Biotechnology in Agriculture and Forestry

Springer
Berlin
Heidelberg
New York
Barcelona
Budapest
Hong Kong
London
Milon
Paris
Tokyo

Volumes already published

Volume 1: Trees I (1986)
Volume 2: Crops I (1986)
Volume 3: Potato (1987)
Volume 4: Medicinal and Aromatic Plants I (1988)
Volume 5: Trees II (1989)
Volume 6: Crops II (1988)
Volume 7: Medicinal and Aromatic Plants II (1989)
Volume 8: Plant Protoplasts and Genetic Engineering I (1989)
Volume 9: Plant Protoplasts and Genetic Engineering II (1989)
Volume 10: Legumes and Oilseed Crops I (1990)
Volume 11: Somaclonal Variation in Crop Improvement I (1990)
Volume 12: Haploids in Crop Improvement I (1990)
Volume 13: Wheat (1990)
Volume 14: Rice (1991)
Volume 15: Medicinal and Aromatic Plants III (1991)
Volume 16: Trees III (1991)
Volume 17: High-Tech and Micropropagation I (1991)
Volume 18: High-Tech and Micropropagation II (1991)
Volume 19: High-Tech and Micropropagation III (1992)
Volume 20: High-Tech and Micropropagation IV (1992)
Volume 21: Medicinal and Aromatic Plants IV (1993)
Volume 22: Plant Protoplasts and Genetic Engineering III (1993)
Volume 23: Plant Protoplasts and Genetic Engineering IV (1993)
Volume 24: Medicinal and Aromatic Plants V (1993)
Volume 25: Maize (1994)
Volume 26: Medicinal and Aromatic Plants VI (1994)
Volume 27: Somatic Hybridization in Crop Improvement I (1994)
Volume 28: Medicinal and Aromatic Plants VII (1994)
Volume 29: Plant Protoplasts and Genetic Engineering V (1994)
Volume 30: Somatic Embryogenesis and Synthetic Seed I (1995)
Volume 31: Somatic Embryogenesis and Synthetic Seed II (1995)
Volume 32: Cryopreservation of Plant Germplasm I (1995)

Volumes in preparation

Volume 33: Medicinal and Aromatic Plants VIII (1995)
Volume 34: Plant Protoplasts and Genetic Engineering VI (1995)
Volume 35: Trees IV
Volume 36: Somaclonal Variation in Crop Improvement II
Volume 37: Medicinal and Aromatic Plants IX
Volume 38: Plant Protoplasts and Genetic Engineering VII

Biotechnology in Agriculture and Forestry 32

Cryopreservation of Plant Germplasm I

Edited by Y.P.S. Bajaj

With 143 Figures

Springer

Professor Dr. Y.P.S. Bajaj
A-137
New Friends Colony
New Delhi 110065, India

ISBN 3-540-57451-4 Springer-Verlag Berlin Heidelberg New York
ISBN 0-387-57451-4 Springer-Verlag New York Berlin Heidelberg

Library of Congress Cataloging-in-Publication Data. Cryopreservation of plant germplasm I/edited by Y.P.S. Bajaj. Includes bibliographical references and index. ISBN 3-540-57451-4 (Berlin: alk. paper). –ISBN 0-387-57451-4 (New York: alk. paper) 1. Crops–Germplasm resources–Cryopreservation. 2. Germplasm resources, Plant–Cryopreservation. I. Bajaj, Y.P.S., 1936– . II. Series. SB123.3.C85 1995 631.5′23′028–dc20 94-46550

Printed in Germany

Typesetting: Thomson Press (India) Ltd., New Delhi

Spin: 10127480 31/3130/SPS – 5 4 3 2 1 0 – Printed on acid-free paper

Dedicated to
Professor Dr. R.G. Butenko
of the Timiryazev Institute of Plant Physiology,
Academy of Sciences Moscow,
whose laboratory I had
the privilege of visiting in 1975

Preface

The germplasm of numerous plant species, especially those of forest trees, some agricultural crops, and medicinal plants, is endangered and threatened with extinction. This depletion of germplasm pools and the shrinkage of naturally occurring genetic resources have caused international concern. Conventionally, the germplasm of plants is conserved through seeds, tubers, roots, corms, rhizomes, bulbs, cuttings, etc. However, the germplasm of a number of trees and plantation crops (such as coconut, cocao, coffee, oil palm, rubber, mango, horse chestnut, etc.) cannot be preserved since their seed are short-lived (recalcitrant). Likewise, germplasm of vegetatively propagated crops (such as potato and cassava) cannot be stored on a long-term basis and has to be grown and multiplied periodically in nurseries and fields. The plants are thus exposed to unpredictable weather conditions and diseases, with the result that instances are known where entire genetic stocks are lost. Therefore, unconventional methods are being developed for the storage and international exchange of germplasm. For this purpose in vitro cultures have been employed, but they can only enable short- to medium-term preservation; moreover, cell cultures upon repeated subculture undergo genetic erosion. In view of the recent developments in the in vitro induction of genetic variability through somaclonal variation, somatic hybridization, recombinant DNA technology, etc., new methods need to be employed for the storage of desirable cultures. In this regard freeze preservation of cells in liquid nitrogen (−196 °C), like that of semen, enables long-term storage, theoretically, for an indefinite period of time.

Considering the aforementioned aspects, while working at the Freie Universität at Berlin in 1974, I regenerated whole plants of carrot and tobacco from cell suspensions frozen at −20, −70, and −196 °C. This work evoked an enthusiastic response when presented at the International Botanical Congress Leningrad in 1975, and then at the International Colloquium Plant Cell and Tissue Culture held at the Ohio State University, Columbus, in 1977. Since then, numerous reports have appeared, and cryopreservation has now become a very fascinating area of research in plant biotechnology, especially for the conservation of (1) germplasm of vegetatively propagated crops, (2) plants with recalcitrant seeds, (3) in vitro induced genetic variability, and (4) somaclones for use in the pharmaceutical industry.

The present book, *Cryopreservation of Plant Germplasm I*, has thus attempted to present the work done, especially on agricultural crops, fruit and forest trees, and medicinal plants, achievements made, to survey the literature, and to highlight the potential of this technology for long-term storage and establishment of gene banks. This book, comprising 35 chapters, has been divided into the following six sections:

I Cryopreservation of Plant Cells, Tissues, Organs, and Seeds;
II Tree Species—*Citrus, Malus, Morus, Pyrus, Juglans*, Date Palm, Oil Palm, Coconut, *Veitchia, Howea, Calamus, Hevea, Larix, Picea*, and *Pinus*;
III Cereals and Grasses—Wheat, Rice, Maize, Sugarcane, *Setaria*, and *Puccinellia*;
IV Legumes and Oilseed Crops—*Arachis hypogaea, Cicer arietinum, Pisum sativum, Medicago, Trifolium* and *Brassica* Species;
V Fruits, Vegetables, Tuber and Root Crops—*Actinidia*, Asparagus, Sugarbeet, Strawberry, Tomato, Banana, Potato, and Cassava;
VI Medicinal, Aromatic, and Ornamental Plants—*Anisodus, Chrysanthemum, Dianthus, Digitalis, Dioscorea*, and *Lavandula.*

New Delhi, March 1995

Professor Y.P.S. Bajaj
Series Editor

Contents

Section I Cryopreservation of Cells, Tissues, Organs, and Recalcitrant Seeds

Section II Cryopreservation of Germplasm of Trees

II.7 Cryopreservation of Date Palm, Oil Palm, and Coconut
F. ENGELMANN, B. ASSY-BAH, S. BAGNIOL, D. DUMET, and N. MICHAUX-FERRIÈRE (With 18 Figures)

II.8 Cryopreservation of Germplasm of Ornamental Palms (*Veitchia*, *Howea*, and *Calamus* Species)
B. KRISHNAPILLAY, H.F. CHIN, and M. MARZALINA (With 4 Figures)

II.9 Cryopreservation of Germplasm of Rubber (*Hevea brasiliensis*)
M.N. NORMAH and H.F. CHIN (With 1 Figure)

II.10 Cryopreservation of Germplasm of *Larix* and *Picea* Species
P.J. CHAREST and K. KLIMASZEWSKA (With 6 Figures)

II.11 Cryopreservation of Scots Pine (*Pinus sylvestris* L.)
A. HOHTOLA (With 1 Figure)

Section III Cereals and Grasses

Section IV Legumes and Oilseed Crops (for Coconut and Oil Palm see Section II)

Section VI Medicinal, Aromatic, and Ornamental Plants

List of Contributors

Assy-Bah, B., IDEFOR/DPO, Station Principale de La Mé, 13 BP 989, Abidjan 13, Ivory Coast, Africa

Bagniol, S., GRFP, Conservatoire National Botanique de Porquerolles, 83400 Ile de Porquerolles, France

Bajaj, Y.P.S., Former Professor of Tissue Culture, Punjab Agricultural University, Ludhiana, India (Present address: A-137 New Friends Colony, New Delhi 110065, India)

Benson, E.E., Department of Molecular and Life Sciences, Dundee Institute of Technology, Bell Street, Dundee, Scotland, DD1 1HG, UK

Berjak, P., Plant Cell Biology Research Unit, Department of Biology, University of Natal, King George V Avenue, Durban, 4001 South Africa

Braun, A.M., American Crystal Sugar Company, Research Center, PO Box 1227, Moorhead, MN 56561-1227, USA

Brison, M., Laboratory of Plant Cell Physiology, University of Bordeaux I, Avenue des Facultés, 33405 Talence Cédex, France

Butenko, R.G., Department of Cell Biology and Biotechnology, K.A. Timiryazev Institute of Plant Physiology, Russian Academy of Sciences, Botanical str. 35, Moscow 127276, Russia

Cachită, C.D., Institute of Biological Research, Republicii Street 48, 3400 Cluj-Napoca, Romania

Charest, P.J., Molecular Genetics and Tissue Culture Group, Petawawa National Forestry Institute, Canadian Forest Service, Chalk River, Ontario, KOJ 1JO, Canada

Chin, H.F., Department of Agriculture, Agricultural University of Malaysia, 43400 Serdang, Selangor, Malaysia

CRĂCIUN, C., Department of Electron Microscopy, Faculty of Biology, "Babeș-Bolyai" University, Clinicilor Street 5-6,3400 Cluj-Napoca, Romania

CRISP, P.C., Crisp Innovar Ltd., Glebe House, Station Road, Reepham, Norfolk NR10 4NB, UK

DE BOUCAUD, M.T., Laboratory of Plant Cell Physiology, University of Bordeaux I, Avenue des Facultés, 33405 Talence Cédex, France

DEREUDDRE, J., Plant Cryobiology Laboratory, Pierre and Marie Curie University (Paris 6), 12 rue Cuvier, 75230 Paris Cédex 05, France

DUMET, D., ORSTOM, 911 Avenue Agropolis, BP 5045, 34032 Montpellier Cédex 1, France

DURAN-VILLA, N., Department of Plant Protection and Biotechnology, Instituto Valenciano de Investigaciones Agrarias, Apartado oficial, 46113 Moncada (Valencia), Spain

ENGELMANN, F., ORSTOM, 911 Avenue Agropolis, BP 5045, 34032 Montpellier Cédex 1, France (Present address: IPGRI, Via delle Sette Chiese 142, 00145 Rome, Italy)

FUKAI, S., Faculty of Agriculture, Kagawa University, Miki-cho, Kita-gun 761-07, Japan

GROUT, B.W.W., Crisp Innovar Ltd., Glebe House, Station Road, Reepham, Norfolk NR10 4NB, UK

GYULAI, G., Department of Genetics and Plant Breeding, University of Agriculture, Páter K. út 1, 2103 Gödöllö, Hungary

HESZKY, L.E., Department of Genetics and Plant Breeding, University of Agriculture, Páter K. út 1, 2103 Gödöllö, Hungary

HOHTOLA, A., Department of Botany, University of Oulu, 90570 Oulu, Finland

HOPE, B., Department of Biology, University of Natal, King George V Avenue, Durban, 4001 South Africa

HUMMER, K.E., National Clonal Germplasm Repository, USDA/ARS 33447 Peoria Road, Corvallis, OR 97333, USA

JEKKEL, Zs., Department of Genetics and Plant Breeding, University of Agriculture, Páter K. út 1, 2103 Gödöllö, Hungary

JIAN, L.C., Laboratory of Cytology, Institute of Botany, Academia Sinica, Beijing, China

KLIMASZEWSKA, K., Molecular Genetics and Tissue Culture Group, Petawawa National Forestry Institute, Canadian Forest Service, Chalk River, Ontario, KOJ 1JO, Canada

KRISHNAPILLAY, B., Seed Technology Laboratory, Forest Research Institute of Malaysia, Kepong, 52109 Kuala Lumpur, Malaysia

LU, T.G., Department of Cell Biology, Institute of Botany, Academia Sinica, Beijing 100044, China

LUO, J.P., Department of Plant Physiology, Kunming Institute of Botany, Academia Sinica, Kunming, Yunnan 650204, China

MARZALINA, M., Seed Technology Laboratory, Forest Research Institute of Malaysia, Kepong, 52109 Kuala Lumpur, Malaysia

MICHAUX-FERRIÈRE, N., CIRAD, BIOTROP, BP 5035, 34032 Montpellier Cédex 1, France

MONETTE, P.L., Agriculture and Agri-Food Canada, Centre for Plant Health, 8801 East Saanich Road, Sidney, British Columbia, V8L 1H3, Canada

MORIGUCHI, T., Division of Breeding, Fruit Tree Research Station, Tsukuba, Ibaraki 305, Japan (Present address: Fruit Tree Research Station, Okitsu Branch, Shimizu, Shizuoka 424-02 Japan)

MYCOCK, D.J., Department of Botany, University of the Witwatersrand Johannesburg, Private Bag 3, Wits, 2050 South Africa

NIINO, T., National Institute of Agrobiological Resources, Shinjo, Yamagata 996, Japan (Present address: Tohoku National Agricultural Experiment Station, Shinjo, Yamagata 996, Japan)

NORMAH, M.N., Department of Botany, Faculty of Life Sciences, Universiti Kebangsaan Malaysia, 43600 Bangi, Selangor, Malaysia

PANIS, B., Laboratory of Tropical Crop Husbandry, Catholic University of Leuven (K.U. Leuven), Kardinaal Mercierlaan 92, 3001 Heverlee, Belgium

PENCE, V.C., Plant Conservation Division, Center for Reproduction of Endangered Wildlife, Cincinnati Zoo and Botanical Garden, 3400 Vine Street, Cincinnati, OH 45220, USA

Popov, A.S., Department of Cell Biology and Biotechnology, Cryopreservation Laboratory, K.A. Timiryazev Institute of Plant Physiology, Russian Academy of Sciences, Botanical str. 35, Moscow 127276, Russia

Reed, B.M., National Clonal Germplasm Repository, USDA/ARS 33447 Peoria Road, Corvallis, OR 97333, USA

Sakai, A., The Institute of Low Temperature Science, Hokkaido University, Sapporo, 060 Japan (Retired, Emeritus Professor) (Present address: 1-5-23 Azabucho, Sapporo, 001 Japan)

Seitz, U., Department of Plant Physiology, University of Tübingen, Auf der Morgenstelle 1, 72076 Tübingen, Germany

Seufferheld, M., Department of Horticulture, Colorado State University, Fort Collins, CO 80523-1173, USA

Stushnoff, C., Department of Horticulture, Colorodo State University, Fort Collins, CO 80523-1173, USA

Sun, C.S., Department of Cell Biology, Institute of Botany, Academia Sinica, Beijing 100044, China

Swennen, R., Laboratory of Tropical Crop Husbandry, Catholic University of Leuven (K.U. Leuven), Kardinaal Mercierlaan 92, 3001 Heverlee, Belgium

Takahashi, J., Discovery Research Laboratories I, Discovery Research Division, Takeda Chemical Industries Ltd., 10 Wadai, Tsukuba, Ibaraki 300-42, Japan

Tanaka, A., Department of Synthetic Chemistry and Biological Chemistry, Faculty of Engineering, Kyoto University, Yoshida, Sakyo-ku, Kyoto 606-01, Japan

Tannoury, M., Plant Cryobiology Laboratory, Pierre and Marie Curie University (Paris 6), 12 rue Cuvier, 75230 Paris Cedex 05, France

Uragami, A., Vegetable Breeding Laboratory, Hokkaido National Agricultural Experiment Station, 1 Hitsujigaoka, Sapporo 062, Japan

Volkova, L.A., Department of Cell Biology and Biotechnology, Cryopreservation Laboratory, K.A. Timiryazev Institute of Plant Physiology, Russian Academy of Sciences, Botanical str. 35, Moscow 127276, Russia

WATT, P., Department of Biology, University of Natal, King George V Avenue, Durban, 4001, South Africa

WESLEY-SMITH, J., Plant Cell Biology Research Unit, Department of Biology & Electron Microscope Unit, University of Natal, King George V Avenue, Durban, 4001 South Africa

ZHENG, G.Z., Department of Plant Physiology, Kunming Institute of Botany, Academia Sinica, Kunming, Yunnan 650204, China

Section I
Cryopreservation of Cells, Tissues, Organs, and Recalcitrant Seeds

I.1 Cryopreservation of Plant Cell, Tissue, and Organ Culture for the Conservation of Germplasm and Biodiversity

Y.P.S. Bajaj[1]

1 Conservation of Germplasm

The indiscriminate clearing of forests and agricultural land has led to the depeletion of the naturally occurring reservoirs of plant germplasm. About 20 000 species of higher plants are believed to be endangered, rare, and threatened with extinction. Moreover, the green revolution no longer appears to be green, as many of the original native cultivars, especially those of wheat, have been discarded and eventually lost. In Greece, for instance, since World War II, 95% of the wheat varieties have been abandoned (Plucknett et al. 1983), and some of them must have been lost, leading to a continuous depletion of genetic variability. On the other hand, the importance of conservation of rice germplasm has been adequately emphasized (Chang and Vaughan 1991), as wild relatives of rice are also disappearing because of various developmental projects. In maize, for instance, the damage caused by *Helminthosporium maydis* in the 1970s in the USA alerted plant breeders to look for unconventional sources of genetic variability to avoid such an eventuality in the future. The germplasm of a number of medicinal plants and forest trees is also threatened with extinction (Hussain 1983; Sakai 1986).

Conventionally, the germplasm of plants is conserved through seeds, tubers, roots, bulbs, corms, rhizomes, buds, cuttings, etc. However, the germplasm of a number of plantation crops and fruit trees cannot be preserved as their seeds are "recalcitrant" and the embryo degenerates early. Moreover, germplasm of vegetatively propagated crops cannot be stored on a long-term basis, it must be grown and multiplied periodically in nurseries or fields. Thus, it is exposed to unpredictable weather and/or diseases and instances are known when entire genetic stocks have been lost.

The depletion of germplasm pools and the reduction of naturally occurring genetic resources have attracted international attention and caused great concern. Thus, unconventional methods are being developed for the storage, maintenance, conservation, and international exchange of germplasm and to establish germplasm banks for rare plant materials. For this purpose, in vitro cultures have been used which are normally stored at 25 °C. However, this requires periodic transfer to fresh media, involving not only man-power and high

[1] Former Professor of Tissue Culture, Punjab Agricultural University, Ludhiana (Punjab), India. *Present address*: A-137 New Friends Colony, New Delhi 110 065, India

Biotechnology in Agriculture and Forestry, Vol. 32
Cryopreservation of Plant Germplasm I (ed. by Y.P.S. Bajaj)

costs, but also the hazards of contamination, and sometimes the loss of the entire material. Moreover, the cell cultures retained on periodic subculture/transfer undergo genetic erosions.

Various methods have been shown to reduce the growth rate of cultures and thus to delay or reduce the frequency of subculture. Some of these methods are:

1. *Minimal Media and Growth Retardants.* The use of minimal media and growth retardants, such as abscisic acid (Henshaw et al. 1978) or the growth of cultures in a sucrose-free medium (Jones 1974) has helped to delay the subculture/transfer period.
2. *Mineral Oil Overlay.* Caplin reported in 1959 that the mineral oil overlay of carrot callus tissue considerably reduced the rate of growth and subsequently delayed the transfer frequency. Later, by using mineral oil overlay, tissue cultures of various medicinal plants were stored for 4–6 months without subculture (Augereau et al. 1986).
3. *Desiccation.* Nitzsche (1980) reported the growth of dried callus after 1 year of storage. Recently (see Gray et al. 1995), extensive studies have been conducted on the desiccation of somatic embryos of carrot, grape, alfalfa, orchard-grass, etc.
4. *Low Pressure/Low Oxygen.* Bridgen and Staby (1981) stored tissue cultures under low atmospheric pressure and low oxygen and achieved fourfold reductions in the growth rate of callus cultures of tobacco, tomato, and chrysanthemum.
5. *Low Temperature Storage.* Of all the above-mentioned methods, low temperature storage has been extensively used. The storage of medicinal plant cell cultures at nonfreezing temperatures and its effects on secondary metabolites have been reviewed (Hiraoka 1988). The storage of cultures at nonfreezing temperatures has been applied to a large number of species (see Bajaj 1991a) and material thus preserved for months to years. The in vitro plantlets of *Chrysanthemum* and *Petunia* stored at 4–5 °C for up to 6 years, with occasional exposure to light in the culture room, flowered after transfer to pots; no abnormalities were observed (Bajaj 1993). However, all the above-mentioned methods enable short- to medium-term conservation. Cryopreservation of in vitro cultures in liquid nitrogen (LN) has enabled plant recovery in a number of species and has been recommended for the long-term storage of germplasm, especially that of the vegetatively propagated crops (Bajaj 1990a). Moreover, the germplasm of rare, elite, and endangered plant species must be conserved. In this case, cryopreservation also holds great promise.

In view of the recent advancements in the in vitro induction of genetic variability (Fig. 1) through somaclonal and gametoclonal variation, somatic hybridization, cybridization, recombinant DNA technology, etc., cryopreservation methods must be employed for the storage of desirable cultures.

Basically, cryopreservation involves bringing the culture to a state of nondivision and zero metabolism, which is achieved by subjecting it to superlow temperatures in the presence of cryoprotectants. The method of freezing varies. After freeze storage the cultures are generally thawed at 35–40 °C, washed, and recultured. Considerable progress has been made with cells, protoplasts, meri-

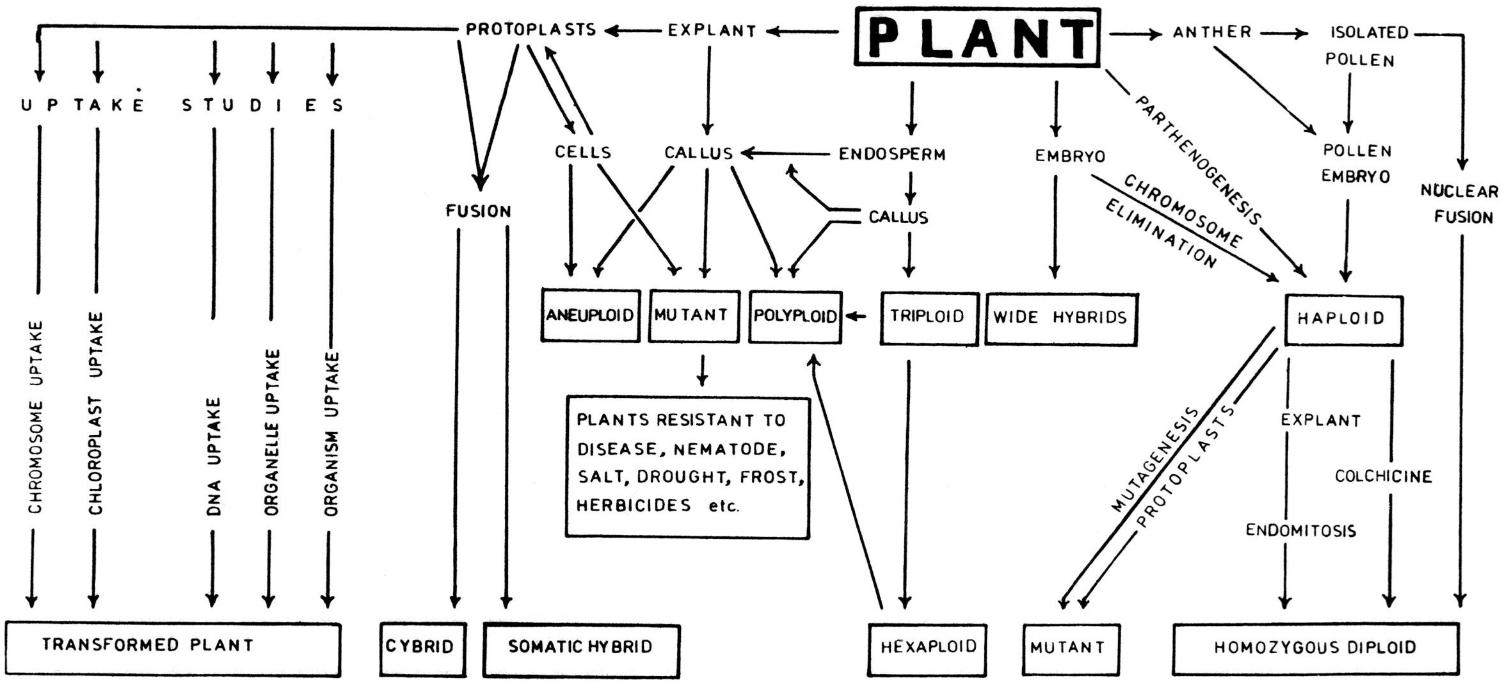

Fig. 1. In vitro induction of genetic variability through culture of cells, tissues, and organs, and the fusion of protoplasts and genetic transformation. (Modified from Bajaj 1979a)

stems, embryos, etc. stored at $-196\,^{\circ}C$ for various lengths of time (Table 1). This introductory chapter, which is an updated version of earlier publications by the author (Bajaj 1979a, b, 1983a, 1986a, 1991a), deals with the general aspects of cryopreservation and its implications in the conservation of in vitro induced genetic variability and the establishment of germplasm banks. For details on cryopreservation studies on individual plant species, the readers are referred to the following chapters of this book.

Table 1. Plant species (alphabetically arranged) in which cryopreservation studies have been conducted on protoplast, cell, tissue, and organ cultures

Botanical name (common name)	Material frozen	Reference
Abies nordmanniana	Cell cultures	Nørg et al. (1993)
Acer pseudoplatanus (sycamore)	Cell suspension	Sugawara and Sakai (1974); Withers (1978)
Actinidia chinensis (kiwifruit)	Stem segment	Jian and Sun (1989)
Aesculus hippocastanum (horse chestnut)	Somatic embryo	Jekkel et al. (1992)
Allium sativa (garlic)	Shoot tips	Niwata (1992)
Anisodus acutangulus	Cell suspension	Zheng Guang-zhi et al. (1983)
Arachis hypogaea (peanut)	Meristems	Bajaj (1979c, 1983b)
	Pollen embryos	Bajaj (1983c)
Arachis villosa	Pollen embryos	Bajaj (1983c)
Araucaria hunsteinii	Desiccated embryo	Pritchart and Prendergast (1986)
Asparagus officinalis	Somatic embryo	Uragami et al. (1989)
	Cell suspension	Nishizawa et al. (1992)
Atropa belladonna (belladonna)	Cell suspension	Nag and Street (1975a)
	Anthers/pollen embryos	Bajaj (1977a, 1978b)
	Protoplasts	Bajaj (1988b)
Beta vulgaris (sugarbeet)	Meristems	Braun (1988)
	Root culture	Benson and Hamill (1991)
Brassica campestris (mustard)	Pollen embryos	Bajaj (1983c)
B. campestris	Cell suspension	Langis et al. (1990a)
B. napus	Anthers/pollen embryos	Bajaj (1983c)
B. napus (rapeseed)	Microspores	Charne et al. (1988); Chen and Beversdorf (1992a, b)
B. napus	Shoot tips	Withers et al. (1988)
B. napus	Microspore embryo	Uragami et al. (1993)
Bromis innermis	Protoplasts	Mazur and Hartmann (1979)
Brunfelsia densifolia	Cell suspension	Pence (1990)
Camellia sinensis (tea)	Embryo axis	Chaudhury et al. (1991); Wesley-Smith et al. (1992)
Cannabis sativa (hemp)	Cell suspension	Jekkel et al. (1989)
Carya illinoinensis (pecan)	Zygotic embryo	Abou Taleb et al. (1992)
Catharanthus roseus (periwinkle)	Cell suspension	Chen et al. (1984)
Chamomilla recutita	Shoot tips	Diettrich et al. (1990)
	Cells	Cellarova et al. (1992)

Table 1. (*Contd.*)

Botanical name (common name)	Material frozen	Reference
Chrysanthemum morifolium	Shoot tip	Fukai (1990); Fukai et al. (1991)
Cicer arietinum (chickpea)	Meristems	Bajaj (1983b)
Cichorium intybus (chicory)	Shoot tips	Demeulemeester et al. (1993)
Cinchona ledgeriana	Callus	Hunter (1986)
Citrus species	Ovule, nucellus, somatic embryo	Bajaj (1984b; Marin and Duran-Vila (1988); Kobayashi et al. (1990); Sakai et al. (1991a, b)
Cocos nucifera (coconut)	Embryo segment	Bajaj (1984b)
	Embryos	Chin et al. (1989); Assy-Bah and Engelmann (1992a, b)
Coffea arabica (coffee)	Somatic embryos	Bertrand-Desbrunais et al. (1988)
	Zygotic embryos	Abdelnour-Esquivel et al. (1992a)
C. canephora	Cell suspension, somatic embryo	Tessereau et al. (1991)
Cucumis melo (melon)	Somatic embryo	Shimonishi et al. (1991)
Datura innoxia	Protoplasts	Bajaj (1988b)
D. stramonium	Cell suspensions	Bajaj (1976a)
Daucus carota (carrot)	Cell suspension	Nag and Street (1975b); Bajaj (1976a); Bhandal et al. (1985)
	Somatic embryos	Bajaj (1976a); Withers (1979); Lecouteux et al. (1991); Dereuddre et al. (1991)
	Protoplasts	Takeuchi et al. (1980); Hauptmann and Widholm (1982)
Dianthus caryophyllus (carnation)	Shoot apices	Seibert (1976); Seibert and Wetherbee (1977); Dereuddre et al. (1988); Uemura and Sakai (1980); Langis et al. 1990; Tannoury et al. (1991)
D. hybrida	Shoot tips	Fukai (1989)
Digitalis lanata	Cell suspension	Dietterich et al. (1982); Seitz et al. (1983); Göldner et al. (1991)
Dioscorea deltoidea (yam)	Cell suspension	Butenko et al. (1984); Popov and Fedorovskii (1992)
Elaeis guineesis (oil palm)	Somatic embryo	Engelmann et al. (1985); Engelmann and Duval (1986); Dumet et al. (1993)
	Zygotic embryo	Grout et al. (1983)
Euphorbia longan	Embryonic axes	Fu et al. (1990)
Fragaria × *ananassa* (strawberry)	Runner apex	Sakai et al. (1978)
	Apical meristem	Kartha et al. (1980)
Glycine max (soybean)	Cell suspension	Bajaj (1976a)
	Protoplasts	Takeuchi et al. (1982); Weber et al. (1983)
Gossypium arboreum (cotton)	Anther culture, callus	Bajaj (1982c)
G. hirsutum	Ovule	Bajaj (1982c)
Grevillea scapigera	Shoot tip	Touchell et al. (1992)

(*Contd.*)

Table 1. (*Contd.*)

Botanical name (common name)	Material frozen	Reference
Haplopappus gracilis	Shoot primordia	Taniguchi et al. (1988)
Hevea brasiliensis (rubber)	Zygotic embryo	Normah et al. (1986)
Hordeum vulgare (barley)	Protoplasts	Takeuchi et al. (1982)
	Callus	Hahne and Lörz (1987)
Howea fosteriana (ornamental palm)	Zygotic embryo	Chin et al. (1988)
Ipomoea species (sweet potato)	Cell suspension	Latta (1971)
	Shoot tip	Towill and Jarret (1992)
Juglans regia (Persian walnut)	Embryonic axis	de Boucaud et al. (1991)
Juglans regia (walnut)	Somatic embryo	de Boucaud et al. (1994)
Landolphia kirkii	Embryonic axis	Vertucci and Berjak (1991)
Lavandula vera (lavender)	Cell suspension	Watanabe et al. (1983)
Larix × eurolepis (larch)	Cell cultures	Klimaszewska et al. (1992)
Linum usitatissimum (flax)	Cell suspension	Quatrano (1968)
Lycopersicon esculentum (tomato)	Meristems	Grout et al. (1978)
Malus (apple)	Alginate-coated shoot tips	Niino and Sakai (1992)
	Callus	Kurola and Nishiyama (1983)
	Shoot tips	Kuo and Lineberger (1985)
	Shoot tips	Niino et al. (1992a)
	Buds	Tyler and Stushnoff (1988); Stushnoff (1991); Stushnoff et al. (1992)
Manihot esculenta (cassava)	Shoot apices	Bajaj (1977c, 1982a, 1983f,1985c); Kartha et al. (1982); Benson et al. (1992)
	Somatic embryos	Sudarmonowati and Henshaw (1990)
	Zygotic embryo	Marin et al. (1990)
Marchantia polymorpha	Protoplasts	Takeuchi et al. (1980, 1982)
Medicago sativa (alfalfa)	Callus	Finkle et al. (1980); Cachita et al. (1991)
Mentha (mint)	Shoot tips	Towill (1990)
Morus alba (mulberry)	Shoot tips	Niino and Sakai (1992); Niino et al. (1992b)
Morus bomycis	Meristems	Yakuwa and Oka (1988)
Musa species (banana)	Zygotic embryos	Abdelnour-Esquivel et al. (1992b)
	Cell suspension	Panis et al. (1990)
Nicotiana tabacum (tobacco)	Cell suspension	Bajaj (1976a); Hauptmann and Widholm (1982); Maddox et al. (1983); Bhandal et al. (1985)
	Anthers/pollen embryos	Bajaj (1977a, 1978b)
	Protoplasts	Bajaj (1988b)
N. plumbaginifolia, *N. sylvestris*	Cell suspension	Maddox et al. (1983)
N. rustica	Root cultures	Benson and Hamill (1991)
Onobrychis viciaefolia (sainfoin)	Callus	Sun and Jian (1990)
Oryza sativa (rice)	Endosperm	Bajaj (1981a)
	Zygotic embryos	Bajaj (1981a, 1984c)
	Pollen embryos	Bajaj (1984c)

Table 1. (*Contd.*)

Botanical name (common name)	Material frozen	Reference
	Pollen	Coulibaly and Demarly (1979)
	Cell suspension/ callus	Sala et al. (1979); Bajaj (1982b), Cella et al. (1982); Finkle and Ulrich (1982); Kuriyama et al. (1990); Watanabe et al. (1990)
Oryza sativa + *Pisum sativum*	Fused protoplasts	Bajaj (1983d)
Oryza sativa + *Sorghum bicolor*	Fused protoplasts	Bajaj (1983e)
Panax ginseng (ginseng)	Cell suspension	Butenko et al. (1984); Seitz and Reinhard (1987)
	Root culture	Yoshimatsu et al. (1995)
Panicum miliaceum	Callus	Sun et al. (1988)
Petunia hybrida	Anthers	Bajaj (1978b)
Phoenix dactylifera (date palm)	Meristem	Bagniol and Engelmann (1991, 1992)
	Callus	Finkle et al. (1980); Tisserat et al. (1981); Mater (1987)
Picea abies (Norway spruce)	Cell masses	Gupta et al. (1987)
Picea glauca (white spruce)	Cell cultures	Kartha et al. (1988)
Picea glauca-engelmanni	Cell cultures	Cyr et al. (1994)
Picea mariana (black spruce)	Cell cultures	Klimaszewska et al. (1992)
Pinus sylvestris (Scots pine)	Buds	Kuoksa and Hohtola (1991)
Pinus caribaea (Caribean pine)	Cell suspension	Laine et al. (1992)
Pinus taeda (loblolly pine)	Cell masses	Gupta et al. (1987)
Pisum sativum (pea)	Protoplasts	Bajaj (1983d)
	Embryonic axis	Mycock et al. (1991)
	Seedlings	Sun (1958)
	Meristem	Haskins and Kartha (1980)
Poncirus trifoliata	Embryonic axis	Radhamani and Chandel (1992)
Populus euramericana (poplar)	Cell suspension	Sakai and Sugawara (1973)
Primula obconica	Androgenic anthers	Bajaj (1981b)
Puccinellia species (halophyte grasses)	Cell suspension	Heszky et al. (1990)
Pyrus communis (pear)	Shoot apices	Moriguchi et al. (1985); Reed (1990); Dereuddre et al. (1990a, b); Niino et al. (1992a)
	Winter buds	Oka et al. (1991)
Quercus petraea (oak)	Embryo	Jörgensen (1990); Gonzales-Benito and Perez-Ruiz (1992)
Ribes species (currants)	Apical meristem	Reed (1992)
Rubus species (brambles)	Apical meristem	Reed and Lagerstedt (1987) Chen et al. (1979)
Saccharum officinarum (sugarcane)	Callus/cell suspension	Bajaj et al. (1987); Jian et al. (1987); Ulrich et al. (1979); Finkle and Ulrich (1979); Gnanapragasam and Vasil (1990); Eksomtramage et al. (1992)
	Meristems	Bajaj et al. (1987); Paulet et al. (1993)
	Plantlets	Bajaj et al. (1987)
Secale cereale (rye)	Protoplasts	Langis and Steponkus (1990)

(*Contd.*)

Table 1. (*Contd.*)

Botanical name (common name)	Material frozen	Reference
Setaria italica (foxtail millet)	Callus/cell suspension	Lu and Sun (1992)
Solanum tuberosum (potato)	Meristems/ axillary buds	Bajaj (1977c, 1978a, 1981c, 1985c); Towill (1984); Fabre and Dereuddre (1990); Harding et al. (1991)
	Protoplast-derived callus	Bajaj (1986b)
	Root tips	Bajaj (1978a, 1987b)
S. goniocalix	Meristem	Grout and Henshaw (1978)
S. etuberosum	Meristems	Towill (1981)
S. phureja	Encapsulated shoot tips	Fabre and Dereuddre (1990)
Theobroma cacao (coco)	Immature zygotic embryos	Pence (1991a)
Trifolium repens (clover)	Apical meristem	Yamada et al. (1991)
Triticum aestivum (wheat)	Cells/callus	Bajaj (1980a); Kendall et al. (1990)
	Zygotic embryos	Bajaj (1984c)
	Pollen embryos	Bajaj (1983c, 1984c)
Triticum aestivum + *Pisum sativum*	Fused protoplasts	Bajaj (1983d)
Vaccinium species	Apical meristems	Reed 1989
Veitchia merrillii (ornamental palm)	Zygotic embryo	Chin et al. (1988)
Vigna mungo × *Vigna radiata*	Hybrid embryos	Bajaj (1990c)
Vitis vinifera (grapes)	Cell suspension	Dussert et al. (1992)
Wasabia japonica (wasabi)	Apical meristems	Matsumoto et al. (1994)
Xanthosoma brasiliense (tanier spinach)	Somatic embryo	Jayos et al. (1993)
Zea mays (maize)	Callus/cell suspension	Withers and King (1979); Jian et al. (1987); Sun and Jian (1989)
	Protoplasts	Zhang et al. (1990)
	Embryos	Delvallée et al. (1989)

2 Prospects for Cryopreservation of In Vitro Cultures

Some of the most obvious prospects for cryopreservation are outlined in Fig. 4 and discussed below:

1. *Conservation of Somaclonal and Gametoclonal Variation.* Plant cell cultures undergo genetic erosions and are thus a rich source of genetic variability, both somaclonal and gametoclonal. They show polyploidy, aneuploidy, mutations, chromosome elimination, gene amplification, translocation, etc. (see Bajaj 1991b). If plant cell cultures are maintained in culture, or if they are periodically subcultured, they undergo changes and their genetic stability cannot be ensured. However, this can be achieved by freeze storage.

2. *Vegetatively Propagated Crops and Genetic Uniformity of Clones.* In vegetatively propagated crops such as potato, cassava, yam, etc., in order to

maintain the genetic stocks (tubers, roots), they have to be stored and grown annually in nurseries. This is not only time-consuming and cumbersome, requiring both space and labor, but the material is also exposed to pests and pathogens and there is always the risk of losing germplasm. Moreover, as yet, no method is known by means of which the germplasm can be stored on a long-term basis. Cryopreservation offers such a possibility (Bajaj 1990a).

3. *Recalcitrant Seeds.* The germplasm is traditionally maintained through seeds (Stanwood 1985), however, many large seeds, especially those of plantation crops and fruit trees, are sensitive to temperature and humidity. Under such conditions the embryo is short-lived and aborts (Stanwood 1985). A number of such recalcitrant seed species, such as coconut, oil palm, coffee, mango, chestnut, walnut, etc. (Pence 1991b), are known in which germplasm cannot be maintained through seeds. Cryopreservation of whole embryos or their segments (Bajaj 1984b) would help to overcome such an eventuality.

4. *Conservation of Rare Germplasm.* It is rightly asserted that extinction is forever. There are over 20 000 species of higher plants that are rare or threatened with extinction in their natural habitats. In addition, rare combinations of germplasm obtained through somatic hybridization, cybridization, genetic transformation, etc. must also be conserved.

5. *Medicinal and Alkaloid-Producing Cell Lines and Hairy Root Cultures.* In view of the importance of cell cultures for the production of medicines and drugs, it is highly desirable to obtain selected cell cultures en masse. Thus, some of the overproducing cell lines or mutant cells can be cryopreserved (Bajaj 1988a) and stored for use when sufficient quantities are available for the isolation of medicines and other products of pharmaceutical importance. Recently, genetically transformed roots have been cryopreserved, and they have retained their biosynthetic potential (Benson and Hamill 1991).

6. *Meristems for Disease-Free Micropropagation and Micrografting.* Meristem culture is an accepted tool for obtaining virus-free stocks, large-scale clonal propagation, short-term storage, and international exchange of germplasm. In addition, micrografting of young meristems onto aging stocks, especially those of old trees, restores juvenility and vigor. Cryopreservation adds another dimension to the long-term storage of germplasm.

7. *Retention of Morphogenetic Potential.* Long-term tissue cultures are known to show a decline or even a complete loss of morphogenetic potential. Freeze storage would help to extend the period of retention of morphogenetic and biosynthetic potential.

8. *Slow Metabolism and Aging.* At superlow temperatures there is total growth arrest, and the cells are in a state of inactive metabolism. This prevents or virtually "stops" the aging process.

9. *Storage of Pollen to Enhance Longevity.* From the plant breeder's point of view, one of the most important aspects is the storage of pollen. Frozen pollen stored in liquid nitrogen for various lengths of time retain their viability. The

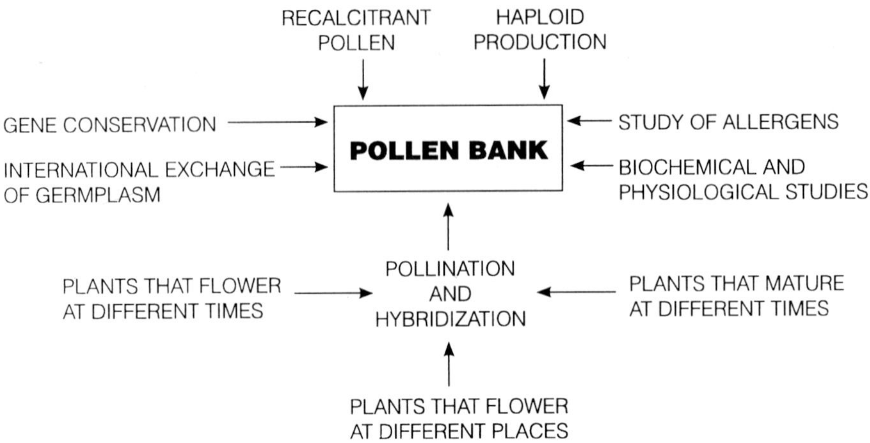

Fig. 2. Importance of pollen storage and the establishment of pollen banks. (Bajaj 1987a)

freeze storage of pollen would enable (1) hybridization between plants that flower at different times; (2) hybridization between plants that grow at different and distant places; (3) reduction in the spread of diseases by pollination vectors; and (4) maintenance of germplasm and enhancement of longevity (Fig. 2).

3 Methods of Cryopreservation

3.1 Conventional Methods

The following basic steps are involved in the freeze preservation of in vitro cultures:

1. Establishment of in vitro cultures or the sterilization of tissues and organs taken directly from the plant.
2. Addition of an appropriate cryoprotectant.
3. Subjection of cultures to superlow temperatures.
4. Storage of frozen cultures in liquid nitrogen.
5. Thawing of retrieved cultures.
6. Removal of cryoprotectant by washing.
7. Determination of viability.
8. Reculture of retrieved material.
9. Induction of growth and regeneration of plants.

Freezing is achieved by the following three methods:

1. *Fast freezing* by direct immersion of the material into liquid nitrogen.
2. *Prefreezing* the cultures in various steps to -20, -50, -70, $-100\,^{\circ}$C, and then $-196\,^{\circ}$C. This cooling is periodic and at defined intervals.

3. *Gradual and slow freezing* at a determined slow rate of cooling (1–5 °C/min), which can be achieved by commercially available cryostats. A large variety of liquid nitrogen freezers, both automatic and manual, are available in which the rate of cooling and thawing can be regulated (see Bajaj and Reinert 1977).

3.2 Recently Developed or Modified Methods

Though conventional methods form the basis of any cryopreservation study, the following modifications and refinements have helped to simplify the protocol and to increase the viability. Moreover, in some cases, the sophisticated and expensive cryostats for controlled freezing rates are not needed as the material is immersed directly into liquid nitrogen.

1. *Vitrification.* Vitrification involves treatment of the culture with higher concentrations of cryoprotectants. The main advantage of vitrification is that controlled-rate cryostats are not required, thus the time needed for cooling is reduced. Complete vitrification eliminates concern for the potentially damaging effects of intra- and extracellular crystallization. Vitrification, in fact, is the formation of a glass-like, noncrystalline solid at the freezing point of an aqueous solution. At sufficiently low temperatures, a highly concentrated cryoprotective solution becomes so viscous that it solidifies into a metastable glass. The goal is to achieve a sufficiently high concentration of solutes to prevent the solution from freezing into crystalline ice when cooled and to ensure its transition into the amorphous or glassy state. Vitrification, originally used in the cryopreservation of animal cells (Fahy et al. 1984), has recently been extended to plant cells.

Langis et al. (1990a) successfully cryopreserved cell suspensions of *Brassica campestris* by vitrification, by treating them with 1.5 M ethyl glycol solution followed by sudden freezing. Almost 80% cell survival, followed by callus regrowth, was observed. Later, Sakai et al. (1990, 1991a, b) used a very highly concentrated PVS_2 solution (containing 30% glycerol, 15% ethylene glycol, and 15% DMSO together with 0.4 M sucrose) for nucellar cells of *Citrus sinensis.*

The somatic embryos (SE) of asparagus treated with a mixture of glycerol (22%), ethylene glycol (15%), propylene glycol (15%), and DMSO (7%) in MS medium enriched with 0.5 M sorbitol showed a viability of about 50% (Uragami et al. 1989). However, 100% survival of SEs of carrot and coffee was observed by using a high concentration of sucrose (0.85–1 M) as the sole source of cryoprotectant (Tesserreau et al. 1991). A combination of high sugar and dehydration also proved successful for zygotic embryos of coconut (Assy-Bah and Engelmann 1992a, b). Similarly, meristems of carnation (Langis et al. 1990b), mint (Towill 1990), clover (Yamada et al. 1991), and apple and pear (Niino et al. 1992a) have been successfully cryopreserved by vitrification. Thus, judging from the simplicity of the protocol for vitrification, it appears to be the method of choice for future studies.

2. *Encapsulation/Alginate Coating.* Carrot SEs first encapsulated in calcium alginate, followed by incubation in 10.2% sucrose (as the sole source of

cryoprotectant), desiccated for 4 h in a laminar flow chamber, and then subjected to rapid freezing in LN, had a viability of about 92%. These embryos grew directly into plants (Dereuddre et al. 1991).

3. *Dehydration/Silica Gel.* Dumet et al. (1993) considerably improved the cryoability of oil palm SEs by additional dehydration by keeping them (for 6–18 h in the dark) in an air-tight box containing 40 g silica gel before immersion in LN. This improved process was successfully applied to seven different clones. The desiccation with silica gel is likely to ensure more reproducible dehydration conditions than those obtained with a laminar flow chamber which can vary, depending on the air humidity and the speed of air flow.

4 General Guidelines and Factors Affecting Cryoability

Though plants have been regenerated from the frozen cultures of a number of plant species (Table 1), the survival and the extent of regeneration of retrieved cultures depend on a number of factors. Some of these are outlined below:

1. Physiological state, age, and nature of the culture to be preserved.
2. Concentration and nature of the cryoprotectants.
3. Method of freezing.
4. Storage temperature.
5. Method of thawing.
6. Method for the determination of viability.

From a survey of the literature (Table 1), for the successful revival of the cells, following general guidelines should be taken into consideration:

1. Periodically transferred cell suspensions in their exponential growth phase, highly cytoplasmic, thin-walled, nonvacuolated, small cells, and cell aggregates are able to withstand freezing much better than large, thick-walled, and vacuolated cells (Fig. 3).
2. For the preservation of plantlets and whole organ or embryos, partial desiccation before freezing is necessary.
3. For the maintenance of clones, especially those of vegetatively propagated crops, meristems should be used; they will also ensure virus-free stocks.
4. Small, young, and immature zygotic, somatic, or pollen embryos show higher cryoability than older or fully differentiated embryos. Thus, globular and heart-shaped embryos are more likely to show higher survival rates.
5. Before freezing, the cells or tissues may be precultured for a few days on a medium containing a low concentration of cryoprotectants (5% DMSO, for instance).
6. During treatment with the cryoprotectants, the cultures should be maintained in ice to avoid deleterious effects.
7. The results on the method of freezing are rather variable. Cell survival has been obtained in cultures subjected to sudden freezing, using both prefrozen

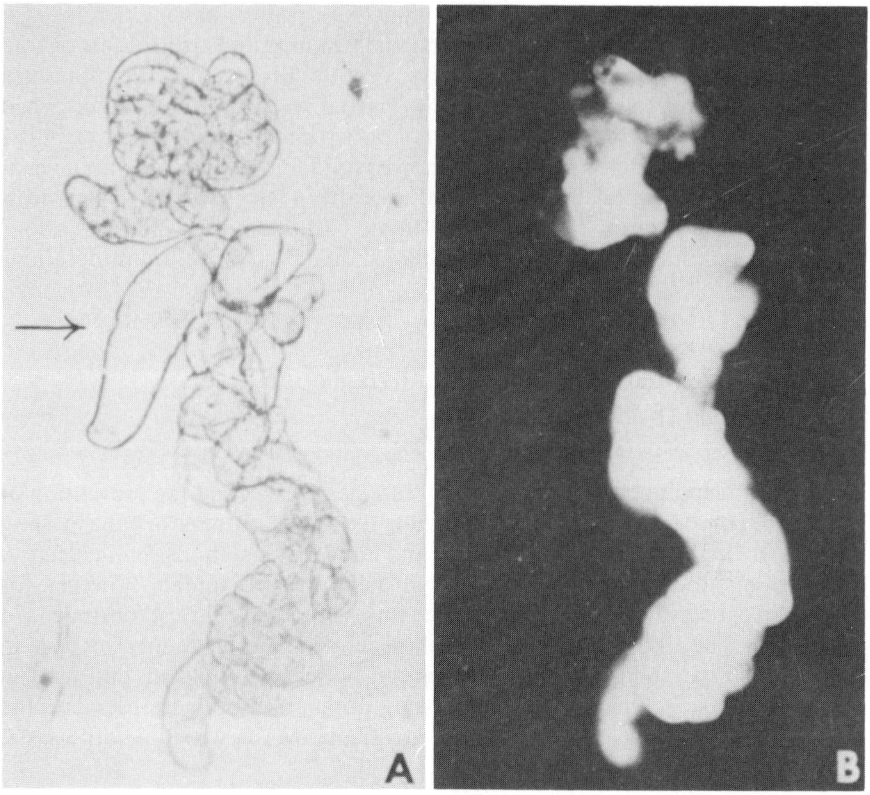

Fig. 3 A, B. Survival of a haploid tobacco cell suspension cooled at a rate of 2 °C/min and subjected to −196 °C in the presence of 7% DMSO. Twin photographs in tungsten (**A**) and ultraviolet light (**B**); note the complete survival (fluorescence) of a highly cytoplasmic clump of cells, and the death of a large vacuolated cell (*arrow* in **A**). (After Bajaj 1976a)

material and cells frozen at a slow and regulated rate of cooling. However, a recently developed and simple method of vitrification, in which the material is subjected to sudden freezing, is preferred.

8. Cultures stored at −20 °C or −70 °C have been observed to undergo deterioration with the passage of time. Thus, for long-term conservation of germplasm, material should be maintained in liquid nitrogen.
9. In most of the studies, fast-thawing at 35–40 °C proved to be better than other methods.
10. A lag phase is due to cryodamage, resulting in the depletion of viable cells, and eventually, selection pressure. This is highly undesirable with respect to the maintenance of clones. Thus, the maximum recovery of cells is of utmost importance.
11. It is pertinent to point out that staining methods alone may not give accurate information about survival. There are always some cells that, although

giving a positive staining reaction immediately after thawing, later die in culture. There are also cells that are partially injured and are in a state of cold shock, and thus do not give a positive reaction. Moreover, with fluorescein diacetate, the cells sometimes show partial or light fluorescence when compared to the control. Thus, it is not possible to determine their viability. It was observed that tobacco cells showed mild fluorescence or gave a negative staining reaction; however, in cultures after a lag phase of 3 to 4 months, they started to grow (Bajaj 1976a). In view of these observations, various growth parameters and plating efficiency should be taken into consideration.

5 Gene Banks and the International Exchange of Germplasm (Fig. 4)

In view of the imminent danger of losing rare germplasm and the prevention of endangered ones from extinction, it is highly desirable to strengthen the existing germplasm banks by incorporating new and innovative methods. In this regard, for short- and medium-term storage, in vitro cultures are suitable; however, for long-term conservation, superlow temperatures are ideal. The establishment of gene banks is necessary since such in vitro generated variability does not exist in nature. Such variability and/or cell lines can be selected, banked, and used in response to research needs. Ever-increasing population pressures demand the improvement and development of new cultivars. However, with this enthusiastic

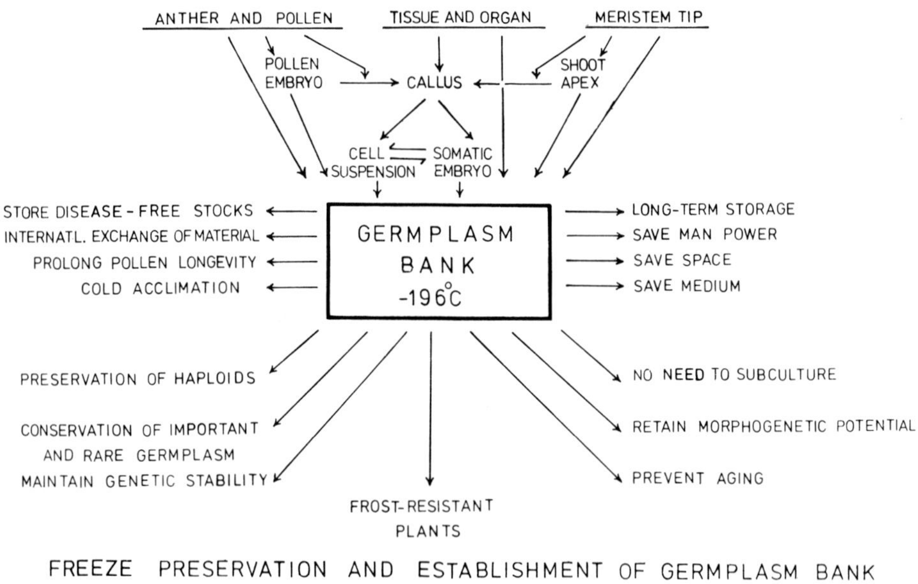

Fig. 4. Diagrammatic representation showing the freeze storage of plant cells, tissues, and organs in liquid nitrogen and their prospects for the establishment of germplasm banks. (Bajaj 1979b)

search for new plants and the rapid increase in their number, it is not only becoming difficult, but at times impossible, to maintain or preserve some of the stocks which, at present, are not needed. Thus, some of the in vitro material, which may not seem to be of importance today, but might be needed in future, is ignored or completely lost. Therefore, in order to preserve such material, cryopreservation facilities are being developed.

Hitherto, mostly meristem cultures have been used for the storage and exchange of germplasm. Now that somatic embryos of many plant species, including those of recalcitrant ones, can be produced in large numbers in bioreactors we can propagate them on a large scale, freeze them for preservation (see Bajaj 1995), and internationally exchange germplasm as such or coated with calcium alginate.

Studies on vegetatively propagated crops, such as potato, cassava, and sugarcane have shown that the methodology has been refined enough to use it for the long-term preservation of their germplasm.

6 Summary

Cryopreservation studies conducted over the last two decades have enabled the regeneration of entire plants from cells, protoplasts, meristems, embryos, endosperm, etc. of a variety of genera belonging to diverse families. Notable examples are the vegetatively propagated crops (potato, cassava), cereal and grasses (rice, wheat, maize, sugarcane), medicinal and aromatic plants (*Atropa*, carnation, chrysanthemum, citrus, coffee, *Digitalis*), legumes and oilseed crops (peanut, chickpea, brassica, oil palm) etc. These studies have demonstrated that, in addition to various other potentials of cryopreservation, the following applications are possible: (1) conservation of germplasm of vegetatively propagated crops; (2) conservation of germplasm of plants with recalcitrant seeds; and (3) storage of cell cultures of desirable somaclones, especially those of high-yielding alkaloids or secondary metabolites of pharmaceutical interest.

In view of the recent advancements in the in vitro induction of genetic variability (through somatic hybridization, cybridization, somaclonal and gametoclonal variation, recombinant DNA technology, etc.), cryopreservation methods must be further refined and simplified for routine use. The quest of agriculture to obtain new cultivars for producing more food and the rapid multiplication of the plants requires the utilization of freeze-preservation methods for the storage of the desired germplasm.

7 Conclusion – Reproducibility of Results

It should be emphasized that cryopreservation involving "freezing – storage – thawing – culture" is a complicated, multiple event. An error at any stage or alteration of factors can result in lethal injury to the system. It is now an established fact that different species, cultivars varieties, clones, lines, hybrids, etc. show variations in their response to the same in vitro culture conditions. The same is true for the reproducibility of results in cryopreservation studies. In addition to other factors, the cryoability of cells or tissues greatly depends not only on the genotype, age, and physiological state of the plant and/or cultures, but also on the conditions under which they are grown prior to freezing. Tropical plants, in general, are presumed to be more sensitive compared to the plants grown under subzero temperatures. It has been reported that due to seasonal variations there is a change in the soluble proteins (Siminovitch and Briggs 1949) and an increase in the level of sugars (Heber 1958; Parker 1959) and polyhydric alcohols (Sakai 1960), which is generally believed to increase the cold resistance of plants. Thus, material taken from the plants in winter might behave differently from plants grown under extreme summer conditions. Furthermore, material taken from in vitro cultures differs from that taken from the field. Harding et al. (1991) observed that potato meristems taken from long-term in vitro cultures showed less cryoability than those taken from recently grown in vitro plantlets. It has been noted that sometimes the results obtained in one laboratory cannot be exactly reproduced in another laboratory working with a different genotype grown or cultured under different conditions. Thus, it is highly pertinent that the above-mentioned points be taken into consideration before drawing hasty conclusions.

References and Further Reading

(Also given are references not cited in the text.)

Abdelnour-Esquivel A, Villalobos V, Engelmann F (1992a) Cryopreservation of zygotic embryos of *Coffea* spp. Cryo-Lett 13: 297–302

Abdelnour-Esquivel A, Mora A, Villalobos V (1992b) Cryopreservation of zygotic embryos of *Musa acuminata* (AA) and *Musa balbisiana* (BB). Cryo-Lett 13: 159–164

Abou Taleb S, Yates I, Wood BW, Fouad MM (1992) Cryogenics and tissue culture for preserving pecan germplasm. HortSci 27 (6): 693

Assy-Bah B, Engelmann F (1992a) Cryopreservation of immature embryos of coconut (*Cocos nucifera* L.). Cryo-Lett 13: 67–74

Assy-Bah B, Engelmann F (1992b) Cryopreservation of mature embryos of coconut (*Cocos nucifera* L.) and subsequent regeneration of plantlets. Cryo-Lett 13: 117–126

Augereau JM, Courtois D, Petiard V (1986) Long-term storage of callus cultures at low temperatures or under mineral oil layer. Plant Cell Rep 5: 372–376

Bagniol S, Engelmann F (1991) Effects of pregrowth and freezing conditions on the resistance of meristems of date palm (*Phoenix dactylifera* L. var. Bou Sthammi noir) to low temperatures and to freezing in liquid nitrogen. Cryo-Lett 12: 279–286

Bagniol S, Engelmann F (1992) Effect of thawing and recovery conditions on the regrowth of meristems of date palm (*Phoenix dactylifera* L.) after cryopreservation in liquid nitrogen. Cryo-Lett 13: 253–260

Bajaj YPS (1976a) Regeneration of plants from cell suspensions frozen at −20, −70, and −196 °C. Physiol Plant 37: 263–268

Bajaj YPS (1976b) Gene preservation through freeze storage of cell, tissue, and organ cultures. Acta Hortic 63: 75–84

Bajaj YPS (1977a) Survival of *Atropa* and *Nicotiana* pollen embryos frozen at −196 °C. Curr Sci 46: 305–307
Bajaj YPS (1977b) Initiation of shoots and callus from potato tuber-sprouts and axillary buds frozen at −196 °C. Crop Improv 4: 48–53
Bajaj YPS (1977c) Clonal propagation and cryopreservation of cassava through tissue culture. Crop Improv 4: 198–204
Bajaj YPS (1978a) Tuberization in potato plants regenerated from freeze-preserved meristems. Crop Improv 5: 137–141
Bajaj YPS (1978b) Effect of superlow temperature on excised anthers and pollen embryos of *Atropa, Nicotiana* and *Petunia*. Phytomorphology 28: 171–176
Bajaj YPS (1979a) Technology and prospects of cryopreservation of germplasm. Euphytica 28: 267–285
Bajaj YPS (1979b) Establishment of germplasm banks through freeze storage of plant tissue cultures and their implications in agriculture. In: Sharp WR, Larsen PO, Paddock PO, Raghavan V (eds) Plant cell and tissue culture – principles and applications. Ohio State Univ Press, Columbus, pp 745–774
Bajaj YPS (1979c) Freeze preservation of meristems of *Arachis hypogaea* and *Cicer arietinum*. Indian J Exp Biol 17: 1405–1407
Bajaj YPS (1980a) Freeze preservation of plant cells – a novel approach to the conservation of germplasm. In: Gupta AK (ed) Genetics and wheat improvement. Oxford & IBH, New Delhi, pp 141–149
Bajaj YPS (1980b) Induction of androgenesis in rice anthers frozen at −196 °C. Cereal Res Commun 8: 365–369
Bajaj YPS (1980c) Freeze-preservation of plant cell and tissue cultures – progress and prospects. In: Rao PS, Heble MR, Chadha MS (eds) Plant tissue culture genetic manipulation and somatic hybridisation of plant cells. Natl Symp BARC, Bombay, pp 50–65
Bajaj YPS (1981a) Growth and morphogenesis in frozen (−196 °C) endosprem and embryos of rice. Curr Sci 50: 947–948
Bajaj YPS (1981b) Regeneration of plants from ultra-low frozen anthers of *Primula obconica*. Sci Hortic 14: 93–95
Bajaj YPS (1981c) Regeneration of plants from potato meristems freeze-preserved for 24 months. Euphytica 30: 141–145
Bajaj YPS (1982a) Cryopreservation of germplasm of potato, cassava, peanut and chickpea. In: Fujiwara A (ed) Plant tissue culture 1982. Maruzen, Tokyo, pp 799–800
Bajaj YPS (1982b) Induction and cryopreservation of genetic variability in rice. In: Rice tissue culture planning conf 1980, IRRI, Los Baños, pp 99–111
Bajaj YPS (1982c) Survial of anther- and ovule-derived cotton callus frozen in liquid nitrogen. Curr Sci 51: 139–140
Bajaj YPS (1983a) Cryopreservation and the international exchange of germplasm. In: Sen SK, Giles KL (eds) Plant cell culture in crop improvement. Plenum, New York, pp 19–41
Bajaj YPS (1983b) Production of normal seeds from plants regenerated from meristems of *Arachis hypogaea* and *Cicer arietinum* cryopreserved for 20 months. Euphytica 32: 425–430
Bajaj YPS (1983c) Regeneration of plants from pollen embryos of *Arachis*, *Brassica* and *Triticum* spp. cryopreserved for one year. Curr Sci 52: 484–485
Bajaj YPS (1983d) Survival of somatic hybrid protoplasts of wheat × pea and rice × pea subjected to −196 °C. Indian J Exp Biol 21: 120–122
Bajaj YPS (1983e) Cryopreservation of germplasm of cereals – progress and prospects. In: Sakamoto S (ed) Proc 6th Int Wheat Genet Symp, Kyoto Univ, Kyoto, pp 565–574
Bajaj YPS (1983f) Cassava plants from meristem cultures freeze-preserved for three years. Field Crop Res 7: 161–167
Bajaj YPS (1984a) Cryopreservation of germplasm of crops – present status and prospects. Proc Int Symp Plant tissue and cell culture application to crop improvement. Czech Acad Sci, Prague, pp 607–616
Bajaj YPS (1984b) Induction of growth in frozen embryos of coconut and ovules of citrus. Curr Sci 53: 1215–1216

Bajaj YPS (1984c) The regeneration of plants from frozen pollen embryos and zygotic embryos of wheat and rice. Theor Appl Genet 67: 525–528
Bajaj YPS (1985a) Cryopreservation of embryos. In: Kartha KK (ed) Cryopreservation of plant cells and organs. CRC, Boca Raton, pp 227–242
Bajaj YPS (1985b) Cryopreservation of plant cell cultures and its prospects in agricultural and forest biotechnology. In: Natesh S et al. (ed) Int Worksh Biotechnology in agriculture: evolving a research agenda for the International Centre of Genetic Engineering and Biotechnology. Oxford & IBH, New Delhi, pp 109–131
Bajaj YPS (1985c) Cryopreservation of germplasm of potato and cassava – viability studies on excised meristems cryopreserved up to four years. Indian J Exp Biol 23: 285–287
Bajaj YPS (1986a) In vitro preservation of genetic resources – techniques and problems. Int Symp Nuclear techniques and in vitro culture for plant improvement IAEA/FAO, Vienna 1985, pp 43–57
Bajaj YPS (1986b) Cryopreservation of potato somaclones. In: Semal J (ed) Proc Symp Somaclonal variations and crop improvement. Nijhoff, Dordrecht, pp 244–251
Bajaj YPS (1986c) In vitro regeneration of diverse plants and the cryopreservation of germplasm in wheat (*Triticum aestivum* L.). Cereal Res Commun 14: 305–311
Bajaj YPS (1987a) Cryopreservation of pollen and pollen embryos, and the establishment of pollen banks In: Giles KL, Prakash J (eds) Pollen development and cytology. Academic Press, London, pp 397–420
Bajaj YPS (1987b) Cryopreservation of potato germplasm. In: Bajaj YPS (ed) Biotechnology in agriculture and forestry, vol 3. Potato. Springer, Berlin Heidelberg New York, pp 472–486
Bajaj YPS (1988a) Cryopreservation and the retention of biosynthetic potential in cell cultures of medicinal and alkaloid-producing plants. In: Bajaj YPS (ed) Biotechnology in agriculture and forestry, vol 4. Medicinal and aromatic plants I. Springer, Berlin Heidelberg New York, pp 169–187
Bajaj YPS (1988b) Regeneration of plants from frozen (−196 °C) protoplasts of *Atropa belladonna* L., *Datura innoxia* Mill. and *Nicotiana tabacum* L. Indian J Exp Biol 26: 289–292
Bajaj YPS (1988c) Biotechnology of the cryopreservation of germplasm of plants. In: Ilahi I, Hughes KW (eds) In vitro selection and propagation of economic plants. Jadoon Printing Press, Peshawar, pp 97–102
Bajaj YPS (1988d) Biotechnology of the conservation of germplasm and its implications in agriculture and forestry. In: Int Symp Genetic manipulation in crops, Beijing 1984. IRRI/Cassell Tycooly, London, pp 125–127
Bajaj YPS (1988e) Biotechnologie kryokonzervace genetickych zdrrojo rostlin. Symp Kryokonzervace v živočišné a rostlinné vyrobě. Nové Město na Moravé Sept 13–15. Dum techniky CSVTS Usti nad Labem, Czechoslovakia, pp 100–105
Bajaj YPS (1989a) Induction and cryopreservation of somaclonal variation in wheat and rice. Int Symp Genetic manipulation in crops. CIMMYT, Mexico, pp 195–203
Bajaj YPS (1989b) Cryopreservation of plant protoplasts. In: Bajaj YPS (ed) Biotechnology in agriculture and forestry, vol 8. Plant protoplasts and genetic engineering I. Springer, Berlin Heidelberg New York, pp 97–106
Bajaj YPS (1990a) Cryopreservation of germplasm of vegetatively propagated crops. Bull Soc Bot Fr 137. Actual Bot (3/4): 99–114
Bajaj YPS (1990b) Cryopreservation of germplasm of wheat. In: Bajaj YPS (ed) Biotechnology in agriculture and forestry, vol 13. Wheat. Springer, Berlin Heidelberg New York, pp 669–681
Bajaj YPS (1990c) Cryopreservation of germplasm of legumes and oilseed crops. In: Bajaj YPS (ed) Biotechnology in agriculture and forestry, vol 10. Legumes and oilseed crops I. Springer, Berlin Heidelberg New York, pp 49–62
Bajaj YPS (1991a) Storage and cryopreservation of in vitro cultures. In: Bajaj YPS (ed) Biotechnology in agriculture and forestry, vol 17. High-tech and micropropagation I. Springer, Berlin Heidelberg New York, pp 361–381
Bajaj YPS (1991b) Somaclonal variation – origin, induction, cryopreservation, and implications in plant breeding. In: Bajaj YPS (ed) Biotechnology in agriculture and forestry, vol 11. Somaclonal variation in crop improvement I. Springer, Berlin Heidelberg New York, pp 3–48

Bajaj YPS (1993) A suggested method for in vitro long-term storage at 4 °C of *Chrysanthemum* and *Petunia* germplasm. Plant Tissue Cult 3(1): 57–58

Bajaj YPS (1995) Cryopreservation of somatic embryos. In: Bajaj YPS (ed) Biotechnology in agriculture and forestry, vol 30. Somatic embryogenesis and synthetic seed I. Springer, Berlin Heidelberg New York, pp 221–229

Bajaj YPS, Reinert J (1975) Regeneration of plants from cells subjected to superlow temperatures. Abstr Int Bot Congr, Leningrad, p 278

Bajaj YPS, Reinert J (1977) Cryobiology of plant cell cultures and establishment of gene banks. In: Reinert J, Bajaj YPS (eds) Applied and fundamental aspects of plant cell, tissue, and organ culture. Springer, Berlin Heidelberg New York, pp 757–777

Bajaj YPS, Sala F (1991) Cryopreservation of germplasm of rice (*Oryza sativa* L.). In: Bajaj YPS (ed) Biotechnology in agriculture and forestry, vol 14. Rice. Springer, Berlin Heidelberg New York, pp 553–571

Bajaj YPS, Korneva SB, Gutierrez R, Maribona RH (1987) Freeze preservation of plantlets, excised meristems and cell cultures of sugarcane (*Saccharum officinarum* L.) at −196 °C. In: Bose A, Sengupta P (eds) Proc Int Conf Cryogenics. McGraw-Hill, New Delhi, pp 222–226

Benson EE, Hamill JD (1991) Cryopreservation and post freeze molecular and biosynthetic stability in transformed roots of *Beta vulgaris* and *Nicotiana rustica.* Plant Cell Tissue Organ Cult 24: 163–172

Benson EE, Harding K, Smith H (1989) Variation in recovery of cryopreserved shoot tips of *Solanum tuberosum* exposed to different pre- and post-freeze light regimes. Cryo-Lett 10: 323–344

Bensen EE, Chabrillange N, Engelmann F (1992) A comparison of cryopreservation methods for the long-term in vitro conservation of cassava (*Manihot esculenta*). Proc Soc Low Temp Biol Meet, Sterling, Scotland

Bertrand-Desbrunais AJ, Fabre F, Engelmann F, Dereuddre J, Charrier A (1988) Reprise de l'embryogenèse adventive d'embryons somatiques de caféier (*Coffea arabica*) après congélation dans l'azote liquide. CR Acad Sci Paris 307, Ser III: 795–801

Bhandal IS, Hauptmann RM, Widholm JM (1985) Trehalose as cryoprotectant for the freeze preservation of carrot and tobacco cells. Plant Physiol 78: 430–432

Braun A (1988) Cryopreservation of sugarbeet germplasm. Plant Cell Tissue Organ Cult 14: 161–168

Bridgen MP, Staby GL (1981) Low pressure and low oxygen storage of plant tissue cultures. Plant Sci Lett 22: 117–186

Butenko RG, Popov AS, Volkova LA, Chernyak ND, Nosov AM (1984) Recovery of cell cultures and their biosynthetic-capacity after storage of *Dioscorea deltoides* and *Panax ginseng* cells in liquid nitrogen. Plant Sci Lett 33: 285–292

Cachita CD, Zapirtan M, Craciun C, Rakosy TL, Vicol A, Cristea V, Varga P, Craciun V (1991) The response of various types of cormophytoinocula to cryopreservation (−196 °C) IVth Natl Symp Plant cell and tissue culture. Cluj-Napoca, Romania, 1989, pp 90–91

Caplin SM (1959) Mineral-oil overlay for conservation of plant tissue cultures. Am J Bot 46: 324–329

Cella R, Colombo R, Galli MG, Nielsen E, Rollo F, Sala F (1982) Freeze-preservation of rice cells: a physiological study of freeze-thawed cells. Physiol Plant 55: 279–284

Cellarova E, Cernicka T, Vranova E, Brutovska R, Lapar M (1992) Viability of *Chamomilla recutita* L. Rauschert cells after cryopreservation. Cryo-Lett 13: 37–42

Chang TT, Vaughan DA (1991) Conservation and potentials of rice genetic resources. In: Bajaj YPS (ed) Biotechnology in agriculture and forestry, vol 14. Rice. Springer, Berlin Heidelberg New York, pp 531–552

Charne DG, Pukachi, Kott LS, Beversdorf WD (1988) Embryogenesis following cryopreservation in isolated microspores of rapeseed (*Brassica napus* L.) Plant Cell Rep 7(6): 407–409

Chaudhury R, Radhamani J, Chandel KPS (1991) Preliminary observations on the cryopreservation of desiccated embryonic axes of tea [*Camellia sinensis* (L.) O. Kuntze] seeds for genetic conservation. Cryo-Lett 12: 31–36

Chen JL, Beversdorf WD (1992a) Cryopreservation of isolated microspores of spring rapeseed (*Brassica napus* L.) for in vitro embryo production. Plant Cell Tissue Organ Cult 31: 141–149

Chen JL, Beversdorf WD (1992b) Production of spontaneous diploid lines from isolated microspores following cryopreservation in spring rapeseed (*Brassica napus* L.). Plant Breeding 108: 324–327

Chen THH, Kartha KK, Leung NL, Kurz WGW, Chatson KB, Constabel F (1984) Cryopreservation of alkaloid-producing cell cultures of periwinkle (*Cathranthus roseus*). Plant Physiol 75: 726–731

Chen THH, Kartha KK, Gusta LV (1985) Cryopreservation of wheat suspension culture and regenerable callus. Plant Cell Tissue Organ Cult 4: 101–109

Chen WH, Cockburn W, Street HE (1979) Preliminary experiments on freeze preservation of sugarcane cells. Taiwania 24: 70–74

Chin HF, Krishanapillay B, Alang ZC (1988) Cryopreservation of *Veitchia* and *Howea* palm embryos: non-development of the haustorium. Cryo-Lett 9: 372–379

Chin HF, Krishnapillay R, Hor YL (1989) A note on the cryopreservation of embryos from young coconuts (*Cocos nucifera* var. MAWA). Pertanika 12(2): 183–186

Coulibaly Y, Demarly Y (1979) Sur les conditions de développment des microspores de *Nicotiana tabacum* et d'*Oryza sativa* soumises à la temperature de l'azote liquide (– 196 °C). CR Acad Sci Paris Ser D 286: 1065–1068

Cyr DR, Lazaroff WR, Grimes SMA, Quian G, Benthune TD, Dunstan DI, Roberts DR (1994) Cryopreservation of interior spruce (*Picea glauca-engelmanni* complex) embryogenic cultures. Plant Cell Rep 13: 574–577

de Boucaud MT, Brison M, Ledoux C, Germain E, Lutz A (1991) Cryopreservation of embryonic axes of recalcitrant seed: *Juglans regia* L. cv. Franquette: Cryo-Lett 12: 163–166

de Boucaud MT, Brison M, Dosba F (1994) Cryopreservation of walnut somatic embryos. Cryo Lett 15: 151–160

Delavallée I, Guillaud J, Beckert M, Dumas C (1989) Cryopreservation of immature maize embryos after freeze-hardening in the ear and in vitro. Plant Sci 60: 129

Demeulemeester MAC, Vandenbussche B, de Profit MP (1993) Regeneration of chicory plants from cryopreserved in vitro shoot tips. Cryo-Lett 14: 57–64

Dereuddre J, Fabre J, Bassaglia C (1988) Resistance to freezing in LN of carnation (*Dianthus caryophyllus* L. var. Eolo) apical and axillary shoot tips. Plant Cell Rep 7: 170–173

Dereuddre J, Scottez C, Arnaud Y, Duron M (1990a) Effects of cold hardening on cryopreservation of axillary pear (*Pyrus communis* L. cv. Beurre Hardy) shoot tips of in vitro plantlets. CR Acad Sci Paris 310, Ser III: 265–272

Dereuddre J, Scottez C, Arnaud Y, Duron M (1990b) Resistance of alginate-coated shoot tips of pear tree (*Pyrus communis* L. cv. Beurre Hardy) in vitro plantlets to dehydration and subsequent freezing in liquid nitrogen: effects of previous cold hardening. CR Acad Sci Paris 310, Ser III: 317–323

Dereuddre J, Hassen N, Blandin S, Kaminski M (1991) Resistance of alginate-coated somatic embryos of carrot (*Daucus carota* L.), to desiccation and freezing in liquid nitrogen. 2. Thermal analysis. Cryo-Lett 12: 135–148

Diettrich B, Popov AS, Pfeiffer B, Neumann D, Butenko RG, Luckner M (1982) Cryopreservation of *Digitalis lanata* cell cultures. Planta Med 46: 82–87

Diettrich B, Donath P, Popov AS, Butenko RG, Luckner M (1990) Cryopreservation of *Chamomilla recutita* shoot tips. Biochem Physiol Pflanzen 186: 63–67

Dumet D, Engelmann F, Chabrillange N, Duval Y (1993) Cryopreservation of oilpalm (*Elaeis guineensis* Jacq.) somatic embryos involving a desiccation step. Plant Cell Rep 12: 352–355

Dussert S, Mauro MC, Engelmann F (1992) Cryopreservation of grape embryogenic cell suspensions: 2. Influence of post-treatment conditions and application to different strains. Cryo-Lett 13: 12–22

Eksomtramage T, Paulet F, Guiderdoni E, Glaszmann JC, Engelmann F (1992) Development of a cryopreservation process for embryogenic calluses of a commercial hybrid of sugarcane (*Saccharum* sp.) and application to different varieties. Cryo-Lett 13: 239–252

Engelmann F, Duval Y (1986) Cryopreservation d'embryons somatiques de palmier á huile (*Elaeis guineenis* Jacq.): resultats et perspectives d'application. Oleagineux 41(4): 169–172

Engelmann F, Duval Y, Dereuddre J (1985) Survie et prolifération d'embryons somatiques de palmier á huile (*Elaeis guineensis* Jacq.) aprés congélation dans l'azote liquide. CR Acad Sci Paris Ser III, 301: 111–116

Fabre J, Dereuddre J (1987) Effects of different sugars (sucrose, glucose, sorbitol and mannitol) on the resistance to deep freezing in liquid nitrogen of meristems from in vitro cultured carnations

(*Dianthus caryophyllus* L., var. Eolo) – cryopreservation. CR Sceances Acad Sci Ser 3, 20: 507–510
Fabre J, Dereuddre J (1990) Encapsulation dehydration: a new approach to cryopreservation of *Solanum* shoot tips. Cryo-Lett 11: 413–426
Fahy GM, MacFarlane DR, Angell CA, Meryman HT (1984) Vitrification as an approach to cryopreservation. Cryobiology 21: 407–426
Finkle BJ, Ulrich JM (1979) Effects of cryoprotectants in combination on the survival of frozen sugarcane cells. Plant Physiol 63: 598–604
Finkle BJ, Ulrich JM (1982) Cryoprotectant removal as a factor in the survival of frozen rice and sugarcane cells. Cryobiology 19: 329–335
Finkle BJ, Ulrich JM, Tisserat B, Rains DW (1980) Regeneration of date palm and alfalfa plants after freezing callus tissues to $-196\,^{\circ}C$ in a combination of cryoprotective agents. Cryobiology 17: 625–626
Fu JR, Zhang BZ, Wang XP, Qiao YZ, Huang XL (1990) Physiological studies on desiccation, wet storage and cryopreservation of recalcitrant seeds of three fruit species and their excised embryonic axes. Seed Sci Technol 18: 743–754
Fukai S (1989) Plant regeneration from shoot tips of *Dianthus hybrida* cryopreserved in liquid nitrogen to 2 years. Plant Tissue Cult Lett 6: 177–178
Fukai S (1990) Cryopreservation of chrysanthemum shoot tips. Sci Hortic 45: 167–174
Fukai S, Goi M, Tanaka M (1991) Cryopreservation of shoot tips of *Chrysanthemum morifolium* and related species native to Japan. Euphytica 54: 201–204
Gnanapragasam S, Vasil IK (1990) Plant regeneration from a cryopreserved embryogenic cell suspension of a commerical sugarcane hybrid (*Saccharum* sp.). Plant Cell Rep 9: 419–423
Göldner EM, Seitz U, Reinhard E (1991) Cryopreservation of *Digitalis lanata* Ehrh. cell cultures: preculture and freeze tolerance. Plant Cell Tissue Organ Cult 24: 19–24
Gonzalez-Benito ME, Perez-Ruiz C (1992) Cryopreservation of *Quercus faginea* embryonic axes. Cryobiology 29: 685–690
Gray DJ, Compton ME, Harrel RC (1995) Somatic embryogenesis and the technology of synthetic seed. In: Bajaj YPS (ed) Biotechnology in agriculture and forestry, vol 30. Somatic embryogenesis and synthetic seed I. Springer, Berlin Heidelberg New York, pp 126–151
Grout BWW, Henshaw GG (1978) Freeze-preservation of potato shoot-tip cultures. Ann Bot 42: 1227–1229
Grout BWW, Westcott RJ, Henshaw GG (1978) Survival of shoot meristems of tomato seedlings frozen in liquid nitrogen. Cryobiology 15: 478–483
Grout BWW, Shelton K, Pritchard HW (1983) Orthodox behaviour of oil palm seed and cryopreservation of the excised embryo for germplasm conservation. Ann Bot 52: 381–384
Gupta PK, Durzan DJ, Finkle BJ (1987) Somatic embryogenesis in embryonal cell masses of *Picea abies* and *Pinus taeda* after freezing in liquid NO_2. Can J For Res 17: 1130–1136
Hahne G, Lörz H (1987) Cryopreservation of embryogenic callus cultures from barley (*Hordeum vulgare* L.) Plant Breeding 99: 330–332
Harding K (1990) Molecular stability of cryopreserved shoot tips of *Solanum tuberosum.* VIIth Int Congr Plant tissue and cell culture, Amsterdam, Abstr, p 376
Harding K, Benson EE, Smith H (1991) The effects of in vitro culture period on the recovery of cryopreserved shoot tips of *Solanum tuberosum.* Cryo-Lett 12: 17–22
Haskins RH, Kartha KK (1980) Freeze-preservation of pea meristems: cell survival. Can J Bot 833–840
Hauptmann RM, Widholm JM (1982) Cryostorage of cloned amino acid analog resistant carrot and tobacco suspension culture. Plant Physiol 70: 30–37
Heber H (1958) Ursachen der Frostresistenz bei Winterweizen I. Die Bedeutung der Zucker für die Frostresistenz. Planta 52: 188–201
Heszky LE, Jekkel Z, Ali AH (1990) Effect of cooling rate, cryoprotectant and holding time at different transfer temperatures on the survival of cryopreserved cell suspension culture [*Puccinellia distans* (L.) Parl.]. Plant Cell Tissue Organ Cult 21: 217–226
Hiraoka N (1988) Nonfrozen storage of cell cultures and its effect on metabolites. In: Bajaj YPS (ed) Biotechnology in agriculture and forestry, vol 4. Medicinal and aromatic plants I. Springer, Berlin Heidelberg New York, pp 157–168

Hunter CS (1986) In vitro propagation and germplasm storage of *Cinchona*. In: Withers LA, Alderson P (eds) Plant tissue culture and its agricultural applications. Butterworth, London, pp 291–301

Hussain A (1983) Conservation of genetic resources of medicinal plants in India. In: Jain SK, Mehra KL (eds) Conservation of medicinal plant resources. NBPGR, New Delhi, pp 110–117

Jayos EE, Hoekstra FA, Staritsky G (1993) Cryopreservation of somatic embryos of tanier spinach [*Xanthosoma brasiliense* (Desf.) Engler]. Plant Cell Tissue Organ Cult

Jekkel ZS, Heszky LE, Ali AH (1989) Effect of different cryoprotectants and transfer temperatures on the survival rate of hemp (*Cannabis sativa* L.) cell suspension in deep freezing. Acta Biol Hung 40: 127–136

Jekkel ZS, Kiss J, Kiss E, Gyulai G, Heszky LE (1992) Freeze preservation of somatic embryos of recalcitrant seed species: horsechestnut (*Aesculus hippocastanum* L.). Proc XIIIth EUCARPIA Congr, Angers, France, July 6–11, 1992

Jian LC, Sun LH (1989) Cryopreservation of the stem segments of Chinese gooseberry. Acta Bot Sin 31: 66–68

Jian LC, Sun DL, Sun LH (1987) Sugarcane callus cryopreservation. In: Li PH (ed) Plant cold hardiness. Alan R Liss, New York, pp 323–337

Jones LH (1974) Long-term survival of embryoids of carrot (*Daucus carota* L.). Plant Sci Lett 2: 221–224

Jörgensen J (1990) Conservation of valuable gene resources by cryopreservation in some forest tree species. J Plant Physiol 136: 373–376

Jullien M (1983) Medium-term preservation of mesophyll cells isolated from *Asparagus officinalis* L.: development of a simple method by storage at reduced temperature. Plant Cell Tissue Organ Cult 2: 305–316

Kartha KK, Leung NL, Pahl K (1980) Cryopreservation of strawberry meristems and mass propagation of plantlets. J Am Soc Hortic Sci 105: 481–484

Kartha KK, Leung NL, Mroginski LA (1982) In vitro growth responses and plant regeneration from cryopreserved meristems of cassava (*Manihot esculenta* Crantz). Z Pflanzenphysiol 107: 133–140

Kartha KK, Fowke LC, Leung NL, Caswell KL, Hakman I (1988) Induction of somatic embryos and plantlets from cryopreserved cell cultures of white spruce (*Picea glauca*). J Plant Physiol 132: 529–539

Katano M, Ishihara RA, Sakai A (1983) Survival of dormant apple shoot tips after immersion in liquid nitrogen. HortScience 18: 707–708

Kendall EJ, Qureshi JA, Kartha KK, Leung N, Chevrier N, Caswell K, Chen THH (1990) Regeneration of freezing-tolerant spring wheat (*Triticum aestivum* L.) plants from cryoselected callus. Plant Physiol 94: 1756–1762

Klimaszewska K, Ward C, Cheliak WM (1992) Cryopreservation and plant regeneration from embryogenic cultures of larch (*Larix* x *eurolepis*) and black spruce (*Picea mariana*). J Exp Bot 43: 73–79

Kobayashi S, Sakai A, Oiyama I (1990) Cryopreservation in liquid nitrogen of cultured navel orange (*Citrus sinensis* Osb.) nucellar cells and subsequent plant regeneration. Plant Cell Tissue Organ Cult 23: 15–20

Kuo CC, Lineberger BD (1985) Survival of in vitro cultured tissue of 'Jonathan' apples exposed to $-196\,^{\circ}C$. HortScience 20: 764–767

Kuoksa T, Hohtola A (1991) Freeze-preservation of buds from Scots pine trees. Plant Cell Tissue Organ Cult 27: 89–93

Kuriyama A, Watanabe K, Ueno S, Mitsuda H (1990) Effect of post-thaw treatment on the viability of cryopreserved *Lavandula vera* cells. Cryo-Lett 11: 171–178

Kurola H, Nishiyama Y (1983) Liquid nitrogen storage of apple callus cultures. Hokkaido Nat Agric Stn Res Bull 136: 15–21

Laine E, Bade F, David A (1992) Recovery of plants from cryopreserved embryogenic cell suspensions of *Pinus caribaea*. Plant Cell Rep 11: 295–298

Langis R, Schnabel B, Earle ED, Steponkus PL, (1990a) Cryopreservation of *Brassica campesteris* L. cell suspensions by vitrification. Cryo-Lett 10: 421–428

Langis R, Schnabel B, Preikstas B, Earle ED, Steponkus PL (1990b) Cryopreservation of carnation shoot tips by vitrification. Cryobiology 27: 657–658

Langis P, Steponkus PL (1990) Cryopreservation of rye protoplasts by vitrification. Plant Physiol 92: 666–671

Latta R (1971) Preservation of suspension cultures of plant cells by freezing. Can J Bot 49: 1253–1254

Lecouteux C, Florin B, Tessereau H, Bollon H, Petiard V (1991) Cryopreservation of carrot embryos using a simplified freezing process. Cryo-Lett 1: 319–328

Lu TG, Sun CS (1992) Cryopreservation of millet (*Setaria italica* L.). J Plant Physiol 139: 295–298

Maddox AD, Gonsalves F. Shields R (1983) Successful preservation of suspension cultures of three *Nicotiana* species at the temperature of liquid nitrogen. Plant Sci Lett 28: 157–162

Marin ML, Duran-Vila N (1988) Survival of somatic embryos and recovery of plants of sweet orange (*Citrus sinensis* L. Osb.) after immersion in liquid nitrogen. Plant Cell Tissue Organ Cult 7: 51–57

Marin ML, Mafla G, Roca WM, Withers LA (1990) Cryopreservation of cassava zygotic embryos and whole seeds in liquid nitrogen. Cryo-Lett 11: 257–264

Mater AA (1987) Production and cryogenic freezing of date palm germplasm and regeneration of plantlets from frozen material. Iraqi J Agric Sci 5: 35–49

Matsumoto T, Sakai A, Yamada K (1994) Cryopreservation of in vitro-grown apical meristems of wasabi (*Wasabia japonica*) by vitrification and subsequent high plant regeneration. Plant Cell Rep 13: 442–446

Mazur RA, Hartmann JX (1979) Freezing of plant protoplasts. In: Sharp WR, Larsen PO, Paddock EF, Raghavan V (eds) Plant cell and tissue culture – principles and applications. Ohio State Univ Press, Columbus, p 876

Moriguchi T, Akihawa T, Kozaki I (1985) Freeze-preservation of dormant pear shoot apices. Jpn J Breed 35: 196–199

Murashige T, Skoog F (1962) A revised medium for rapid growth and bioassays with tobacco tissue cultures. Physiol Plant 15: 473–497

Mycock DJ, Watt MP, Berjak P (1991) A simple procedure for the cryopreservation of hydrated embryonic axes of *Pisum sativum.* J Plant Physiol 138: 728–733

Nag KK, Street HE (1975a) Freeze-preservation of cultured plant cells, I. The pretreatment phase. Physiol Plant 34: 254–260

Nag KK, Street HE (1975b) Freeze-preservation of cultured plant cells. II. The freezing and thawing phases. Physiol Plant 34: 261–265

Niino T, Sakai A (1992) Cryopreservation of alginate-coated in vitro-grown shoot tips of apple, pear and mulberry. Plant Sci 87: 199–206

Niino T, Sakai A, Yakuwa H, Nojiri K (1992a) Cryopreservation of in vitro-grown shoot tips of apple and pear by vitrification. Plant Cell Tissue Organ Cult 28: 261–266

Niino T, Sakai A, Enomoto S, Magoshi J, Kato S (1992b) Cryopreservation of in vitro-grown shoot tips of mulberry by vitrification. Cryo-Lett 13: 303–312

Niino T, Sakai A, Yakuwa H (1992c) Cryopreservation of dried shoot tips of mulberry winter buds and subsequent plant regeneration. Cryo-Lett 13: 51–58

Nishizawa S, Sakai A, Amano Y, Matsuzawa T (1992) Cryopreservation of asparagus (*Asparagus officinalis*) embryogenic suspension cells and subsequent plant regeneration by new vitrification method. Cryo-Lett 13: 379–388

Nitzsche W (1980) One year storage of dried carrot callus. Z Pflanzenphysiol 100: 269–271

Niwata E (1992) Cryopreservation of shoot tips of garlic by vitrification. Jpn J Breed 42 (Suppl 1): 326

Nørg JV, Baldursson S, Krogstrup P (1993) Genotypic differences in the ability of embryogenic *Abies nordmanniana* cultures to survive cryopreservation. Silvae Genet 42: 93–97

Normah MN, Chin HF, Hor YL (1986) Desiccation and cryopreservation of embryonic axes of *Hevea brasiliensis* Muell.-Arg. Pertanika 9(3): 299–303

Oka S, Yakuwa H, Sato K, Niino T (1991) Survival and shoot formation in vitro of pear winter buds cryopreserved in liquid nitrogen. HortScience 26: 65–66

Panis B, Withers LA, De Langhe E (1990) Cryopreservation of *Musa* suspension cultures and subsequent regeneration of plants. Cryo-Lett 11: 337–350

Parker J (1959) Seasonal variations in sugars of conifers with some observations on cold resistance. For Sci 5: 56–63

Paulet F, Engelmann F, Glaszmann JC (1993) Cryopreservation of apices of in vitro plantlets of sugarcane (*Saccharum* sp. hybrids) using encapsulation/dehydration. Plant Cell Rep 12: 525–529
Pence VC (1990) In vitro collection, regeneration, and cryopreservation of *Brunfelsia densifolia*. Abstr VIIth Int Congr Plant tissue and cell culture. Amsterdam, p 377
Pence VC (1991a) Cryopreservation of immature embryos of *Theobroma cacao*. Plant Cell Rep 10: 144–147
Pence VC (1991b) Cryopreservation of seeds of Ohio native plants and related species. Seed Sci Technol 19: 235–251
Plucknett DL, Smith NJH, Williams JT, Murthy Anishetty N (1983) Crop germplasm conservation and developing countries. Science 220: 163–169
Popov AS, Fedorovskii DN (1992) Injuries to the plasmalemma of *Dioscorea* cells cultured in vitro incurred in the process of their cryopreservation. Sov Plant Physiol 39: 211–216
Pritchard HW, Prendergast FG (1986) Effects of desiccation and cryopreservation on the in vitro viability of embryos of the recalcitrant seed species *Araucaria hunsteinii* K. Shum. J Exp Bot 37: 1388–1397
Quatrano RS (1968) Freeze-preservation of cultured flax cells utilizing dimethyl sulfoxide. Plant Physiol 43: 2057–2061
Radhamani J, Chandel KPS (1992) Cryopreservation of embryonic axes of trifoliate orange (*Poncirus trifoliata* L.) RAF. Plant Cell Rep 11: 372–374
Reed BM (1989) The effect of cold hardening and cooling rate on the survival of apical meristems of *Vaccinium* species frozen in liquid nitrogen. Cryo Lett 10: 315–322
Reed BM (1990) Survival of in vitro-grown apical meristems of *Pyrus* following cryopreservation. HortScience 25(1): 111–113
Reed BM (1992) Cryopreservation of *Ribes* apical meristem. 29th Annu Meet Cryobiology, p 41
Reed BM, Lagerstedt HB (1987) Freeze preservation of apical meristems of *Rubus* in liquid nitrogen. HortScience 22(2): 302–303
Reuff I, Seitz U, Ulrich B, Reinhard E (1988) Cryopreservation of *Coleus blumei* suspension and callus cultures. J Plant Physiol 133: 414–418
Sakai A (1960) The frost-hardening process of woody plants. VIII. Relation of polyhydric alcohols to frost hardiness. Low Temp Sci Ser B 18: 15–22
Sakai A (1986) Cryopreservation of germplasm of woody plants. In: Bajaj YPS (ed) Biotechnology in agriculture and forestry, vol 1. Trees I. Springer, Berlin Heidelberg New York, pp 113–129
Sakai A, Sugawara Y (1973) Survival of poplar callus at superlow temperatures after cold acclimation. Plant Cell Physiol 14: 1202–1204
Sakai A, Kobayashi S, Oiyama Y (1990) Cryopreservation of nucellar cells of navel orange (*Citrus sinensis* Osb. var. *brasiliensis* Tanaka) by vitrification. Plant Cell Rep. 9: 30–33
Sakai A, Kobayashi S, Oiyama I (1991a) Survival by vitrification of nucellar cells of navel orange (*Citrus sinensis* var. *brasiliensis* Tanaka) cooled to −196 °C. J Plant Physiol 137: 465–470
Sakai A, Kobayashi S, Oiyama I (1991b) Cryopreservation of nucellar cells of navel orange (*Citrus sinensis* Osb.) by a simple freezing method. Plant Sci 74: 243–248
Sakai A, Yamakawa M, Sakato D, Harada T, Yakuwa T (1978) Development of a whole plant from an excised strawberry runner apex frozen to −196 °C. Low Temp Sci Ser B, Biol Sci 36: 31–38
Sala F, Cella R, Rollo F (1979) Freeze-preservation of rice cells. Physiol Plant 45: 170–176
Seibert M (1976) Shoot initiation from carnation shoot apices frozen to −196 °C. Science 191: 1178–1179
Seibert M, Wetherbee PJ (1977) Increased survival and differentiation of frozen herbaceous plant organ cultures through cold treatment. Plant Physiol 59: 1043–1046
Seitz U, Reinhard E (1987) Growth and ginsenoside pattern of cryopreserved *Panax ginseng* cell cultures. J Plant Physiol 131: 215–223
Seitz U, Alfermann AW, Reinhard E (1983) Stability of biotransformation capacity in *Digitalis lanata* cell cultures after cryogenic storage. Plant Cell Rep 2: 273–276
Shimonishi K, Ishikawa M, Suzuki S, Cosawa K (1991) Cryopreservation of melon somatic embryos by desiccation method. Jpn J Breed 41: 347
Siminovitch D, Briggs DR (1949) The chemistry of the living bark of the black locust tree in relation to frost hardiness. I. Seasonal variations in protein content. Arch Biochem 23: 8–17

Stanwood PC (1985) Cryopreservation of seed germplasm for genetic conservation. In: Kartha KK (ed) Cryopreservation of plant cells and organs. CRC Press, Boca Raton, pp 199–226

Stushnoff C (1991) Cryopreservation of fruit crop genetic resources. HortScience 26: 518–522

Stushnoff C, Remmele RL, Esensee V (1992) Use of phase diagrams to assess cryopreservability of dormant buds from several apple cultivars and from *Amelanchier alnifolia* Nutt. 'SMOKY', CRYO-92, 41 pp

Sudarmonowati E, Henshaw GG (1990) Cryopreservation of cassava somatic embryos. VIIth Int Congr Plant tissue and cell culture, Amsterdam, Abstr, p 378

Sugawara Y, Sakai A (1974) Survival of suspension-cultured sycamore cells cooled to the temperature of liquid nitrogen. Plant Physiol 54: 772–774

Sun CN (1958) The survival of excised pea seedlings after drying and freezing in liquid nitrogen. Bot Gaz 119: 234–236

Sun DL, Sun LH, Jian LC (1988) Plant regeneration from cryopreserved *Panicum miliaceam* calli. Chin Acad Sci 5: 230–239

Sun LH, Jian LC (989) The cryopreservation of maize (*Zea mays* L.) calli. Chinese Bull Bot 6: 28–30

Sun LH, Jian LC (1990) Cryopreservation of sainfoin tissue cultures and their ultrastructural observation. Acta Bot Sin 32: 262–267

Takeuchi M, Matasushima H, Sugawara Y (1980) Long-term freeze-preservation of protoplasts of carrot and Marchantia. Cryo-Lett 1: 519

Takeuchi M, Matasushima H, Sugawara Y (1982) Totipotency and viability of protoplasts after long-term freeze preservation. In: Fujiwara A (ed) Plant tissue culture 1982. Maruzen, Tokyo, pp 797–798

Taniguchi K, Tanaka R, Ashitani N, Miyagawa H (1988) Freeze preservation of tissue cultured shoot primordia of the annual *Haplopappus gracilis* (2n = 4). Jpn J Genet 63: 267–272

Tannoury M, Ralambosoa J, Kaminski M, Dereuddre J (1991) Cryoconservation par vitrification d'apex enrobés d'oeillet (*Dianthus caryophyllus* L.) cultive in vitro. CR Acad Sci Paris, Ser III 313: 633–638

Tessereau H, Lecouteux C, Florin B, Schlienger C, Petiard V (1991) Use of a simplified freezing process and dehydration for the storage of embryogenic cell lines and somatic embryos. Rev Cytol Biol Végétal Bot 14: 297–310

Tisserat B, Ulrich JM, Finkle BJ (1981) Cryogenic preservation and regeneration of date palm tissue HortSci. 16: 47–48

Touchell DH, Dixon KW, Tan B (1992) Cryopreservation of shoot-tips of *Grevillea scapigera* (Proteaceae) from western Australia. Australian J Bot 40: 305–310

Towill LE (1981) *Solanum etuberosum*: a model for studying the cryobiology of shoot-tips in the tuber bearing *Solanum* species. Plant Sci Lett 20: 315–324

Towill LE (1984) Survival at ultra-low temperature of shoot tips from *Solanum tuberosum* groups andigena, phureja, stenotomum, tuberosum, and other tuber-bearing *Solanum* species. Cry Lett 5: 319–326

Towill LE (1990) Cryopreservation of isolated mint shoot tips by vitrification. Plant Cell Rep 9: 178–180

Towill E, Jarret RL (1992) Cryopreservation of sweet potato [*Ipomoea batatas* (L.) Lam.] shoot tips by vitrification. Plant Cell Rep 11: 175–178

Tyler NJ, Stushnoff C (1988) The effects of prefreezing and controlled dehydration on cryopreservation of dormant vegetative apple buds. Can J Plant Sci 68: 1163–1167

Uemura M, Sakai A (1980) Survival of carnation (*Dianthus caryphyllus* L.) shoot apices frozen to the temperature of liquid nitrogen. Plant Cell Physiol 21: 85–94

Ulrich JM, Finkle BJ, Moore PH, Ginoza H (1979) Effect of a mixture of cryoprotectants in attaining liquid nitrogen survival of callus cultures of a tropical plant. Cryobiology 16: 550–556

Uragami A, Sakai A, Nagai M, Takahashi T (1989) Survival of cultured cells and somatic embryos of *Asparagus officinalis* cryopreserved by vitrification. Plant Cell Rep 8: 418–421

Uragami A, Sakai A, Nagai M (1990) Cryopreservation of dried axillary buds from plantlets of *Asparagus officinalis* L. grown in vitro. Plant Cell Rep 9: 328–331

Uragami A, Lucas MO, Ralambosa J, Renard M, Dereuddre J (1993) Cryopreservation of microspore embryos of rapeseed *Brassica napus* L. by dehydration in air with or without alginate encapsulation. Cryobiology 2: 83–90

Vertucci CW, Berjak P (1991) Cryopreservation of embryonic axes of an homeohydrous (recalcitrant) seed in relation to calorimetric properties of tissue water. Cryo-Lett 12: 339–350

Watanabe K, Mitsuda H, Yamada Y (1983) Retention of metabolic and differentiation potential of green *Lovandula vera* callus after freeze-preservation. Plant Cell Physiol 24: 119–122

Watanabe K, Ueno S, Mitsuda H (1990) Differences in viability after freeze-preservation in liquid nitrogen among rice callus lines. Jpn Agric Biol Chem 54: 1295–1296

Weber G, Roth EJ, Schweiger HG (1983) Storage of cell suspensions and protoplasts of *Glycine max* (L.) Merr., *Brassica napus* (L.), *Datura innoxia* (Mill.), and *Daucus carota* (L.) by freezing. Z Pflanzenphysiol 109: 29–39

Wesley-Smith J, Vertucci CW, Berjak P, Pammenter NW, Crane J (1992) Cryopreservation of desiccation-sensitive axes of *Camellia sinenis* in relation to dehydration, freezing rate and the thermal properties of tissue water. J Plant Physiol 140: 596–604

Withers LA (1978) The freeze-preservation of synchronously dividing cultured cells of *Acer pseudoplatanus* L. Cryobiology 15: 87–92

Withers LA (1979) Freeze-preservation of somatic embryos and clonal plantlets of carrot (*Daucus carota* L.). Plant Physiol 63: 460–467

Withers LA, King PJ (1979) Proline – a novel cryoprotectant for the freeze-preservation of cultured cells of *Zea mays* L. Plant Physiol 64: 675–678

Withers LA, Benson EE, Martin M (1988) Cooling rate/medium interactions in the survival and structural stability of cryopreserved shoot tips of *Brassica napus.* Cryo-Lett 9: 114–119

Yakuwa H, Oka S (1988) Plant regeneration through meristem culture from vegetative buds of mulberry (*Morus bombycis* Koidz) stored in liquid nitrogen. Ann Bot 62: 79–82

Yamada T, Sakai A, Matsumura T, Higuchi S (1991) Cryopreservation of apical meristems of white clover (*Trifolium repens* L.) by vitrification. Plant Sci 78: 81–87

Yoshimatsu K, Yamaguchi H, Shimomura K (1995) Cold storage and cryopreservation of hairy roots of *Panax ginseng.* Plant Cell Rep (in press)

Zhang SB, Jian LC, Qian YQ, Bajaj YPS (1994) Cryopreservation of germplasm of maize. In: Bajaj YPS (ed) Biotechnology in agriculture and forestry, vol 25. Maize. Springer, Berlin Heidelberg New York, pp 619–628

Zhang SB, Jian LC, Kuo CS, Qu GP, Qian YQ (1990) Bud and root differentiation of maize (*Zea mays* L.) protoplasts after cryopreservation. Acta Biol Exp Sin 23: 117–121

Zheng Guang-zhi, He Jing-bo, Wang Shi-ling (1983) Cryopreservation of calli and their suspension culture cells of *Anisodus acutangulus.* Acta Bot Sin 25: 512–517 (in Chinese)

I.2 Cryopreservation of Recalcitrant Seeds

V.C. Pence[1]

1 Introduction

Long-term ex situ storage of plant germplasm is of increasing importance, both for maintaining the genetic diversity of species with current human use, and for the preservation of species threatened with extinction in the wild. Seed banks have long been used for this purpose, taking advantage of a natural plant preservation mechanism: the dry seed. Most agricultural species have desiccation-tolerant, or *orthodox*, seeds, which can often remain viable for many years. Longevity can be increased further by storing the seeds at low temperatures (e.g., Ellis and Roberts 1980), and orthodox seeds are generally adaptable to cryopreservation (storage in liquid nitrogen, LN) (Stanwood 1985).

In contrast, an unknown number of species have desiccation-sensitive, or *recalcitrant*, seeds (Roberts 1973; Table 1). A number of tropical timber, fruit, and plantation crops, as well as species from several threatened habitats, fall into this category, and aspects of their physiology and development have recently been reviewed (Farrant et al. 1988; Berjak et al. 1989, 1990). Recalcitrant seeds are generally short-lived and cannot be stored intact using the traditional methods of drying. Nontraditional storage methods, though, such as cryopreservation, offer the possibility of maintaining tissues from recalcitrant seeds in a stable state for very long periods of time. This chapter will describe methods which are being developed for utilizing LN storage for the long-term preservation of tissues from recalcitrant seeds, as well as some prospects for the application of this technology.

2 Recalcitrant Seeds

Recalcitrant seeds are defined as seeds which are sensitive to desiccation and cannot be dried below a certain critical moisture level. This level differs with different species, but is significantly higher than the 5–10% moisture levels

[1]Center for Reproduction of Endangered Wildlife, Cincinnati Zoo and Botanical Garden, 3400 Vine Street, Cincinnati, Ohio 45220, USA

Biotechnology in Agriculture and Forestry, Vol. 32
Cryopreservation of Plant Germplasm I (ed. by Y.P.S. Bajaj)

Table 1. Some species considered to be or likely to be recalcitrant[a]

Family	Species	Common name
Aceraceae	*Acer pseudoplatanus*	Plane tree maple
	A. saccharinum	Silver maple
Amaryllidaceae	*Scadoxus membranaceus*	
Anacardiaceae	*Bouea macrophylla*	Gandaria
	Mangifera caesia	
	M. foetida	
	M. indica	Mango
	M. odorata	
	Spondias dulcis	Golden apple
Annonaceae	*Stelechocarpus burahol*	
Apocynaceae	*Landolphia kirkii*	
Araceae	*Anthurium* spp.	Anthurium
	Dieffenbachia spp.	Dieffenbachia
	Monstera tenuis	Monstera
Araucariaceae	*Araucaria angustifolia*	Parana pine
	A. araucana	Hardy monkey puzzle tree
	A. bidwilii	Monkey puzzle tree
	A. heterophylla	Norfolk Island pine
	A. hunsteinii	Klinkii pine
Arecaceae	*Areca catechu*	
	Chameodorea spp.	
	Chrysalidocarpus lutescens	Butterfly palm
	Cocos nucifera	Coconut
	Geonoma spp.	
	Welfia georgii	Welfia palm
Bombacaceae	*Durio kutejensis*	Lai
	D. zeylanicum	Durian
	D. zibethinus	Durian
	Pachira aquatica	
Brassicaceae	*Wasabia japonica*	Wasabi
Combretaceae	*Conocarpus erecta*	Buttonwood mangrove
	Laguncularia racemosa	White mangrove
Cycadaceae	*Cycas* spp.	Cycads
	Zamia spp.	Cycads
Dipterocarpaceae	*Anisoptera marginata*	
	Dipterocarpus spp.	Dipterocarp
	Dryobalanops aromatica	Borneo camphor
	D. lanceolata	
	Hopea spp.	
	Neobalanocarpus heimii	
	Parashorea spp.	
	Shorea spp.	
	Stemonoporua canaliculatus	
	Vatica spp.	
Ebenaceae	*Diospyros kaki*	Japanese persimmon
	D. marmorata	
	D. virginiana	Persimmon
Erythroxylaceae	*Erythroxylum coca*	Coca
Euphorbiaceae	*Baccaurea dulcis*	Ketupa
	B. motleyana	Rambai
	Hevea brasiliensis	Rubber

Table 1. (*Contd.*)

Family	Species	Common name
Fabaceae	*Castanospermum australe*	Moreton bay chestnut
	Inga spp.	
	Mora megistosperma	Mora
	Pentaclethra macroloba	Gavilan
	Prioria copaifera	
Fagaceae	*Castanea* spp.	Chestnuts
	Lithocarpus densiflorus	Tanoak
	Quercus spp.	Oaks
Flacourtiaceae	*Casearia corymbosa*	Cerito
Ginkgoaceae	*Ginkgo biloba*	Ginkgo
Gnetaceae	*Gnetum* spp.	
Guttiferae	*Garcinia dulcis*	Mandu
	G. mangostana	Mangosteen
Hippocastanaceae	*Aesculus* spp.	Buckeyes, horsechestnut
Lauraceae	*Cinnamomum camphora*	Camphor
	C. zibethinus	Cinnamon
	Persea americana	Avocado
Lecythidaceae	*Barringtonia asiatica*	
	Bertholletia excelsa	Brazil nut
	Couroupita guianensis	Canonball tree
	Lecythis spp.	
Meliaceae	*Carapa guianensis*	Caobilla
	Lansium domesticum	Langsat
	Sandoricum koetjape	
	Swietenia mahagoni	Mahogany
Moraceae	*Artocarpus altilis*	Breadfruit
	A. heterophyllus	Jackfruit
	A. integer	Cempedak
	Brosimum spp.	
	Poulsenia spp.	
	Pseudolmedia spp.	
	Trophis spp.	
Myristicaceae	*Myristica fragrans*	Nutmeg
	Virola sebifera	Wild nutmeg
Myrtaceae	*Syzygium aquaeum*	Watery rose apple
	S. aromaticum	Clove
	S. malaccense	
Oxalidaceae	*Averrhoa carambola*	Carambola
Pellicieraceae	*Pelliciera rhizophorae*	Tea mangrove
Plantaginaceae	*Plantago cordata*	Heartleaf plantain
Poaceae	*Porteresia coarctata*	
	Spartina anglica	
	Zizania palustris[b]	Wild rice
Podocarpaceae	*Podocarpus henkelii*	
	P. makoyi	
Rhizophoraceae	*Carallia* spp.	
	Rhizophora harrisonii	Red mangrove
	R. mangle	
	R. mucronata	
Rubiaceae	*Posoqueria latifolia*	Boca de vieja
	Sickingia maxonii	Guaytil colorado

(*Contd.*)

Table 1. (*Contd.*)

Family	Species	Common name
Rutaceae	*Poncirus trifoliata*	Hardy orange
Salicaceae	*Salix* spp.	Willows
Sapindaceae	*Euphoria longan*	Longan
	Litchi chinensis	Lychee
	N. lappaceum	Rambutan
	N. mutabile	
Sapotaceae	*Chrysophyllum cainito*	Star apple
	Manilkara achras	Sapodilla
Sterculiaceae	*Cola nitida*	Kola nut
	Theobroma cacao	Cacao
Surianaceae	*Guilfoylia monostylis*	
Theaceae	*Camellia sinensis*	Tea
Trapaceae	*Trapa natans*	Water chestnut
Verbenaceae	*Avicennia bicolor*	Black mangrove
	A. germinans	Black mangrove
Zingiberaceae	*Elettaria cardamomum*	Cardamon

[a]In some cases, species are included for which recalcitrance has been strongly suggested by a description of seed characteristics in the literature, but for which desiccation testing data to confirm this are not yet available. Sources include: King and Roberts (1979); Becwar et al. (1983); Janzen (1983); Ellis (1984); Nkang and Chandler (1989); Probert and Longley (1989); Mycock and Berjak (1990); Janzen and Vàzquez-Yanes (1991); Pammenter et al. (1991); Young and Young (1992); Attree and Fowke (1993); Matsumoto et al. (1994); Pence (unpubl.).
[b]May be dried under certain conditions.

routinely tolerated by orthodox seeds. Recalcitrant seeds are shed from the plant in a naturally hydrated state and remain hydrated until germination or deterioration occurs. Many recalcitrant seeds from tropical areas are also cold-sensitive, thus limiting the potential for traditional seed storage even further.

Desiccation-sensitive species generally fall into three categories:

1. Large-seeded temperate trees such as *Quercus*, *Aesculus*, and *Castanea*. These can be stored in a cool, hydrated state for several months or years, but if not germinated, they succumb to deterioration, and cannot be maintained in the long-term.
2. Species from moist tropical forests where many species germinate readily from newly shed seeds, and maintain their germplasm in a seedling, rather than a soil seed bank.
3. Wetland species, such as *Zizania*, *Spartina*, and *Porteresia* (Probert and Longley 1989). The extent of recalcitrance among wetland and tropical species is not known, and surveys of these groups are needed to clarify this point.

In some cases, species have been misidentified as being recalcitrant; this topic has been reviewed in detail (Roberts et al. 1984; Ellis 1991). Seeds which are dried may change in their requirements for germination, and if these are not met, the ungerminated seeds may be classified as nonviable and thus, sensitive to desiccation. In addition, some species have been shown to be desiccation-tolerant, but not fully orthodox. Species of *Juglans*, once identified as recalcitrant, can survive careful drying (Ellis et al. 1985b). They are still not long-lived in storage, probably due to their high lipid content, and have been designated *suborthodox* seeds (Bonner 1986). Still other seeds have been found to be tolerant of desiccation, but sensitive to low temperatures when dried. Seeds of *Coffea arabica* (Ellis et al. 1991b), *Carica papaya* (Ellis et al. 1991a), *Elaeis guineensis*, and *Oreodoxa regia* (Ellis et al. 1991c) show this characteristic and have been termed *intermediate* species.

It is clear that species which have been classified as recalcitrant do not form a homogeneous group. This review will discuss species which are classified as desiccation-sensitive, but also species which have traditionally been regarded as recalcitrant, but for which recent evidence indicates drying and/or LN tolerance under certain conditions.

3 Factors Affecting Cryopreservation of Recalcitrant Seeds

3.1 Seed Size and the Use of Isolated Embryos

Recalcitrant seeds are often large, with the embryo or embryo axis comprising a smaller percentage of the total dry weight of the seed than in orthodox seeds (Chin et al. 1989b). This presents difficulties in applying uniform and rapid drying and cooling procedures to the embryo. Tissues within the seed may also lose water at different rates, as with *Elaeis guineensis*. When the seed is dried to a level of 5% moisture overall, the embryo contains 15% moisture. This originally led to the misidentification of oil palm as a recalcitrant seed, although the excised embryos can be dried to a moisture level of 10% and successfully cryopreserved (Grout et al. 1983).

3.2 Desiccation Tolerance

Although by definition, recalcitrant seeds are intolerant of drying, they do in fact vary in the level of moisture they must retain to remain viable. Recalcitrant seeds undergo desiccation damage when moisture levels drop below a *critical moisture level*. However, if levels are too high during LN exposure, ice will form and damage the tissue. Thus, there may be only a narrow "window" at which maximum survival through LN exposure is obtained, as with *Aesculus hippocastanum* axes (Pence 1992; Fig. 1). This has been observed in several species (Grout et al. 1983; Normah et al. 1986; Pritchard and Prendergast 1986).

When seed tissues are dried, freezable water is lost first, followed by nonfreezable water, which is associated with macromolecules. When freezable water is present in the seed tissues, potentially damaging phase transitions measured by differential scanning calorimetry (DSC) (Vertucci 1989) are observed during cooling as water freezes. At the *threshold moisture level*, all freezable water is removed and only nonfreezable water is present. This point differs with different tissues and is related to lipid content (Vertucci and Leopold

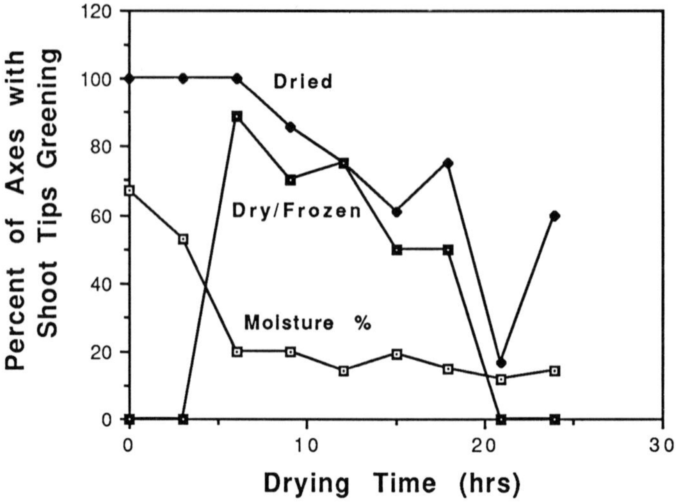

Fig. 1. Desiccated and dry/frozen *Aesculus hippocastanum* axes from various harvest times which showed shoot tips greening after 6 days in culture

1987; Vertucci and Roos 1990). Among species of *Araucaria*, for example, seeds with high starch levels were more sensitive to desiccation than those containing high levels of lipids (Tompsett 1984).

In contrast to orthodox seeds, it appears that recalcitrant seeds do not tolerate the loss of nonfreezable water (Becwar et al. 1983; Pammenter et al. 1991; Berjak et al. 1992), but they may also suffer desiccation damage at even higher moisture levels (Hor et al. 1990). Depending on the species and the procedures used, tissues may survive LN exposure with low levels of freezable water present, while some tissues with no freezable water may still be damaged by LN (Vertucci 1989). It has been shown that very rapid rates of cooling, which inhibit the formation of large ice crystals during freezing, can increase survival in tissues with relatively high levels of freezable water (Wesley-Smith et al. 1992).

The ability to survive desiccation does not guarantee that cryodamage will be avoided. Even when the loss of nonfreezable water can be tolerated, there still may be a relatively narrow range of moisture at which LN exposure is tolerated, as with the "intermediate" *Coffea arabica* seeds (Normah and Vengadasalam 1992), or LN may not be tolerated, even though there is survival at $-20\,^{\circ}C$, as with wild rice seeds (Vertucci et al. 1993). In most cases, though, the critical and threshold moisture levels of the tissue will determine whether desiccation can be used as a protection against LN damage.

3.3 Tissue Type

Tissues within the recalcitrant seed can also differ in their desiccation tolerance, as in *Quercus robur*, where axes and cotyledons differ in their critical moisture

Table 2. Percent of *A. hippocastanum* axes responding after drying 0–8 h, with and without LN freezing. Measurements were taken after 2 weeks on BW medium, excluding root data. After 3 weeks, embryos were transferred to MS medium lacking growth regulators, and root initiation after 12 weeks was recorded. (Pence 1992)

Axis treatment	No. of axes	Moisture (%)	Axes Responding (%)				
			Shoot greening	Leaf growth	Hypocotyl elongation	Bud formation	Root initiation
Dried							
t=0 h	12	69	100	100	92	58	88
2.5	12	30	100	100	75	58	92
5	10	22	90	90	80	50	75
8	12	19	100	75	42	42	60
Frozen							
t=0 h	8	69	0	0	0	0	0
2.5	12	30	75	100	0	0	0
5	7	22	100	57	43	14	25
8	15	19	93	67	40	67	73

levels (Finch-Savage 1992). Even within isolated embryos, tissues may differ in their desiccation sensitivity. Below the critical moisture level, embryos of *Araucaria hunsteinii* increasingly showed damage, but damage was greater in shoots than in roots (Pritchard and Prendergast 1986). This is similar to that observed with *Quercus rubra*, *Q. robur*, and *Castanea* embryo axes (Grout 1986; Pritchard 1991; Pence 1992; Poulsen 1992). With *Aesculus hippocastanum* axes there was a similar survival of root and shoot meristems through drying, but roots were inhibited by freezing at higher overall moisture levels (30%) than shoots (Table 2). This suggests that the actual moisture levels of the root and shoot meristems within the same embryo may differ during the course of drying and that this may result in differential sensitivity to LN exposure. Similarly, in *Veitchia* and *Howea* palm embryos at 10% moisture, the haustorium was sensitive to LN damage even though plants could be grown in culture from cryopreserved embryos (Chin et al. 1988).

3.4 Developmental Stage

The application of desiccation at the proper development time can be critical to maximizing survival of tissues through LN exposure. In orthodox seeds maturation drying acts as a trigger to turn off events associated with embryo development and, upon rehydration, turn on events associated with germination (Kermode et al. 1989). Once germination begins, the orthodox seeds become sensitive to desiccation.

In contrast, events associated with germination in recalcitrant seeds are not as clearly delineated from maturation and appear to begin near the time of shedding from the plant (Farrant et al. 1992). It has been shown for *Avicennia marina* that as the seed moves toward germination, the embryo becomes more sensitive to desiccation, i.e., the critical moisture level increases (Farrant et al. 1986). Because of this, if drying occurs early and quickly, the embryos can tolerate much lower moisture levels than if drying is done slowly. Other recalcitrant and intermediate species have been shown to change in their level of tolerance to desiccation through various stages of development and germination (Hong and Ellis 1990; Ellis et al. 1991b; Berjak et al. 1992; Tompsett and Pritchard 1993).

Recalcitrant seeds differ, however, in how readily germination progresses and how long the seeds remain viable after shedding, and this can affect the response to drying and freezing. Many tropical species germinate rapidly, such as *Theobroma cacao* and *Mangifera indica*, but some, such as *Cocos nucifera* and *Bertholletia excelsa*, may remain dormant for several months or years. Temperate species, such as *Quercus* and *Zizania*, normally overwinter as moist seeds, but some, such as species of *Salix*, germinate readily. Seed tissues may exhibit maximal tolerance of desiccation (although not to the levels of orthodox seeds) before or shortly after shedding, or for a longer period, depending on the species. Sensitivity of embryo axes to desiccation did not increase with storage with seeds of *Quercus robur* (Finch-Savage 1992), whereas seeds of *Zizania palustris* can tolerate orthodox levels of desiccation, but only for a short-time after harvest (Kovach and Bradford 1992; Vertucci et al. 1993). In several cases, it has been noted that embryos survive desiccation and freezing better when they are fresh (Pence 1990; deBoucaud et al. 1991).

4 Application of Cryopreservation to Recalcitrant Seeds

Because of the variety among recalcitrant seeds, it appears that no one method will be ideal for all species. Two general approaches have been directed at the cryopreservation of these tissues. In some cases, seeds or parts of seeds are exposed to LN in a desiccated or partially desiccated state. This is done when the tissue can be dried sufficiently to avoid freezing damage. When this is not possible, tissues can be frozen while hydrated or partially hydrated, using cryoprotectants and controlled cooling to minimize freezing damage. In some cases, these two approaches have been combined. A summary of this work is given in Table 3.

4.1 Tissues

In some cases, whole seeds have been directly exposed to LN. With large seeds, such as *Quercus* spp., this has been generally less successful than axis freezing (Ahuja 1991; Gonzalez-Benito and Perez-Ruiz 1992), although with seeds of intermediate species, such as *Carica papaya* and *Coffea liberica*, success with whole seeds has been reported (Chin and Krishnapillay 1989; Normah and Vengadasalam 1992). Studies on recalcitrant species with smaller seeds, such as

Table 3. A summary of research on the survival of nonorthodox seed tissues through LN exposure[a]

Species	Tissue	Method[b]	Results	Reference
Acer saccharinum	Seeds	Air drying, rapid cooling	N.S.[c]	Becwar et al. (1983)
Aesculus glabra	Axes	Rapid drying to 31%, rapid cooling	80% survival, 20% shoot growth	Pence (1992)
Aesculus hippocastanum	Immature embryos	10% DMSO, 5% sucrose, slow cooling	No growth, but also no browning	Jorgensen (1990)
	Axes	Rapid drying to 10%, rapid cooling	Multiple shoot buds from hypocotyl	Pence (1990)
	Axes	Rapid drying to 12%, rapid cooling	72% germination	Pence (1992)
Aesculus octandra	Axes	Rapid drying, rapid cooling	80% hypocotyl elongation	Pence (1990)
Aesculus parviflora	Axes	Rapid drying, rapid cooling	N.S.	Pence (1990)
Araucaria hunsteinii	Embryos	Rapid drying to 13%, rapid cooling	80% embryo survival, root growth	Pritchard and Prendergast (1986)
Artocarpus integer	Axes	Drying to 41–65%, rapid cooling	N.S.	Hor et al. (1990)
Camellia sinensis	Axes	Rapid drying, rapid cooling	95% survival	Chaudhury et al. (1990)
	Seeds	LN storage	93% emergence	Ma (1991)
	Axes	Rapid drying to 52–62%, very rapid cooling	100% survival and germination	Wesley-Smith et al. (1992)
	Seeds	LN storage	20% germination	Guo and Shi (1990)
Carica papaya	Seeds	Drying to 9–10%, rapid cooling	Viable	Becwar et al. (1983)
	Seeds	Drying to <10%, rapid cooling	Viable	Chin and Krishnapillay (1989)
Castanea mollissima	Axes	Rapid drying, rapid cooling	Callus	Pence (1990)
Castanea sativa	Axes	Rapid drying to 9%, rapid cooling	Callus, roots	Pence (1990)
Chrysalidocarpus lutescens	Seeds	Air drying, rapid cooling	N.S.	Becwar et al. (1983)
Cinnamomum camphora	Seeds	LN storage	20% germination	Guo and Shi (1990)

(*Contd.*)

Table 3. (*Contd.*)

Species	Tissue	Method[b]	Results	Reference
Cocos nucifera	Embryos, whole or cut	Partial drying, 7% DMSO, 7% sucrose, rapid cooling	17–25% elongation or cell proliferation	Bajaj (1984)
	Immature embryos	1-h drying, 10% glycerol, 10% DMSO, rapid cooling	Growth of 1 embryo	Chin et al. (1989a)
	Immature embryos	4h on 60% glucose, 15% glycerol, rapid cooling	Up to 43% survival, 1 plantlet obtained	Assy-Bah and Engelmann (1992a)
	Embryos	4-h drying, 11–20 h on 60% glucose, 15% glycerol, rapid cooling	33–93% survival and growth, depending on variety	Assy-Bah and Engelmann (1992b)
Coffea arabica	Seeds	Air drying to 17%, slow cooling to −80 °C	60% survival	Becwar et al. (1983)
	Somatic embryos	0.75 M sucrose overnight, 5% DMSO, slow cooling	Lag phase, somatic embryogenesis	Bertrand-Desbrunais et al. (1988)
	Embryos	Rapid drying to 16%, rapid cooling	96% survival and growth	Abdelnour-Esquivel et al. (1992)
Coffea canephora	Embryos	Rapid drying to 29%, rapid cooling	41% survival and growth	Abdelnour-Esquivel et al. (1992)
Coffea liberica	Seeds	Rapid drying to 16%, rapid cooling	53% survival	Normah and Vengadasalam (1992)
	Axes	Rapid drying to 15%, cooling	86% survival	Normah and Vangadasalam (1992)
Durio zibethinus	Axes	Drying to 41–65%, rapid cooling	N.S.	Hor et al. (1990)
Elaeis guineensis	Embryos	Rapid drying to 10%, rapid cooling	No loss of viability	Grout et al. (1983)
	Somatic embryos	0.75 M sucrose for 7 days, slow and rapid cooling	20% survival	Engelmann et al. (1988)
	Somatic embryos	0.75 M sucrose for 7 days, drying to 19–35%, rapid cooling	13–53% survival	Dumet et al. (1993)

Euphoria longan	Axes	Rapid drying to 19%, rapid cooling	80% survival	Fu et al. (1990)
Fagus grandiflora	Axes	Rapid drying, rapid cooling	100% survival and germination	Pence (1990)
Fagus sylvatica	Immature embryos	10% DMSO, 5% sucrose, slow cooling	90–100% survival	Jorgensen (1990)
	Seeds	Slow and rapid cooling	N.S.	Ahuja (1991)
	Axes	Rapid cooling	Growth	Ahuja (1991)
Hevea brasiliensis	Axes	Rapid drying to 14–20%, rapid cooling	20–69% survival and germination	Normah et al. (1986)
Juglans nigra	Axes	Rapid drying to 6%, rapid cooling	90% survival and germination	Pence (1990)
Juglans regia	Axes	5 M propanediol, 20% sucrose, drying to 20%, rapid cooling	75% regrowth	de Boucaud et al. (1991)
	Axes	Rapid drying, rapid cooling	100% survival, shoot greening and callus	Pence (1990)
Landolphia kirkii	Axes	Rapid drying to 23–31%, moderately slow cooling to −70 °C	90% survival	Vertucci et al. (1991)
Nephlium lappaceum	Axes	Drying to 13–45%, rapid cooling	N.S.	Hor et al. (1990)
Poncirus trifoliata	Axes	Rapid drying to 14%, rapid cooling	Germination	Radhamani and Chandel (1992)
Quercus alba	Axes	Rapid drying, rapid cooling	Callus	Pence (1990)
Quercus faginea	Axes	Rapid drying to 20%, rapid cooling	60% germination	Gonzalez-Benito and Perez-Ruiz (1992)
	Axes	15% DMSO, rapid cooling	Callus	Gonzalez-Benito and Perez-Ruiz (1992)
Quercus marilandica	Axes	Rapid drying, rapid cooling	N.S.	Pence (1990)
Quercus muhlenbergii	Axes	Rapid drying, rapid cooling	Callus	Pence (1990)
Quercus petraea	Immature embryos	10% DMSO, 5% sucrose, slow cooling	90–100% survival	Jorgensen (1990)
	Seeds	Rapid cooling	N.S.	Jorgensen (1990)
	Seeds	Slow and rapid cooling	N.S.	Ahuja (1991)

(*Contd.*)

Table 3. (*Contd.*)

Species	Tissue	Method[b]	Results	Reference
Quercus robur	Axes	Partial drying to 25%, slow or rapid cooling	N.S.	Poulsen (1992)
	Seeds	Slow and rapid cooling	N.S.	Ahuja (1991)
Quercus rubra	Seeds	Slow and rapid cooling	N.S.	Ahuja (1991)
Salix nigra	Seeds	Rapid drying, or vitrification	Up to 100% germination	Pence (in prep.)
Theobroma cacao	Immature embryos	Preculture with sucrose and ABA, rapid drying and rapid cooling	Callus, one with somatic embryogenesis	Pence (1991b)
	Immature embryos	10% DMSO, 0.5 M sucrose, slow freezing	Somatic embryogenesis in up to 29%	Pence (1991b)
Zizania palustris	Seeds	Slow, drying, above 25 °C to 6–8%, freezing at −20 °C, slow imbibition,	40–90% germination	Kovach and Bradford (1992)
	Seeds	Drying, −20 °C and LN freezing	Survival only at −20 °C freezing	Vertucci et al. (1993)

[a] In a few cases, information on freezing to temperatures warmer than −196 °C has been included.
[b] Unless otherwise indicated, percentages given indicate moisture content of the tissue used for freezing.
[c] N.S. = No survival.

Zizania palustris and *Salix nigra*, have also utilized the entire seed (Kovach and Bradford 1992; Pence, in prep.).

With large recalcitrant seeds, the embryo or embryo axis is often excised for cryopreservation. This, in turn, requires the use of in vitro techniques for subsequent germination. Embryos may be damaged during the excision process, and this may affect subsequent growth (Marin et al. 1990). For some species, conditions for growing isolated embryos or axes have not been optimized, and this will limit the ability to detect cryodamage (e.g., Fu et al. 1990; Assy-Bah and Engelmann 1992a; Poulsen 1992). Procedures for improving embryo culture in recalcitrant species will also improve cryopreservation technologies.

For in vitro germination, the embryos must be removed aseptically from sterilized seeds or fruit, or they must be sterilized after excision. Embryos of *Araucaria hunsteinii* survived desiccated LN exposure better when they were aseptically excised than when directly surface sterilized (Pritchard and Prendergast 1986). In some cases, embryos have been sterilized after cryostorage (Normah and Vengadasalam 1992; Wesley-Smith et al. 1992), allowing cooling and rewarming procedures to be done without aseptic considerations. Sterilizing solutions generally consist of dilutions of commercial sodium hypochlorite, but $HgCl_2$ and hibitane (Wesley-Smith et al. 1992) have also been used. Embryos of recalcitrant species often harbor fungal contaminants which make in vitro culture difficult (Mycock and Berjak 1990). In some cases, LN exposure reduces the growth of these contaminants (Pence, unpubl.), but it has also been noted that the stress of LN exposure may release endogenous bacteria, resulting in higher contamination rates than controls (Ahuja 1991).

Less emphasis has been placed in cryopreservation research on the freezing of hydrated, intact seeds. Work in which the cryoprotectant DMSO has been applied to small hydrated, orthodox seeds, however, has suggested that in some species the potential for germination can be maintained through freezing using this method (Grout 1979; Pence 1991a). Seeds of *Salix nigra* have also been frozen without drying using a cryoprotectant solution (Pence, in prep.)

4.2 Pretreatments

Pretreating embryos and embryo axes can also affect their response to desiccation and cryopreservation. Preculture media containing elevated levels of sucrose and/or ABA may improve survival through desiccation and/or freezing, as has been observed for somatic and zygotic embryos of orthodox species (Shimonishi et al. 1991; Lecouteux et al. 1992; Kendall et al. 1993). Tolerance to drying improved in axes of *Quercus robur* when they were pretreated with polyethylene glycol (PEG) (Pritchard 1991), while survival of somatic embryos of *Elaeis guineensis* through desiccated freezing was improved by preculture on 0.75 M sucrose for 7 days (Dumet et al. 1993). Similarly, tolerance to desiccated LN exposure was slightly improved by preculture of immature embryos of *Theobroma cacao* on 21% sucrose and 10 μM ABA, although there was no detectable effect on the tolerance of hydrated freezing (Pence 1991b). It is not known whether recalcitrant species possess the mechanisms to respond to these treatments to the extent of orthodox species.

4.3 Desiccation

There are no absolute definitions of the rates of slow and rapid or flash drying used for desiccating seed tissues for cryostorage, but generally slow drying is accomplished by air drying (Becwar et al. 1983), while rapid drying is done under a moving air stream, such as that supplied by a laminar flow hood or a flash drying chamber (Pammenter et al. 1991). This has also been achieved using drying in a 30 °C oven (Normah and Vengadasalam 1992). In some cases, drying is slowed by the restrictions of the seed coat, as with *Zizania* (Kovach and Bradford 1992), the seeds of which were dried over saturated salt solutions. A controlled atmosphere of 15% relative humidity and 15 °C can also be used for drying (Pritchard and Prendergast 1986).

In order to determine the optimum level of desiccation for LN exposure, tissues are harvested at various time intervals during drying and tested for viability, for viability after freezing, and for moisture content. The latter is determined by weighing before and after oven drying (Ellis et al. 1985a).

The rate at which seed tissues are dried may also be critical for maximum survival of the tissue, and this in turn will affect maximum survival through LN exposure. When embryo axes of *Landolphia kirkii* were dried slowly, the critical moisture level was 41–47%[1], whereas when they were dried rapidly, the critical moisture level was 23%, just above the threshold moisture level of 22% (Pammenter et al. 1991). In the case of *Avicennia marina*, seeds became more desiccation-sensitive with increasing storage time so that rapid drying was beneficial (Farrant et al. 1986). Survival of whole seeds of *Mangifera indica* was improved by rapid drying (Wang and Fu 1991). Similarly, in a study of three recalcitrant species, the ones most tolerant of desiccation underwent the most rapid drying (Farrant et al. 1989). In some cases, this may be related to species differences in seed size (Tompsett 1984, 1987), but as tissues of *Landolphia kirkii* became more desiccation-sensitive, their drying rate also slowed (Berjak et al. 1992).

4.4 Cryoprotection

Depending on their composition, cryoprotectant solutions are thought to stabilize membranes and to dehydrate the tissues, removing water which can cause freezing damage (Finkle et al. 1985; Assy-Bah and Engelmann 1992a). A common cryoprotectant solution for slow freezing protocols is 5–10% dimethylsulfoxide (DMSO) and/or 5–25% sucrose (e.g., Jörgensen 1990). In some cases, 10–15% glycerol has been used (Chin et al. 1989a). These may be

[1] All moisture levels are given on a wet weight basis.

combined with preculture (days or weeks) or pretreatment (hours) with elevated levels of sucrose (up to 21%, Pence 1991b) or glucose (60%, Assy-Bah and Engelmann 1992a).

In some cases, cryoprotectant solutions or preculture have been used before drying or partial drying, increasing the ability of the tissue to vitrify during LN exposure (Chin et al. 1989a; de Boucaud et al. 1991; Pence 1991b; Assy-Bah and Engelmann 1992b). When *Juglans regia* embryo axes were dried before cryoprotection, however, the cryoprotectant did not affect survival through freezing of embryos with 5% moisture, but did improve survival of embryos with 20% moisture (de Boucaud et al. 1991).

A related method is that of encapsulation, a technique which has been applied to somatic embryos of orthodox species (Dereuddre et al. 1991a; Uragami et al. 1993). Over 90% survival was obtained with torpedo stage carrot embryos encapsulated in alginate beads, incubated in 0.3 M sucrose, and dried to less than 20% moisture before immersion into LN. The high levels of sucrose, concentrated further by drying, promote the formation of glasses, stabilizing them through LN exposure (Dereuddre et al. 1991b).

The use of very high concentrations of cryoprotectants also promotes vitrification of tissues during rapid LN exposure, and mixtures of concentrated cryoprotectants have been used for vitrification in a number of plant tissues (Yamada et al. 1991). Seeds of *Salix nigra* have survived hydrated LN exposure using this procedure (Pence, in prep.). Work with imbibed pea axes suggests the applicability of vitrification procedures to similar tissue in recalcitrant seeds (Mycock et al. 1991). While these procedures have not been used widely with recalcitrant seed tissues at this point, some relatively high levels of cryoprotectants, such as 5 M propanediol plus 20% sucrose and 60% glucose plus 15% glycerol have been combined with drying, and these probably allow for vitrification (de Boucaud et al. 1991; Assy-Bah and Engelmann 1992b).

4.5 LN Exposure

Dry embryos can be contained in cryovials for exposure to LN, particularly if aseptic conditions are required. In some cases, direct freezing of the embryos without a container has been reported, permitting extremely rapid cooling rates (Wesley-Smith et al. 1992). The tissues are surface sterilized after thawing. Rapid freezing can increase survival of tissues which are relatively hydrated. Embryos of *Camellia sinensis* survived LN exposure with a moisture level of 52% when cooled rapidly by direct immersion into subcooled LN. Only drier tissues were able to survive slower cooling rates.

When cryoprotectant solutions are used, tissues are generally frozen in small cryovials, or similar containers. Slow, controlled freezing of tissues has been used in some cases, which removes water from the tissues during the freezing process (Meryman and Williams 1985). Rates of freezing have been in the range of 0.4–1 °C/min down to temperatures of −30 to −50 °C, at which point the sample is plunged into LN (Jörgensen 1990; Pence 1991b; Poulsen 1992). Work with somatic embryos of carrot has shown that sucrose pretreatment combined with precooling in a conventional −20 °C freezer can also provide protection through LN exposure for small, immature embryos (Lecouteux et al. 1991). Vitrification and desiccation procedures have employed rapid freezing, obtained by plunging the cryovial directly into LN.

4.6 Thawing

Thawing of the desiccated embryos is accomplished by equilibrating them with room temperature or by placing them directly into liquid medium. The temperature at which thawing occurs may be important, as with embryos of *Euphoria longan*, which survived warming at 30 °C better than at lower temperatures (Fu et al. 1990). Rehydration of dried tissues can be accomplished by equilibrating the embryos in 100% humidity, by placing them in a liquid medium, or by placing them directly on recovery medium.

For tissues frozen in a cryoprotectant solution, rapid thawing is commonly used and is achieved by immersing the cryovial in a 40 °C water bath for 1–2 min, until the cryoprotectant solution is melted. The tissues may or may not be washed of cryoprotectant before being moved to a semisolid medium for recovery and growth.

4.7 Recovery

Embryos are put onto a tissue culture medium for recovery and growth, and this may also serve to rehydrate desiccated embryos. *Coffea arabica* embryos, pretreated with 0.75 M sucrose before freezing, were recovered on the same medium, but transferred daily, decreasing the sucrose to 0.1 M (Bertrand-Desbrunais et al. 1988). Isolated embryos or axes may be capable of normal germination, and for this a basal medium and sucrose are often sufficient. Growth regulators are often not needed and can inhibit normal germination (Pence 1992), but they can be useful in identifying the survival of cells through the freezing process, by stimulating callus growth, even if the embryo is too damaged to germinate (Pence 1990). Ideally, such a damaged embryo can be stimulated to produce shoots or somatic embryos from surviving cells, as has been done with frozen immature embryos of *Theobroma cacao* (Pence 1991b). Studies on embryo culture and regeneration may be needed before freezing, in order to determine the optimal recovery medium.

5 Conclusions and Prospects

As research on the cryopreservation of recalcitrant seeds continues, the techniques best adapted to the size and true desiccation tolerance of each species will be adopted. In recent years, the number of species reported as surviving cryopreservation has increased significantly, and this is likely to continue. Other tissues of recalcitrant species may be targets of cryopreservation, such as cultured shoot tips or cell cultures, but seed tissue offers the potential for maintaining more easily a greater portion of the genetic diversity of a species than clonal material.

A large portion of the work on recalcitrant seed freezing has centered on the use of desiccation as a cryoprotective measure and drying the tissues to the maximum tolerable limit before LN exposure. Other work has used more traditional cryoprotective approaches, originally developed for the freezing of hydrated animal tissues. This has involved the use of cryoprotectant solutions and slow freezing, but more recently has moved to higher amounts of cryoprotectant and fast freezing, adapting the techniques of vitrification. Desiccation in combination with cryoprotectants may also allow vitrification of the tissues. Along these lines, encapsulation has been demonstrated as a feasible technique for hydrated embryos but has yet to be widely applied.

The need to use isolated embryos in some large-seeded species makes the banking of recalcitrant seed materials more labor-intensive and expensive than orthodox seed banking, because of the effort involved in extracting embryos for cryostorage and in germinating them in vitro. The cost should compare favorably, however, with the cost of maintaining "living germplasm collections" in plantations and gardens. The use of somatic embryo systems as well as hydrated axes of orthodox species can give some assistance in providing materials for experimentation and the development of techniques. Each recalcitrant species is likely to be somewhat different, however, requiring specialized refinement of procedures.

As more species with recalcitrant seeds are examined in detail, the definition of recalcitrance becomes increasingly qualified. Even for species which are now known to be desiccation-tolerant, cryopreservation may still be a very useful tool. For species such as *Juglans* and *Salix*, which are not long-lived even in the dry state in traditional cold storage, LN should increase longevity dramatically. For several species designated as "intermediate", which are tolerant of desiccation, but are damaged by storage in the dry state at cool temperatures (4, $-20\,^{\circ}$C), good survival has been obtained after LN exposure. On the other hand, there are some species, such as *Zizania palustris*, which can be dried but which appear to be sensitive to LN, but not to $-20\,^{\circ}$C. Further work with such species will determine whether modifications in freezing protocols can overcome this sensitivity to cryopreservation.

Many recalcitrant species, particularly large-seeded tropical species, still lack adequate protocols for long-term maintenance through cryopreservation, and their large embryos present particular problems for the application of any drying and freezing protocols. For some of these, freezing of immature embryos and the subsequent production of somatic embryos, or maturation and germination in vitro may be necessary. In preserving germplasm of endangered wild species, however, practical problems may arise in collecting immature embryos, particularly from tall, climax rainforest species.

The protection of recalcitrant tissues through LN exposure, whether by cryoprotectants, desiccation, or combinations of the two, should allow for the preservation of germplasm hitherto unadaptable to long-term storage. Ex situ preservation of this germplasm has thus far been maintained in gardens and plantations. Only a few species have the economic interest to support large germplasm collections under these conditions. For most recalcitrant species grown in gardens, space severely limits the amount of genetic representation which can be preserved. Cryopreservation can offer the opportunity of preserving seeds or embryos in a manner analogous to a traditional seed bank, allowing a large amount of genetic variation to be conserved in a very small space, and to be easily "backed up" in several geographic locations, irrespective of climate. As more habitats become threatened and as the expense of maintaining living collections increases, cryostorage can offer a supporting technique for protection against the loss of genetic diversity of species with recalcitrant seeds.

6 Selected Protocols

6.1 Cryopreservation of Desiccated Embryo Axes

Used for *Juglans*, *Aesculus*, and other large-seeded temperate tree species (Pence 1990, 1992).

1. Isolate embryo axes from seeds. Collect on moist paper over ice.
2. Surface sterilize embryos in a 5% dilution of commercial sodium hypochlorite for 5 min, followed by 2 rinses in sterile, distilled water.
3. Blot axes dry on sterile filter paper and transfer to a sterile dish with dry filter paper. Leave in open dish under the air flow of the laminar flow hood.
4. To determine the optimal drying time for a new species, harvest embryos at various time intervals, including $t = 0$, for moisture determination, culture on recovery medium, and LN exposure, thawing and culture on recovery medium.
5. For cryopreservation, transfer embryos from drying dish to 2 ml polypropylene sterile cryovials. Plunge these directly into LN.
6. Thaw by equilibrating vials with ambient temperature.
7. Remove embryos and transfer to a recovery medium (basal Murashige and Skoog 1962; Driver and Kuniyuki 1984; Lloyd and McCown 1980; etc. medium, depending on the species).

6.2 Cryopreservation of Desiccated Embryo Axes

Used for *Camellia sinensis*. (Wesley-Smith et al. 1992).

1. Excise embryos and collect on moist paper.
2. Dry in air flow of flash drying chamber (Pammenter et al. 1991).
3. Harvest at various time intervals and process as above to determine optimum drying time.
4. Immerse axes directly into supercooled LN.
5. Thaw and rehydrate by transferring axes to liquid medium at room temperature for 30 min.
6. Surface sterilize axes in 1% sodium hypochlorite and 0.2% hibitane for 10 min, followed by 2 rinses in sterile, distilled water.
7. Transfer to recovery medium.

6.3 Cryopreservation of Hydrated Embryos by Slow Freezing

Used for immature embryos of *Theobroma cacao* (Pence, 1991b).

1. Aseptically excise embryos from fruit surface sterilized with 70% ethanol. Collect in dish with sterile liquid medium.

2. Transfer embryos to a known volume of sterile liquid medium with 3% sucrose. Add slowly, over a period of 45 min, an equal volume of the same medium with 1 M sucrose and 20% DMSO, to give a final cryoprotectant concentration of 0.5 M sucrose and 10% DMSO.
3. Cool to 4 °C for 30 min. Transfer to sterile, 2 ml polypropylene cryovials.
4. Freeze at a rate of 0.4 °C/min down to −35 °C. Transfer to LN.
5. Thaw rapidly by immersing the cryovial in a 40 °C water bath for 2 min.
6. Transfer embryos to recovery medium. (In the case of immature *T. cacao* embryos, this was a medium which would promote somatic embryogenesis.)

References

Abdelnour-Esquivel A, Villalobos V, Engelmann F (1992) Cryopreservation of zygotic embryos of *Coffea* spp. Cryo Lett 13: 297–302

Ahuja MR (1991) Application of biotechnology to preservation of forest tree germplasm. In: Ahuja MR (ed) Woody plant biotechnology. Plenum Press, New York, pp 307–313

Assy-Bah B, Engelmann F (1992a) Cryopreservation of immature embryos of coconut (*Cocos nucifera* L.). Cryo Lett 13: 67–74

Assy-Bah B, Engelmann F (1992b) Cryopreservation of mature embryos of coconut (*Cocos nucifera* L.) and subsequent regeneration of plantlets. Cryo Lett 13: 117–126

Attree SM, Fowke LC (1993) Embryogeny of gymnosperms: advances in synthetic seed technology of conifers. Plant Cell Tissue Organ Cult 35: 1–35

Bajaj YPS (1984) Induction of growth in frozen embryos of coconut and ovules of citrus. Curr Sci 53: 1215–1216

Becwar MR, Stanwood PC, Leonhardt KW (1983) Dehydration effects on freezing characteristics and survival in liquid nitrogen of desiccation-tolerant and desiccation-sensitive seeds. J Am Soc Hortic Sci 108: 613–618

Berjak P, Farrant JM, Pammenter NW (1989) The basis of recalcitrant seed behavior. Cell biology of the homoiohydrous seed condition. In: Taylorson RB (ed) Recent advances in the development and germination seeds. Plenum Press, New York, pp 89–108

Berjak P, Farrant JM, Mycock DJ, Pammenter NW (1990) Recalcitrant (homoiohydrous) seeds: the enigma of their desiccation-sensitivity. Seed Sci Technol 18: 297–310

Berjak P, Pammenter NW, Vertucci C (1992) Homoiohydrous (recalcitrant) seeds: development status, desiccation sensitivity and the state of water in axes of *Landolphia kirkii* Dyer. Planta 186: 248–261

Bertrand-Desbrunais A, Fabre J, Engelmann F, Dereuddre J, Charrier A (1988) Reprise de l'embryogenèse adventive à partir d'embryons somatiques de cafeier (*Coffea arabica* L.) après leur congélation dans l'azote liquide. CR Acad Sci Paris Ser III 307: 795–801

Bonner F (1986) Technologies to maintain tree germplasm diversity. In: Technologies to maintain biological diversity, vol 2, part D. Office of Technology Assessment, Washington, DC, pp 630–672

Chaudhury R, Lakhanpaul S, Chandel KPS (1990) Germination and desiccation tolerance of tea (*Camellia sinensis* (L.) O. Kuntze) seeds and feasibility of cryopreservation. Sri Lanka J Tea Sci 59: 89–94

Chin HF, Krishnapillay B (1989) Cryogenic storage of some horticultural species. Acta Hortic 253: 107–112

Chin HF, Krishnapillay B, Alang ZC (1988) Cryopreservation of *Veitchia* and *Howea* palm embryos: nondevelopment of the haustorium. Cryo Lett 9: 372–379

Chin HF, Krishnapillay B, Hor YL (1989a) A note on the cryopreservation of embryos from young coconuts (*Cocos nucifera* var. MAWA). Pertanika 12: 183–186

Chin HF, Krishnapillay B, Stanwood PC (1989b) Seed moisture: recalcitrant vs. orthodox seeds. In: Stanwood PC, McDonald MB (eds) Seed moisture. Crop Science Society of America Spec Publ No 14, Madison, Wisconsin, pp 15–22

de Boucaud MT, Brison M, Ledoux C, Germain E, Lutz A (1991) Cryopreservation of embryonic axes of recalcitrant seed: *Juglans regia* L. cv. Franquette, Cryo Lett 12: 163–166
Dereuddre J, Blandin S, Hassen N (1991a) Resistance of alginate-coated somatic embryos of carrot (*Daucus carota* L.) to desiccation and freezing in liquid nitrogen. 1. Effects of preculture. Cryo Lett 12: 125–134
Dereuddre J, Hassen N, Blandin S, Kaminski M (1991b) Resistance of alginate-coated somatic embryos of carrot (*Daucus carota* L.) to desiccation and feeezing in liquid nitrogen. 2. Thermal analysis. Cryo Lett 12: 135–148
Driver JA, Kuniyuki AH (1984) In vitro propagation of paradox walnut rootstock. Hort Sci 19: 507–509
Dumet D, Engelmann F, Chabrillange N, Duval Y (1993) Cryopreservation of oil palm (*Elaeis guineensis* Jacq.) somatic embryos involving a desiccation step. Plant Cell Rep 12: 352–355
Ellis RH (1984) Revised table of seed storage characteristics. Plant Genet Resour Newsl 58: 16–33
Ellis RH (1991) The longevity of seeds. HortScience 26: 1119–1125
Ellis RH, Roberts EH (1980) The influence of temperature and moisture on seed viability period in barley (*Hordeum distichum* L.). Ann Bot 45: 31–37
Ellis RH, Hong RD, Roberts EH (1985a) Handbook of seed technology for genebanks, vol 1. International Board for Plant Genetic Resources, Rome
Ellis RH, Hong RD, Roberts EH (1985b) Handbook of seed technology for genebanks, vol 2. International Board for Plant Genetic Resources, Rome
Ellis RH, Hong TD, Roberts EH (1991a) Effect of storage temperature and moisture on the germination of papaya seeds. Seed Sci Res 1: 69–72
Ellis RH, Hong TD, Roberts EH (1991b) An intermediate category of seed storage behavior? II. Effects of provenance, immaturity, and imbibition on desiccation-tolerance in coffee. J Exp Bot 42: 653–657
Ellis RH, Hong TD, Roberts EH, Soetisna U (1991c) Seed storage behaviour in *Elaeis guineensis*. Seed Sci Res 1: 99–104
Engelmann F, Duval Y, Pannetier C (1988) Use of cryoprotection for setting up a bank of oil palm (*Elaeis guineensis* Jacq.) somatic embryos. Oleagineux 43: 327–328
Farrant JM, Pammenter NW, Berjak P (1986) The increasing desiccation sensitivity of recalcitrant *Avicennia marina* seeds with storage time. Physiol Plant 67: 291–298
Farrant JM, Pammenter NW, Berjak P (1988) Recalcitrance – a current assessment. Seed Sci Technol 16: 155–166
Farrant JM, Pammenter NW, Berjak P (1989) Germination-associated events and the desiccation sensitivity of recalcitrant seeds – a study of three unrelated species. Planta 178: 189–198
Farrant JM, Pammenter NW, Berjak P (1992) Development of the recalcitrant (homoiohydrous) seeds of *Avicennia marina*: anatomical, ultrastructural and biochemical events associated with development from histodifferentiation to maturation. Ann Bot 70: 75–86
Finch-Savage WE (1992) Embryo water status and survival in the recalcitrant species *Quercus robur*. Evidence for a critical moisture content. J Exp Bot 43: 663–669
Finkle BJ, Zavala ME, Ulrich JM (1985) Cryoprotective compounds in the viable freezing of plant tissues. In: Kartha KK (ed) Cryopreservation of plant cells and organs. CRC Press, Boca Raton, pp 75–113
Fu JR, Zhang BZ, Wang XP, Qiao YZ, Huang XL (1990) Physiological studies on desiccation, wet storage and cryopreservation of recalcitrant seeds of three fruit species and their excised embryonic axes. Seed Sci Technol 18: 743–754
Gonzalez-Benito ME, Perez-Ruiz C (1992) Cryopreservation of *Quercus faginea* embryonic axes. Cryobiology 29: 685–690
Grout BWW (1979) Low temperature storage of imbibed tomato seeds: a model for recalcitrant seed storage. Cryo Lett 1: 71–76
Grout BWW (1986) Embryo culture and cryopreservation for the conservation of genetic resources of species with recalcitrant seed. In: Withers LA, Alderson PG (eds) Plant tissue culture and its agricultural applications . Butterworth, London, pp 303–309
Grout BWW, Shelton K, Pritchard HW (1983) Orthodox behavior of oil palm seed and cryopreservation of the excised embryo for germplasm conservation. Ann Bot 52: 381–384

Guo CG, Shi SX (1990) Conservation of tea and camphor seeds at very low temperatures. Crop Gen Res 2: 16

Hong TD, Ellis RH (1990) A comparison of maturation drying, germination, and desiccation tolerance between developing seeds of *Acer pseudoplatanus* L. and *Acer platanoides* L. New Phytol 116: 589–596

Hor YL, Stanwood PC, Chin HF (1990) Effects of dehydration on freezing characteristics and survival in liquid nitrogen of three recalcitrant seeds. Pertanika 13: 309–314

Janzen DH (1983) Costa Rican natural history. The University of Chicago Press, Chicago

Janzen DH, Vàzquez-Yanes C (1991) Aspects of tropical seed ecology of relevance to management of tropical forested wildlands. In: Gómez-Pompa A, Whitmore TC, Hadley M (eds) Rain forest regeneration and management. UNESCO, Paris, and The Parthenon Publishing Group, Park Ridge, NJ, pp 137–157

Jörgensen J (1990) Conservation of valuable gene resources by cryopreservation in some forest tree species. J Plant Physiol 136: 373–376

Kendall EJ, Kartha KK, Qureshi JA, Chermak P (1993) Cryopreservation of immature spring wheat zygotic embryos using an abscisic acid pretreatment. Plant Cell Rep 12: 89–94

Kermode AR, Oishi MY, Bewley JD (1989) Regulatory roles for desiccation and abscisic acid in seed development: a comparison of the evidence from whole seeds and isolated embryos. In: Stanwood PC, McDonald MB (eds) Seed moisture. CSSA Spec Publ 14, Crop Science Society of America, Madison, WI, pp 23–50

King MW, Roberts EH (1979) The storage of recalcitrant seeds – achievements and possible approaches. International Board for Plant Genetic Resources Secretariat, Rome

Kovach DA, Bradford KJ (1992) Imbibitional damage and desiccation tolerance of wild rice (*Zizania palustris*) seeds. J Exp Bot 43: 747–757

Lecouteux C, Florin B, Tessereau H, Bollon H, Petiard V (1991) Cryopreservation of carrot somatic embryos using a simplified freezing process. Cryo Lett 12: 319–328

Lecouteux C, Tessereau H, Florin B, Courtois D, Petiard V (1992) Preservation of carrot (*Daucus carota* L.) somatic embryos by dehydration after pretreatment with sucrose. C R Acad Sci Paris Ser III 314: 423–428

Lloyd G, McCown B (1981) Commercially feasible micropropagation of mountain laurel, *Kalmia latifolia*, by use of shoot-tip culture. Int Plant Prop Soc 30: 421–427

Ma YS (1991) Advances in conservation technology for crop germplasm resources in China. Crop Gen Res 3: 1–3

Marin ML, Mafla G, Roca WM, Withers LA (1990) Cryopreservation of cassava zygotic embryos and whole seeds in liquid nitrogen. Cryo Lett 11: 257–264

Matsumoto T, Sakai A, Yamada K (1994) Cryopreservation of in vitro-grown apical meristems of wasabi (*Wasabia japonica*) by vitrification and subsequent high plant regeneration. Plant Cell Rep 13: 442–446

Meryman HT, Williams RJ (1985) Basic principles of freezing injury to plant cells; natural tolerance and approaches to cryopreservation. In: Kartha KK (ed) Cryopreservation of plant cells and organs. CRC Press, Boca Raton, pp 13–47

Murashige T, Skoog F (1962) A revised medium for rapid growth and bioassays with tobacco tissue cultures. Physiol Plant 15: 473–497

Mycock DJ, Berjak P (1990) Fungal contaminants associated with several homoiohydrous (recalcitrant) seed species. Phytophylactica 22: 413–418

Mycock DJ, Watt MP, Berjak P (1991) A simple procedure for the cryopreservation of hydrated embryonic axes of *Pisum sativum*. J Plant Physiol 138: 728–733

Nkang A, Chandler G (1989) Changes during germination in rainforest seeds with orthodox and recalcitrant viability characteristics. J Plant Physiol 134: 9–16

Normah MN, Vengadasalam M (1992) Effects of moisture content on cryopreservation of *Coffea* and *Vigna* seeds and embryos. Cryo Lett 13: 199–208

Normah MN, Chin HF, Hor YL (1986) Desiccation and cryopreservation of embryonic axes of *Hevea brasiliensis* Muell-Arg. Pertanika 9: 299–303

Pammenter NW, Vertucci CW, Berjak P (1991) Homoiohydrous (recalcitrant) seeds: dehydration, the state of water and viability characteristics in *Landolphia kirkii*. Plant Physiol 96: 1093–1098

Pence VC (1990) Cryostorage of embryo axes of several large-seeded temperate tree species. Cryobiology 27: 212–218

Pence VC (1991a) Cryopreservation of seeds of Ohio native plants and related species. Seed Sci Technol 19: 235–251

Pence VC (1991b) Cryopreservation of immature embryos of *Theobroma cacao*. Plant Cell Rep 10: 144–147

Pence VC (1992) Desiccation and the survival of *Aesculus*, *Castanea*, and *Quercus* embryo axes through cryopreservation. Cryobiology 29: 391–399

Poulsen KM (1992) Sensitivity to desiccation and low temperature (−196 °C) of embryo axes from acorns of the pedunculate oak (*Quercus robur* L.). Cryo Lett 13: 75–82

Pritchard HW (1991) Water potential and embryonic axis viability in recalcitrant seeds of *Quercus rubra*. Ann Bot 67: 43–50

Pritchard HW, Prendergast FG (1986) Effects of desiccation and cryopreservation on the in vitro viability of the recalcitrant seed species *Araucaria hunsteinii* K. Schum. J Exp Bot 37: 1388–1397

Probert RJ, Longley PL (1989) Recalcitrant seed storage physiology in three aquatic grasses (*Zizania palustris*, *Spartina anglica* and *Porteresia coarctata*). Ann Bot 63: 53–63

Radhamani J, Chandel KPS (1992) Cryopreservation of embryonic axes of trifoliate orange (*Poncirus trifoliata* [L.] RAF.). Plant Cell Rep 11: 204–206

Roberts EH (1973) Predicting the storage life of seeds. Seed Sci Technol 1: 499–514

Roberts EH, King MW, Ellis RH (1984) Recalcitrant seeds: their recognition and storage. In: Holden JHW, Williams JT (eds) Crop genetic resources: conservation and evaluation. George Allen & Unwin, London, pp 38–52

Shimonishi K, Ishikaura M, Suzuki S, Oosawa K (1991) Cryopreservation of melon somatic embryos by desiccation method. Jpn J Breed 41: 347–351

Stanwood PC (1985) Cryopreservation of seed germplasm for genetic conservation. In: Kartha KK (ed) Cryopreservation of plant cells and organs. CRC Press, Boca Raton, pp 199–226

Tompsett PB (1984) Desiccation studies in relation to the storage of *Araucaria* seed. Ann Appl Biol 105: 581–586

Tompsett PB (1987) Desiccation and storage studies of *Dipterocarpus* seeds. Ann Appl Biol 110: 371–380

Tompsett PB, Pritchard HW (1993) Water status changes during development in relation to the germination and desiccation tolerance of *Aesculus hippocastanum* L. seeds. Ann Bot 71: 107–116

Uragami A, Lucas MO, Ralambosoa J, Renard M, Dereuddre J (1993) Cryopreservation of microspore embryos of oilseed rape (*Brassica napus* L.) by dehydration in air with or without alginate encapsulation. Cryo Lett 14: 83–90

Vertucci CW (1989) Relationship between thermal transitions and freezing injury in pea and soybean seeds. Plant Physiol 90: 1121–1128

Vertucci CW, Leopold AC (1987) Water binding in legume seeds. Plant Physiol 85: 224–231

Vertucci CW, Roos EE (1990) Theoretical basis of protocols for seed storage. Plant Physiol 94: 1019–1023

Vertucci CW, Berjak P, Pammenter NW, Crane J (1991) Cryopreservation of embryonic axes of an homeohydrous (recalcitrant) seed in relation to calorimetric properties of tissue water. Cryo Lett 12: 339–350

Vertucci CW, Crane J, Oelke EA (1993) The effect of moisture content and temperature on the viability of *Zizania* embryos. Plant Physiol 102 (Suppl): 76

Wang XF, Fu JR (1991) Studies on desiccation and storage of mango seeds. Acta Bot Sin 33: 118–123

Wesley-Smith J, Vertucci CW, Berjak P, Pammenter NW, Crane J (1992) Cryopreservation of desiccation-sensitive axes of *Camellia sinensis* in relation to dehydration, freezing rate and the thermal properties of tissue water. J Plant Physiol 140: 596–604

Yamada T, Sakai A, Matsumura T, Higuchi S (1991) Cryopreservation of apical meristems of white clover (*Trifolium repens* L.) by vitrification. Plant Sci 78: 81–87

Young JA, Young CG (1992) Seeds of woody plants in North America. Dioscorides Press, Portland, OR

Section II
Cryopreservation of Germplasm of Trees

II.1 Cryopreservation of Germplasm of Woody Plants

A. Sakai[1]

1 Introduction

The conservation of clones of rare, elite, and other important trees is of great importance in clonal propagation. The recent demonstration that embryogenic conifer cultures are amenable to cryopreservation indicates their obvious application for long-term storage of conifer tissues for clonal forestry programs (Gupta et al. 1987; Kartha et al. 1988; Klimaszewska et al. 1992). These in vitro practices will undoubtedly play an important role in large-scale production and improvement of fruit and forest trees, and eventually in achieving an increased production of biomass energy.

Cryopreservation is based on the reduction and subsequent arrest of metabolic functions of biological materials, while maintaining viability (see Bajaj 1991). At the temperature of liquid nitrogen (LN) almost all the metabolic activities of cells are at a standstill and they can be preserved in such a state for extended periods. However, one of the basic requirements prior to cryopreservation is the availability or development of efficient and reliable protocols for cryopreservation and faithful regeneration of plants from cryopreserved cells or meristems. This consideration obviously narrows the choice of plant material currently available for cryopreservation unless research efforts are extended to include many other plant species as yet untreated.

Conventional cryopreservation has generally been accomplished by slow freezing to about $-40\,^{\circ}C$ prior to plunging into liquid nitrogen. This slow freezing method is basically the most important protocol, but needs expensive controlled freezing equipment and a complicated cryoprotective procedure. Recently, two novel cryogenic protocols – the complete vitrification method (Sakai et al. 1990, 1991b), and the alginate-coated dehydration technique (Dereuddre et al. 1990b) – have been developed, and thus the number of woody species to be cryopreserved has increased within a few years.

This chapter is an updated version of an earlier account (Sakai 1986) and primarily focusses on recent work on cryopreservation using vitrification method.

[1]Hokkaido University, Sapporo, 060 Japan

Biotechnology in Agriculture and Forestry, Vol. 32
Cryopreservation of Plant Germplasm I (ed. by Y.P.S. Bajaj)

2 Survival Mechanism of Very Hardy Tissues or Twigs After Cooling to the Temperature of Liquid Nitrogen (−196 °C)

To maintain survival after cooling to the temperature of liquid nitrogen, it is essential to avoid lethal intracellular freezing, which occurs during rapid cooling into LN. Thus, cells and meristems have to be sufficiently dehydrated to avoid lethal intracellular freezing prior to plunging into liquid nitrogen. Extracellular freezing is an effective method of dehydrating hardy cells. Sakai (1956, 1960) reported that very hardy twigs prefrozen at −30 °C remained alive after immersion into LN even when thawed slowly in air at 0 °C.

Winter-hardy mulberry cortical tissues were mounted with water and prefrozen at different temperatures prior to plunging into LN. Prefrozen cells below −20 °C remained alive after rapid cooling in liquid nitrogen, regardless of warming rate (Fig. 1). However, in the prefrozen cells at temperatures above −15 °C, the warming rate seriously influenced survival. In the prefrozen cells at −5 °C, all cells remained alive by rapid warming in water at 30 °C, but were killed by slow warming in air at 0 °C, indicating that the prefrozen cells at −5 °C vitrified upon rapid cooling into LN and that these vitrified cells were killed due to intracellular crystallization when warmed slowly in air. The temperature range in which vitrified cells are injured during slow warming was determined. The prefrozen tissues at −10 °C were plunged into LN and then warmed in air at 0 °C. During the slow warming process, the tissue sections were rapidly warmed in water at 35 °C, immediately after reaching a definite temperature. As shown in

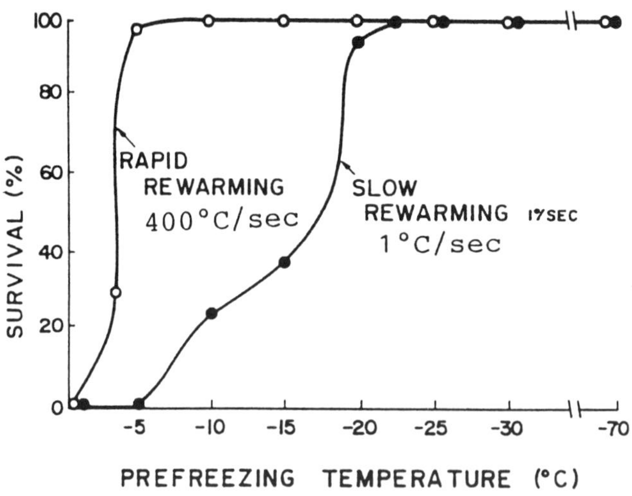

Fig. 1. Effect of warming rate on the survival of mulberry cortical cells immersed into liquid nitrogen following prefreezing at various temperatures. Cortical tissue from winter twig were mounted between cover glasses with 0.05 ml water. Tissue sections were warmed rapidly in water at 30 °C (400 °C s), or slowly in air 0 °C (1 °C s). (Sakai 1966)

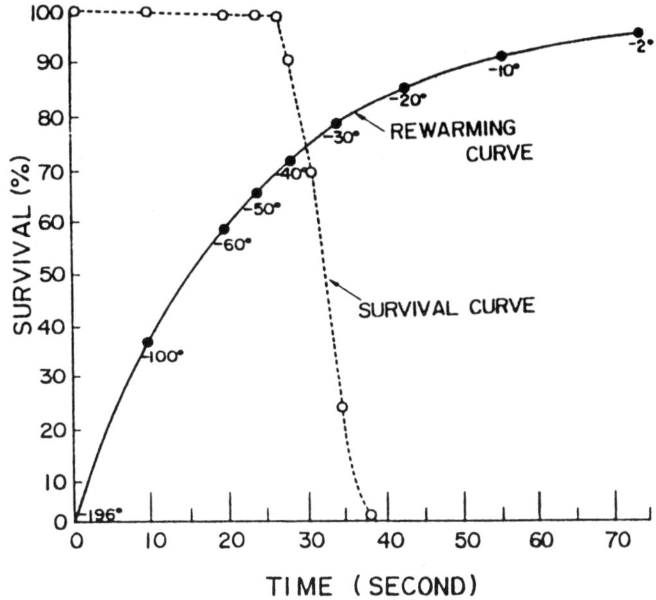

Fig. 2. Effect of temperature range on cell injury in the process of warming following removal from liquid nitrogen. The mulberry winter cortical tissues prefrozen at −10 °C and immersed into LN were slowly warmed (2.6 °C s) in air at 0 °C. In the process of warming following removal from LN, tissue sections were rapidly warmed at various temperatures by immersion in water at 30 °C. Survival of the cells in these sections warmed rapidly at various temperatures is shown on the survival curve. Temperature was determined with a 0.1-mm copper-constantan thermocouple and was recorded by an oscilloscope. (Sakai and Yoshida 1967)

Fig. 2, almost all the vitrified cells were killed within 6 s at the temperature range between −40 and −30 °C during the slow warming process (Sakai and Yoshida 1967).

To elucidate further the survival mechanism of very hardy materials, winter twigs of willow (0.5 cm in diameter, 15 cm long) prefrozen at different temperatures were plunged into liquid nitrogen and then warmed rapidly in water at 35 °C or in air at 0 °C (Table 1). Twigs prefrozen at −5 °C were killed after immersion in LN when warmed either rapidly or slowly. Interestingly, the twigs prefrozen at −10 °C for 16 h survived immersion into liquid nitrogen when warmed rapidly, but were killed after slow warming. A possible interpretation is that the twigs prefrozen at −10 °C vitrify during rapid cooling and are killed due to crystallization during slow warming. The twigs prefrozen at −15 °C survived immersion into LN regardless of warming methods. On the other hand, the twigs prefrozen at temperatures from −20 to −70 °C or below survived immersion in LN only when warmed slowly in air at 0 °C. It was confirmed that twigs prefrozen below −20 °C were killed as a result of rapid thawing without cooling in liquid nitrogen (Sakai 1965a). These results indicate that partially freeze dehydrated cells or twigs [residual water content (fr. wt.): 18 to 25%] remained alive after immersion in LN by vitrification when warmed rapidly, while in fully freeze

Table 1. Survival of prefrozen twigs rewarmed rapidly or slowly without or after immersion in liquid nitrogen (Sakai 1973)

Prefreezing temperatures (°C)	*Salix sachalinensis*			*Betula platyphylla*		
	Rewarming conditions		Resistance to rapid thawing[a]	Rewarming conditions		Resistance to rapid thawing[a]
	In air (0 °C)	In water 35 °C)		In air (0 °C)	In water (35 °C)	
−5	●	●	○	●	●	○
−10	●	△	○	●	○	○
−15	○	○	○	○	○	○
−20	○	●	●	○	△	△
−25	○	●	●	○	●	●
−30	○	●	●	○	●	●

○ = Normal; △ = injured; ● = killed.
Twigs immersed in liquid nitrogen following prefreezing to various temperatures in 5 °C decrements at hourly intervals were rewarmed slowly in air at 0 °C or rapidly in water at 35 °C (150 °C s). These twigs survived freezing down to −120 °C when cooled and rewarmed slowly.
[a]Prefrozen twigs were rewarmed rapidly by direct immersion in water at 35 °C without immersion in liquid nitrogen.

dehydrated cells or twigs, slow warming is necessary to maintain a high level of survival. Prefreezing temperatures below which twigs survive subsequent immersion in LN followed by slow warming (maximum quenching temperature) differed by species and varied in the range from −15 to about −30 °C. The hardier samples required less prefreezing (Sakai 1965a, b). The maximal quenching temperature was −15 °C for the extremely hardy winter twigs of willow, poplar, and white birch. In the same species, the temperature varied with seasonal fluctuations in hardiness. Hirsh et al. (1985) provided a definite explanation for the resistance of very hardy woody plants to the LN temperature even if quench cooling begins at temperatures as high as −20 °C and warming is very slow. Their evidence for vitrification in the extracellular frozen cells was based on the absence of ice crystals in freeze fractured samples and on the direct observation of second-order transitions in the differential scanning calorimeter.

3 Cryopreservation of Winter Woody Plant Materials

Winter willow twigs prefrozen at −30 °C were successfully cryopreserved in LN for 1 year and subsequently developed into entire plants (Sakai 1960). This was the first experiment in the cryopreservation of nondesiccated plant materials. This prefreezing method was successfully applied to several hardy fruit trees. Prefrozen apple buds stored in liquid nitrogen for 1 year developed normal shoots after grafting on root stocks (Sakai and Nishiyama 1978). Tyler and

Stushnoff (1988) demonstrated that winter-hardy apple twigs, which were previously dehydrated at −4 °C prior to prefreezing to −30 °C, produced high levels of survival for the majority of cultivars after immersion into LN when thawed slowly in air at 2 °C. These cryopreserved apple buds were successfully grafted on root stocks. Recently, Niino et al. (1992a) found that in winter mulberry buds attached to small twig segments, partial dehydration (water content ca. 40% fr. wt. base) at 25 °C prior to a prefreezing to −20 °C in 5 °C steps at daily intervals improved the shoot formation (about 70%) through meristem culture from frozen-thawed buds.

Another possible approach for cryopreservation of winter-hardy buds would be excessive dehydration by air drying. Excised shoot tips from winter mulberry buds were air-dried to 19% (fr. wt.) on a sterilized filter paper in a parafilm-sealed Petri dish containing 25 g dry silica gel at 25 °C and then cooled in liquid nitrogen. These dehydrated shoot tips expressed about 70% shoot formation through meristem culture (Niino et al. 1992a). It was also demonstrated that

Table 2. Cryopreservation of winter hardy twigs or bud segments cooled to −196 °C and subsequent plant regeneration

Species or cultivar	Procedure	Freezing regime	Warming	Regrowth	Reference
Camellia sinensis (tea plant) embryogenic axes	Drying	Rapid	Slow	MC	Chaudhury et al. (1991)
Malus domestica (apple), twigs	Freezing	5 °C/day PF: −30	Slow	Graft	Sakai and Nishiyama (1978)
Malus domestica 15 cultivars	Drying + freezing	10 °C/h PF: −30	Slow	Graft	Tyler and Stushnoff (1988)
Morus bombysis (mulberry)	Drying (part) + freezing	5 °C/day PF: −20	Rapid	MC	Niino et al. (1992a)
Morus bombysis shoot tips	Drying	Rapid	Slow	MC	Niino et al. (1992a)
Morus bombysis shoot tips	Encapsulation dehydration	Rapid	Slow	MC	Niino et al. (1992a)
Pinus sylvestris (Scots pine)	Freezing	1 °C/min PF: −39 °C	Rapid	MC	Kuoksa and Hohtola (1991)
Poplus nigra (poplar), twig	Freezing	5 °C/day PF: −30	Slow	Cutt	Sakai (1960, 1965)
Ribes grossulavia (gooseberry)	Freezing	5 °C/day PF: −30	Slow	Cutt	Sakai and Nishiyama (1978)
Rubus sp. cv. (raspberry)	Freezing	5 °C/day PF: −30	Slow	Cutt	Sakai and Nishiyama (1978)
Salix sachalinensis (willow), twig	Freezing	5 °C/h PF: −30	Slow	Cutt	Sakai (1960, 1965a)
Salix koriyanagi twig	Freezing	5 °C/h PF: −30	Slow	Cutt	Sakai (1960, 1965a)

PF: prefreezing temperature; MC: meristem culture from shoot tips excised from frozen or dehydrated buds; Cutt: cutting; Graft: grafting. Slow warming: in air at 0 °C; Rapid warming in water at 36 to 40 °C.

mulberry winter shoot tips trapped into alginate-coated beads containing 0.5 M sucrose were precultured on a medium supplemented with 1 M sucrose for 1 day and then dehydrated up to 25% (fr. wt.) followed by immersing in LN. Approximately 70% shoot formation was achieved (Niino et al. 1992a). Cryopreservation of dormant winter buds provides several advantages as a method of conserving hardy woody germplasm: (1) clonal integrity is maintained; (2) clonal propagation in vitro from cryopreserved materials is easy; (3) it does not require cold hardening and preculturing to increase hardiness; (4) the method for cryopreservation is reliable; (5) it avoids possible toxic effects of chemical cryoprotectants. Cryogenic protocols which enabled hardy woody plants to survive cooling to the temperature of LN are listed in Table 2.

4 Cryopreservation of Cultured Cells and In Vitro Grown Meristems

4.1 Cryogenic Strategies

Successful cryopreservation procedures for cultured cells and meristems can be divided into four different categories based on the dehydration method prior to rapid cooling.

1. Conventional slow freezing procedure prevents intracellular freezing by dehydrating cells sufficiently to allow them to undergo vitrification upon rapid cooling.
2. Alternative cryogenic strategies reduce or eliminate the need for cellular dehydration during slow cooling by osmotically dehydrating cells and meristems prior to freezing at −30 °C (simple freezing method).
3. An ideal cryogenic approach could be extensive dehydration using a highly concentrated vitrification solution (complete vitrification) at 25 or 0 °C prior to direct immersion in LN. The vitrification procedure permits cells and meristems to be cryopreserved by a direct transfer to LN.
4. Another possible approach to avoid intracellular freezing would be extensive dehydration by air drying. However, the induction or modification of drought tolerance is vital to successful cryopreservation when using this approach.

4.2 Conventional Slow Prefreezing Method

The prefreezing method is widely used as the routine method for cryopreservation to less hardy cultured cells and meristems, with the prior application of cryoprotectants. The conventional slow prefreezing method is depicted in Fig. 3.

Cultured cells and meristems to be cryopreserved are generally sensitive to freezing. Thus, they are previously treated with cryoprotectants so that they can survive freezing below −30 °C. As a successful cryoprotectant for cultured cells

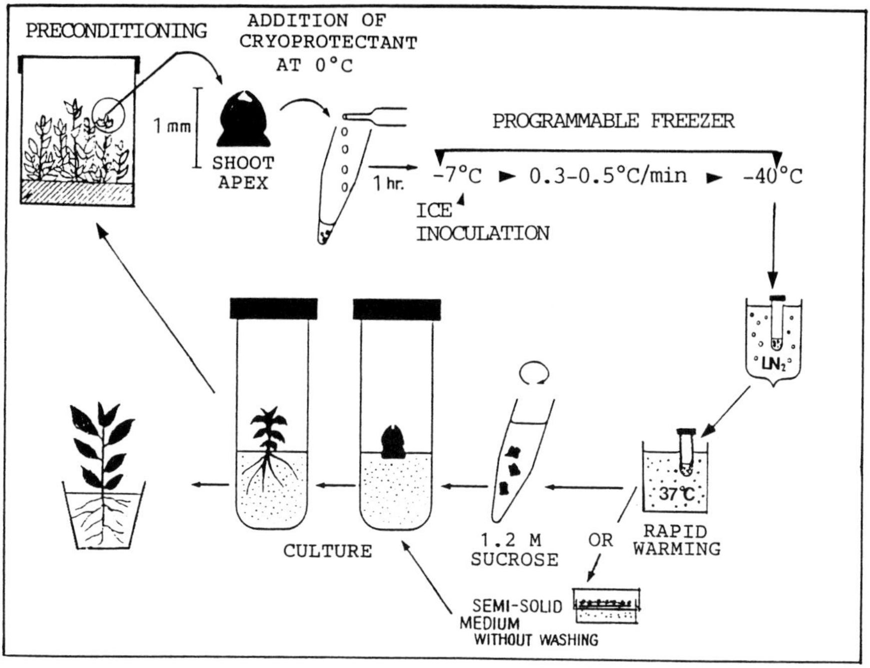

Fig. 3. General slow prefreezing method for apical meristems (Sakai 1984, modified)

and meristems, a mixture of 5 to 10% DMSO and 5 to 15% sugar or sorbitol has often been employed (Table 3). To decrease toxicity, precooled cryoprotectants were gradually added to a cell suspension or meristems in an ice bath and then held for 1 h at 0 °C to allow equilibration.

Most cultured cells showed high survival rates when prefrozen at the rate of 0.3 to 0.5 °C/min to ca. −40 °C before being immersed in LN, and subsequently thawed rapidly (Table 2). Slow freezing following ice inoculation to −40 °C produces sufficient concentration in the unfrozen fraction of the suspending solution and in the cytosol to be capable of vitrifying upon rapid cooling in liquid nitrogen. The survival of the vitrified cells depends on the fate of the glassy cytoplasm during the warming process. When warming is slow enough to permit sufficient time for crystallization of the cytoplasm, low survival was observed. Thus, the vitrified cells following freeze dehydration to −40 °C must be warmed rapidly to produce a high level of survival. On the other hand, in most cultured cells and meristems, slow deep cooling from −40 to −70 °C gives low survival (Sugawara and Sakai 1974; Uemura and Sakai 1980). Thus, slow freezing of specimens to ca. −40 °C prior to plunging into LN and subsequent rapid warming has been the optimal method for cryopreservation of cultured cells and meristems, unlike very hardy winter twigs.

Preculture (Sakai 1984) and post-thaw handling (Chen et al. 1984) are vital to successful cryopreservation. A high level of sugar (Dereuddre et al. 1988) or sorbitol (Chen et al. 1984) during preculture was reported to be important in

Table 3. Cryopreservation of cultured cells and meristems of woody plants by slow prefreezing method and subsequent plant regeneration

Species	Preculture	Cryoprotectant	Freezing regime	Reference
Citrus sinensis (navel orange) nucellar cells	None	5% DMSO + 1.2 M sucrose	0.5 °C/min PF: −40 °C	Kobayashi et al. (1990)
Coffea arabica (somatic embryos)	0.75 M suc 1 day	5% DMSO + 0.75 M sucrose	0.5 °C/min PF: −40 °C	Bertrand-Desbrunais et al. (1988)
Larix × eurolepis (larch)[a]	0.4 M suc 1 day	0.4 M sorbitol + 10% DMSO	0.33 °C/min PF: −40 °C	Klimaszewska et al. (1992)
Phoenix dactylifera (date palm)	None	PGD[b]	1 °C/min PF: −30 °C	Ulrich et al. (1982)
Picea abies (Norway spruce)[a]	None	PGD	1 °C/min PF: −30 °C	Gupta et al. (1987)
Picea glauca (white spruce)[a]	0.2 M sor + 0.4 M sor (each 1 day)	5% DMSO + 0.4 M sorbitol	0.3 °C/min PF: −40 °C	Kartha et al. (1988)
Picea mariana (black spruce)[a]	0.4 M sor 1 day	5% DMSO + 0.4 M sorbitol	0.33 °C/min PF: −40 °C	Klimaszewska et al. (1992)
Pinus taeda (Loblolly pine)[a]	None	PGD	1.0 °C/min PF: −30 °C	Gupta et al. (1987)
Pyrus sp. (pear) meristem	Hardening Prec: 5% DMSO 2 days	PGD	0.8 °C/min PF: −40 °C	Reed (1990)
Vitis vinifera (grape)[a]	None	5% DMSO + 0.25 M maltose	0.5 °C/min PF: −40 °C	Dussert et al. (1990)

[a]Embryogenic cells, [b]PGD: PEG 10%, glucose 10%, DMSO 10% (each w/v%); PF: prefreezing temperature; Prec: preculture; suc: sucrose, sor: sorbitol.

improving survival of cryopreserved cells and meristems (Table 3). Special attention has been given to the deleterious effects of washing with culture medium after thawing. Withers and King (1979) suggested that in addition to triggering injury by rapid deplasmolysis, post-thaw washing, especially in hypotonic culture medium, may cause damage by removing vital solutes lost by the specimens. Chen et al. (1984) achieved the regrowth of cryopreserved cells by transferring them without washing onto filter paper disks over nutrient medium solidified with agar. In our procedure for cryopreservation, frozen and thawed cells or meristems were placed in 1.2 M sucrose solution dissolved in culture medium, and then dispensed onto a double layer filter paper without washing (Kobayashi et al. 1990; Yamada et al. 1991).

The conventional slow prefreezing method is still the most important protocol for cryopreservation of cultured cells of woody plants (Kartha et al. 1988; Klimaszewska et al. 1992) (Table 3). However, the method needs controlled freezing equipment, and the complicated cryoprotective procedure is time-consuming.

4.3 A Simple Freezing Method

In this method (Sakai et al. 1991b; Nishizawa et al. 1992), cells or meristems were cryoprotected with a mixture of 2 M glycerol and 0.4 M sucrose at 25 °C for 10 min and then spontaneously frozen by placing the 1.8-ml cryotube in a freezer at −30 °C for 1 h prior to plunging into liquid nitrogen. The cell suspension initiated freezing spontaneously at about −15 °C and then cooled to −30 °C at the rate of about 2 °C/min (Sakai et al. 1991b). It was confirmed that the prefrozen cells at −30 °C by direct transfer from 25 °C produced nearly the same level of survival as those cooled at 0.5 °C/min to −30 or −40 °C by a programmable freezer. In the simple freezing method, osmotic partial dehydration by a mixture of 2 M glycerol and 0.4 M sucrose at 25 °C for 10 min led to high levels of survival and simplified complicated cryoprotective and freezing procedures.

The main advantages of this procedure for cryopreservation are that (1) the time required for slow freezing and the time used to equilibrate cells in the cryoprotective solution at 0 °C are significantly reduced, and (2) controlled-rate freezing equipment and ice seeding of the extracellular medium are not necessary. The temperature of −30 °C corresponds to that of ordinary freezers. This method has been successfully applied to many cultured cells of herbaceous plants.

4.4 Vitrification Method

In the vitrification method, cells or meristems must be sufficiently dehydrated with a highly concentrated vitrification solution at about 25 or 0 °C without causing injury prior to a direct transfer to the temperature of LN. We used a glycerol-based, less toxic vitrification solution designated PVS2, which contains 30% (w/v) glycerol, 15% (w/v) ethylene glycol, and 15% (w/v) DMSO in culture medium containing 0.4 M sucrose (pH 5.8) (Sakai et al. 1990, 1991a). This solution easily supercooled below −70 °C when cooled rapidly, and solidified into metastable glass at about −115 °C. On subsequent slow warming, the vitrified PVS2 showed a glass transition (Tg) at about −115 °C followed by exothermic devitrification (crystallization) and endothermic melting (Sakai et al. 1990). The one-step vitrification method is presented in Sakai et al. (1990, 1991a). As shown in Fig. 4, cells or meristems are treated with PVS2 at 25 or 0 °C for a definite time and then directly plunged into LN. Cell or meristem suspensions in 0.5-ml straws or 1.8-ml cryotubes were warmed in a water bath at 25 °C and then placed in 2 ml of a diluting solution containing 1.2 M sucrose in culture medium at 25 °C for 10 min. Exposure to PVS2 at 25 °C produced time-dependent survival of nucellar cells of navel orange after cooling to −196 °C (Fig. 5). Highest survival was obtained in the cells treated with PVS2 for 3 min. The cells treated with PVS2 for up to 5 min still maintained high levels of survival (80 to 90%) after cooling to −196 °C, but the survival decreased rapidly after longer exposure. However, the cells treated with PVS2 at 0 °C for up to 20 min showed high levels of survival (70–75%). Vitrified and warmed nucellar cells of navel

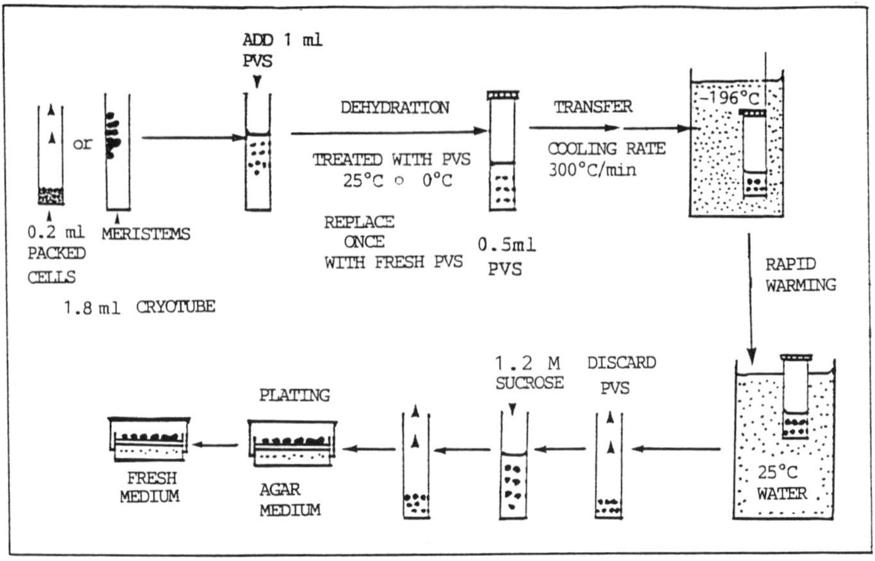

Fig. 4. Vitrification procedure for nucellar cells and meristems (Sakai et al. 1991)

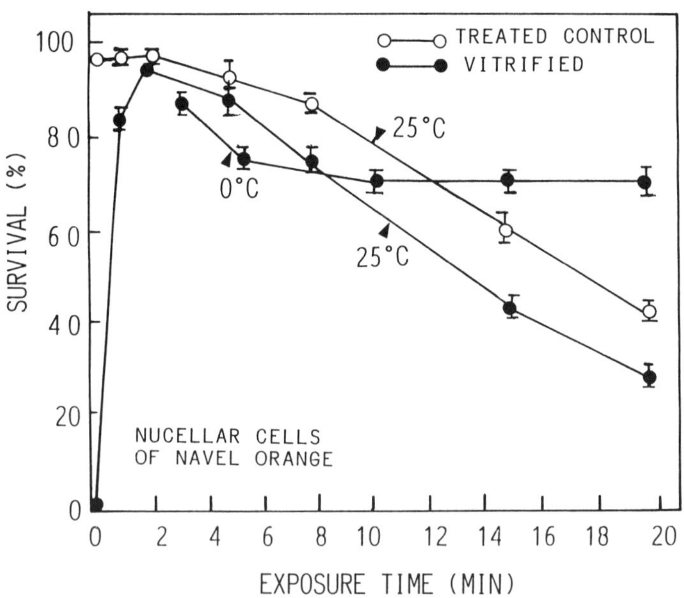

Fig. 5. Effect of exposure time to PVS2 at 25 or 0 °C on the survival of vitrified nucellar cells of navel orange. Cells treated with PVS2 at 25 or 0 °C for various lengths of time were placed in a 1.8-ml cryotube or loaded into a 0.5-ml plastic straw and then directly plunged into LN for 30 min. After rapid warming, the cell suspension was expelled into 2 ml of MT medium containing 1.2 M sucrose. Treated control: same as vitrified cells without cooling to −196 °C. *Vertical bars* SE (n = 2). (After Sakai et al. 1991b)

Fig. 6. Regenerated plant via embryogenesis of the nucellar cells of navel orange cooled to −196 °C by vitrification. Nucellar cells were treated with PVS2 for 3 min at 25 °C and then directly plunged into liquid nitrogen for 1 day. Rapidly warmed cells were placed in MT medium supplemented with 1.2 M sucrose for 10 min and then plated on paper disks onto agar medium. One month after transferring in a pot. *Bar* 1 cm (Photo Kobayashi)

orange were grown on filter paper disks over agar medium. Control and vitrified cells started growth within 3 days after plating, but the regrowth rate of vitrified cells was lower than that of controls until 12 days after plating: it then followed a growth pattern similar to that of the controls. Cotyledonary embryoids were produced from the vitrified cells and then developed into whole plants after 2 or 3 months (Fig. 6). The nucellar cells reestablished from cryopreservation for 1 year by vitrification maintained embryogenic potential identical to that of the nontreated controls, and plants regenerated from cryopreserved cells were morphologically uniform (Sakai et al. 1991b).

The vitrification procedure was successfully applied to the nucellar cells of three other citrus plants. High levels of survival (above 90%) were obtained in these plants cryopreserved for 320 days. Survival of nucellar cells of navel orange cooled to −196 °C was compared to three different cryogenic protocols (Table 4). The highest survival was obtained in the vitrified cells treated with PVS2 at 25 °C for 3 min.

This vitrification method was successfully applied to in vitro-grown shoot tips of some woody plants. Excised apple shoot tips (about 2 mm long, 1.5 mm base diameter) from cold-hardened plantlets at 0 °C for 3 weeks were precultured on a solidified agar medium supplemented with 0.7 M sucrose for 1 day at 5 °C. Following preculture, shoot tips were treated with PVS2 for various lengths of time at 25 °C prior to a plunge into LN_2 (Fig. 7). Shoot formation of vitrified shoot tips increased gradually with increasing time of exposure to PVS2 and reached a miximum at about 80 min exposure in apple, pear, and mulberry. The incubation time in PVS2, and thus meristem dehydration time, may necessarily increase with size of the excised meristems. Recently, we confirmed that the optimal exposure time for the shoot tips (0.5 to 1 mm long) was 20 to 25 min at 25 °C.

Table 4. Survival of nucellar cells of navel orange cooled to −196 °C by different cryogenic protocols

Method	Survival (%)
Conventional slow prefreezing[a]	73.0 ± 1.4
Simple freezing method[b]	65.1 ± 1.6
Vitrification[c]	90.0 ± 1.1

[a]Cooled to −40 °C at 0.3 °C/min in the presence of 5% DMSO and 1.2 M sucrose. (Kobayashi et al. 1990).
[b]Cells were treated with a mixture of 2 M glycerol and 0.4 M sucrose at 25 °C for 10 min and then directly transferred to a freezer at −30 °C for 1 h prior to plunging into LN. (Sakai et al. 1991a).
[c]Treated with PVS2 at 25 °C for 3 min prior to a plunge into LN. Values are mean of duplicates ± SE. (Sakai et al. 1991b).

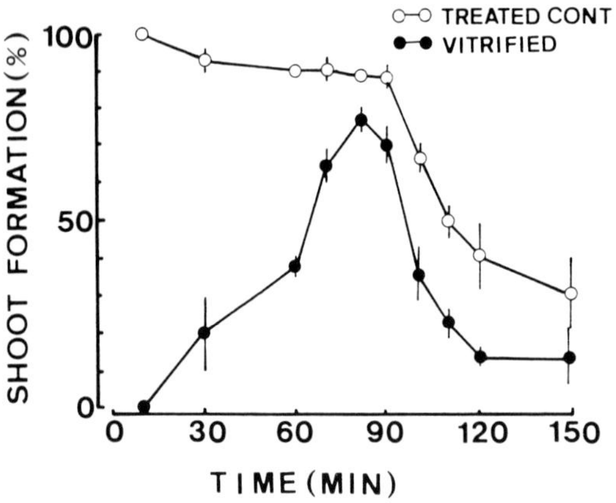

Fig. 7. Effect of exposure time to PVS2 at 25 °C on the shoot formation of apple shoot tips cooled to −196 °C by vitrification. Materials: *Malus domestica* cv. Fuji. Cold-hardened, precultured shoot tips were treated with PVS2 for various lengths of time at 25 °C and then directly plunged into LN for 1 day. *Treated control* Same as vitrified meristems without cooling to −196 °C. Cold hardening was carried out 3 weeks at 5 °C (8 h/day photoperiod). Shoot tips excised from cold-hardened in vitro plantlets were precultured on agar MS medium supplemented with 0.7 M sucrose for 1 day at 5 °C. Approximately 20 shoot tips were tested for each of two replicates. (Niino et al. 1992b)

Successfully vitrified apple shoot tips resumed growth within about 3 days after culture, and started to develop shoots within 2 weeks without intermediary callus formation. This vitrification method using PVS2 was successfully applied to in vitro-grown shoot tips of 5 apple species or cultivars, 8 pear cultivars (Niino et al. 1992b), 13 mulberry species or cultivars (Niino et al. 1992c), tea plant (Kuranuki and Sakai 1992), *Ribes* (Reed 1992) (Table 6), and many herbaceous plants.

Table 5. Effects of cold hardening and preculturing on the shoot formation of apple shoot tips cooled to −196 °C by vitrification (Niino et al. 1992b)

Period of preculture (days)	Shoot formation (% ± SE)	
	Hardening	Nonhardening
0	5.0 ± 2.9	0
1	60.0 ± 4.1	0
2	32.5 ± 7.5	0

Material: *Malus domestica* cv. Fuji
Cold hardening: 3 weeks at 5 °C (8 h/day photoperiod); preculturing: at 5 °C on MS agar medium supplemented with 0.7 M sucrose; shoot tips were treated with PVS2 for 80 min at 25 °C and then directly plunged into liquid nitrogen. Shoot formation (%): percent of shoot tips producing about 40 days after plating.

Cold hardening, preculturing and post-thaw handling of apical meristems from in vitro plantlets of woody plants are essential to successful cryopreservation by any other cryogenic protocol (Dereuddre et al. 1990a; Niino et al. 1992b,c). In vitro grown apple plantlets, precultured at 5 °C for 1 day following cold hardening, improved the recovery rate of hardened shoot tips cooled to −196 °C by vitrification (Table 5). Reed (1990) also demonstrated that cold hardening significantly improved the recovery rates of apical meristems of all four pear cultivars tested. In Table 6, successful cryopreservation of woody plants by vitrification along with the encapsulation-dehydration technique are listed.

It is particularly important that cryopreserved meristems be capable of producing plants identical to the nontreated phenotype. Haskins and Kartha (1980) demonstrated in frozen-thawed pea meristems by the conventional freezing method that renewed growth was usually from tissues other than those of the original meristem dome, and that renewed growth led to differentiation and whole-plant regeneration. Recovery of proliferated structures from surviving cells by secondary embryogenesis or adventitious buds was also reported. A callus phase prior to shoot formation is undesirable, since callusing potentially increases the frequency of genetic variants. However, successfully vitrified and warmed meristems generally produced direct shoot formation in many materials tested in both herbaceous and woody plants (Langis et al. 1990; Yamada et al. 1991; Niino et al. 1992a,b; Schnabel et al. 1992).

Recently, the vitrification procedure has been applied to a wide range of cultured meristems of herbaceous plants. However, this method did not produce high levels of survival for cultured cells except for some limited species or cultivars. This may suggest that direct exposure of less resistance-cultured cells to a highly concentrated vitrification solution may lead to harmful effects due to osmotic stress or chemical toxicity. Thus, to improve survival, we presented a two-step vitrification method for cultured cells (Nishizawa et al. 1993). In this method, cells were loaded with a mixture of 2 M glycerol and 0.4 M sucrose at

Table 6. Cryopreservation of cultured cells and meristems of woody-plants by vitrification or encapsulation-dehydration techniques and subsequent plant regeneration

Species	Cryogenic protocol	Reference
Citrus sinensis (navel orange) nucellar cells	Vitrification	Sakai et al. (1990)
Citrus paradisi (grapefruit)	Vitrification	Sakai et al. (1991a)
Camellia sinensis (tea plant)[a]	Vitrification	Kuranuki and Sakai (1992)
Malus 5 species (or cvs. (apple)[a]	Vitrification	Niino et al. (1992b)
Malus 3 species or cvs.	Encapsulation-dehydration	Niino and Sakai (1992)
Morus 13 sp. or cvs. (mulberry)	Vitrification	Niino et al. (1992c)
Morus bombycis 3 sp. or cvs.[a]	Encapsulation-dehydration	Niino and Sakai (1992)
Pyrus 6 sp. or cvs. (bear)[a]	Vitrification	Niino et al. (1992b)
Pyrus 3 sp. or cvs.	Encapsulation-dehydration	Niino and Sakai (1992)
Pyrus communis[a]	Encapsulation-dehydration	Dereuddre et al. (1990)
Ribes[a] 3 sp.	Vitrification	Reed (1992)

[a] In vitro grown meristems.

25 °C for 10 min, and then treated with PVS2 at 0 °C for 20 min prior to plunging into LN. Another approach for successful vitrification of cultured cells is a gradual addition of PVS2 (20, 65, and 100% at 0 °C) (Van Iren, unpubl.). It is necessary to extend the applicability of vitrification techniques to a wide range of cultured cells.

4.5 Drying Method

Partially-desiccated tissues of tree species, such as coconut and citrus subjected to freezing have survived (Bajaj 1984), however another possible approach for cryopreservation is the extensive dehydration by air drying. However, the induction or modification of drought tolerance of in vitro grown shoot tips or embryogenic cell masses may be the key to successful cryopreservation in this approach. To induce drought tolerance, in vitro plantlets are cold-hardened and/or precultured with high concentrations of sugar (Dereuddre et al. 1990b; Fabre and Dereuddre 1990; Niino and Sakai 1992; Table 6). The encapsulation-dehydration technique is easy to handle and alleviates the dehydration process. In this method, resistance to dehydration and deep freezing was induced by preculturing encapsulated shoot tips in medium enriched with sucrose before dehydration. Alginate-coated shoot tips of in vitro-grown pear (Dereuddre et al. 1990b), apple, and mulberry (Niino and Sakai 1992) were successfully cryopreserved following air drying.

In the encapsulation-dehydration technique, sucrose molarity increased drastically during drying and reached or exceeded the saturation point of sucrose solution and underwent glass transition during cooling (Dereuddre et al. 1991). Recently, a revised method for the encapsulated-dehydration technique was presented (Niino and Sakai 1992). To induce dehydration tolerance, cold-hardened apple apical meristems were precultured before being trapped into alginate-coated beads by progressive transfer daily to media containing 0.1, 0.4, and 0.7 M sucrose. The precultured shoot tips, trapped in alginate-coated beads containing 0.5 M sucrose, were treated in a medium supplemented with 1 M sucrose for 16h at 5 °C. The beads were dehydrated on sterile silica gel at 25 °C before being immersed in LN. This revised method was successfully applied to four apple, one mulberry, and three pear species or cultivars (Niino and Sakai 1992) (Table 6). The encapsulation-dehydration technique seems promising as a practical method for cryopreservation of embryogenic cell masses, meristems, and somatic embryos.

5 Summary and Conclusions

During the last decade, there has been increased interest in mass propagation and biotechnology in woody plants, and cryopreservation of embryogenic cell masses, meristems, and somatic embryos has become increasingly important. Cryopreservation of plant materials has proven to be a potentially ideal method for preserving plant germplasm using a minimum of space and maintenance. The conventional slow freezing method is still a basically important protocol for embryogenic cell cultures. However, two novel cryogenic protocols such as complete vitrification and encapsulation-dehydration techniques have been developed and the number of species or varieties to be cryopreserved has considerably increased within a few years.

References

Bajaj YPS (1984) Induction of growth in frozen embryos of coconut and ovules of citrus. Curr Sci. 53: 1215–1216

Bajaj YPS (1991) Storage and cryopreservation of in vitro cultures. In: Bajaj YPS (ed) Biotechnology in agriculture and forestry, vol 17. High-tech and micropropagation I. Springer, Berlin Heidelberg New York, pp 361–381

Bertrand-Desbrunais A, Fabre J, Engelmann F, Dereuddre J, Charrier A (1988) Adventive embryogenesis recovery from coffee (*Coffea arabica* L.) somatic embryos after freezing in liquid nitrogen. CR Acad Sci Paris 307 (III): 795–801

Chaudhury R, Radhamani J, Chandel KPS (1991) Preliminary observations on the cryopreservation of desiccated embryonic axes of tea [*Camellia sinensis* (L.) O. Kuntze] seeds for genetic conservation. Cryo Lett 12: 31–36

Chen THH, Kartha KK, Leung NL, Kurz WGW, Chatson KB, Constabel F (1984) Cryopreservation of alkaloid-producing cell cultures of periwinkle (*Catharanthus roseus*). Plant Physiol 75: 726–731

Dereuddre J, Fabre J, Basaglia C (1988) Resistance to freezing in liquid nitrogen of carnation (*Dianthus caryophyllus* L. var Eolo) apical and axially shoot tips excised from different aged in vitro plantlets. Plant Cell Rep 7: 170–173

Dereuddre J, Scottez C, Arnaud Y, Duron M (1990) Resistance of alginate-coated shoot tips of pear tree (*Pyrus communis* L. cv Beurré Hardy) in vitro plantlets to dehydration and subsequent freezing in liquid nitrogen: effects of previous cold hardening. CR Acad Sci Paris 310 (III): 317–323

Dussert S, Mauro MC, Deloire A, Hamon S, Engelmann F (1990) Cryopreservation of grape embryogenic cell suspensions. 1-Influence of pretreatment, freezing and thawing conditions. Cryo Lett 12: 287–298

Fabre J, Dereuddre J (1990) Encapsulation-dehydration: a new approach to cryopreservation of *Solanum* shoot tips. Cryo Lett 11: 413–426

Gupta PK, Durzan DJ, Finkle BJ (1987) Somatic polyembryogenesis in embryogenic cell masses of *Picea abies* (Norway spruce) and *Pinus taeda* (loblolly pine) after thawing from liquid nitrogen. Can J For Res 17: 1130–1134

Haskins RH, Kartha KK (1980) Freeze preservation of pea meristems: cell survival. Can J Bot 58: 833–840

Hirsh AG, Williams RJ, Meryman HT (1985) A novel method of natural cryoprotection. Plant Physiol 79: 41–56

Kartha KK, Fowke LC, Leung NL, Caswell KL, Hawkman I (1988) Induction of somatic embryos and plantlets from cryopreserved cell cultures of white spruce (*Picea glauca*). J Plant Physiol 132: 529–539

Klimaszewska K, Ward C, Cheliak WM (1992) Cryopreservation and plant regeneration from embryogenic cultures of larch (*Larix* × *eurolepis*) and black spruce (*Picea mariana*). J Exp Bot 43 (246): 73–79

Kobayashi S, Sakai A, Oiyama I (1990) Cryopreservation in liquid nitrogen of cultured navel orange (*Citrus sinensis* osb.). Plant Cell Tissue Organ Cult 23: 15–20

Kuoksa T, Hohtola A (1991) Freeze-preservation of buds from Scots pine trees. Plant Cell Tissue Organ Cult 27: 89–93

Kuranuki, Sakai A (1992) Effect of cold-hardening on the survival of shoot tips of in vitro-grown tea plants cooled to $-196\,^{\circ}C$ by vitrification. Jpn J Breed 42 (Suppl 1): 330–331

Langis R, Schnabel B, Preikstas B, Earle ED, Steponkus PL (1990) Cryopreservation of carnation shoot tips by vitrification. Cryobiology 27 (6): 657–659

Monod V, Poissonnier M, Paques M, Dereuddre J (1992) Cryopreservation of shoot tips of in vitro plantlets of *Eucalyptus* after encapsulation and air dehydration. Cryobiology 29 (6): 737–738

Niino T, Sakai A (1992) Cryopreservation of alginate-coated in vitro grown shoot tips of apple, pear and mulberry. Plant Sci 87: 199–206

Niino T, Sakai A, Yakuwa H (1992a) Cryopreservation of dried shoot tips of mulberry winter buds and subsequent plant regeneration. Cryo Lett 13: 51–58

Niino T, Sakai A, Yakuwa H, Nojiri K (1992b) Cryopreservation of in vitro-grown shoot tips of apple and pear by vitrification. Plant Cell Tissue Organ Cult 28: 261–266

Niino T, Sakai A, Enomoto S, Magoshi J, Kato S (1992c) Cryopreservation of in vitro-grown shoot tips of mulberry by vitrification. Cryo Lett 13: 303–312

Nishizawa S, Sakai A, Amano Y, Matuzawa T (1992) Cryopreservation of asparagus (*Asparagus officinalis* L.) embryogenic suspension cells and subsequent plant regeneration by a simple freezing method. Cryo Lett 13: 379–388

Nishizawa S, Sakai A, Amano Y, Matuzawa T (1993) Cryopreservation of asparagus (*Asparagus officinalis* L.) embryogenic suspension cells and subsequent plant regeneration by vitrification. Plant Sci 91: 67–73

Reed B (1990) Survival of in vitro-grown apical meristems of *Pyrus* following cryopreservation. Hort Sci 25: 111–113

Reed B (1992) Cryopreservation of *Ribes* apical meristem. Cryobiology 29(6): 740
Sakai A (1956) Survival of plant tissue at super-low temperature. Low Temp Sci Ser B 16: 17–23
Sakai A (1960) Survival of the twigs of woody plants at −196 °C. Nature 185: 393–394
Sakai A (1965a) Survival of plant tissue at super-low temperatures III. Relation between effective prefreezing temperatures and the degree of frost hardiness. Plant Physiol 40: 882–887
Sakai A (1965b) Determining the degree of frost-hardiness in highly hardy plants. Nature 206: 1064–1065
Sakai A (1966) Survival of plant tissue at super-low temperatures IV. Cell survival with rapid cooling and rewarming. Plant Physiol 41: 1050–1054
Sakai A (1973) Characteristics of winter hardiness in extremely hardy twigs of woody plants. Plant Cell Physiol 14: 1–9
Sakai A (1984) Cryopreservation of apical meristem. Hortic Rev 6: 357–372
Sakai A (1986) Cryopreservation of germplasm of woody plants. In: Bajaj YPS (ed) Biotechnology in agriculture and forestry, vol I. Trees I. Springer, Berlin Heidelberg New York, pp 113–129
Sakai A, Nishiyama Y (1978) Cryopreservation of winter vegetative buds of hardy fruit trees in liquid nitrogen. Hort Sci 13: 225–227
Sakai A, Yoshida S (1967) Survival of plant tissue at super-low temperature. VI. Effects of cooling and rewarming rates on survival. Plant Physiol 42: 1695–1701
Sakai A, Kobayashi S, Oiyama I (1990) Cryopreservation of nucellar cells of navel orange (*Citrus sinensis* Osb. var. *brasiliensis* Tanaka) by vitrification. Plant Cell Rep 9: 30–33
Sakai A, Kobayashi S, Oiyama I (1991a) Survival by vitrification of nucellar cells of navel orange (*Citrus sinensis* Osb. var. *brasiliensis* Tanaka) cooled to −196 °C. J Plant Physiol 137: 465–470
Sakai A, Kobayashi S, Oiyama I (1991b) Cryopreservation of nucellar cells of navel orange (*Citrus sinensis* Osb.) by a simple freezing method. Plant Sci 74: 243–248
Sugawara Y, Sakai A (1974) Survival of suspension-cultured sycamore cells cooled to the temperature of liquid nitrogen. Plant Physiol 54: 722–724
Tyler NJ, Stushnoff C (1988) The effects of prefreezing and controlled dehydration on cryopreservation of dormant vegetative apple buds. Can J Plant Sci 68: 1163–1167
Uemura M, Sakai A (1980) Survival of carnation (*Dianthus caryophyllus* L.) shoot apices frozen to the temperature of liquid nitrogen. Plant Cell Physiol 21: 85–94
Ulrich JM, Finkle BJ, Tisserat BH (1982) Effects of cryogenic treatment on plantlet production from frozen and unfrozen date palm callus. Plant Physiol 69: 624–627
Yamada T, Sakai A, Matsumura T, Higuchi S (1991) Cryopreservation of apical meristems of white clover (*Trifolium repens* L.) by vitrification. Plant Sci 78: 81–87
Withers LA, King PJ (1979) Proline: a novel cryoprotectant for the freeze preservation of cultured cells of *Zea mays* L. Plant Physiol 64: 675–678

II.2 Cryoconservation of Germplasm of Citrus

N. Duran-Vila[1]

1 Introduction

1.1 Plant Distribution and Its Important Species

Citrus rank second after grapes as the major fruit crop. Commercial citrus are grown in tropical and subtropical climates, mainly in a uniform belt from 35°N to 35°S. Under mild environmental conditions such as those found in the Mediterranean regions, citrus may be grown as far as 44°N. About 90 countries produce citrus, most of which are grown for local consumption. Only eight countries produce 50% of the total world production with five accounting for most of the exported citrus as fresh fruit and processed juices (FAO 1992).

Sweet orange accounts for 70% of the world production, followed by mandarins (12%), lemons and limes (10%), and grapefruits and pummelos (7%). The remaining 1% includes other species processed in the jam industry or as spices and essential oils. Since most commercial citrus are propagated as grafted scions, the availability of a wide range of rootstocks is essential to overcome soil problems, and to grow citrus in areas where certain diseases are endemic. Therefore, rootstock species have to be taken into consideration.

The taxonomy of *Citrus* remains controversial, with two elaborate systems of classification proposed by Tanaka (1961) with 159 genera and Swingle (1943) with 16. There is a general agreement about the classification of citrus relatives, but the classification of true citrus is still controversial, and alternative systems have also been proposed (Carpenter and Reece 1969).

Citrus are included in the family Rutaceae, subfamily Aurantioideae, tribe Citreae. True citrus fruits and their hybrids fall into three genera (*Citrus, Poncirus*, and *Fortunella*). However, taxonomists also include the genera *Microcitrus* and *Eremocitrus*, which are thought to be citrus ancestors, and more recently the less known genus *Clymenia* (Swingle and Reece 1967; Vardi and Spiegel-Roy 1978). Commercial citrus fruits (sweet oranges, mandarins, lemon, limes, grapefruits, and pummelos) are all included in the genus *Citrus. Poncirus* is a monotypic genus with a single species, *P. trifoliata*, which is of commercial interest as a rootstock; *Fortunella* includes only the kumquats which are of limited commercial interest, principally as ornamentals (Hodgson 1967).

[1]Instituto Valenciano de Investigaciones Agrarias Apartado Oficial, 46113 Moncada, Valencia, Spain

Biotechnology in Agriculture and Forestry, Vol. 32
Cryopreservation of Plant Germplasm I (ed. by Y.P.S. Bajaj)

The classification of the genus *Citrus* is hindered by the high frequency of mutations and the natural occurrence of interspecific and intergeneric hybrids. Bitters (pers. comm.) has recognized that at least half of the 33 genera of the Rutaceae hybridize with *Citrus*. It is generally accepted that a number of commercial citrus including sweet orange and grapefruit may have arisen as natural hybrids of unknown origin (Barrett and Rhodes 1976; Scora et al. 1982). In addition, hybrids have also been artificially produced by conventional plant breeding, and more recently by somatic hybridization. Hybrids produced in conventional plant breeding programs include the citranges (*Citrus sinensis* L. Osb. × *Poncirus trifoliata* L. Raf.) and citrumelos (*C. paradisi* Macf. × *Poncirus trifoliata* L. Raf.), which are used as rootstocks, and hybrid mandarins like tangelos (*C. paradisi* Macf. × *C. tangerina* Hort. ex. Tan.) and tangors (*C. sinensis* L. Osb. × *C. tangerina* Hort. ex. Tan.) as commercial varieties. With protoplast fusion technologies, a large number of interspecific (Ohgawara et al. 1985a; Kobayashi et al. 1988; Grosser et al. 1989; Tusa et al. 1990) and intergeneric somatic hybrids have also been obtained. Successful intergeneric hybrids include sweet orange and mandarin hybrids with *Poncirus trifoliata* (Ohgawara et al. 1985b; Grosser et al. 1988a), *Severinia disticha* (Grosser et al. 1988b), *Fortunella crassifolia* (Deng et al. 1992), and *Citropsis gilletiana* (Grosser et al. 1990).

More recently, with the advent of taxonomy based on numerical and molecular biology technologies (Barrett and Rhodes 1976; Scora et al. 1982; Potvin et al. 1983; Green et al. 1986; Roose 1988), the classification of common citrus into only three species, *C. medica, C. reticulata* and *C. maxima*, has been proposed. The latter grouping includes sweet orange, sour orange, lemon, and grapefruit. Current genetic analysis also supports evidence that sweet oranges, sour oranges, lemons, mandarins, and grapefruits have a relatively narrow genetic basis, even though they are recognized as distinct commodities by plant breeders, growers, and consumers. Advances in the field of molecular genetics will certainly clarify some of the controversies and incongruities inherent in the classification systems available today.

Citrus cultivars and rootstocks can be grouped as shown in Table 1. The botanical names follow the nomenclature proposed by Carpenter and Reece (1969) which is more practical in grouping commercial citrus into distinct horticultural groups.

1.2 Methods for Storage of Citrus Germplasm

Traditionally, citrus have been conserved as living collections in field plantings maintained in botanical gardens or at research facilities (Fig. 1A). Like most woody species, plants held in the field are desirable for evaluation and characterization. However, to prevent loses due to biological and climatic hazards, it is advisable to conserve selected materials in screened houses or greenhouses (Fig. 1B), where they can be easily protected from pests and insect borne diseases. In addition, when the plants are grown in containers with artificial soil mixtures, soil borne diseases and nutrition can be controlled.

Table 1. Main *Citrus* species grown as commercial varieties and rootstocks

Generic name	Common name	Botanical species
Oranges	Sweet orange	*Citrus sinensis* (L.) Osbeck.
Mandarins	Clementine	*C. reticulata* Hort. ex Tan
	Common mandarin	*C. deliciosa* Ten
	Dancy	*C. tangerina* Hort. ex Tan
	Ponkan	*C. reticulata* Blanco
	Satsuma	*C. unshiu* (Mak.) Marc
Lemons	Lemon	*C. limon* (L.) Burm. f.
	Meyer lemon	*C. meyery* Y. Tan.
Limes	Bearss lime	*C. latifolia* Tan.
	Mexican lime	*C. aurantifolia* (Christm.) Swing
	Sweet lime	*C. limetta* Risso
Grapefruits	Grapefruit	*C. paradisi* Macf.
Pummelos	Pummelo	*C. grandis* (L.) Osb.
Rootstocks	Alemow	*C. macrophylla* Wester
	Cleopatra mandarin	*C. reshni* Hort. ex Tan.
	Rangpur lime	*C. limonia* Osb
	Rough lemon	*C. jambhiri* Lush
	Sour orange	*C. aurantium* L.
	Trifoliata orange	*Poncirus trifoliata* (L.) Raf.
	Volcamer lemon	*C. volkameriana* Tem Pasq.

Fig. 1A, B. Citrus germplasm maintained at the Instituto Valenciano de Investigaciones Agrarias: **A** field collection; **B** collection in screened house. (Juárez and Duran-Vila, unpubl.)

There are numerous citrus collections conserved as field plantings all over the world (IBPGR 1986). However, not all the countries with unique genetic resources have the required land, equipment, and personnel to protect the plant materials from biological and climatic hazards (IBPGR 1982). Virtually all the collections are subjected to the risks imposed by epidemics like blight, which is becoming increasingly important in South America and Florida; greening, which is devastating in some areas of Southeast Asia and South Africa; and other diseases like tristeza and root rot which are present in practically all regions where citrus are grown commercially. In addition, some collections are located in areas like Orlando and Riverside (USA) and in Mediterranean countries which are periodically subjected to freezing hazards.

The International Board of Plant Genetic Resources (IBPGR 1982) in consultation with a Citrus Working Group, and after taking into consideration the areas of possible origin, the present distribution of cultivated citrus, and the location of major citrus collections, encouraged a number of countries to maintain regional collections. Spain was invited to house a collection of cultivars from the Mediterranean region. This collection is located in the facilities of the Instituto Valenciano de Investigaciones Agrarias and, at present, includes 340 accessions of local cultivars and materials imported from other countries (Navarro et al. 1988). The collection is disease-free and the plants are grown in containers in insect-proof, screened houses (Fig. 1B) where they are subjected to a preliminary characterization. Selected cultivars are also maintained in the field (Fig. 1A) for further agronomic evaluation.

The screened-in collections are still vulnerable to a number of potential hazards, in addition to the considerable expenses which limit the number of specimens that can be conserved. Therefore, research efforts are presently being devoted to the development of alternative methods which include in vitro collections and cryconservation procedures.

In vitro collections are based on the periodic subculture of proliferating structures. The approach, proposed for a number of woody species, involves the culture of shoot tips and buds that are maintained under standard or minimal growth conditions by periodic subculture (Styer and Chin 1983; Bajaj 1991; Withers 1992). However, it has a limited application for citrus, because the proliferating shoots and buds fail to grow after a few subcultures (Marín 1988). Therefore, an alternative method, based on the establishment of primary nodal cultures to recover in vitro plants followed by the maintenance of successive cycles of secondary cultures, was developed for several species and varieties (Marín and Duran-Vila 1991).

1.3 Need for Conservation and Cryopreservation of Endangered *Citrus* Species and Relatives

The centers of origin of the major cultivated citrus are southern China, Southeast Asia, South Vietnam, northern India, Indonesia, and the Philippines. The secondary centers include Japan for mandarins, and the Mediterranean region for sweet orange and lemon. Grapefruits are believed to be the result of a bud mutation or a natural hybrid of pummelo occurring in the West Indies before 1750. With the exception of sour orange, the major, cultivated citrus are not found wild.

It is difficult to assess the magnitude of the actual risk of genetic erosion through the replacement of a large number of unselected citrus varieties by a small number of improved cultivars. However, in many countries where wild aurantioids are found, dense populations have increased the pressure on natural vegetation for agricultural purposes (Sastrapradja 1975). Although citrus-collecting missions have been initiated (Akihama 1984), and some local collections have been established in Indonesia (Sastrapradja and Soetjiko 1984), West Malaysia (Jones and Ibrahim 1984), Papua New Guinea (Tarepe 1984), the Philippines (Espino 1984), India (Dass 1984), and Thailand (Ratanadaros and Chomchalow 1984), genetic erosion has undoubtedly occurred and is still occurring.

Most citrus research has been carried out on species adapted to the warm temperate zones, subtropics, and drier parts of some tropical areas. With the diversity of commercial cultivars, it has been possible to develop citrus which are adapted to various environmental conditions and cultural practices. However, the lack of effective means of control of certain diseases like greening, witches'-broom, blight, and tristeza limits the citrus industry in certain regions. Pest control is not a limiting factor but relies on the periodic application of chemicals with the subsequent contamination of the environment. Therefore, the further search for better means of inducing resistance or tolerance to the major pests and diseases is highly desirable. In addition, commercial citrus are presently grown on a very limited number of rootstocks which are not always well adjusted to some soil problems like salinity, high calcium content, drought, Phytophthora root rot, and nematodes. The limited number of rootstocks also increases the vulnerability of the citrus industry to potential hazards. Therefore, there is an urgent need to develop new rootstocks to cover present and future demands. A number of disease and frost resistance characteristics, as well as some desirable horticultural traits, have already been identified in citrus and citrus relatives (Ford and Feder 1960; Bitters et al. 1969, 1977; Giacometti and Van Sloten 1984), whereas other desirable traits require further screening of unexploited wild material.

On the other hand, only limited research has focused on species grown in the humid tropics where citrus production is restricted to a few types like pummelos, limes, and calamondins which are well adapted or can be grown under minimal agricultural conditions. In such cases, the desirable traits necessary for potential utilization in citrus breeding have probably not been as yet identified.

From all the considerations outlined above it can be concluded that a broad collection of germplasm including wild, primitive, and currently cultivated species would be highly desirable. The task of collecting and conserving citrus requires international cooperation and coordination of the agencies involved. As recommended by the Citrus Working Group to IBPGR (IBPGR 1982), special support must be given to the conservation of: (1) the wild Aurantioideae in East Asia, Southeast Asia, and Central Africa; (2) the old cultivars in East Asia and Southeast Asia; and (3) cultivars in the Mediterranean and other regions.

2 Cryoconservation of Citrus

2.1 General Account

Several research groups have devoted much effort to the development of alternative methods to field collections for the genetic conservation of citrus.

Table 2. *Citrus* explants which survived after being subjected to $-196\,^{\circ}C$

Explant	Species	Reference
Seeds	Lemon	Mumford and Grout (1979)
Ovules	Orange	Bajaj (1984)
Embryos	Sweet orange	Marín and Duran-Vila (1988); Marín et al. (1993)
Embryonic axes	Trifoliate orange	Radhamani and Chandel (1992)
Cells	Sweet orange	Kobayashi et al. (1990); Sakai et al. (1990, 1991a, b); Pérez et al. (1993)
	Grapefruit, Sudachi, Tangor	Sakai et al. (1991a)
	Cleopatra mandarin	Pérez et al. (1993)
	Common mandarin	Pérez et al. (unpubl.)

Most of the methods proposed are based on the exploitation of in vitro culture technologies of micropropagation, ovule culture, embryo culture, cell culture, and somatic embryogenesis.

Different specimens have been demonstrated to be suitable for cryoconservation, including seeds, ovules, embryos, and cells (Table 2). The proposed cryoconservation protocols cover a wide range of approaches: desiccation followed by fast cooling has been used for seeds (Mumford and Grout 1979), ovules (Bajaj 1984), and excised embryonic axes (Radhamani and Chandel 1992); dehydration and cryoprotection followed by slow cooling for somatic embryos (Marín and Duran-Vila 1988; Marín et al. 1993) and embryogenic cell lines (Kobayashi et al. 1990; Sakai et al. 1991b); and vitrification and fast cooling for embryogenic cell lines (Sakai et al. 1990, 1991a).

Citrus seeds, which do not remain viable when stored for extended periods of time, were considered recalcitrant for many years. Following the report of Mumford and Grout (1979), who demonstrated that the seeds of a single citrus species, *Citrus limon*, were tolerant to desiccation, citrus seeds are now considered to be orthodox (Chin 1988). However, tolerance to desiccation is limited and drying below 5% moisture content may damage most citrus seeds (King et al. 1981). Other species like *Poncirus trifoliata* are even more sensitive to reduced moisture contents below 20% (Radhamani and Chandel 1992). Therefore, the methods recommended for the storage of orthodox seeds are not generally applicable to citrus, thus alternative methods including the use of cryogenic techniques should be explored (King et al. 1981; Chin 1988).

Somatic embryos and embryogenic cell lines of polyembryonic species, have also been subjected to cryogenic treatments with success (Table 2).

Recovery of whole plants from cryopreserved partially dehydrated segments of young ovules (Bajaj 1984) and other citrus tissues was reported by Marín and Duran-Vila (1988). Somatic embryos of Washington navel sweet orange (*C. sinensis*), derived from in vitro cultured ovules, were subjected gradually to a final cryoprotectant solution of 10% v/v DMSO and slow cooling to $-42\,^{\circ}C$ at an estimated cooling rate of $0.5\,^{\circ}C/min$ and stored in liquid nitrogen. Only 5% survival was achieved when the embryos were thawed by slow warming at room temperature. However, when the embryos were thawed by immersion in a water bath at 37 °C, an average survival of 30% was achieved (Marín et al. 1993). The survival of these tissues did not demonstrate the survival of whole organ but the recovery of proliferating cells followed by secondary embryogenesis. Although this was initially considered to be a limitation, the subsequent observation and analysis of greenhouse plants indicated that true-to-type plants were probably recovered (Marín et al. 1993).

Cryopreservation of embryogenic cell lines of navel sweet orange (*C. sinensis*) has also been reported. Cell cultures derived from in vitro cultured ovules were immersed directly in liquid nitrogen after a vitrification procedure using a solution containing glycerol, ethylene glycol, DMSO, and sucrose (Sakai et al. 1990). The same method was also successfully used to freeze embryogenic cell lines of grapefruit (*Citrus paradisi*), sudachi (*Citrus sudachi* hort. ex Shirai), and Murcott tangor, with survival rates estimated at about 90% (Sakai et al. 1991a). Embryogenic cell lines of three varieties of sweet orange and Cleopatra mandarin have also been successfully cryopreserved by more conventional treatments with DMSO and slow cooling to −40 °C at a cooling rate of 0.1–0.5 °C/min and storage in liquid nitrogen. The survival achieved was always high and the surviving cells regenerated embryos and produced apparently normal plants (Kobayashi et al. 1990; Pérez et al. 1993). Simplification of the freezing method by placing the specimens in a freezer at −30 °C for 20–30 min prior to immersion in liquid nitrogen gave comparable results (Sakai et al. 1991b).

From the above it can be concluded that embryos and embryogenic cell lines present high survival rates and regenerate new embryos and whole plants after being subjected to a variety of cryogenic treatments. The treatments assayed covered the two basic principles of cryconservation (Mazur 1969): slow cooling rates and an adequate cryoprotectant concentration so as to produce sufficient dehydration of the plant cells; and ultrafast cooling through a vitrification process to prevent the formation of ice crystals. Considering the response shown by sweet orange cells to all these treatments, and the fact embryogenic cells can also be produced from other species and varieties, the cryoconservation of embryogenic tissues seems to be a logical approach to the conservation of citrus.

2.2 Methodology/Protocol

2.2.1 Production of Somatic Embryos and Embryogenic Cell Lines

Preparation of Explants. The method is based on previous reports (Button and Bornman 1971; Navarro et al. 1979). Ovules excised from flowers collected prior to anthesis and from immature fruits collected 2 to 8 weeks after anthesis are used to initiate the cultures. Ovaries and fruits are collected from field- or greenhouse-grown trees and surface-disinfected by soaking for 20 min in a 1% sodium hypochlorite solution containing 0.1% wetting agent (Tween 20) (Fig. 2A). They are cut open under the microscope under sterile conditions (Fig. 2B). The ovules are excised with the aid of the scalpel tip and placed on the surface of the culture medium without further treatment (Fig. 2C).

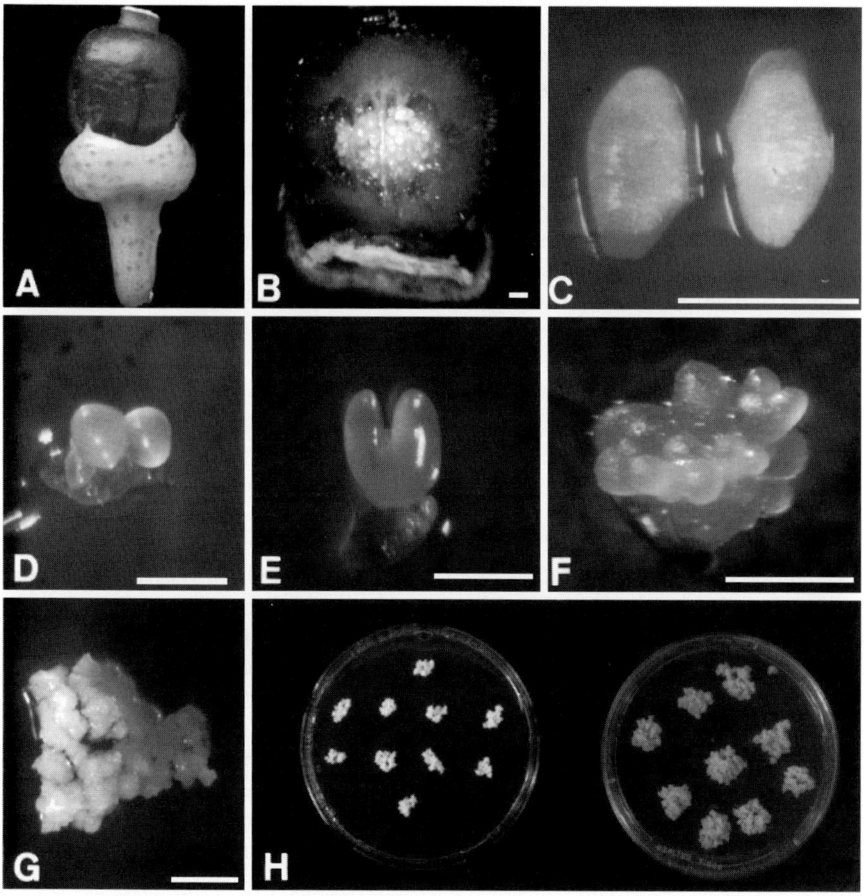

Fig. 2A–H. Ovule culture and production of somatic embryos and embryogenic cell lines: immature fruits (**A**) collected 2–8 weeks after anthesis are surface-disinfested and cut open before ovule excision (**B**). The ovules are placed on the surface of culture medium (**C**) and incubated for 8–20 weeks. Cultured ovules produce spherical or slightly elongated structures termed 'pseudobulbi' (**D**), heart-shaped embryos (**E**), and complex structures containing pseudobulbi, embryos, and small clusters of disorganized cells (**F**). These cell clusters are transferred to fresh medium and proliferate into more cells (**G**) which can be eventually maintained and/or increased by monthly subculturing (**H**). During routine subcultures, ca. 10 calli (20–30 mg each) per dish are transferred to fresh medium (*left*) and continue to grow over the following 4 weeks, reaching 3–5 times the original size (*right*); *bar* = 5 mm. (Duran-Vila, unpubl.)

Culture Medium. The culture medium contains the inorganic salts of Murashige and Skoog (1962), 100 mg/l *i*-inositol, 0.2 mg/l thiamine-HCl, 1 mg/l pyridoxine-HCl, 1 mg/l nicotinic acid, 50 g/l sucrose, 500 mg/l malt extract, and 10 g/l agar (Difto Bacto agar). The pH is adjusted to 5.7 ± 0.1 with 1 N NaOH prior to autoclaving. The medium is sterilized at 121 °C for 15 min. Twenty-five ml of autoclaved medium is dispensed into 9-cm-diameter disposable Petri dishes.

Culture Conditions. An average of 15 ovules is cultured per dish. The dishes are sealed with Parafilm and incubated in the culture room at 27 °C, with a 16-h day length and a photosynthetic photon flux (PPF) of 43 μmol/m^2/s.

Growth of Cultured Ovules. For most species, growth of cultured ovules takes place after 3–4 weeks in culture, whereas slow-growing species take up to 20 weeks. The ovules that grow produce spherical bodies termed "Pseudobulbi" which are believed to be proembryos (Fig. 2D) and embryos (Fig. 2E). Only a small number of cultures produce directly from the nucellar tissues, small clusters of disorganized cells which remain loose and friable (Galiana et al. 1993). In general, these cells grow very slowly among clustered embryos and pseudobulbi (Fig. 2F).

Production of Somatic Embryos for Cryoconservation Assays. Pseudobulbi and somatic embryos developing directly from the nucellar tissues can be used for cryoconservation assays. These primary embryos can also be transferred to fresh medium, producing more embryos through a budding process.

Production of Embryogenic Cell Lines for Cryoconservation Assays. The small clusters of disorganized cells which grow among embryos and pseudobulbi (Fig. 2F) are excised from the original explant and transferred to fresh medium. After several monthly subcultures, during which disorganized cells are selected from the proliferating embryos, stable cell lines virtually devoid of embryos can be established (Fig. 2G). As recommended by other authors (J.W. Grosser, pers. comm.) when callus grows poorly, the medium can be supplemented with growth regulators (1 mg/l benzylaminopurine and 1 mg/l indole acetic acid have been recommended) for a few subculture cycles. During routine transfers, ca. ten calli (20–30 mg each) per dish (Fig. 2H, left) are subcultured in the same medium and incubated under the same environmental conditions, reaching 50–100 mg each after 4 weeks (Fig. 2H, right).

2.2.2 Freezing-Thawing Procedures

Pretreatment and Cryoprotection. Somatic embryos or callus cells (150–200 mg) are transferred to 2-ml screw-cap, cryogenic, polypropylene tubes with 1.8 ml of culture medium and 10% v/v DMSO. The tubes are sealed with silicon washers and kept in ice for 30 min before being subjected to the freezing protocol.

Freezing Procedures

Slow Cooling. The ampoules containing the specimens are placed in a stainless steel basket and immersed into a cooling bath. Using pure methanol as coolant, an estimated cooling rate of about 0.5 °C/min down to −40 °C can be consistently achieved. The cooling bath (Heto CB10) has shown to be extremely reliable.

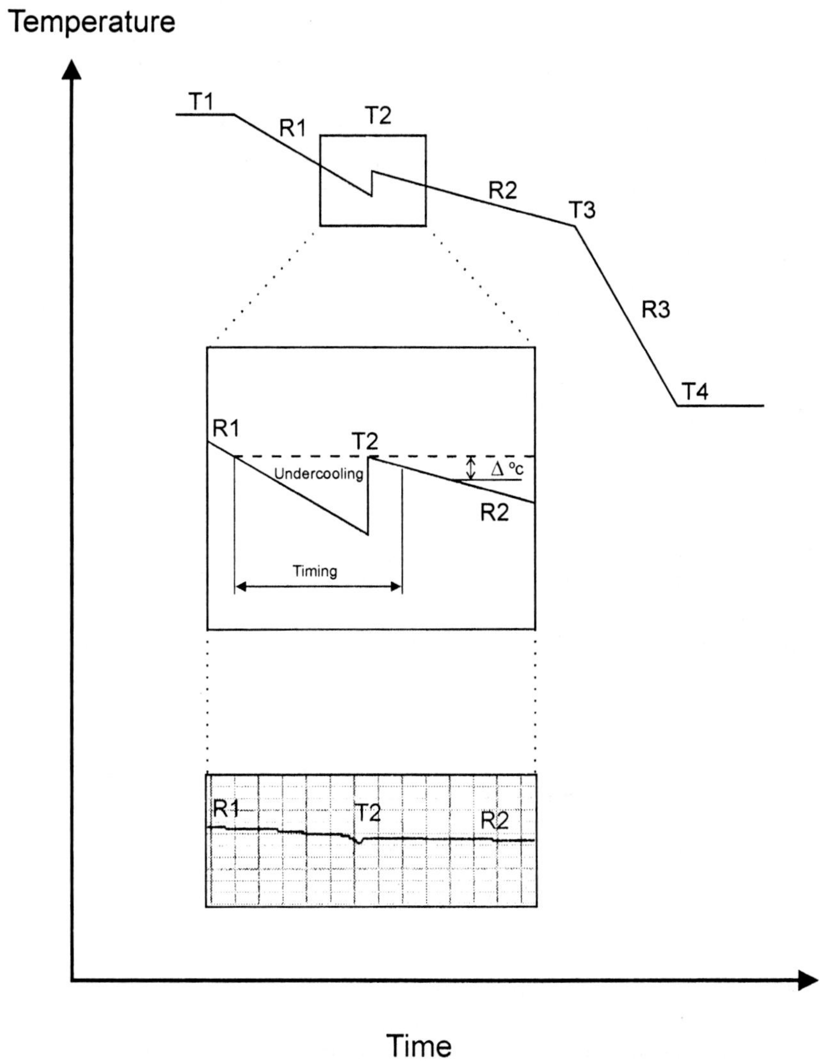

Fig. 3. *Above* Schematic representation of the time-temperature course of freezing of an aqueous solution using a programmable freezer; *Center* the freezing point. *Below* Cooling curve at the freezing point of a 10% DMSO solution subjected to the selected freezing protocol with R_1 and R_2 both set at 0.5 °C/min

Controlled Cooling. Cooling is achieved by a programmable freezing unit which controls the automatic injection of liquid nitrogen from an adjacent vessel connected with a transfer line into the freezing chamber. The cooling rate is programmable and depends on the amount and periodicity of liquid nitrogen injected into the freezing chamber. The freezing program is defined by several parameters (Fig. 3):

1. starting temperature T_1 at which the chamber is held before introducing the samples;
2. freezing point T_2 which must be determined for each cryoprotectant solution;
3. liquid-phase cooling rate R_1;
4. cooling power W needed to overcome the latent heat of fusion, which is applied as soon as the temperature of the liquid-phase sample falls below T_2;
5. timing during which the cooling power W is applied;
6. temperature difference Δ between T_2 and the point at which the cooling rate R_2 is applied;
7. solid-phase cooling rate R_2;
8. temperature T_3 at which the cooling rate can be changed;
9. time at which T_3 is held;
10. cooling rate R_3 between T_3 and the final temperature T_4.

The starting temperatures T_1 is set at 4 °C. The freezing point T_2 has been determined to be −6 °C, and T_3 and T_4 are at −40 °C and −150 °C, respectively. The cooling power W is set at W = 5 and applied for 40s, providing a smooth cooling curve (Fig. 3, bottom) with Δ equal to zero. The cooling rates R_1 and R_2 are set between 0.1 and 0.5 °C/min and the final cooling R_3 at 20 °C/min. The cryogenic tubes containing the specimens are placed in the rack provided in the freezing chamber of the apparatus, and the cooling curve is always confirmed with a temperature sensor in a test sample.

Fast Cooling. The tubes containing the specimens are placed in a stainless steel basket and immersed directly into an LN container.

Storage. After reaching the final temperature provided by the cooling system chosen, the ampoules are placed into an LN container. In routine assays the specimens are kept in liquid nitrogen overnight.

Thawing Procedure

Thawing by fast warming is performed by immersing the cryogenic tubes for 5 min in a water bath kept at 37 °C. Slow warming is achieved by leaving the tubes for 15 min at room temperature.

Recovery of Surviving Tissues and Whole Plants. After thawing, the cryoprotection solution is removed from the tubes, and the specimens are washed three times with 1.8 ml of culture medium. The specimens are then cultured and kept for several weeks under the same environmental conditions described above. Growth of the specimens after freezing and thawing is considered the only reliable viability test. Growth of surviving cultures is observed after 15 days (frozen cells), or after 30–60 days (frozen embryos). Surviving cultures produce globular embryos which show normal development into torpedo, heart-shaped, and finally into cotyledonary stage embryos and whole plantlets.

2.3 Results

2.3.1 Production of Somatic Embryos and Embryogenic Cell Lines

The growth of ovules cultured in vitro showed great variation among samples and among species. Some cultures proliferated after 4–5 weeks, whereas others needed as much as 20 weeks. The number and kind of proliferating structures including pseudobulbi, embryos, and disorganized cells varied considerably among species. The pseudobulbi and primary embryos were used for cryogenic assays or subcultured to fresh medium. Subcultured embryos germinated into whole plants and/or presented secondary embryogenesis, producing more pseudobulbi, embryos, and cells.

The production of disorganized cells and callus was generally restricted to only a few cultures and growth proceeded very slowly among clustered embryos and pseudobulbi. Upon selection and periodic subculture, stable cell lines virtually devoid of organized structures were eventually established (Fig. 4A; Galiana et al. 1993). With the exception of grapefruit, all cell lines were obtained by periodic selection and subculturing on standard medium. Grapefruit cell lines were obtained after supplementing the culture medium with 10 mg/l benzyladenine (Galiana et al., unpubl.).

Embryogenesis of cell lines was induced by supplementing the culture medium with maltose or lactose instead of sucrose (Button 1978; Kochba et al. 1982; Fig. 4B). The embryos were excised from the callus and transferred to fresh medium, and germinated into whole plants (Fig. 4C; Galiana et al. 1993, unpubl. results).

2.3.2 Cryoconservation of Somatic Embryos

Initial assays on the cryoconservation of somatic embryos were performed with Washington navel sweet orange embryos using a cooling bath. Only embryos

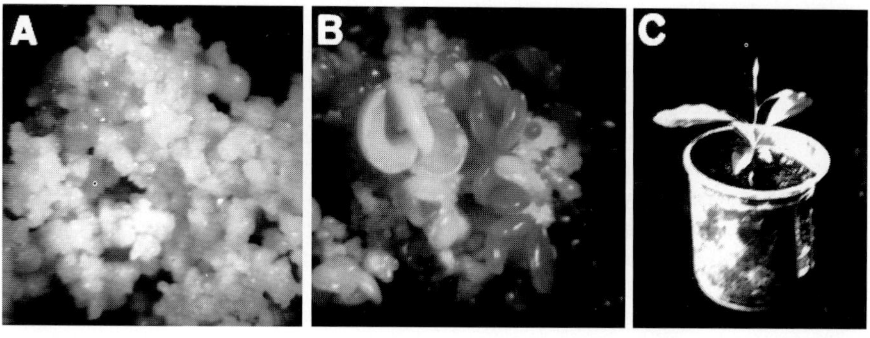

Fig. 4. A Embryogenic cell lines virtually devoid of organized structures obtained by selection and periodic subculture. **B** Embryogenesis induced by supplementing the culture medium with maltose or lactose. **C** whole plants are obtained by germination of somatic embryos. (Galiana and Pérez, unpubl.)

Table 3. Survival and growth of somatic embryos and embryogenic cells of sweet orange after being subjected to − 196 °C

Cryoprotection	Treatment Freezing (cooling rate)	Thawing (warming rate)	Survival (frequency) Somatic embryos[a]	Embryogenic cells[b]
+	−	−	100	100
−	−	−	99.2	100
+	Fast	Slow	0	0
+	Fast	Fast	0	0
+	Slow	Slow	3.7	6.2
+	Slow	Fast	30.5	100

[a] Marín and Duran-Vila (1988); Marín et al. (1993).
[b] Pérez et al. (1993).

subjected to slow cooling survived (Marín and Duran-Vila 1991; Marín et al. 1993). As summarized in Table 3, survival ranged from 3.7 to 30.5%. The assays performed using a programmable freezing unit gave similar results with survival rates averaging 36% (Pérez et al., unpubl. results). Regardless of the freezing protocol, the surviving cultures did not refer to the whole embryo but to proliferating tissue patches which produced new embryos by secondary embryogenesis. These results were not improved by selecting uniform embryo populations or by modifying the cooling rates. Salustiana sweet orange embryos subjected to cryoconservation gave similar results.

2.3.3 Cryoconservation of Embryogenic Cell Lines

Initial assays on the cryoconservation of embryogenic cell lines were performed with Salustiana sweet orange cells using a cooling bath. Growth was observed 2–6 weeks after the freezing/thawing treatments. As in the case of embryos, only cells subjected to slow cooling survived. As summarized in Table 3, 100% survival was achieved when the cultures were thawed by fast warming. The assays performed using a programmable freezing unit indicated that cooling rates R_1 and R_2 may range from 0.1 to 0.5 °C with no effect on cell survival (Pérez et al. 1993). Increasing R_2 to 1 °C also gave similar results, but a further increase to 5 °C caused a reduction in cell survival (Pérez et al., unpubl. results). A freezing program with R_1 and R_2 both set at 0.5 °C is now routinely used for the cryoconservation of embryogenic cell lines. As illustrated in Table 2, the protocol has been successfully used for the cryoconservation of three sweet orange varieties and three mandarin types. In all the cases after thawing, the cultures spontaneously produced embryos which germinated into plantlets without further treatment.

2.4 Discussion

Cryoconservation may be useful for the preservation of plant materials from which cell lines are already available. This is the case for a number of commercial cultivars which are presently being subjected to plant breeding technologies based on protoplast isolation and genetic engineering (Vardi and Galun 1989; Gmitter et al. 1992). However, the production of embryogenic cell lines simply as a preliminary step for storage of genetic material may not be practical. In such cases, cryoconservation of the embryos developing during early stages of ovule culture may provide an alternative. An even simpler and desirable approach would be the cryoconservation of ovules (Bajaj 1984).

Cryoconservation of embryogenic cell lines and somatic embryos has only been demonstrated for polyembryonic citrus. Initial efforts to culture ovules of two monoembryonic species, clementine and citron, were unsuccessful (Juárez and Navarro, pers. comm.), and the development of somatic embryos and cells was restricted to the culture of nucellus isolated from immature seeds (Rangan et al. 1968; Esan 1973). However, many of the plants recovered from eight clementine cultivars (Juárez et al. 1976) were not true to type (Navarro et al. 1985). Therefore, the cryoconservation of embryos and embryogenic cell lines derived from the nucellus of monoembryonic citrus cannot be used for genetic conservation.

The cryoconservation of embryogenic cell lines has the advantage of preserving the citrus genome in a form readily applicable to the modern technologies of protoplast culture and genetic transformation (Gmitter et al. 1992). However, the recovered plants are always juvenile and several years must elapse before adult and productive trees are available. Therefore, additional technologies for the preservation of tissues capable of producing adult plants need to be investigated.

In spite of the limitations described above, the cryoconservation of embryogenic cell lines and somatic embryos is a valid conservation approach. In principle, the method may be applied to a wide range of polyembryonic species and cultivars. However, its implementation for the conservation of wild species collected during expeditions may not be feasible. In such cases, the conservation of ovules and seeds would be a broader and more desirable approach.

3 Summary and Conclusions

It has been demonstrated that somatic embryos and embryogenic cell lines of polyembryonic citrus survived and regenerated whole plants after being subjected to and stored at the temperature of liquid nitrogen. The survival rates were high and the regenerated plants did not show any phenotypic abnormalities, thus opening new perspectives for the conservation of genetic resources of citrus. The cryoconservation protocol described here can be readily used for the conservation of polyembryonic species and cultivars from which cell lines are available.

Acknowledgments. This work would not have been possible without the cooperation of Dr. M.L. Marín, R.M. Pérez, V. Greño, A. Galiana, and C. Ortega, who have been involved in the maintenance of embryogenic cell lines and performed the cryoconservation assays. The helpful comments and suggestions of Dr. L. Navarro and J. Juárez are also acknowledged.

References

Akihama T (1984) Report of the IBPGR citrus collecting mission in Thailand. IBPGR Newsl 8: 19–23

Bajaj YPS (1984) Induction of growth in frozen embryos of coconut and ovules of citrus. Curr Sci 53: 1215–1216

Bajaj YPS (1991) Storage and cryopreservation of in vitro cultures. In: Bajaj YPS (ed) Biotechnology in agriculture and forestry, vol 17. High-tech and micropropagation I. Springer, Berlin Heidelberg New York, pp 361–381

Barrett HC, Rhodes AM (1976) A numerical taxonomy study of affinity relationships in cultivated *Citrus* and its close relatives. Syst Bot 1: 105–136

Bitters, WP, Cole DA, Brusca JA (1969) The citrus relatives as citrus rootstocks. In: Chapman HD (ed) Proc Int Citrus Symp 1, Riverside, CA, pp 411–415

Bitters WP, Cole DA, McCarty CD (1977) Citrus relatives are not irrelevant as dwarfing stocks or interstocks of citrus. In: Grierson W (ed) Proc Int Soc Citric 2, Lake Alfred, FL, pp 561–567

Button J (1978) The effects of some carbohydrates on the growth and organization of citrus ovular callus. Z Planzenphysiol 88: 61–68

Button J, Bornman CM (1971) Development of nucellar plants from unpollinated and unfertilized ovules of Washington navel orange in vitro. J S Afr Bot 37: 127–134

Carpenter JB, Reece PC (1969) Catalog of genera, species and subordinate taxa in the subfamily Aurantioideae (Rutaceae). Agricultural Research Service, USDA, Beltsville, MD, pp 34–106

Chin HF (1988) Recalcitrant seeds – a status report. IBPGR, Rome

Dass HC (1984) Genetic resources of citrus in India. IBPGR Newsl 8: 23–26

Deng X, Grosser JW, Gmitter FG (1992) Intergeneric somatic hybrid plants from *Fortunella crassifolia* cultivar 'Meiwa' with *Citrus sinensis* cultivar 'Valencia'. Sci Hortic 49: 55–62

Esan EB (1973) A detailed study of adventive embryogenesis in the *Rutaceae*. PhD Diss, University of California, Riverside

Espino RRC (1984) Citrus germplasm in the Philippines. IBPGR Newsl 8: 14–15

FAO (1992) Citrus fruit, annual statistics 1990. FAO, Rome

Ford HW, Feder WA (1960) *Citropsis gilletiana*, a citrus relative resistant to the burrowing nematode in laboratory tests. Proc Fla State Hortic Soc 73: 60–64

Galiana A, Pérez RM, Navarro L, Duran-Vila N (1993) Obtención de líneas celulares embriogénicas de cítricos. Actas Hortic (SECH) 10: 947–953

Giacometti DC, Van Sloten DH (1984) Citrus genetic resources. Proc Int Soc Citric 1: 1–4

Gmitter FG, Grosser JW, Moore GA (1992) Citrus. In: Hammerschlag FA, Litz RE (eds) Biotechnology of perennial crops. Biotechnology in agriculture, vol 8. CAB International, Wallingford, Oxon, UK, pp 335–370

Green, RM, Vardi A, Galun E (1986) The plastome of *Citrus*: Physical map, variation among citrus cultivars and species and comparison with related genera. Theor Appl Genet 72: 170–177

Grosser JW, Gmitter FG, Chandler JL (1988a) Intergeneric somatic hybrid plants of *Citrus sinensis* cv. Hamlin and *Poncirus trifoliata* cv. flying dragon. Plant Cell Rep 7: 5–8

Grosser JW, Gmitter FG, Chandler JL (1988b) Intergeneric somatic hybrid plants from sexually incompatible woody species *Citrus sinensis* and *Severinia disticha*. Theor Appl Genet 75: 397–401

Grosser JW, Moore GA, Gmitter FG (1989) Interspecific somatic hybrid plants from the fusion of 'Key' lime (*Citrus aurantifolia*) with 'Valencia' sweet orange (*Citrus sinensis*) protoplasts. Sci Hortic 39: 23–29

Grosser JW, Gmitter FG, Tusa N, Chandler JL (1990) Somatic hybrid plants from sexually incompatible woody species: *Citrus reticulata* and *Citropsis gilletiana*. Plant Cell Rep 8: 656–659

Hodgson RW (1967) Horticultural varieties of citrus. In: Reuther W, Webber J, Batchelor LD (eds) The citrus industry, vol I. Division of Agricultural Science, University of California, Berkeley, pp 431–591

IBPGR (1982) Genetic resources of citrus. AGPG: IBPGR/82/7 Report, Rome, 13 pp

IBPGR (1986) Citrus species and *Poncirus trifoliata*. In: Genetic resources of tropical and subtropical fruits and nuts. IBPGR, Rome, pp 105–115

Jones DT, Ibrahim AZ (1984) Citrus cultivation and germplasm collection in West Malaysia. IBPGR Newsl 8: 9–12
Juárez J, Navarro L, Guardiola JL (1976) Obtention de plantes nucellaires de divers cultivars de clémentiniers au moyen de la culture de nucelle in vitro. Fruits 31: 751–762
King MW, Doetisna U, Roberts EH (1981) The dry storage of citrus seeds. Ann Bot 48: 865–872
Kobayashi S, Ohgawara T, Ohgawara E, Oiyama I, Ishii S (1988) A somatic hybrid plant obtained by protoplast fusion between navel orange (*Citrus sinensis*) and satsuma mandarin (*Citrus inshiu*). Plant Cell Tissue Organ Cult 14: 63–69
Kobayashi S, Sakai A, Oiyama I (1990) Cryoconservation in liquid nitrogen of cultured navel orange (*Citrus sinensis* Osb.) nucellar cells and subsequent plant regeneration. Plant Cell Tissue Organ Cult 23: 15–20
Kochba J, Spiegel-Roy P, Neumann H, Saad S (1982) Effect of carbohydrates on somatic embryogenesis in subcultured nucellar callus of *Citrus* cultivars. Z Pflanzenphysiol 105: 359–368
Marín ML (1988) Conservación de germoplasma de cítricos por cultivo in vitro. PhD Diss, Universidad de Oviedo, Spain
Marín ML, Duran-Vila N (1988) Survival of somatic embryos and recovery of plants of sweet orange [*Citrus sinensis* (L.) Osb.] after immersion in liquid nitrogen. Plant Cell Tissue Organ Cult 14: 51–57
Marín ML, Duran-Vila N (1991) Conservation of citrus germaplasm in vitro. J Am Soc Hortic Sci 116: 740–746
Marín ML, Duran-Vila N (1992) Cryopreservation of somatic embryos of 'Washington Navel' sweet orange. Proc Int Soc Citricult (in press)
Marín ML, Gogorcena Y, Ortiz J, Duran-Vila N (1993) Recovery of whole plants of sweet orange from somatic embryos subjected to freezing thawing treatments. Plant Cell Tissue Organ Cult 34: 17–33
Mazur P (1969) Freezing injury in plants. Annu Rev Plant Physiol 20: 419–448
Mumford PM, Grout WW (1979) Desiccation and low temperature (– 196 °C) tolerance of *Citrus limon* seed. Seed Sci Technol 7: 407–410
Murashige T, Skoog F (1962) A revised medium for rapid growth and bioassays with tobacco tissue cultures. Physiol Plant 15: 473–497
Navarro L, Juárez J, Ballester JF, Pina JA, Ortega C (1979) Obtención de plantas nucelares libres de virus de diversaas variedades de agrios del grupo Navel [*C. sinensis* (L.) Osbeck] por cultivo de óvulos in vitro. Anales de INIA, Serie Protección vegetal, No 12, 19 pp
Navarro L, Ortiz JM, Juárez J (1985) Aberrant citrus plants obtained by somatic embryogenesis of nucelli cultured in vitro. HortScience 20: 214–215
Navarro L, Juárez J, Pina JA, Ballester JF, Arregui JM (1988) The citrus variety improvement program in Spain after eleven years. In: Timmer LW, Garnsey SM, Navarro L (eds) Proc 10th Conf IOCV, Riverside, CA, pp 400–406
Ohgawara T, Kobayashi S, Ishii S, Yoshinaga K, Oiyama I (1985a) Somatic hybridization in *Citrus*: navel orange (*C. sinensis* Osb.) and grapefruit (*C. paradisi* Macf.). Theor Appl Genet 78: 609–619
Ohgawara T, Kobayashi S, Ohgawara E, Uchimaya H, Ishii S (1985b) Somatic hybrid plants obtained by protoplast fusion between *Citrus sinensis* and *Poncirus trifoliata*. Theor Appl Genet 71: 1–4
Pérez RM, Navarro L, Duran-Vila N, Máss O (1993) La crioconservación de lineas celulares como alternativa para la conservación de recursos genéticos de citricos. Actas Horticult (SECH) 10: 940–946
Potvin C, Bergerson Y, Simon JP (1983) A numerical taxonomy study of selected *Citrus* species (Rutaceae) based on biochemical characters. Syst Bot 8: 127–133
Radhamani J, Chandel KPS (1992) Cryoconservation of embryonic axes of trifoliate orange [*Poncirus trifoliata* (L.) Raf.] Plant Cell Rep 11: 204–206
Rangan TS, Murashige T, Bitters WP (1968) In vitro initiation of nucellar embryos in monoembryonic *Citrus*. HortScience 3: 226–227
Ratanadaros A, Chomchalow N (1984) Citrus germplasm in Thailand. IBPGR Newsl 8: 16–19
Roose ML (1988) Isozymes and DNA restriction fragment Iength polymorphisms in citrus breeding and systematics. Proc Int Citrus Congr 1: 225–229

Sakai A, Kobayashi S, Oiyama I (1990) Cryoconservation of nucellar cells of navel orange (*Citrus sinensis* Osb. var. *brasiliensis* Tanaka) by vitrification. Plant Cell Rep 9: 30–33

Sakai A, Kobayashi S, Oiyama I (1991a) Survival of nucellar cells of navel orange (*Citrus sinensis* var. *brasiliensis* Tanaka) cooled to $-196\,^{\circ}C$. J Plant Physiol 137: 465–470

Sakai A, Kobayashi S, Oiyama I (1991b) Cryoconservation of nucellar cells of navel orange (*Citrus sinensis* Osb.) by simple freezing method. Plant Sci 74: 243–248

Sastrapradja S (1975) Tropical fruit germplasm in South East Asia. In: Williams JT, Lamoureux CH, Wulijarni-Soetjipto N (eds) South East Asian plant genetic resources. IBPGR/SEAMEO/BIOTROP/LIPI, Bogor, pp 33–46

Sastrapradja S, Soetjipto N (1984) Citrus in Indonesia. IBPGR Newsl 8: 6–8

Scora RW, Kumamoto J, Soost RK, Nauer EM (1982) Contribution to the origin of the grapefruits, *Citrus paradisi* (Rutaceae). Syst Bot 7: 170–177

Styer DJ, Chin CK (1983) Meristem shoot-tip culture for propagation, pathogen elimination and germplasm preservation. Hortic Rev 5: 221–227

Swingle WT (1943) The botany of *Citrus* and its wild relatives. In: Webber HJ, Batchelor LC (eds) The citrus industry, vol I. Division of Agricultural Science, University of California, Berkeley, pp 129–474

Swingle WT, Reece PC (1967) The botany of citrus and its wild relatives. In: Reuther W, Webber J, Batchelor LD (eds) The citrus industry, vol I. Division of Agricultural Science, University of California, Berkeley, pp 190–430

Tanaka T (1961) Citrilogia, semicentennial commemoration papers on citrus studies. Citrologia Supporting Foundation. Univ Osaka Prefecture, Sasaki, Osaka. Japan, 114 pp

Tarepe TN (1984) Citrus germplasm in Papua New Guinea. IBPGR Newsl 8: 13–14

Tusa N, Grosser JW, Gmitter FG Jr (1990) Plant regeneration of 'Valencia' sweet orange and 'Feminello' lemon, and the interspecific somatic hybrid following protoplast fusion. J Am Soc Hortic Sci 115: 1043–1046

Vardi A, Galun E (1989) Isolation and culture of citrus protoplasts. In: Bajaj YPS (ed) Biotechnology in agriculture and forestry 8. Plant protoplasts and genetic engineering I. Springer, Berlin Heidelberg New York, pp 147–159

Vardi A, Spiegel-Roy P (1978) Citrus breeding, taxonomy and the species problem. In: Cary PR (ed) Proc Int Soc Citricult Griffith, Australia, pp 51–57

Withers LA (1992) In vitro conservation. In: Hammerschlag FA, Litz RE (eds) Biotechnology of perennial crops. Biotechnology in agriculture 8. CAB International, Wallingford, Oxon, UK, pp 169–200

II.3 Cryopreservation of Apple (*Malus* Species) Genetic Resources

C. Stushnoff and M. Seufferheld[1]

1 Introduction

1.1 Plant Distribution and Important Species

Apples are grown in virtually all temperate climates and with modern storage and transportation are found in every country, essentially year-round. The origin of modern cultivars can be traced to a few centers where genetic diversity flourished because of environmental conditions conducive to the evolution and sustenance of diverse genotypes. Global exploration and immigration over the centuries included apple culture and cultivar development as an integral component of developing civilizations. Centers of origin for 32 primary species have been identified in several countries on three different continents (Table 1).

The taxonomy of *Malus* has been studied extensively. Although as many as 74 species and hundreds of interspecific hybrids have been reported by several authors, Way et al. (1990) listed 33 primary species in a review on genetic resources of *Malus*. Yunnan Province in China is a particularly rich region of biodiversity for *Malus*, where 24 indigenous species have been identified (Zhang et al. 1993).

1.2 Methods for the Conservation of Apple Germplasm

Apples are genetically heterozygous, driven by a strong self-incompatibility system. Accordingly, a wealth of genetic diversity has enabled selection and breeding to develop, based upon rich sources of natural variability from the centers of origin. Historically, the centers of origin have represented the most important means of germplasm conservation. As civilizations became globally mobile, selection and growing of open-pollinated seedlings in regions remote from the centers of origin generated another important source of diversity and conservation. More recently, however, centralized marketing and distribution have fostered production of a largely monocultivar system relegating hundreds, if not thousands, of cultivars to relict status. Many of these are being saved in private collections as sources of gene conservation.

National as well as local government and public institutions have recognized the need to conserve the gene resources of this valuable crop by initiating

[1]Department of Horticulture, Colorado State University, Fort Collins, CO 80523, USA

Biotechnology in Agriculture and Forestry, Vol. 32
Cryopreservation of Plant Germplasm I (ed. by Y.P.S. Bajaj)

Table 1. Distribution of primary species of *Malus* by centers of origin. (Westwood, in Way et al. 1990)

Species	Center of origin
M. pumila Miller *M. sylvestris* Miller *M. florentina* (Zuccagni) C. Schneider	Europe
M. pumila Miller *M. trilobata* (Poiret) C. Schneider	Asia Minor
M. sikkimensis (Wenzig) Koehne	Himalaya
M. prattii (Hemsley) C. Schneider *M. yunnanensis* (Franchet) C. Schneider	SW China
M. micromalus Makino	SE China
M. hupehensis (Pampan) Rehder *M. honanensis* Rehder	Central China
M. sieversii (Ledeb.) M. Roemer *M. kansuensis* (Batalin) C. Schneider	NW China
M. prunifolia (Willd.) Borkh. *M. asiatica* Nakai *M. baccata* (L.) Borkh.	N & NE China
M. doumari (Bois) A. Chev.	Taiwan
M. baccata *M. halliana* Koehne *M. sieboldii* (Regel) Rehder *M. sargentii* Rehder *M. tschonoskii* (Maxim.) C. Schneider	Japan
M. prunifolia *M. asiatica* *M. micromalus* *M. baccata* *M. sieboldii* (Regel) Rehder	Korea
M. fusca (Raf.) C. Schneider *M. ioensis* (Alph. Wood) Britton *M. angustifolia* (Aiton) Michaux *M. coronaria* (L.) Miller	North America

and maintaining, with various degrees of success, field gene banks or clonal repositories for future use. Because clonal integrity is the keystone to a successful conservation program for the apple, maintenance of gene resources presents special and unique problems and opportunities not found in seed-propagated crops.

Germplasm of apple and other clonally propagated fruit crops can be stored as field-grown plants, as potted plants in screenhouses or greenhouses, as in vitro culture collections, and recently as cryopreserved dormant buds. At present, tissue culture collections and cryopreserved dormant buds are used to back up a collection, and not as the primary collection. Pollen has not yet become a conventional vehicle for germplasm preservation, but has potential for preservation of biodiversity in *Malus* (Akihama and Omura 1986; Way et al. 1990). In the USA, clonal apple collections are stored in field gene banks in the National Clonal Apple Repository, within the National Plant Germplasm System (NPGS) at Geneva, New York (Forsline 1983).

Two approaches have been used to conserve in vitro derived genetic resources: short to medium range storage and long-term storage. Short and medium storage techniques rely on the principles of slow growth, e.g., low light, nutrients, and cool temperatures (Towell and Roos 1989). Apple shoots have been stored successfully for 12 months in the dark at 1 and 4 °C (Sakai and Nishiyama 1978; Lundergan and Janick 1979; Orlikowska 1991). Druart (1985) reported up to 2 years of survival for defoliated apple shoots stored at 2 °C.

Long-term preservation by cryopreservation of plant tissue cultures has been studied with various tissues. In vitro shoot tips of apple (Katano et al. 1983; Kuo and Lineberger 1985; Niino et al. 1992) and callus cultures (Kuroda and Nishiyama 1983) have been cryopreserved.

Dormant vegetative buds have also been successfully recovered from cryopreservation (Stushnoff 1991; Forsline et al. 1993). A pilot project designed to optimize handling methods and to test recovery of different taxa following short (1–5 years) and long-term (10–25 years) storage in liquid nitrogen is in progress at the National Seed Storage Lab (NSSL), Fort Collins, CO, USA, in collaboration with the National Clonal Apple Repository, Geneva, NY.

1.3 Need for Conservation/Cryopreservation

Accessibility to a diverse gene pool of the world's germplasm is essential to enable continual progress in developing new cultivars to serve the industry. Unfortunately, the world's gene pool of fruit crops is not expanding, but is more often than not under pressure and threatened with loss of diversity because of the activities of mankind. The need to develop technology to preserve existing resources and make them more available is more pressing than ever.

Although there are no documented cases in the literature of one or more species of apple becoming extinct, there is evidence that some species are known to exist as only one or a few clones, e.g., *M. sargentii* and *M. tschonoski* (Maxim.) C. Schneider (Way et al. 1990). Aside from the obvious anthropogenic threats of human encroachment, genetic dilution of wild *Malus* populations bordering advanced selected cultivars is of concern. Dilution of wild populations is strongly driven by self-incompatibility, forcing generation of hybrids with nearby cultivars which usually carry neither genes for resistance to diseases such as apple scab nor many other valuable traits. Thus, native populations which have taken centuries to evolve are potentially diluted. It will be very important to preserve not only *Malus* species from their centers of origin, but also land races that have been used for centuries and become wild and acclimated to different ecosystems, for example, apple cultivars introduced to the Patagonian Andes of Argentina. Although a noble concept, in situ conservation alone is for the same reasons which have historically reduced biodiversity of many other perennial species, unlikely to adequately serve our interests to conserve valuable germplasm. World politics and environmental pressures cannot be controlled, and risking the consequences of inaction is intolerable. Accordingly, it is in our best interests to foster conservation of valuable germplasm by as many methods as practical. Recent developments in cryopreservation promise to provide another alternative. One of the most compelling reasons for preserving the apple vegetatively is to maintain desirable gene combinations which have taken centuries to evolve. Maintenance of clonal integrity cannot be accomplished by storing seed, thus the job becomes more challenging, and cryopreservation of asexual propagules or tissues with vegetative reproduction potential offers an attractive option.

2 Rationale and Potential Application for Cryopreservation of Apple

For several reasons, conservation of germplasm from vegetatively propagated species such as apple is best served by storing unique gene combinations through preservation of vegetative propagules in place of, or in addition to seeds. Field gene banks have been assembled as clonal repositories at many locations around the world, and some attempts have been made to protect valuable resources in their natural sites, but neither approach is free of hazards. The practice of using in vitro culture as a means of storing plant materials is about 20 years old. The use of cryopreservation as a long-term storage technique is the most recent development, little more than a decade old and currently undergoing intense scientific research (see Bajaj 1991).

Cryopreservation of mature dormant vegetative buds is a promising method of conserving clonal base collections since it is inexpensive, space-efficient and simple. It maintains clonal integrity and requires low maintenance, with reduced vulnerability to environmental and pathological agents. After post-thaw recovery and grafting, dormant buds can rapidly produce flowering plants for plant breeding purposes (Sakai 1986; Chen and Kartha 1987; Stushnoff 1991). Dormant buds from *Malus, Amelanchier, Sorbus*, and *Prunus* have been cryopreserved, resulting in 80 to 100% viability and recovery. Cultivars and species can be classified in terms of response to desiccation, seasonal adaptation, and cryopreservation viability (Stushnoff 1987, 1991; Tyler and Stushnoff 1988a, b).

2.1 Mechanisms Controlling Tolerance to Desiccation and Freezing Temperatures

Knowledge of biochemical and biophysical mechanisms controlling tissue responses to moisture stress and low temperatures simplifies the understanding of cryopreservation technology. Responses to low temperatures fall into two general categories, *cold tender* tissues and *cold hardy* tissues. In addition, it is useful to classify tissues as *desiccated* or *hydrated* to facilitate development of cryopreservation protocols.

Dormant buds of many temperate-zone woody plant species (including apple), desiccation-tolerant seeds, excised embryo axes, and pollen of some species can be desiccated to very low moisture levels. In order to survive cryopreservation, vegetative tissues and organs must tolerate freeze-induced desiccation, as well as temperatures of liquid nitrogen (−196 °C). The natural capability of some species of the Rosaceae and other families to acquire cold hardiness (Sakai 1973; Forsline 1983; Way et al. 1990) provides a useful strategy to cryopreserve *Malus* germplasm (Stushnoff 1987, 1991, 1992).

Second-order transitions to the glassy phase occur when aqueous solutions become viscous and form an amorphous noncrystalline solid (MacKenzie 1977; Franks 1985). Glass transitions can be detected by differential scanning calorimetry. With this information phase diagrams can be constructed to enable studies of the thermodynamic behavior of dormant apple buds under freezing and desiccation conditions (Stushnoff et al. 1992). As dehydration is increased in cold hardy species, the concentration of the intracellular aqueous solution increases until the glass transition temperature is attained. At this temperature – concentration point, viscous solutions become rubbery then undergo a transition to an amorphous solid. Plant tissues are hypothesized to be very stable at cryogenic temperatures in the amorphous phase because rotational and translational molecular motion are effectively inhibited or slowed, preventing molecular reactions including water loss (Burke 1987). In complex tissues such as dormant apple buds, however, it is likely that microdomains of ice crystals exist in and among tissues which have undergone transition to the glassy state.

In nature endogenous changes occur, permitting dehydration without injury and tolerance to freezing temperatures. These changes permit cryopreservation and recovery without injury (Figs. 1 and 2). Dormant buds and associated

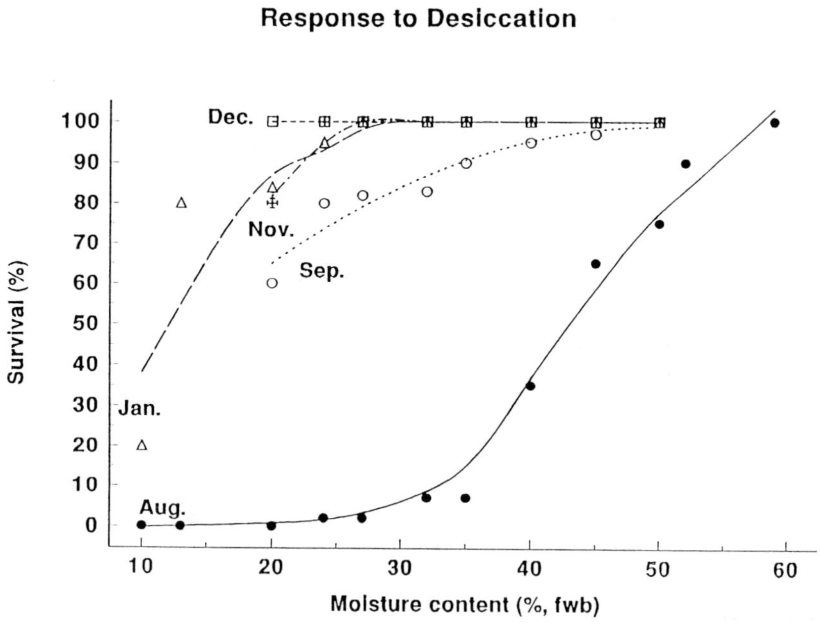

Fig. 1. Response of seasonal physiological development to desiccation tolerance in cold hardy apples. (Stushnoff 1991)

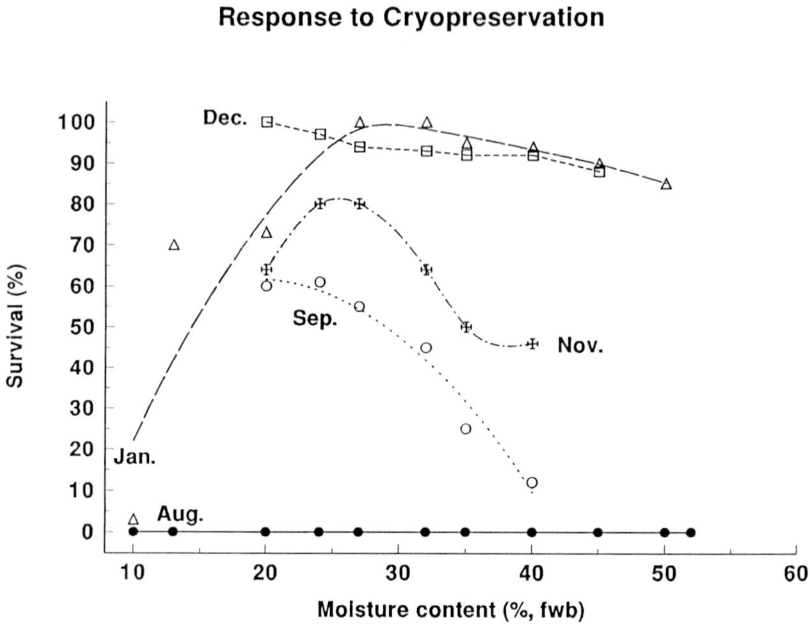

Fig. 2. Response of seasonal physiological development to cryopreservability in cold hardy apples. (Stushnoff 1991)

cortical tissues can be desiccated from a field moisture content of approximately 50% to 20–30% moisture (fresh weight basis, fwb) for cryopreservation purposes. A remarkable transition in cold tolerance, from only a few degrees below 0 °C to survival without injury at liquid nitrogen temperatures (−196 °C) is accompanied, among other biochemical changes, by the endogenous production of products such as sugars and polyols (Sakai and Yoshida 1968; Raese et al. 1978; Stushnoff et al. 1993). These compatible osmolytes can stabilize macromolecules in solutions (Remmele et al. 1994) and invoke a tendency to produce intracellular glasses in dormant apple buds (Stushnoff et al. 1988).

2.2 Cryopreservation Strategies for Dormant Buds

Three general strategies have been used to cryopreserve dormant apple buds:

1. The *first strategy* is applicable to taxa which naturally tolerate desiccation and freezing to −30 °C or colder at maximum hardiness. Cooling at a slow rate (1–2 °C/h) to −20 or −30 °C followed by 12 to 24 h storage at these temperatures is commonly used in cryopreservation protocols with cold hardy dormant apple buds (Sakai and Nishiyama 1978; Tyler et al. 1988). A slow cooling rate allows water movement through membranes to extracellular spaces, effectively setting up a freeze-desiccated condition in critical meristematic tissues.

A complete understanding of the water relations during cold acclimation in woody species is crucial for the development of successful cryopreservation protocols. Freezing tolerance in vegetative buds is a function of both the ability to survive dehydration and the ability to maintain a certain level of water. Calorimetric and NMR studies suggest that changes occur in both the level of desiccation tolerance and the amount of unfreezable water during acclimation and during progression of the cold season. Consequently, no injury occurs when the freezing of water does not desiccate tissues to damaging levels. In the apple, injury occurs when the level of unfrozen water is greater than the critical moisture level for desiccation damage (Tyler et al. 1988; Vertucci and Stushnoff 1992).

If tissues are appropriately treated at subzero temperatures by "prefreezing" for a period of time, they survive −196 °C in liquid nitrogen (Sakai and Nishiyama 1978; Tyler and Stushnoff 1988a ,b). Studies using pulsed nuclear magnetic resonance spectroscopy of dormant apple buds have shown that prefreezing temperatures had a greater effect than the duration of the prefreezing treatment on reducing the quantity of liquid water.

Although cryopreservation of buds from very cold hardy taxa is possible using only a prefreezing treatment before immersion in liquid nitrogen, controlled desiccation treatments enhance viability, and extend both sampling interval and the range of cultivars that can be cryopreserved (Tyler and Stushnoff 1988b). Dormant bud desiccation tolerance is dependent upon tissue maturity and on prevailing climatic conditions. Buds collected just after defoliation, but before substantial cold acclimation, are intolerant of desiccation (Fig. 1). Also, in years with warm periods during winter, buds are less tolerant to desiccation. In early autumn dormant buds are intolerant of desiccation, but by November and December they are fully tolerant to a loss of 50% of their moisture content (Stushnoff 1987, 1991; Tyler and Stushnoff 1988a, b).

Different procedures can be used to dehydrate apple buds. One method is to expose the twigs to 0 to −4 °C in a temperature-controlled freezer, using a fan to circulate the cold air to remove moisture while avoiding dehardening. Our research has shown that temperatures above 0 °C can rapidly deacclimate dormant bud twigs, but at 0 °C or a few degrees below 0 °C hardiness levels are generally stabilized. Another method is to expose the bud stems to low relative humidity over glycerol or various saturated salts in hermetic containers at −4 to 0 °C.

The optimal moisture content for cryopreservation, after dehydration from a field moisture level of 45 to 55%, is generally between 20 to 30% (fwb). Moisture levels below about 15%, depending on the taxa, usually result in tissue injury, even during the period of maximum desiccation tolerance.

After controlled dehydration, a prefreezing treatment is essential. Dormant bud twigs survive a regime of cooling at 2 °C/h to −30 °C, holding for 24 h at −30 °C, followed by a direct plunge into liquid nitrogen without warming. Slow cooling to −20 or −30 °C, depending on initial tissue hardiness, encourages extracellular ice formation and further reduces the amount of liquid water within the cells (Tyler et al. 1988). The second cooling process, in which tissues are rapidly cooled from −30 to −196 °C, is considered to have two functions: (1) to prevent additional dehydration of the cells by extracellular ice growth and (2) to favor vitrification, preventing intracellular ice formation.

Recovery of the buds from liquid nitrogen involves thawing and rehydration in moist peat moss before grafting the bud onto 1 to 1.5 cm diameter rootstocks under greenhouse conditions. Buds recovered in this manner have produced flowers within 2 years after cryopreservation in liquid nitrogen (Stushnoff 1987, 1991).

2. The *second strategy* used for cryopreservation of dormant buds involves pretreatment with cryoprotective sugars and the addition of cryoprotectants to enhance survival of taxa that are less cold hardy or do not acclimate sufficiently (Seufferheld et al. 1991).

Attempts to cryopreserve both cold tender and cold hardy cultivars prior to development of maximum cold hardiness usually result in survival of 10% or less. Accordingly, a modified vitrification solution (MPVS) was developed for apple, consisting of 7% (w/v) DMSO, 15% (w/v) ethylene glycol, 15% (w/v) propylene glycol, and 15% (w/v) glycerol in 0.5 M sorbitol (Seufferheld et al. 1991). This solution permeabilizes cellular membranes and promotes the formation of amorphous glasses, using appropriate concentrations and freezing rates. Survival of cold tender apple buds has also been enhanced by application of alginate gel encapsulation prior to desiccation, similar to the method used by Dereuddre et al. (1990) for in vitro derived tissues. Alginate gel encapsulation plus the use of MPVS has resulted in survival of cold tender taxa, not only at the stage of maximum hardiness in samples taken in December and January, but also in samples from October, November and February (Seufferheld et al. 1991). Furthermore, this was achieved with some cold tender samples of *Corylus*, *Ribes*, *Pyrus*, and *Vaccinium* taxa. Interestingly, the xylem ray parenchyma tissues that are normally killed at −40 °C survived −196 °C after treatment with MPVS. Data from differential thermal analysis show that the nucleation temperature was lowered by at least 10 °C (Seufferheld et al. 1991). These results illustrate that the freezing response of nonhardy tissues can be altered significantly to facilitate survival during cryopreservation.

Survival of MPVS-vitrified buds was tested by forcing buds to develop into shoots in warm humid environments after recovery from liquid nitrogen. Although largely successful, there are some problems yet to be resolved. For example, there is some residual toxicity of the MPVS, which interferes with grafting by patch budding, even though the buds can be forced to develop into emerging shoots on the cryopreserved bud sticks during recovery.

3. The *third strategy* is to manipulate cold hardiness by controlling fluctuating temperature regimes. Research on cold hardening processes suggests that fluctuating temperatures can be used to induce cold hardiness in nonacclimated tissues. This suggests that instead of relying only on nature to induce optimal conditions prior to harvesting buds for cryopreservation, buds may be hardened by controlled methods. We are attempting to determine the best sequence of environmental conditions for inducing maximum hardiness and suitability for cryopreservation. In the future, it may be best to collect samples in October or November and subject them to appropriate controlled inductive conditions, rather than waiting

for natural conditions to induce hardiness and desiccation resistance outdoors. The sampling period could thus be expanded, and potentially damaging climatic freeze events that occur in mid-winter might be avoided. Preliminary evidence shows that a regime subjecting nonhardy stems to 12-h diurnal fluctuation from 0 to 10 °C for 30 days is more effective in inducing hardiness than treatment at −5 to 10 °C, or constant exposure to 10 or −5 °C.

2.3 Viability Assays

The most common methods of determining viability in apple buds after freezing are as follows:

1. *Visual Evaluation of Tissue Browning*. This method consists of incubation of the buds at 100% relative humidity, in plastic bags with moist peat moss at room temperature for at least 5 days. The most common injuries are to the connective tissues of the bud to the cortex, the meristematic region of the bud, and the cortical tissues. Injury is rated subjectively by visual scores according to the degree of browning and type of tissue affected.

2. *Conductivity Methods*. These procedures allow establishment of the degree of damage by measuring electrical conductivity of leachate of cellular electrolytes that diffuse out of the cells because membrane semipermeability properties have been affected by freezing stress. Percent of injury can be calculated as an aid to evaluation of various treatments.

3. *Colorimetric*. This method is based on the reduction of colorless triphenyl tetrazolium chloride (TTC) to the red form (formazan) by living tissues. Apple twigs are placed in TTC solution and after a short vacuum are incubated 24 h at room temperature. It is important to standardize the size of the twigs in order to avoid excessive variability. After that the red formazan is extracted and absorbance is measured with a spectrophotometer at 550 nm (Quamme and Stushnoff 1983). The regression of percentage damage can be used to obtain the slope and intercept to determine the lethal temperature at which 50% of the samples were killed (LT50).

4. *Recovery by Grafting Cryopreserved Buds*. The buds can be grafted by patch budding to potted rootstocks in a greenhouse. This process requires several weeks, but it is the most reliable and most positive indicator of survival.

3 Cryopreservation of In Vitro Systems with Apple

Many fruit crops which are to be entered into cryogenic preservation are tissue culture-derived, and materials regenerated after cryogenic storage are done so in vitro. Shoot tips are better able to regenerate true-to-type plants than are other in vitro systems such as callus or suspension cultures. With this technology, it is possible to produce rapid clonal multiplication. Shipment for international exchange is simplified. Additionally, there is good potential for low-temperature in vitro acclimation to accommodate cryopreservation (Caswell et al. 1986; Niino et al. 1992; Scottez et al. 1992), as well as growth reduction (Stushnoff and Fear 1985; Withers 1991; Bajaj 1991).

Meristematic shoot tips have been used to cryopreserve apples (Katano et al. 1983; Kuo and Lineberger 1985; Chang et al. 1992; Niino et al. 1992). In vitro systems provide an acceptable way to conserve clonal integrity. However, there is some question about genetic stability (Towill 1988). One problem with cryogenic storage is the maintenance of organization within the tip and subsequent regeneration without callus formation, since callus increases the frequency of somaclonal variations. With current procedures, apple shoot tips exposed to liquid nitrogen regenerate without intermediary callus formation (Niino et al. 1992). Stability of the shoot tips depends on the cryopreservation protocol and regrowth media (Towill and Roos 1989; Withers 1991). More important concerns about in vitro systems are the need for considerable manipulation of cultural procedures, the need for regular rejuvenation, and the specific media requirements of each species and cultivar. As well, several years may be necessary for recovery and development of adult flowering plants.

3.1 Pretreatment of Plants or Shoot Tips

Taxa with some ability to cold acclimate survive better than those which cannot be acclimated to cold. Tissue cultures of apple can be cold hardened by exposure to low-temperature inductive conditions. Exogenous treatments with sucrose are also very beneficial for increasing tolerance to desiccation and freezing stresses. Growing apple shoots in an elevated sucrose concentration (3 to 14%) results in a decreased shoot moisture content and increased hardiness up to 6 °C (Caswell et al. 1986). Temperature and light regimes can enhance hardening. For example, apple shoot tips survive better at −196 °C when cold hardened for 8 weeks at 5 °C and short days (Niino et al. 1992), or at 4 °C in the dark (Kuo and Lineberger 1985).

Meristems, shoots, and dormant buds at 70 to 90% moisture (fwb) do not tolerate dehydration without the addition of an anhydroprotectant. Cryopreservation of these tissues also requires the use of exogenous agents which help to change physical properties of the tissues, such as increasing the tendency to undergo phase transitions to the glassy state or providing stabilization of membranes and biomolecules (Stushnoff 1992). Chemical and other protocols are needed to achieve these goals (Katano et al. 1983; Kuo and Lineberger 1985; Seufferheld et al. 1991; Niino et al. 1992).

Hydrated tissues generally do not tolerate extreme osmotic stress or freezing. Desiccation has to be done in a way which does not damage the tissues. As dehydration takes place there are changes in osmotic relations within the tissues, as well as modifications of the freezing response. In addition, desiccation behavior can be modified by improving the physical conditions for desiccation, for example, by encapsulating tissues in alginate beads (Dereuddre et al. 1990).

3.2 Cooling

The strategies for survival of in vitro systems cooled to −196 °C can be summarized as follows. DMSO and sucrose have been used frequently as cryoprotectants during prefreezing. The concentration of DMSO has been varied from 5 to 15%, and sucrose from 0.2 to 1.2 M. Also, combinations of DMSO with glycerol, ethylene glycol, or propylene glycol have been used. First, the sample is cooled slowly at 0.5–2 °C/min to −20 to −40 °C, in order to achieve freeze dehydration. Tissues are plunged in liquid nitrogen (LN) to form an intracellular glass. Slow thawing must be avoided to prevent ice formation (resumption of crystallization during warming) and subsequent death of tissue (Sakai 1986). This approach and several modifications have been used to cryopreserve apple and pear (Katano and Ishihara 1983; Kuo and Lineberger 1985; Chang et al. 1992).

The vitrification procedure involves loading the cells with enough cryoprotectant to promote transition to an amorphous state as a glass. This protocol may avoid injury due to ice formation, and subsequent solution effects occurring under conventional cooling procedures. The duration of exposure to cryoprotectants must be adjusted according to the size of the plant material, cryoprotectant toxicity, and temperature (Finkle et al. 1985). In general, a low concentration of vitrification solution is added at room temperature with a stepwise increase in concentration up to 100% and a decrease in temperature (0 °C) to minimize toxicity. After rapid thawing, the vitrification solution is removed by diluting with MS media containing 1–1.5 M sucrose (Niino and Sakai 1992; Scottez et al. 1992). In addition, encapsulation in alginate beads coupled with cold hardening of the shoot tips has been shown to increase desiccation tolerance of apple and pear shoot tips (Dereuddre et al. 1990; Niino and Sakai 1992).

4 Summary and Conclusions

Valuable genetic diversity in apple germplasm has emerged from several centers of origin, over many centuries of selection pressure, driven by climatic and environmental conditions in regions that have been ideal for the genus *Malus*. One or two specific species are threatened by anthropogenic forces, as is the diversity of indigenous species.

Various problems associated with seed storage of a genus not normally reproduced by seed, as well as the need to maintain clonal integrity and to preserve valuable gene combinations, have necessitated the development of conservation methods that rely on vegetative propagation. Such methods include field gene banks of mature trees, in vitro systems, and cryopreservation of dormant buds. There is a need to provide backup and alternative storage methods to field gene banks, and cryopreservation is a promising means of meeting this need.

Conservation technology, enabling cryopreservation of dormant buds and in vitro shoots, has developed to a stage which warrants testing as a routine

Table 2. Summary of research on cryopreservation of apple genetic resources

Author (year)	Species/cultivar	Explant used	Results/remarks
Sakai and Nishiyama (1978)	McIntosh, Red Gold, Jonathan, Starking Delicious, Golden Delicious	Dormant buds	Buds, cortex survived LN; xylem killed
Katano et al. (1983)	Fuji	Dormant shoot tips	Dormant apical buds recovered in vitro
Kuroda and Nishiyama (1983)	Not reported	Callus cultures	Hardened callus, prefrozen to −30 °C survived (16%) LN
Kuo and Lineberger (1985)	Jonathan	In vitro shoot tips	Callus but no organized growth recovered
Akihama and Omura (1986)	*M. pumila* var. *domestica*	Pollen	Low germinability after 8 years in LN
Caswell et al. (1986)	MM106	In vitro shoot tips	Low temperature and high sucrose enhance cold hardiness
Tyler and Stushnoff (1988a, b)	Bottle Greening, Dolgo, Edith Smith, Exeter, G. Delicious, McIntosh, McLean, Mn 1665, park Patterson, Red Delicious, Rescue, Winter, Queen	Dormant buds	Dehydration to 20–30% enhanced survival
Tyler et al. (1988)	McIntosh, Patterson, Red Delicious, Rescue	Dormant buds	NMR shows 0.2–0.3 g H_2O/g dry wt. remains unfrozen in buds surviving LN
Stushnoff et al. (1988, 1992)	Golden Delicious, Dolgo	Dormant buds	Glass transitions detected in buds during cryopreservation
Seufferheld et al. (1991)	Anna, Cox's Orange, Dolgo, Ein Shemer, Empire, G. Delicious, Jonagold, K14, MM106, Mutsu, Wealthy, *M. baccata*, *M. coronaria*, *M. fusca*, *M. kansuensis*, *M.* x *robusta*	Dormant buds	All tissues remain uninjured with encapsulation and use of a vitrification solution
Chang et al. (1992)	Fuji, Jnguang	In vitro shoot tips	Cold hardening, cryoprotectant, slow freezing and rapid warming result in 90% survival
Niino and Sakai (1992) Niino et al. (1992)	Fuji	In vitro shoot tips	Cold hardening, alginate encapsulation, and high sucrose result in 80% survival
Forsline et al. (1993)	Many cultivars and species	Dormant buds	Large-scale pilot project for backup conservation, successful recovery after 5 years in LN

method in germplasm systems. A progression of research discoveries suggests that the practical application of cryopreservation technology should be tested and implemented (Table 2). Dormant buds fall into two categories, cold hardy and cold tender. Cold hardy tissues can be subjected to controlled drying conditions to enable removal of field moisture from approximately 50 to 20–30% (fwb). Freezing at slow rates (1–2 °C/h) and holding tissues at −30 °C for 12 to 24 h followed by a direct plunge into liquid nitrogen provides a high level of recovery following storage. Cold tender tissues require pretreatment involving sucrose or a vitrification solution to achieve satisfactory recovery, but dormant buds of essentially all taxa tested have now been recovered from liquid nitrogen storage.

Cryopreservation of tissue-cultured meristematic shoot tips from apple has been achieved by a variety of procedures, most of which include imbibition in molar concentrations of sucrose prior to freezing; a cold-hardening step; slow and/or rapid cooling rates depending on the investigators and the use of DMSO plus one or more sugars and polyols as cryoprotectants at various concentrations; and recovery by thawing at rapid rates. High recovery has been attained, but development of specific procedures for each taxon is likely to be necessary.

Acknowledgments. The generous financial support provided by USDA, Cooperative Agreement No. 58-5402-0-004, NSSL, Fort Collins, CO, is gratefully acknowledged. The authors also very much appreciate collaboration on various phases of work reported herein by Leigh Towill, Loren Weisner, and John Waddell of USDA, National Seed Storage Laboratory, and by Phillip Forsline, USDA, National Clonal Apple Repository, Geneva, NY. Support from the Colorado Agricultural Experiment Station Project 690 is also gratefully acknowledged.

References

Akihama T, Omura M (1986) Preservation of fruit pollen. In: Bajaj YPS (ed) Biotechnology in agriculture and forestry, vol 1. Trees I. Springer, Berlin Heidelberg New York, pp. 101–112

Bajaj YPS (1991) Storage and cryopreservation of in vitro cultures. In: Bajaj YPS (ed) Biotechnology in agriculture and forestry, vol 17. High-tech and micropropagation I. Springer, Berlin Heidelberg New York, pp 361–381

Burke, MJ (1987) The glassy state and survival of anhydrous systems. In: Leopold C (ed) Membranes, metabolism and dry organisms. Cornell University Press, Ithaca, NY, pp 359–365

Caswell KL, Tyler NJ, Stushnoff C (1986) Cold hardiness of in vitro apple and Saskatoon shoot cultures. HortScience 21(5): 1207–1209

Chen TTH, Kartha KK (1987) Cryopreservation of woody species. In: Bonga JM, Durzan DJ (eds) Cell and tissue culture in forestry, vol 2. Martinus Nijhoff, Dordrecht, pp 305–319

Chang Y, Chen S, Zhao Y, Zhang D (1992) Studies on the cryopreservation of apple shoot tip in vitro. Changli Institute of Pomology, Hebei Academy of Agric Sci 103: 457–465

Dereuddre J, Scottez C, Arnaud Y, Duron M (1990) Resistance of alginate coated axillary tips of pear tree (*Pyrus communis* L. cv. BH) in vitro plantlets to dehydration and subsequent freezing in liquid nitrogen: effects of previous cold hardening. CR Acad Sci Paris 310: 317–323

Druart P (1985) In vitro germplasm preservation technique for fruit trees. In: Schafer-Menuhr A (ed) In vitro techniques: propagation and long-term storage. Martinus Nijhoff, Boston, pp 167–171

Finkle BJ, Zabala ME, Ulrich JM (1985) Cryoprotective compounds in the viable freezing of plant tissues. In: Kartha KK (ed) Cryopreservation of plant cells and organs. CRC Press, Boca Raton, pp 75–113

Forsline PL (1983) Winter hardiness in common New York apple varieties. Proc NY State. Hortic Sci 128: 20–42

Forsline PL, Stushnoff C, Towill LE, Waddell J, Lamboy WF (1993) Pilot project to cryopreserve dormant apple (*Malus* sp.) buds. HortScience 28(5): 478

Franks F (1985) The properties of aqueous solutions at subzero temperatures. In: Franks F (ed) Water – a comprehensive treatise. Plenum Press, New York, pp 215–334

Katano M, Ishihara A, Sakai A (1983) Survival of dormant apple shoot tips after immersion in liquid nitrogen. HortScience 18(5): 707–708

Kuo CC, Lineberger BD (1985) Survival of in vitro culture tissues of Jonathan apples exposed to −196 °C. HortScience 20: 764–767

Kuroda H, Nishiyama Y (1983) Liquid nitrogen storage of apple callus cultures. Hokkaido Natl Agric Stn Res Bull 136: 15–21

Lundergan C, Janick J (1979) Low temperature storage of in vitro apple shoots germplasm preservation. HortScience 14(4): 514

MacKenzie, AP (1977) Non-equilibrium freezing behavior of aqueous systems. Philos Trans R Soc Lond Ser B 278: 167–189

Niino T, Sakai A (1992) Cryopreservation of alginate coated in vitro grown shoot tips of apple, pear and mulberry. Plant Sci 87: 199–206

Niino T, Sakai A, Yakuwa H, Nojiri K (1992) Cryopreservation of in vitro grown shoot tips of apple and pear by vitrification. Plant Cell, Tissue Organ Cult 28: 261–266

Orlikowska AT (1991) Effect of in vitro storage at 4 °C on survival and proliferation of apple rootstocks. Acta Hortic 289: 251–253

Quamme HA, Stushnoff C (1983) Resistance to environmental stress. In: Moore JN, Janick J (eds) Methods in fruit breeding. Purdue University Press, West Lafayette, IN, pp 242–266

Raese JT, Williams MW, Billingsley HD (1978) Cold hardiness, sorbitol and sugar levels of apple shoots as influenced by controlled temperature and season. J Am Soc Hortic Sci 103(6): 796–801

Remmele R Jr, Stushnoff C, Carpenter JF (1994) Real time infrared spectroscopic analysis of lysozyme during lyophilization: structure/hydration behavior and influence of sucrose. In: Cleland J, Langer R (eds) Protein formulations and delivery. ACS Symp Ser ACS Pub, Washington DC 567: 170–192

Sakai A (1965) Survival of plant tissues at super-low temperatures. Relation between effective prefreezing temperatures and the degree of frost hardiness. Plant Physiol 40: 882–887

Sakai A (1966) Seasonal variation in the amounts of polyhydric alcohol and sugars in fruit trees. J Hortic Sci 41: 207–213

Sakai A (1973) Characteristic of winter hardiness in extremely hardy twigs of woody plants. Plant Cell Physiol 14: 1–9

Sakai A (1986) Cryopreservation of germplasm of woody plants. In: Bajaj YPS (ed) Biotechnology in agriculture and forestry, vol 1. Trees I. Springer, Berlin Heidelberg New York, pp 113–129

Sakai A, Nishiyama Y (1978) Freezing storage of winter vegetative buds of apple trees in liquid nitrogen. In: Akihama T, Nakajima K (eds) Long-term preservation of favorable germ plasm in arboreal crops. Okusai Print Service, Tokyo, pp 64–70

Sakai A, Yoshida S (1968) The role of the sugar and related compounds in variation of freezing resistance. Cryobiology 5: 160–174

Scottez C, Chevreau E, Godard N, Arnaud Y, Duron M, Dereuddre J (1992) Cryopreservation of cold-acclimated shoot tips of pear in vitro cultures after encapsulation-dehydration. Cryobiology 29: 691–700

Seufferheld MJ, Fitzpatrick J, Walsh T, Stushnoff C (1991) Cryopreservation of dormant buds from cold tender taxa using a modified vitrification procedure. Cryobiology 28: 576

Stushnoff C (1987) Cryopreservation of apple genetic resources. Can J Plant Sci 67: 1151–1154

Stushnoff C (1991) Cryopreservation of fruit crop genetic resources. HortScience 26(5): 518–522

Stushnoff C (1992) Strategies for cryopreservation based on tissue hydration. Cryobiology 29: 745–746

Stushnoff C, Fear C (1985) The potential use of in vitro storage for temperate fruit germplasm: a status report. FAO, International Board for Plant Genetic Resources, Rome, 21 pp

Stushnoff C, Vertucci CW, Towill LE (1988) Assessment of water status by differential scanning calorimetry for cryopreservation of dormant buds. Proc North Am Thermal Anal Soc 17th Annu Conf 1: 246–251

Stushnoff C, Remmele RL Jr, Esensee V (1992) Use of phase diagrams to assess cryopreservability of dormant buds from several apple cultivars and from *Amelanchier alnifolia* 'Smoky'. Cryobiology 29: 740

Stushnoff C, Remmele R Jr, Esensee V, McNeil M (1993) Low temperature induced biochemical mechanisms; implications for cold acclimation and de-acclimation. In: Jackson MB, Black CR (eds) Interacting stresses on plants in a changing climate. NATO ASI Ser vol I, 16: 647–657

Towill LE (1988) Genetic considerations for germplasm preservation of clonal materials. HortScience 23: 91–95

Towill LE (1990) Cryopreservation. In: Dodds JH (ed) In vitro methods for conservation of plant genetic resources. Chapman and Hall, London, pp 41–69

Towill LE, Roos E (1989) Techniques for preserving of plant germplasm. In: Knutson L, Stoner AK (eds) Biotic diversity and germplasm preservation, global imperatives. Kluwer, Dordrecht, pp 379–403

Tyler NJ, Stushnoff C (1988a) Dehydration of dormant apple buds at different stages of cold acclimation to induce cryopreservability in different cultivars. Can J Plant Sci 68: 1169–1176

Tyler NJ, Stushnoff C (1988b) The effect of prefreezing and controlled dehydration on cryopreservation of dormant vegetative apple buds. Can J Plant Sci 68: 1163–1167

Tyler NJ, Stushnoff C, Gusta LV (1988) Freezing of water in dormant vegetative buds in relation to cryopreservation. Plant Physiol 87: 201–205

Van Stolen DH, Holle M (1988) Temperate fruit genetic resources and the IBPGR. HortScience 23: 73–74

Vertucci CW, Stushnoff C (1992) The state of water in acclimation from *Malus* and *Amelanchier* and its relationship to winter hardiness. Physiol Plant 86: 503–511

Way RD, Aldwinckle HS, Lamb RC, Rejman A, Sansavini S, Shen T, Watkins R, Westwood MN, Yoshida Y (1990) Apples (*Malus*). In: Moore JN, Ballington JR Jr (eds) Genetic resources of temperate fruit and nut crops. International Society for Horticultural Science, Wageningen, The Netherlands, pp 3–62

Withers LA (1991) In vitro conservation. Biol J Linn Soc 43: 31–42

Zhang W, Zhang J, Hu X (1993) Distribution and diversity of *Malus* germplasm resources in Yunnan, China. HortScience 28(10): 978–980

II.4 Cryopreservation of Germplasm of Mulberry (*Morus* Species)

T. NIINO[1]

1 Introduction

1.1 Distribution and Important Species of *Morus*

Mulberry (*Morus*) is an important woody feed crop for rearing silkworm (*Bombyx mori* L.). It occurs in temperate and subtropical regions in Asia, the Middle East, West Africa, and North and Central America. However, it has been commercially cultivated mainly in China, India, Russia, and Japan, where most of the world's cocoons have been produced. In other countries such as Thailand, Iran, Indonesia, Brazil, etc., sericulture is now being progressively introduced.

The genus *Morus* is usually classified into 24 species and 1 variety based on the morphological characters of the flowers (length of style and shape of stigma) and leaves, as well as geological distribution (Koizumi 1917). However, it seems difficult to apply strictly the concept of "biological species" to *Morus*, because most of the *Morus* species can be easily hybridized by pollination. The important species for silkworm rearing are *M. alba* L., *M. acidosa* Griff., *M. bombycis* Koidz., *M. multicaulis* Perr., and *M. mongolica* C.K. Schn., and for edible fruit, *M. nigra* L. and *M. rubra* L.

1.2 Various Methods for the Storage of Germplasm of Mulberry

Mulberry being a vegetatively propagated crop is preserved in the form of whole plants in the "field gene bank". More than 1000 collections of mulberry varieties are now being preserved in the fields in some countries (Katagiri 1989).

Another approach for the storage of germplasm is to use dormant buds, cuttings, scions, and saplings. Recently, micropropagation of vegetatively propagated crops based on in vitro meristem culture has been widely used. The area of germplasm preservation falls into two categories; short- to medium-term preservation and long-term preservation by cryogenic methods (Bajaj 1986). In vitro storage of plant materials, reduced growth, or minimum conditions of growth have been achieved in several ways. The advantage of using in vitro

[1]National Institute of Agrobiological Resources, Shinjo, Yamagata 996, Japan (*Present Address*: Tohoku National Agricultural Experiment Station, Shinjo, Yamagata 996, Japan)

Biotechnology in Agriculture and Forestry, Vol. 32
Cryopreservation of Plant Germplasm I (ed. by Y.P.S. Bajaj)

storage techniques are enormous, especially for germplasms which are vegetatively propagated woody plants. Despite the advantages, in vitro storage methods can ensure only short-medium-term needs. Problems arise from loss of vigor in plantlets, progressive changes in the genome, microbial contaminations, and the task of maintaining all valuable in vitro collections. Thus, it is highly desirable to have techniques available which permit long-term preservation of germplasm.

At present, genetic resources are generally preserved by a comprehensive combination of the field gene bank and the in vitro gene bank for extra security.

2 Cryopreservation

Sakai (1960) first demonstrated that winter-hardy twigs survived immersion into liquid nitrogen (LN), if they were previously freeze-dehydrated to −20 or −30 °C. Prefrozen willow twigs stored in LN for 1 year retained their viability, and developed roots and shoots (Sakai 1960). This prefreezing method was successfully applied to several winter-hardly fruit trees. It was also demonstrated that prefrozen apple buds stored in LN for 1 year developed normal shoots after being grafted onto rootstocks (Sakai and Nishiyama 1978). Yokoyama and Oka (1983) reported that the shoot tips of prefrozen winter buds attached to a 25-cm shoot survived immersion in liquid nitrogen, but did not sprout by either grafting or cutting.

Yakuwa and Oka (1988) developed this prefreezing method and succeeded in regenerating plants through meristem culture from winter buds cooled to −196 °C. In thier study, winter buds attached to 1–1.5-cm shoot segments were prefrozen to −20 °C at 10 °C increments daily before being immersed in LN. After rapid thawing, apical shoots were aseptically excised from cryopreserved buds and cultured on Murashige and Skoog (1962) (MS) medium. The highest shoot formation amounted to about 50%.

In this chapter, four cryogenic protocols, such as prefreezing, vitrification, air drying, and the alginate-encapsulation dehydration technique, are presented for mulberry winter buds and in vitro-grown shoot tips.

2.1 Cryopreservation of Winter Buds

Cryopreservation of dormant winter buds provides several advantages as a method of conserving mulberry gene resources: (1) clonal integrity is maintained; (2) the method is simple, reliable, and space-efficient; (3) it requires no expensive and sophisticated cooling apparatus; (4) very long stroage is theoretically possible; (5) clonal propagation in vitro from cryopreserved materials is easy; (6) it avoids possible toxic effects of chemical cryoprotectants which are required in conventional methods. Two protocols of cryopreservation using winter buds are given here.

2.1.1 *Slow Prefreezing Method*

Shoots were collected in winter, when buds were still in a state of "quiescence". The axillary buds were removed from the stem with about 5 mm of vascular tissues attached to the buds. Buds in a polyethylene bag were placed in a stainless steel freezing canister and kept at 0 °C for 1 day before freezing. Buds were cooled in 5 °C steps at daily intervals from 0 to −20 °C by transferring the containers to successively lower temperatures to −20 °C, and held there for 1 day prior to plunging in LN. After storage in LN, buds were rapidly thawed in a water bath at 37 °C.

Buds were sterilized with 70% ethanol for 30 s followed by sodium hypochlorite solution (effective chlorine concentration: 0.5%) for 15–30 min. After rinsing in sterilized water, shoot tips consisted of the meristem and five to eight leaf primordia were dissected. The shoot tips were then cultured on MS medium containing 2% fructose, 0.8% agar, and supplmented with 1 mg/l 6-benzylaminopurine. The medium was adjusted to pH 6.0 before autoclaving at 120 °C for 15 min (Oka 1985). Shoot tips were grown on 25 ml of medium in 100-ml glass flasks. The cultures were maintained at 25 °C under a 16-h photoperiod (3000 lx).

Shoot formation of excised shoot tips from winter buds prefrozen to −20 °C at 5 or 10 °C increments daily prior to plunging in LN, is shown in Table 1. Prefreezing at daily 5 °C increments produced higher shoot formation than that of 10 °C/day treatment (Niino et al. 1992b). Recently, we found that partial

Table 1. Shoot formation of excised shoot tips from winter buds cooled to −196 °C at different cooling regimes. (Niino et al. 1992b)

Cultivar	Shoot formation (% ± SE)	
	10 °C/day	5 °C/day
Akagiichibei	21 ± 10	46 ± 15
Aobanezumi	23 ± 8	80 ± 10
Fusounishiki	30 ± 0	53 ± 7
Magozaemon	34 ± 5	73 ± 5
Akaunetaoshi	40 ± 7	46 ± 10
Honyamato	55 ± 15	69 ± 7
Fushimagari	48 ± 8	68 ± 3
Kenmochi	55 ± 10	75 ± 10

Winter buds were cooled to −20 °C at 10 °C/day or 5 °C/day prior to a plunge in LN, and then thawed rapidly in a water bath at 37 °C. After sterilization, shoot tips were excised and plated on MS medium. Shoot formation of cryopreserved shoot tips was defined as percent of shoot tips producing normal shoots 40 days after plating. Approximately 20 buds were treated for each set of duplicates. Significance between 10 °C/day and 5 °C/day treatment (t-test): significant at $p = 0.01$.

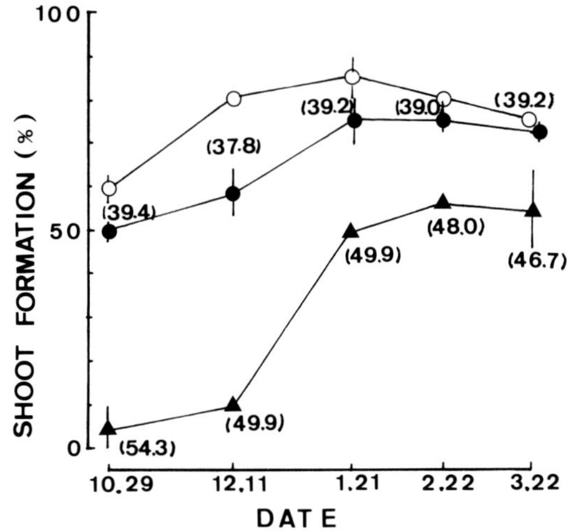

Fig. 1. *Morus bombycis* cv. Kenmochi. Effects of sampling date and dehydration on shoot formation of winter buds after immersion in LN. *Numerals in parentheses* indicate water content of buds before or after partial dehydration for 2 to 3 h at 25 °C. Winter buds were prefrozen to −20 °C at 5 °C increments daily prior to a plunge into LN. *Vertical bars* SE (Niino et al. 1991) ○—○ Unfrozen control; buds were partially dehydrated at 25 °C for 2 to 3h; ●—● cooled to −196 °C after partial dehydration; ▲—▲ cooled to −196 °C without partial dehydration

dehydration at 25 °C prior to prefreezing to −20 °C improved the recovery rates of shoot tips cooled to −196 °C. The optimal water content of dehydrated buds was about 38.5% and the highest rate of shoot formation amounted to about 70% (Fig. 1). This improved procedure enabled us to use the dormant buds collected in late autumn to early spring for cryopreservation (Niino et al. 1991).

It is very important that cryopreserved shoot tips are capable of maintaining viability during storage and of producing plants identical to the nontreated phenotype. The shoot formation from excised shoot tips of winter buds stored at various temperatures for different periods are shown in Table 2 (Niino et al.,

Table 2. Shoot formation of excised shoot tips from winter buds stored at various temperatures for different lengths of duration. (Niino et al. 1993)

Duration	Shoot formation (% ± SE)			
	Temperature (°C)			
	−40	−70	−135	−196
1 month	0±0	2.5±2.5	72.5±2.5	75.0±10.0
6 months	0±0	12.5±2.5	75.0±5.0	72.5± 2.5
1 year	–	12.5±2.5	77.5±2.5	70.0± 0
3.5 years	–	–	75.0±5.0	75.0± 5.0

Material: *Morus bombysis* Koiz., cv Kenmochi. Winter buds were slowly frozen to −30 °C at 10 °C/day prior to storage at various temperatures. Winter buds were thawed rapidly in a water bath at 37 °C. After sterilization, shoot tips were excised and plated on MS medium. Shoot formation of cryopreserved shoot tips was defined as percent of shoot tips producing normal shoots 40 days after plating. Approximately 20 buds were treated for each set of duplicates.

1993). The rate of shoot formation of the shoot tips stored in LN for 3.5 years was 75%. The results also demonstrated the feasibility of cryopreserving shoot tips at −135 °C for at least 3.5 years with little or no decrease in the rate of shoot development compared with those stored in LN.

2.1.2 Air Drying Method

Excised Tips. Excised shoot tips (2 mm long) from winter buds were placed on a sterilized filter paper in a parafilm-sealed Petri dish (9 cm diameter) containing 25 g dry silica gel at 25 °C. After exposure to dehydration for various lengths of time, shoot tips were placed in a 2-ml cryotube and plunged into LN. The shoot tips were then rewarmed by placing the cryotubes in water at 37 °C. Cryopreserved shoot tips were cultured as described previously.

Shoot formation of dried shoot tips before and after exposure to LN at various water contents is shown in Fig. 2. Shoot formation increased greatly with decreasing water content from 30 to 20%, and reached a maximum at about 19% water content (fresh weight base). Subsequently, shoot formation rapidly decreased due to desiccation injury (Niino et al. 1992b).

Alginate Gel. Recently, alginate-coated shoot tips of in vitro-grown pear were successfully cryopreserved following air drying (Dereuddre et al. 1990). Alginate beads protect shoot tips during dehydration and freeze-thaw treatment, and alleviate rapid dehydration of the samples.

Alginate gel beads containing shoot tips were prepared according to the technique of Bapat et al. (1987) and Dereuddre et al. (1990). Shoot tips were suspended in calcium-free culture medium supplemented with 3% Na-alginate solution and 0.5 M sucrose. The mixture was dispensed from a dispenser with sterile tips into 50 ml of culture medium containing 100 mM calcium chloride and

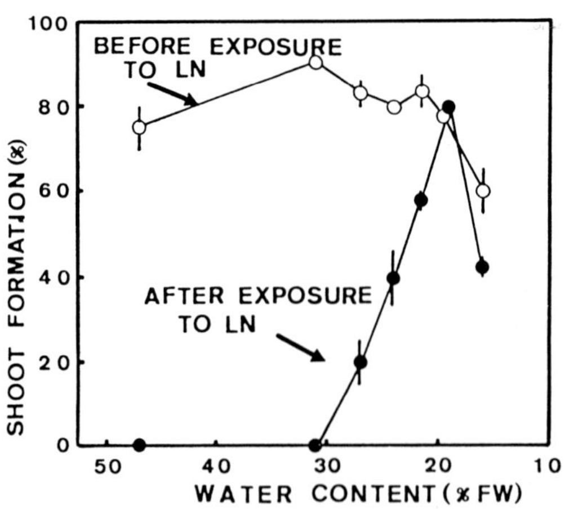

Fig. 2. *Morus bombycis* cv. Kenmochi. Shoot formation of excised shoot tips from winter buds before and after exposure to LN at various water content. Dehydrated shoot tips were placed in a cryotube and then immersed into LN. After 1 day, the cryotube was rapidly transferred into a water bath at 37 °C. Shoot formation (%): percent of shoot tips producing normal shoots 40 days after plating. Approximately 20 shoot tips were treated for each of 2 replicates. (Niino et al. 1992b)

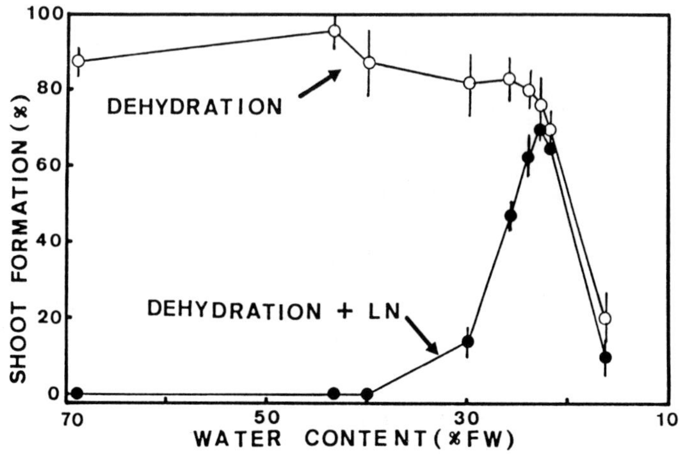

Fig. 3. *Morus bombycis* cv. Kenmochi. Shoot formation of alginate-coated winter shoot tips cooled to −196 °C after dehydration. Shoot tips were encapsulated into alginate-gel beads including 0.5 M sucrose and then precultured in MS medium supplemented with 1 M sucrose at 5 °C for 16h. The alginate-coated shoot tips were subjected to dehydration for different lengths of time at 25 °C prior to a plunge in LN. *Shoot formation* (%) Percent of shoot tips producing normal shoots 40 days after plating. Approximately 20 shoot tips were treated for each of two replicates. (Niino et al. 1992b)

0.5 M sucrose and held for 30 min. Beads of about 5 mm in diameter containing one shoot tip were precultured in MS medium supplemented with 1 M sucrose for 16h at 5 °C. After preculturing, beads were wiped off the surface solution and then subjected to dehydration in parafilm-sealed Petri dishes (9 cm in diameter) containing 40 g dry silica gel held at 25 °C. After dehydration about ten dried beads were placed in a 2-ml cryotube and then plunged into LN. Samples were then warmed by placing the cryotubes in a water bath at 25 °C. Regrowth of shoot tips was achieved by transfer onto MS medium under standard culture conditions described previously. After 7 days of planting, shoot tips were removed from the beads and transferred to fresh MS medium.

Shoot formation of alginate-coated shoot tips cooled to −196 °C following dehydration are shown in Fig. 3. The maximum shoot formation (about 70%) was obtained at the water content of about 22 to 25%. Shoot formation then decreased rapidly with decreasing water content, most likely as a result of desiccation injury (Niino et al. 1992b). Figures 4 and 5 show alginate-coated shoot tips and plantlets developed from the shoot tips cooled to −196 °C following dehydration.

2.2 Cryopreservation of In Vitro-Grown Shoot Tips

The cryogenic strategies for cultured cells and meristems are divided into three categories depending on the method for dehydration: freeze-dehydration, vitrification, and the drying method.

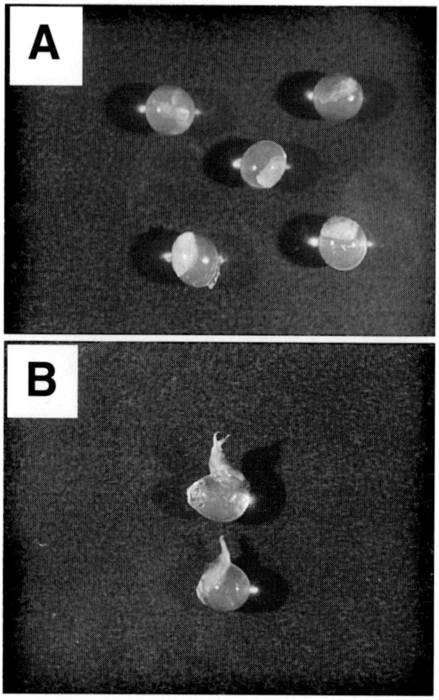

Fig. 4A, B. *Morus bombycis* cv. Kenmochi. Alginate-coated shoot tips cooled to −196 °C following dehydration. **A** After plating on MS medium. **B** 14 days after plating. (Niino et al. 1992b)

Fig. 5. *Morus bombycis* cv. Kenmochi, Plantlet developed from alginate-coated shoot tips cooled to −196 °C following dehydration. (Niino et al. 1992b)

Stock cultures of mulberry plants were maintained on MS medium containing 1 mg/l BA, 2.5% sucrose, and 0.8% agar at pH 5.6. The stock cultures were subcultured every 50 days. All the tissue-cultured plantlets were grown on 25 ml medium in 100-ml glass flasks under white fluorescent light (3000 lx), 16-h photoperiod at 25 °C.

2.2.1 Conventional Slow-Freezing Method

In vitro-grown mulberry shoots were hardened at 5 °C for 20–40 days. Shoot tips (1–1.5 mm) excised from the hardened shoots were dipped in 0.5 M sorbitol solution for 1 day at 5 °C. Ten shoot tips were transferred to 0.3 ml of a cryoprotectant solution containing 10% dimethylsulfoxide (DMSO) and 0.5 M sorbitol in a 2-ml cryotube at 5 °C. After 1 h treatment, the cryotubes were held at 0 °C for 30 min. The cryotubes were placed in the freezing chamber of a programmable freezer precooled at −6 °C. Freezing was induced by dipping the bottom of the cryotube in an ethanol bath at −40 °C. The shoot tips were slowly frozen 0.5 °C/min to −42 °C, and then plunged into LN. After rapid thawing in a water bath at 37 °C, shoot tips were rinsed with sterile water and then plated on modified MS medium. The rate of shoot formation of cryopreserved shoot tips was about 50% (Niino 1990; Niino and Oka 1990). This conventional slow-freezing method requires controlled freezing equipment and a complicated procedure.

2.2.2 Vitrification Methods

Vitrification procedure eliminates the need for controlled slow freezing and permits meristems to be cryopreserved by a direct plunge into LN (Yamada et al. 1991). Three-week-old plantlets were cold-hardened at 5 °C for 3 weeks under 8-h photoperiods. The cold-hardened shoot tips (1.5–2 mm long, 1 mm base diameter) were precultured at 5 °C for 1 day under 8-h photoperiod on agar-MS medium supplemented with 0.7 M sucrose.

The vitrification solution PVS2 (Sakai and Kobayashi 1990) contains 30% (w/v) glycerol, 15% (w/v) ethylene glycol, and 15% (w/v) dimethylsulfoxide in MS basal medium containing 0.4 M sucrose (pH 5.8). Following preculturing, ten shoot tips were transferred to 1 ml of PVS2 in a 2-ml plastic cryotube at 25 °C and were treated for various lengths of time. During this treatment, PVS2 was replaced twice with fresh solution. Shoot tips were finally suspended in 0.6 ml of fresh PVS2 immediately before being directly plunged into and held in LN. After rapid warming in a water bath at 25 °C, PVS2 was drained from the cryotubes and replaced twice with liquid MS medium containing 1.2 M sucrose. After blotting the surface solution, the shoot tips were transferred onto 0.8% agar-MS medium and grown under standard conditions as described above.

Shoot formation of vitrified shoot tips increased considerably from 30 to 80 min exposure and reached a maximum (70%) at 90 min exposure (Fig. 6). The rates of shoot formation of 13 cultivars from 7 species following vitrification are

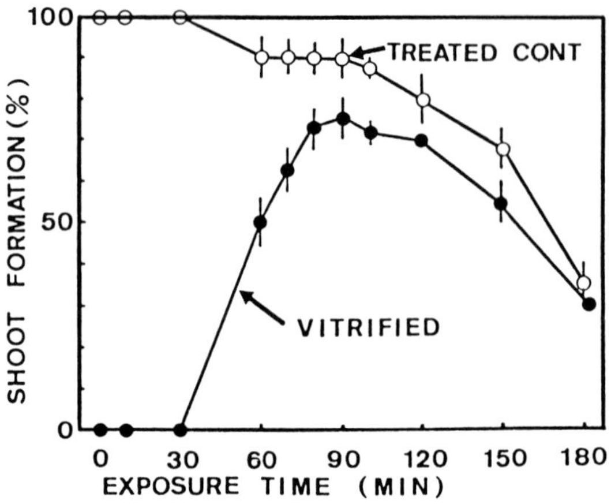

Fig. 6 *Morus bombycis* cv. Kenmochi. Effect of exposure time to PVS2 at 25 °C on the shoot formation of mulberry shoot tips cooled to −196 °C by vitrification. Cold-hardened, precultured shoot tips were treated with PVS2 for various lengths of time at 25 °C and then directly plunged into LN for 1 day. *Treated cont* Same as vitrified shoot tips without cooling to −196 °C. Cold hardening was carried out at 5 °C for 3 weeks with an 8-h photoperiod. Shoot tips excised from cold-hardened in vitro plantlets were precultured on agar MS medium supplemented with 0.7 M sucrose for 1 day at 5 °C. Approximately 20 shoot tips were treated for each of two replicates. (Niino et al. 1992a)

shown in Table 3. A high level of shoot formation was obtained for 12 out of 13 cultivars. Even less or nonhardy cultivars in Southeast Asia showed moderate or high shoot formation (Niino et al. 1992a).

2.2.3 The Air-Drying or Alginate-Encapsulation Dehydration Technique

The air-drying is a simple and reliable method for cryopreservation of in vitro-grown shoot tips as in vitrification. Cold-hardened shoot tips of in vitro-grown mulberry were precultured at 5 °C on agar MS medium supplemented with 0.7 M sucrose for 1 day, and then dehydrated in Petri dishes containing 50 g dry silica gel for various lengths of time. The shoot tips dehydrated up to about 24% water content were immersed in LN. Regrowth was achieved by plating dehydrated shoot tips on MS medium. This air-drying method produced approximately 50% shoot formation, which was not an acceptable rate.

The alginate-encapsulation dehydration technique is another drying method, which is easy to handle and simplifies the dehydration process. Cold-hardened shoot tips were precultured at 5 °C by successive daily transfer of the shoot tips onto MS media supplemented with 0.1, 0.4, and 0.7 M sucrose. Alginate gel beads containing the shoot tips were prepared according to the technique described above. Beads of about 5 mm in diameter containing one shoot tip were treated in MS medium supplemented with 1 M sucrose for 16h at 5 °C. Then the beads were subjected to dehydration in Petri dishes containing 50 g dry silica gel for 7 h (water content: about 33%) at 25 °C prior to plunging in

Table 3. Shoot formation of mulberry shoot tips cooled to −196 °C by vitrification. (Niino et al. 1992a)

Cultivar and/or species	Country of cultivation	Shoot formation (% ± SE)
cv. Yukishinogi (*M. bombycis*)	Japan	81.3 ± 1.3
cv. Kenmochi (*M. bombycis*)	Japan	70.0 ± 0
cv. Ohataro (*M. alba* L.)	Japan	60.8 ± 0.7
cv. Ohaoyama (*M. alba* L.)	Japan	71.4 ± 4.6
cv. Tokiyutaka (*M. alba* L.)	Japan	60.0 ± 5.0
cv. Athubamidori (*M. latiforia*)	Japan	70.0 ± 3.0
cv. Shimaguwa M-28 (*M. australis* Poir)	Japan	55.0 ± 3.0
M. nigra L.	Iran	57.0 ± 3.0
cv. Sulawesi (*M. australis* Poir)	Indonesia	70.0 ± 4.0
M. multicaulis	Indonesia	65.0 ± 5.0
cv. Poo (*M. rotundiloba* Koidz)	Thailand	40.0 ± 4.0
cv. Som (*M. rotundiloba* Koidz)	Thailand	60.0 ± 12.0
cv. Dapang (*M. rotundiloba* Koidz)	Thailand	50.0 ± 8.0

Cold hardening: 3 weeks at 5 °C (8 h/day photoperiod); Preculturing: at 5 °C on agar MS medium supplemented with 0.7 M sucrose; shoot tips were treated with PVS2 for 90 min at 25 °C except *M. multicaulis*, and then directly plunged into LN. The shoot tips of *M. multicaulis* were treated with PVS2 for 50 min at 25 °C. Shoot formation (%): percent of shoot tips showing shoot formation 40 days after plating. Approximately ten shoot tips were tested for each of four replicates.

LN. Samples were warmed by placing the cryotubes in a water bath at 25 °C. Regrowth of alginate-coated shoot tips was achieved by transfer onto the MS medium under standard culture conditions described above. The rate of shoot formation after warming was about 65% (Niino and Sakai 1992).

3 Summary and Conclusions

Winter buds of mulberry prefrozen at 5 °C/day to −20 °C prior to plunging into LN produced high rates of shoot formation (about 50–80%) when rewarmed rapidly. Partial dehydration at 25 °C prior to prefreezing to −20 °C improved the recovery rates of shoot tips of winter buds cooled to −196 °C.

Another possible approach for cryopreservation of winter buds was extensive dehydration by air-drying of shoot tips excised from winter buds. Cryopreserved shoot tips following air-drying (water content fr. wt. base: 17 to 19%) produced high levels of normal shoot formation without intermediary callus formation. However, the alginate-coated dehydration technique is more easy to handle and alleviates the dehydration process than the air-drying method. Thus, the slow prefreezing method or alginate-coated dehydration technique seems promising as a practical method for cryopreservation of mulberry winter buds.

Recently, we confirmed that the winter buds (cultivar Kenmoch) cryopreserved in LN or $-135\,^{\circ}$C for 3.5 years regenerated plantlets after plating on culture medium. No morphological abnormalities were observed in the plants developed from cryopreserved shoot tips. All the regenerated plants were normal diploids with 28 chromosomes (Niino et al. 1993).

In vitro-cultured plants are a preferable source of materials for germplasm preservation. Vitrification procedure and the encapsulation-dehydration technique using in vitro-grown shoot tips would eliminate the need for controlled slow freezing and permit meristems to be cryopreserved by a direct transfer to LN. These two cryogenic methods appear promising as a routine method for cryopreserving in vitro-grown shoot tips of mulberry.

In our Institute, about 420 mulberry cultivars are maintained in the field. Concurrently, winter buds of the same cultivars have been cryopreserved in a deep refrigeration at $-135\,^{\circ}$C for more than 3.5 years. Cryopreservation using in vitro-grown shoot tips of mulberry started 3 years ago.

References

Bajaj YPS (1986) In vitro preservation of genetic resources – techniques and problems. Int Symp Nuclear techniques and in vitro culture for plant improvement. IAEA/FAO, Vienna 1985, pp 43–57

Bapat VA, Mhater M, Rao PS (1987) Propagation of *Morus indica* L. (mulberry) by encapsulated shoot buds. Plant Cell Rep 6: 393–395

Dereuddre J, Scottez C, Arnaud Y, Duron M (1990) Resistance of alginate-coated axillary shoot tips of pear tree (*Pyrus communis* L. Beurre Hardy) in vitro plantlets to dehydration and subsequent freezing in liquid nitrogen: effect of previous cold hardening. CR Acad Sci Paris 310 Ser III: 317–323

Katagiri K (1989) 7. Mulberry. In: Matsuo T (ed) Collected data of plant genetic resources. Kodansha, Japan, pp 691–696 (in Japanese)

Koizumi G (1971) Classification of *Morus*. Bull Seric Exp Stn Jpn 3: 1–62 (in Japanese)

Murashige T, Skoog F (1962) A revised medium for rapid growth and bioassays with tobacco tissue cultures. Physiol Plant 15: 473–497

Niino T (1990) Establishment of the cryopreservation procedure of mulberry shoot tips from in vitro-grown shoot. J Seric Sci Jpn 59: 135–142 (in Japanese with English Summary)

Niino T, Oka S (1990) Plant regeneration from in vitro-grown mulberry shoot tips frozen in liquid nitrogen. J Seric Sci Jpn 59: 111–117 (in Japanese with English Summary)

Niino T, Sakai A (1992) Cryopreservation of alginate-coated in vitro-grown shoot tips of apple, pear and mulberry. Plant Sci 87: 199–206

Niino T, Yakuwa H, Nojiri K, Sakai A (1991) Improved procedures for the cryopreservation of mulberry using winter buds. J Seric Sci Jpn 60: 394–399 (in Japanese with English Summary)

Niino T, Sakai A, Enomoto S, Magosi J, Kato S (1992a) Cryopreservation of in vitro-grown shoot tips of mulberry by vitrification. Cryo Lett 13: 303–312

Niino T, Sakai A, Yakuwa H (1992b) Cryopreservation of dried tips of mulberry winter buds and subsequent plant regeneration. Cryo Lett 13: 51–58

Niino T, Koyama A, Shirata K, Ohuchi S, Suzuki M, Sakai A (1993) Long-term storage of mulberry winter buds by cryopreservation. J Seric Sci Jpn 62: 431–434

Oka S (1985) In vitro culture of isolated buds and organ formation in mulberry. Bull Seric Exp Stn Jpn 29: 747–852 (in Japanese with English Summary)

Sakai A (1960) Survival of the twigs of woody plants at −196 °C. Nature 185: 393–394

Sakai A, Kobayashi S (1990) A simple and efficient procedure for cryopreservation of nucellar cells of navel orange by vitrification. Cryobiology 27: 657

Sakai A, Nishiyama Y (1978) Cryopreservation of winter vegetative buds of hardy fruit trees in liquid nitrogen. Hortic Sci 13: 225–227

Yakuwa H, Oka S (1988) Plant regeneration through meristem culture from vegetative buds of mulberry (*Morus bombycis* Koidz) stored in liquid nitrogen. Ann Bot 62: 79–82

Yamada T, Sakai A, Matsumura T, Higuchi S (1991) Cryopreservation of apical meristems of white clover (*Trifolium repens* L.) by vitrification. Plant Sci 78: 81–87

Yokoyama T, Oka S (1983) Survival of mulberry winter buds at super low temperatures. J Seric Sci Jpn 52: 363–364 (in Japanese with English Summary)

II.5 Cryopreservation and Minimum Growth Storage of Pear (*Pyrus* Species)

T. Moriguchi[1]

1 Introduction

1.1 Plant Distribution and Important Species

Pear (genus *Pyrus*) belongs to the family Rosaceae. It is believed to have arisen during the Tertiary period in the mountainous regions of western China. Dispersal followed the mountain chains towards both the east and west. In the west, two secondary centers, where many populations are differentiated, are formed. The central Asiatic region includes Tadjikistan, Uzbekistan, India, Afghanistan, and the Tian-Shai Mts. In this region, a crossing of *Pyrus communis* with *P. heterophylla*, *P. boissieriana*, or *P. korshinskyi* has been carried out. Another dispersal area is the near-eastern region including the Caucasus Mts. and Asia Minor. Domesticated forms of *P. communis* are believed to have originated in the near-eastern regions. In the east, a secondary center (Chinese center) is the continent of China, where Asian pears are believed to have developed (Asian pea pears and Asian large-fruited pears) (Vavilov 1951).

P. communis is the main commercial species in Europe, North America, South America, Africa, and Australia. In Europe, *P. nivalis* Jacq. (snow pear) is also grown for making perry. In southern and central China and Japan, *P. pyrifolia*, (sand pear) is the main cultivated species. In northern China and Japan, *P. ussuriensis* Max., *P. bretschneideri* Rehd., and *P. betulaefolia* Bge. are grown. In southern China and northern India, selections of *P. pashia* D. Don. are cultivated. Other species are used for rootstock and ornamental purposes (Layne and Quamme 1975).

World production of pear fruit was 9 840 000 t in 1990 (FAO 1990). China is the major producer of pear fruits, with about 30% of the total. In Europe, pear fruits are produced in Italy, Spain, Germany, and France. America, Argentina, and Japan also produced 870 000, 210 000, and 461 000 t, respectively.

1.2 Various Methods for the Storage of Pear Germplasm

In general, maintenance and preservation in the field are traditional for fruit trees including pear (Fig. 1). However, other forms of storage which are labor-reduced and safe were examined (Table 1).

[1]Division of Breeding, Fruit Tree Research Station, Tsukuba, Ibaraki, 305 Japan

Biotechnology in Agriculture and Forestry, Vol. 32
Cryopreservation of Plant Germplasm I (ed. by Y.P.S. Bajaj)

Fig. 1. Maintenance and preservation of pear germplasm in the orchard (Moriguchi, unpubl.)

Table 1. Various methods for storage of pear germplasm

Tissue	Storage method	Storage duration	Reference
Scion	$-1 \sim -3\,^{\circ}C$	More than 1 year	Kozaki (1975a)
	$1 \sim -5\,^{\circ}C$	4 years	Omura et al. (1978a)
Pollen	Freeze drying	More than 10 years	Kozaki (1975b)
	Freeze drying	6 years	Akihama et al. (1978)
Seed	$-3 \sim -5\,^{\circ}C$	7 years	Solovieva (1966)
	Drying	3 years	Omura et al. (1978b)

The storage of fruit tree scions was developed because of its practical benefit for grafting and cutting. Usually, scions are collected, wrapped in a polyethylene bag, stored for a few months in the refrigerator, and used for propagation in spring. However, this is not a preferable method for long-term preservation because of budding, rot, and desiccation during storage, so that, preservation of pear scions at lower temperatures was attempted to prevent deterioration. Kozaki (1975a) showed that the scions of pear could be preserved for more than 1 year at -1 to $-3\,^{\circ}C$ without significant loss of viability. Omura et al. (1978a) also examined the effect of storage temperature and fungicide on pear scions. They showed that scions of Alexandrine Douillard could be preserved well at $1\,^{\circ}C$, while those of Bartlett preferred a lower (-1 to $-3\,^{\circ}C$) temperature (Fig. 2). Survival rate of Chojuro scions was about 80% at $-1\,^{\circ}C$ for 2.5 years, but decreased at 4 years at all temperatures examined.

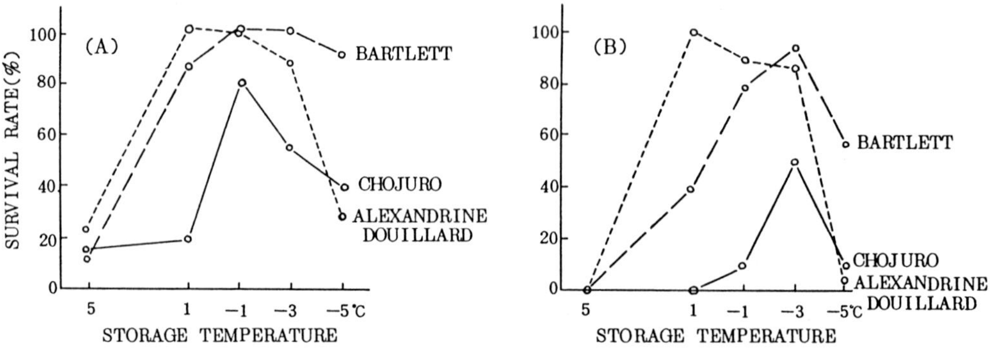

Fig. 2A,B. Effect of storage temperature on survival rate of pear scions in cultivars. Storage duration: **A** 2.5 years and **B** 4 years. (Omura et al. 1978a)

Storage of pollen provides benefits in pollination, breeding, and shipment, and recently for genetic resources. Kozaki (1975b) investigated freeze drying of pear pollen and suggested that freeze drying combined with freeze storage (−20 °C) was the safest and most economical method for periods of more than 10 years. Furthermore, Akihama et al. (1978) showed that both vacuum-dried and the freeze-dried pollen prefrozen at −24, −40, or −196 °C were germinable, and the rates of fruit set when pollinated with pollen stored for 6 years were nearly as high as those with fresh pollen.

In storage of seeds, Crocker and Barton (1931) found that seeds of pear could be dried and were able to germinate through stratification. The seeds could be preserved for 7 years at −3 to −5 °C (Solovieva 1966). Omura et al. (1978b) also found that the fully dried seeds (moisture content 4.3%) survived for 3 years at various temperatures such as room temperature, 0, −25, and −196 °C and the stored seeds germinated after stratification of 4 to 5 months.

1.3 Need for Conservation Using in Vitro Technique

Maintenance and preservation of germplasms, representing genetic diversity of cultivated pears and their relatives, are considered to be essential for breeding. Recently, pear germplasms have been threatened with extinction by environmental disruptions on a worldwide scale. Furthermore, once new cultivars with good cultivation and yield qualities are released, the present major cultivars replaced by them may be lost. Vegetatively propagated crops, including pear, are conventionally maintained in the field. However, in addition to being labor- and cost-intensive, this approach might often result in loss of valuable genetic materials due to disease outbreak, pest infection, environmental perturbation, and other unforeseen problems. Thus, in view of these problems, alternative strategies must be sought, and in this context the application of in vitro techniques must be examined. Meristem cultures have provided not only clonal propagation, but also production of disease-free plants, especially those infected with viruses. Furthermore, since meristems are genetically stable, regenerated shoots maintain genetic stability. These features provide us with ideal factors for the preservation of pear germplasm. Detailed instructions for techniques of both cryopreservation and minimum growth storage of pear are presented here.

2 Cryopreservation

2.1 Brief Review of Cryopreservation of Pear

Since the successful cryopreservation of carnation meristems was reported (Seibert 1976), this technique has been attempted in many species (Bajaj 1991). A summary of cryopreservation of pear is listed in Table 2. Sakai and Nishiyama (1978) reported cryopreservation in liquid nitrogen (LN) of buds from pear trees growing in the orchard after prefreezing at −40 to −70 °C. They tested four pear cultivars, and except for Bartlett, the remaining cultivars showed high survival (Flemish Beauty 100%; Mishirazu 100%; Hosui 78%). Injury of xylem was severe in all cultivars, but if the cortical part was intact, the shoots survived and proceeded to sprout buds. Cryopreservation of excised dormant shoot apices was also examined, in contrast to dormant shoots (Moriguchi et al. 1985; Suzuki et al. 1987; Oka et al. 1991). Suzuki et al. (1987) identified the factors to which dormant pear shoot apices are subjected during cryopreservation as well as fluctuations of freezing hardiness. Pretreatment by immersing the apices in freezing solution (0 to 20% dimethylsulfoxide) (DMSO) at 20 °C for 2 h resulted in an increased survival rate in early September-collected samples frozen at −5 to −30 °C. In this case, 4 to 8% (v/v) of DMSO was optimal, and the following survival rates were obtained in LN immersion after prefreezing at −40 °C in January-collected samples: Flemish Beauty 100%; Mishirazu 92%; Chojuro 75%; Bartlett 17%. Oka et al. (1991) showed that a few whole plants were recovered from the cryopreserved buds of Senryo (10 to 50% shoot formation).

Cryopreservation using in vitro grown apical meristems was achieved by Reed (1990). Apical meristems of four species were tested for their ability to survive immersion in LN. Cultures were grown at 25 °C or cold-hardened for 1 week at −1 °C before cooling to −40 °C, followed by immersion in LN with a mixture of polyethylene glycol (PEG), glucose, and DMSO as cryoprotectant. Regeneration ranged from 0 to 51% for plants grown at 25 °C and from 5 to 95% for cold-hardened plants. Dereuddre et al. (1990a) reported that pear axillary shoot tips from in vitro plantlets survived after enhancement by cold treatment at 0 °C for 3 to 8 weeks. After preculture on medium containing 0.75 M sucrose and incubation with 15% DMSO for 2 h, the apices were cooled progressively (0.5 °C/min) to −30 or −40 °C before immersion in LN. About 60% of apices developed normal shoot tips after rapid rewarming. Furthermore, the cryopreservation of alginate-coated axillary shoot tips from in vitro pear plantlets was successful Dereuddre et al. (1990b). Shoot tips, trapped in alginate gel beads, were precultured in a medium supplemented with 0.75 M sucrose and dehydrated

Table 2. Summary of cryopreservation studies on pear

Materials	Freezing and rewarming methods	Viability	Reference
Hardy shoot	Prefreezing at −40 ~ −50 °C Slow rewarming	78 ~ 100% Regrowth	Sakai and Nishiyama (1978)
Dormant shoot apices	Prefreezing below −40 °C Rapid or slow rewarming	More than 80% Survival	Moriguchi et al. (1985)
	8% DMSO prefreezing at −20 or −40 °C Rapid rewarming	17 ~ 100% Survival	Suzuki et al. (1987)
	Prefreezing at −20 ~ −40 °C Rapid or slow rewarming	Max. 50% Shoot formation	Oka et al. (1991)
In vitro plant material	Preculture for 1 week at −1 °C Prefreezing at −40 °C with mixture of PEG, glucose, and DMSO Rapid rewarming	5 ~ 95% Survival	Reed (1990)
	Preculture for 3 to 8 weeks at 0 °C, followed by incubation with 0.75 M sucrose and 15% DMSO Prefreezing at −30 or −40 °C Rapid rewarming	60% Shoot formation	Dereuddre et al. (1990a)
	Preculture of alginate-coated shoot tips with 0.75 M sucrose, followed by dehydration in an airflow cabinet for 2 to 6 h Direct immersion Slow rewarming	80% Survival, 40% Shoot formation	Dereuddre et al. (1990b)
	Preculture with 0.4 M sorbitol and 1.25 M glycerol for 18 h at 5 °C Prefreezing at −30 °C Rapid rewarming	33 ~ 70% Shoot formation	Niino et al. (1991)
	Preculture with 0.7 M sucrose for 1 day at 5 °C, followed by 30% glycerol, 15% EG, and 15% DMSO for 90 min at 25 °C Direct immersion Rapid rewarming	40 ~ 72.5% Shoot formation	Niino et al. (1992)

in an airflow cabinet for 2 to 6h. The beads were cooled rapidly by direct immersion in LN and rewarmed slowly at room temperature. High survival rates (80%) were obtained after dehydration for 3 h and 40% of the survived shoot tips produced new plantlets after removal of the beads and subculture on a medium. Niino et al. (1991) proposed a simple freezing method for cryopreservation of in vitro grown pear shoots. The shoot tips were cryopreserved with a solution of 0.4 M sorbitol and 1.25 M glycerol at 5 °C for 18 h, followed by placing in a freezer at −30 °C for 1 h prior to plunging in LN. The average rate of shoot formation was about 70%. They also reported the cryopreservation of in vitro grown pear plants by vitrification (Niino et al. 1992). The shoot tips were subjected to treatment for cold hardiness (at 5 °C for 3 weeks under an 8-h photoperiod), and then precultured with 0.7 M sucrose (at 5 °C for 1 day). These shoot tips were hardened further with 30% glycerol, 15% ethylene glycol (EG), and 15% DMSO prior to a direct plunge into LN. Shoot formation rates of 40 to 72.5%, depending on the cultivars, were obtained.

2.2 Protocol

Dormant vegetative buds of a Japanese pear (*P. pyrifolia* Nakai; Kosui) were collected from the orchard in December, and stored at 5 °C until used. Sample apices excised from these buds were used for all experiments. Buds were sterilized for 15 min in 0.5% sodium hypochlorite containing a few drops of Tween 20, and were then rinsed three times in distilled water. Shoot apices, about 1.5 mm long, with five to six leaf primordia were excised from the sterilized buds and sunk in 0.25 ml distilled water in screw-capped vials. An equal volume of DMSO, glucose, or sorbitol solution at twice the final desired concentration was gradually added over a period of 30 min. After being kept at 0 °C for 1 h, the samples were cooled at −40 °C in a freezer. The cooling technique was very similar to that described by Maddox et al. (1982/1983), but a jar was used instead of a polystyrene block. For the evaluation of prefreezing effects on survival, the shoot apices were pretreated at the selected temperatures ranging from 0 to −70 °C, and then immersed in LN for 20 min. To investigate the longevity of shoot apices in LN, shoot apices with 10% DMSO solution were prefrozen at −40 °C, and then immersed in LN for 1, 5, 10, 20, 30, and 40 days.

When using in vitro culture, in vitro grown plants of *P. communis* L. La France were cultured on MS medium (Murashige and Skoog 1962) containing 3% (w/v) sucrose, 1 μM benzyladenine (BA), and 0.7% (w/v) agar, pH 5.8 at 28 °C under a 16-h photoperiod regime. Cold hardening of in vitro plants was induced by treatment at 5, 10, and 20 °C combined with 8, 16, and 24h light for 35 days before excision for meristems. Dissected meristems with mixtures of DMSO and glucose or DMSO alone were immersed in LN for 15 to 20 min after applying the same methods as for dormant pear shoot apices.

The frozen shoot apices were rewarmed rapidly in water at 38 °C or slowly in the air at 0 °C. After rewarming, the shoot apices were placed on MS culture solution containing 3% (w/v) sucrose, 1 μM BA, 0.1% (w/v) agar, pH 5.8, and cultured at 28 °C under a 16-h photoperiod regime. Thus, the apices which developed and showed a green color at 14 days after incubation were considered to be tolerant to the treatment.

2.3 Results and Discussion

The cryoprotective effects of DMSO, glucose, or sorbitol were compared (Table 3). In all treatments, about 95% of the shoot apices appeared to survive after being rewarmed rapidly. The shoot apices from dormant buds showed a survival rate of about 95% even without cryoprotectant. In dormant pear shoot apices, it was not necessary to use a cryoprotectant to maintain their viability during exposure to LN. The effects of prefreezing temperatures on survival were investigated (Fig. 3). Shoot apices were immersed in LN for 20 min after

Table 3. Cryoprotective effects of DMSO, glucose, and sorbitol on the survival of pear shoot apices frozen in liquid nitrogen[a]. (Moriguchi et al. 1985)

Treatment	Concentration (%)						
	0	5	10	15	20	25	30
	Survival (%)						
DMSO	95.3	97.1	97.4	85.7	100	100	100
Glucose	95.3	97.1	89.2	94.2	100	97.1	89.2
Sorbitol	95.3	91.2	100	100	97.2	97.1	94.3

[a]Shoot apices prefrozen at −40 °C were rewarmed rapidly in water at 38 °C.

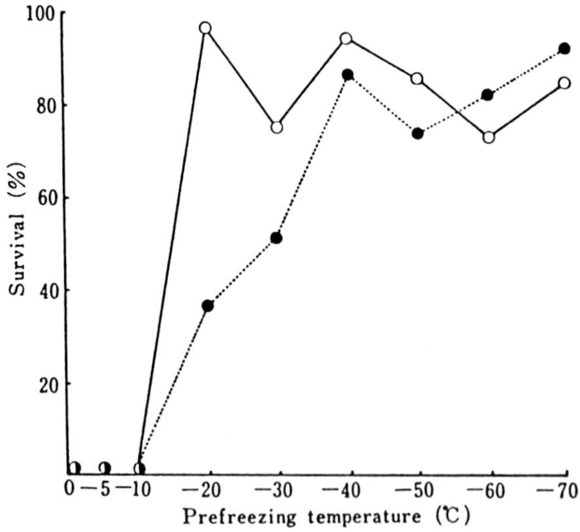

Fig. 3. Effects of prefreezing temperature and rewarming procedure on survival of pear shoot apices. Shoot apices in distilled water were immersed in liquid nitrogen for 20 min and rewarmed rapidly in water at 38 °C (○) or slowly in the air at 0 °C (●). (Moriguchi et al. 1985)

Table 4. Preservation of pear shoot apices in liquid nitrogen[a]. (Moriguchi et al. 1985)

Duration of storage (days)	1	5	10	20	30	40
Survival (%)	95.2	100	100	90.0	85.0	95.2

[a]Shoot apices treated with 10% DMSO were prefrozen at −40 °C. After storage in liquid nitrogen up to 40 days, they were rewarmed rapidly in water at 38 °C.

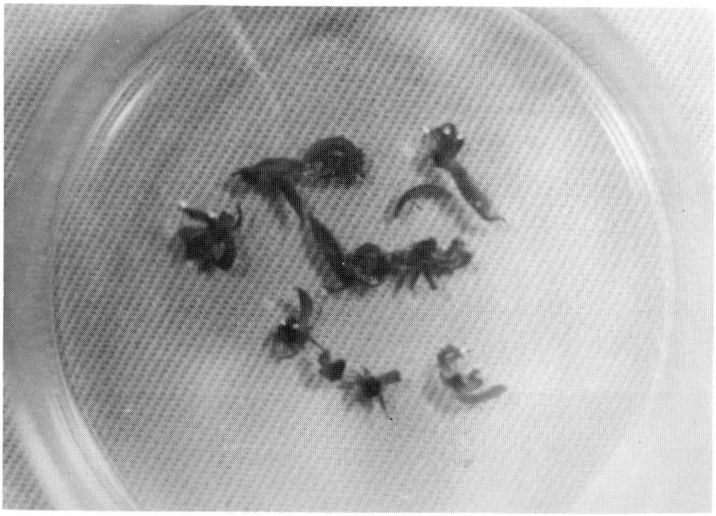

Fig. 4. Development of pear shoot apices frozen in liquid nitrogen. Shoot apices 3 weeks after rewarming. (Moriguchi et al. 1985)

prefreezing at temperatures ranging from 0 to −70 °C in distilled water. Thereafter, they were rewarmed either rapidly or slowly. All apices that had been prefrozen at 0, −5, and −10 °C were killed regardless of the rewarming procedure applied. In apices prefrozen at temperatures ranging between −20 and −30 °C, rapid rewarming improved their survival rate. However, the shoot apices prefrozen below −40 °C survived immersion in LN in all the rewarming procedures. The survival rates of the shoot apices were investigated after storage for 1, 5, 10, 20, 30, and 40 days in LN (Table 4). After 40-day storage, 95.2% of the apices still survived (Fig. 4). In our study, pear shoot apices from dormant buds survived in LN without cryoprotectant. Shoot apices prefrozen at temperatures ranging from 0 to −10 °C were incompletely dehydrated and were killed regardless of the rewarming procedure applied. In apices partially dehydrated by prefreezing at temperatures ranging between −20 and −30 °C, the size of ice crystals may increase when slow rewarming is applied, resulting in death of the

Table 5. Effect of preculture on survival of in vitro cultured pear shoot apices (La France) frozen in liquid nitrogen following prefreezing at $-40\,^{\circ}C^{a}$. (Moriguchi, unpubl.)

Temperature (°C)	Light duration (h)		
	8	16	24
	Survival (%)		
5	57.0	—[b]	54.5
10	0	15.4	0
20	21.2	3.8	0

[a]Shoot apices precultured for 35 days were treated with a mixture of 10% DMSO and glucose or DMSO alone, and immersed in liquid nitrogen for 15–20 min. Rapid rewarming in water at 38 °C was applied.
[b]Not conducted.

Fig. 5. Survival and regeneration of in vitro shoots frozen in liquid nitrogen after prefreezing at −40 °C. (Moriguchi, unpubl.)

tissues. Shoot apices properly dehydrated at −40 °C or below survived regardless of the rewarming procedures.

In vitro cultured plants pretreated at 5 °C combined with 8 and 24 h light showed a survival rate of 57.0 and 54.5%, respectively (Table 5). Some plants could regenerate and propagate vigorously after cryopreservation (Fig. 5). Pretreatments such as low temperature and cryoprotectant appeared to be adequate for inducing freezing hardiness before immersion in LN.

The dormant shoot apices could survive and expand new leaves, but rarely regenerated shoots. This was probably due to the injury of the apical dome itself (leaf primordia appeared to be resistant to freezing compared with the apical dome) in addition to the difficulty of regenerating dormant buds in Japanese pear even without freezing (Oka et al. 1991). Niino et al. (1991, 1992) reported relatively high shoot formation after cryopreservation in LN, although the number of cultivars examined was limited. In addition, the successful levels differed with pear cultivars. Thus, methods which are suitable for many *Pyrus* species must be developed in order to apply cryopreservation routinely.

3 Minimum Growth Storage

3.1 Brief Review

Minimum growth storage has been considered an ideal method for short- to medium-term germplasm preservation. Minimum growth storage may become possible if in vitro shoot-tip culture is established. It has been applied successfully to many fruit species. The growth reduction in culture has been achieved in several ways by applying some stresses to cultures such as: (1) using less than optimal physical or nutrient conditions; (2) addition of growth retardants to the medium. Among the stresses, low temperature was simple and easy to handle, and was adopted for pear species (Table 6). Wanas et al. (1986) reported that the shoots of *P. communis* Caucasica could survive at 4 °C for 18 months under a 16-h photoperiod. Oka and Niino (1989) also reported that cultures of Senryo (probably *P. bretschneideri*) could survive at 5 °C under an 8-h photoperiod and at 10 °C under a 16-h photoperiod, but not at 0 °C in continuous darkness. However, it remains to be determined whether these methods can be applied to Japanese pear cultures.

3.2 Protocol

The shoot-tip cultures of Japanese pear (*P. pyrifolia* Nakai; Hosui, Shinchu, Kosui, Shinusi, Nijisseiki, and Hakataao) and European pear (*P. communis* L.;

Table 6. Summary of minimum storage studied on pear

Species	Storage method and duration		Reference
Pyrus communis	4 °C 16-h photoperiod	18 Months	Wanas et al. (1986)
P. bretschneideri	5 °C 8-h photoperiod or 10 °C 16-h photoperiod	64 Weeks	Oka and Niino (1989)
P. communis	5 °C 16-h photoperiod or 1 °C darkness	12 Months	Moriguchi et al. (1990)
P. pyrifolia	1 °C darkness	20 Months	

La France, Bartlett, and their hybrid Bartlett x La France) were compared. Shoot-tip cultures of European and Japanese pear from orchard-grown trees were established according to the method of Hirabayashi et al. (1987). The shoots were maintained on MS basal medium containing 1 μM BA, 3% (w/v) sucrose, and 0.7% (w/v) agar at 28 °C under a 16-h photoperiod with fluorescent light (3000 lx illumination). These shoot cultures were used as the source of stock for all the experiments. Unless otherwise stated, the storage medium was the same as previously described. The shoots, about 30 mm in length, were cut and placed horizontally in 300-ml culture bottles containing 40 ml of storage medium with five shoots per culture bottle. Bottles were then sealed with screw caps and placed in incubators. The four treatments in the first experiment included incubation at 5, 10, and 15 °C under a 16-h photoperiod at 1 °C in continuous darkness for 12 months each. In the second experiment, there were five treatments. The addition of abscisic acid (ABA) at 1 or 10 μM was compared with media without ABA, but containing sucrose at 3, 6, or 10%. The incubation temperature was 5, 10, and 15 °C under a 16-h photoperiod for all treatments. Shoots were transferred to fresh medium after storage and cultured at 28 °C under a 16-h photoperiod for another month to evaluate recovery growth. The cultures were considered to have survived if the leaves were green and the cultures produced new axillary shoots.

3.3 Results and Discussion

The survival rates after 12 months of La France, Bartlett, and their hybrid Bartlett x La France treated at various temperatures varied with the cultivars (Table 7). The survival rate at 5 °C under a 16-h photoperiod and 1 °C continuous darkness was high in all cultivars. In addition, these shoots survived for 20 months without subculturing, and produced many new shoots after storage (Fig. 6A). In a preliminary experiment, 5 °C combined with continuous darkness was also adequate for these cultivars. However, at 10 and 15 °C combined with a 16-h photoperiod, the survival rate of La France and Bartlett decreased, and sometimes the cultures died. In contrast, the shoots of their hybrids were able to

Table 7. Survival of European pear shoots after storage for 12 months[a]. (Moriguchi et al. 1990)

Treatment		Cultivar		
Photoperiod (h)	Temp. (°C)	La France	Bartlett	Bartlett x La France
		Survival (%)		
0	1	60 ab	90 b	100 a
16	5	100 b	100 b	80 a
16	10	50 a	10 a	80 a
16	15	50 a	0 a	100 a

[a]Values with a column followed by the same letter do not differ significantly ($P = 0.05$).

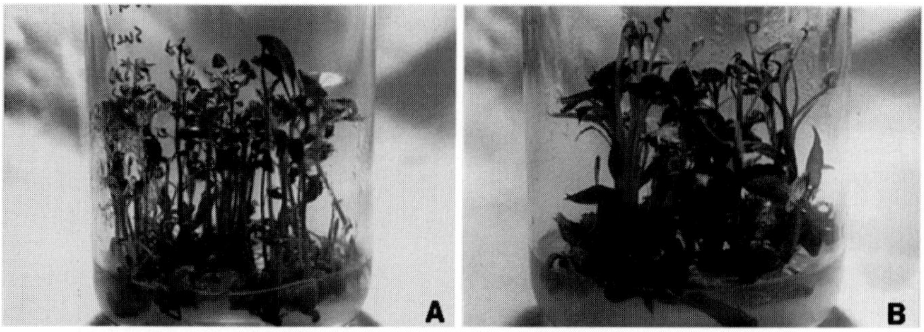

Fig. 6A,B. Formation of a large number of shoots of Bartlett (**A**) and Hosui (**B**) after storage at 1 °C in continuous darkness for 20 and 12 months, respectively. (Moriguchi et al. 1990)

Table 8. Survival of Japanese pear shoots after storage for 12 months[a]. (Moriguchi et al. 1990)

Treatment		Cultivar		
Photoperiod (h)	Temp. (°C)	Hosui	Shinchu	Kosui[b]
		Survival (%)		
0	1	100 b	90 b	100 b
16	5	0 a	0 a	0 a
16	10	0 a	0 a	0 a
16	15	0 a	0 a	0 a

[a]Values with a column followed by the same letter do not differ significantly ($P = 0.05$).
[b]Storage duration; 9 months.

survive at 10 and 15 °C as well as at 5 °C. There were clear differences in the temperature response of Japanese pear compared with European pear. No shoots of Hosui, Shinchu, and Kosui survived at 5, 10, and 15 °C under a 16-h photoperiod (Table 8). All the shoots tended to be compact in form and turned brown early in the storage period. However, when the shoots of these cultivars were stored at 1 °C in continuous darkness, the survival rate of the shoots was greatly improved. They produced many new shoots when transferred to fresh medium after storage for 12 months (Fig. 6B). The other Japanese pear cultivars, Shinsui, Nijisseiki, and Hakataao, also showed a high survival rate after 12 months of storage at 1 °C in continuous darkness (Table 9).

ABA and sucrose treatments did not affect the survival rate at 5 °C (Table 10). The survival rate was significantly higher than that of the control when the cultures were stored at 10 to 15 °C with 1 μM ABA. However, at higher concentrations of 10 μM ABA or 10% sucrose at 10 or 15 °C, the survival rate decreased.

European pear could survive for 20 months of storage at 1 and 5 °C. Wanas et al. (1986) also suggested that 4 °C was suitable for European pear. However,

Table 9. Survival of Japanese pear shoots after storage for 12 months at 1 °C in continuous darkness. (Moriguchi et al. 1990)

Cultivar		
Shinsui	Nijisseiki	Hakataao
Survival (%)		
90	100	95

Table 10. Effect of ABA or sucrose on survival of La France after storage for 12 months[a]. (Moriguchi et al. 1990)

Treatment		Temperature (°C)		
Factor	Conc.	5	10	15
ABA[b]	0	100 a	20 a	50 bc
	1 μM	100 a	70 b	70 c
	10 μM	90 a	20 a	0 a
Sucrose	3%	100 a	20 a	50 bc
	6%	100 a	100 b	50 bc
	10%	90 a	0 a	20 ab

[a]Shoots were stored under 16-h photoperiod.
Values within a column followed by the same letter do not differ significantly ($P = 0.05$).
[b]Contained 3% sucrose.

Japanese pear cultures could not survive at 5 °C, but survived at 1 °C. Such differences may be responsible for the difference in the tolerance to low temperature among *Pyrus* species, as also observed in *Vitis* spp. (Barlass and Skene 1983). Therefore, it is important to identify the most suitable storage temperatures for each cultivar. Many cultivars may die if exposed to the same low temperature during long-term storage.

Growth reduction in the cultures was also achieved by the addition of growth retardants or osmotica such as sucrose and mannitol to the medium. In this experiment, the effect of ABA and sucrose appeared to depend on the concentration and on the storage temperature. The addition of ABA was not necessary, provided the shoots were stored at 5 °C, and a high survival rate was observed with 3% sucrose. Low temperature seemed to be the main factor affecting the survival of pear shoots, and ABA or sucrose concentration caused only a moderate growth inhibition.

4 Summary and Conclusions

Leaf primordia of pear shoot apices could survive after storage in LN, but the regeneration of the shoots rarely occurred or the rate of shoot formation was not always satisfactory in both dormant shoot apices and apical meristems of in vitro-grown plants. Thus, low rates in shoot formation after storage had to be overcome. Pretreatment for inducing cold hardiness appeared to be adequate (Reed 1990; Niino et al. 1991). Recently, vitrification has also been examined (Niino et al. 1992). Pretreatment of freezing methods which are suitable for many *Pyrus* species should be developed. On the other hand, minimum growth storage is simple and easy to handle.

In addition to the method for germplasm preservation, meristem culture is also applied in the development of virus-free plants (Moriguchi et al. 1989). Thus, in the near future, in vitro preservation of pear germplasm will promote the development of virus-free plants and contribute to the exchange of germplasm among nations.

References

Akihama T, Omura M, Kozaki I (1978) Further investigation of freeze-drying for deciduous fruit tree pollen. In: Akihama T, Nakajima K (eds) Long-term preservation of favourable germ plasm in arboreal crops. Fruit Tree Res Stn, Minist Agric For Jpn, pp 1–7

Bajaj YPS (1991) Storage and cryopreservation of in vitro cultures. In: Bajaj YPS (ed) Biotechnology in agriculture and forestry, vol 17. High-tech and micropropagation I. Springer, Berlin Heidelberg New York, pp 361–381

Barlass M, Skene KGM (1983) Long-term storage of grape in vitro. FAO/IBPGR Plant Genet Res Newsl 53: 19–21

Crocker W, Barton LV (1931) After-ripening, germination and storage of certain rosaceous seeds. Contrib Boyce Thompson Inst 3: 385–404

Dereuddre J, Scottez C, Arnaud Y, Duron M (1990a) Effects of cold hardening on cryopreservation of axillary pear (*Pyrus communis* L. cv. Beurré Hardy) shoot tips of in vitro plantlets. CR Acad Sci Paris Ser III 310: 265–272

Dereuddre J, Scottez C, Arnaud Y, Duron M (1990b) Resistance of alginate-coated axillary shoot tips of pear tree (*Pyrus communis* L. cv. Beurré Hardy) in vitro plantlets to dehydration and subsequent freezing in liquid nitrogen: effects of previous cold hardening. CR Acad Sci Paris Ser III 310: 317–323

FAO (1990) Production year book, FAO, Rome

Hirabayashi T, Moriguchi T, Kozaki I, Yamamoto Y, Matsuzaki S (1987) In vitro propagation of *Pyrus* shoot tips. Bull Fruit Tree Res Stn A14: 9–16

Kozaki I (1975a) Storage of fruit tree scions. In: Matsuo T (ed) JIBP synthesis, gene conservation – exploration, collection, preservation and utilization. Univ Tokyo Press, Tokyo, pp 84–88

Kozaki I (1975b) Storage method of pollen. In: Matsuo T (ed) JIBP synthesis, gene conservation – exploration, collection, preservation and utilization. Univ Tokyo Press, Tokyo, pp 89–94

Layne REC, Quamme HA (1975) Pears. In: Janick J, Moore N (eds) Advances in fruit breeding. Purdue Univ Press West Lafayette, IN, pp 38–70

Maddox AD, Gonsalves F, Shields R (1982/1983) Successful preservation of suspension cultures of three *Nicotiana* species at the temperature of liquid nitrogen. Plant Sci Lett 28: 157–162

Moriguchi T, Akihama T, Kozaki I (1985) Freeze-preservation of dormant pear shoot apices. Jpn J Breed 35: 196–199

Moriguchi T, Ohtsu Y, Takanashi K (1989) Development of pear plants free from apple stem grooving virus (ASGV). Bull Fruit Tree Res Stn A16: 25–30

Moriguchi T, Kozaki I, Yamaki S, Sanada T (1990) Low temperature storage of pear shoots in vitro. Bull Fruit Tree Res Stn 17: 11–18

Murashige T, Skoog F (1962) A revised medium for rapid growth and bioassays with tobacco tissue cultures. Physiol Plant 15: 473–497

Niino T, Yakuwa H, Sato K, Sakai A (1991) A simple freezing method for cryopreservation for in-vitro-grown pear shoots. Plant Tissue Cult Lett 8: 185–189 (in Japanese with English Summary)

Niino T, Sakai A, Yakuwa H, Nojiri K (1992) Cryopreservation of in vitro-grown shoot tips of apple and pear by vitrification. Plant Cell Tissue Organ Cult 28: 261–266

Oka S, Niino T (1989) Long term storage of pear shoots cultured in vitro by minimum growth method. Abstr Jpn Soc Hortic Sci Spring Meet 58 (Suppl 1): 120–121 (in Japanese)

Oka S, Yakuwa H, Sato K, Niino T (1991) Survival and shoot formation in vitro of pear winter buds cryopreserved in liquid nitrogen. HortScience 26: 65–66

Omura M, Akihama T, Kozaki I (1978a) Long term preservation of fruit tree scions. In: Akihama T, Nakajima K (eds) Long-term preservation of favourable germ plasm in arboreal crops. Fruit Tree Res Stn, Minist Agric For Jpn, pp 84–88

Omura M, Sato Y, Seike K (1978b) Long term preservation of Japanese pear seeds under extra-low temperatures. In: Akihama T, Nikajima K (eds) Long term preservation of favourable germ plasm in arboreal crops. Fruit Tree Res Stn Minist Agric For Jpn, pp 26–30

Reed BM (1990) Survival of in vitro-grown apical meristems of *Pyrus* following cryopreservation. HortScience 25: 111–113

Sakai A, Nishiyama Y (1978) Cryopreservation of winter vegetative buds of hardy fruit trees in liquid nitrogen. HortScience 13: 225–227

Seibert M (1976) Shoot initiation from carnation shoot apices frozen to −196 °C. Science 191: 1178–1179

Solovieva MA (1966) Long-term storage of fruit seeds. Proc 17th Int Hortic Congr, Maryland, USA

Suzuki T, Harada T, Yakuwa T (1987) Studies on freeze-preservation of fruit tree germplasm. I. Survival of pear leaf bud apices frozen in liquid N_2. Mem Fac Agric Hokkaido Univ 15: 118–123 (in Japanese with English Summary)

Vavilov NI (1951) The origin, variation, immunity and breeding of cultivated plants. Ronald Press, New York

Wanas WH, Callow JA, Withers LA (1986) Growth limitation for the conservation of pear genotypes. In: Withers LA, Anderson PG (eds) Plant tissue culture and its agricultural application. Butterworths, London, pp 285–290

II.6 Cryopreservation of Germplasm of Walnut (*Juglans* Species)

M.T. DE BOUCAUD and M. BRISON[1]

1 General Account

1.1 Important Species and Their Distribution

English or Persian walnut (*Juglans regia*) and eastern black walnut (*Juglans nigra*) are commercially the most important species. *Juglans regia* is known for its wood and edible nuts. At present, *J. regia* represents a major nut crop with an annual worldwide production of 800 000 t (Germain 1990). *Juglans regia* has long been prized by timber and horticultural industries. Its wood is used for veneer, luxury cabinet work, and other items when quality wood is required. Walnut is a large deciduous tree which can reach 30 m in height under favourable conditions. It is monoecious, and the flowers, produced in spring, are wind-pollinated. Walnut appears to be self- and cross-fertile, though some loss of fruit production may occur when male and female flowers do not mature simultaneously. Therefore, a commercially successful production contains a range of cultivars representing different times of pollen shedding.

Juglans regia, the major commercial species, is native to regions of north-central Asia, the western Himalayas, parts of Iran, Turkey, Transcaucasus, and possibly the Balkans. Native forests of *Juglans regia* grow on moist well-drained soils up to 3300 m above sea level, though they may occur at 4000 m altitude in Tibet (Leslie and McGranahan 1988). Wild *Juglans regia* forests are distributed mainly throughout southeastern Europe and western Asia from Turkey to northern India and China. Introduced to Europe from ancient Greece, walnut spread to Italy and then to other European countries (becoming widely distributed from Sweden to Spain) and northern Africa during the Roman period. Today, France, Italy, and Greece are important walnut producers. It was then introduced to the Americas by Spanish explorers and missionaries, and is now found in South America. Its culture has increased rapidly in the western United States (mainly California) which is now the largest world producer, superseding China (Germain 1990). As *Juglans regia* is widely cultivated, the distinction between native and naturalized populations is not always evident. *Juglans nigra* forms forest masses together with conifers in the eastern United States, and was introduced into Europe in the early 17th century (Rodriguez et al. 1989).

[1]Laboratorie de Physiologie Cellulaire Végétale, Université Bordeaux, 1 Avenue des Facultés, 33405 Talence Cédex, France

Biotechnology in Agriculture and Forestry, Vol. 32
Cryopreservation of Plant Germplasm I (ed. by Y.P.S. Bajaj)

1.2 Method of Storage of *Juglans* Germplasm

1.2.1 Germplasm Variability Maintenance

Seeds. For a long time, seed utilization was the main method for walnut multiplication. Although this practice has been virtually abandoned, seed germplasm remains the most convenient system available for the prolonged storage of genetic information (Stushnoff 1991). Walnut conservation is dependent upon harvesting and drying procedures. Rapid nut collection after fall is essential to avoid fungal contamination. Conservation of material, following artificial drying to 10–8% moisture content at 25–30 °C in a special dryer is, at best, only for a few months (Germain 1990).

Pollen. Pollen is easily collected on an annual basis from plants in clonal repositories and can be shipped to breeders. Its use broadens the possibilities for species crossing, though its storage even at subfreezing temperatures has been impossible for many years (Griggs et al. 1971).

1.2.2 Clonal Germplasm Preservation

In Situ Conservation. In situ conservation involves the use of natural sites for preservation of the material. It allows the continued evolution within a species for particular characteristics, including resistance or tolerance to disease or stress conditions (Towill 1988). However, loss of germplasm in natural populations may occur rapidly and so it is necessary to consider alternatives for the preservation of clonal germplasm.

Ex Situ Conservation. This is the system usually adopted for clonal maintenance in the greenhouse, naturalized orchards, and repositories. Interesting cultivars of walnut have been selected and hybridized on the basis of various characteristics (Yadrov 1990) such as time of beginning of vegetative growth, flowering in staminate and pistillate flowers, and fruit maturation; fruit bearing in lateral buds; early start of fruit bearing; optimal nut mass; seed protein content; cold resistance and higher disease resistance.

Some characters can be multiplied by grafting on seedling rootstock for individuals acquired by hybridization. In California, the most used rootstocks are *Juglans hindsii* jeps ex R.E. Smith (northern California black walnut) and Paradox, a hybrid between *J. regia* and *J. hindsii*. *Juglans regia*, *J. nigra*, and their hybrids are used in Europe. However, commercial plantations of *J. nigra* are usually of seedling origin, though they may be improved by grafting to increase vigour and uniformity. Therefore, ex situ collections offer immediate accessibility to diversity for the walnut breeder with, however, the disadvantage of an increased disease vulnerability.

1.2.3 In Vitro Plant Preservation

Maintaining vegetatively propagated plants under field conditions may result in disastrous consequences, i.e., loss of valuable genetic material due to diseases such as bacteriosis caused by *Xanthomonas juglandis* and to anthracnose caused by the fungus *Gonomonia leptostyla* on leaves and fruits. Without treatment, 80–100% of the harvest may be lost. Abundant rainfall in spring is the most important factor responsible for the disease. Furthermore, walnut trees growing in water-saturated soil are attacked by several species of *Phytophthora* directly implicated in the development of the disease and/or the death of *J. regia* cultivation in California. The Paradox hybrid has a high resistance to various species of *Phytophthora* (Rodriguez et al. 1989). Otherwise, all species used as rootstocks are hypersensitive to the Cherry Roll Virus (CLRV) which is transmitted via pollen, seeds, and scions from infected trees (Germain 1990, 1992). Thus, to resolve the various problems in vitro storage is now being used.

In recent years, the potential of tissue culture techniques for the propagation of lignified plant material has been largely described (Bajaj 1986). In addition to clonal propagation, which ensures maintenance of genetic stability, tissue culture techniques have been used to eliminate virus and to maintain virus-free stocks.

Attempts at in vitro regeneration were reported for *Juglans nigra* (Cummins and Ashby 1969; Lenartowicz and Millikan 1977; Somers 1982) and for *Juglans regia* (Chalupa 1981; Rodriguez and Sanchez-Tames 1981; Rodriguez 1982a, b). Steffan and Millikan (1983) tried to culture black walnut meristems from mature trees though growth did not exceed 8 weeks, presumably due to the oxidation of endogenous phenolic compounds. However, Meynier (1985) succeeded in culturing hybrid walnut meristems excised from shoots harvested in spring from annually pollarded walnut trees. Some meristems gave rise to shoots which were multiplied, rooted, and acclimatized. Heile-Sudholt et al. (1986) cultured in vitro embryonic axes and seedling shoot tips of *J. nigra* and obtained micropropagated plants which were subsequently raised in a greenhouse. Despite these results, the rooting of multiplied shoots remains a difficult step in the micropropagation of *Juglans*.

Somatic embryogenesis potentially avoids problems associated with the rooting of multiplied shoots, since it is theoretically possible to produce whole juvenile plants directly with an apical bud and a main root tip. Somatic embryogenesis has been obtained from immature cotyledons of *Juglans regia*, *J. hindsii*, *J. nigra*, *J. major*, and interspecific *J. nigra* x *J. regia* (Tulecke and McGranahan 1985; Cornu 1988, 1989). However, the low efficiency or embryo germination and transformation into a plantlet is a problem that has been examined and partly resolved by de Deng and Cornu (1992). Recently various aspects of micropropagation (Leslie and McGranahan 1992) and somatic embryogenesis (Tulecke et al. 1995) have been reviewed.

1.3 Need for Cryopreservation

Conventional storage is not entirely satisfactory for the germplasm of *Juglans*, since large losses in seed viability occur progressively over a few months even under optimal storage conditions (4 °C and 10% seed moisture content). Continuous deterioration of seeds during storage, risks of low seed yield, contamination, and shifts in the genetic composition of the original gene pool make it necessary to find another method for preserving seed germplasm. Storage in liquid nitrogen (LN) at −196 °C may be a suitable method.

This chapter describes the different techniques used for cryopreservation of *Juglans* and outlines the problems encountered, progress made to date and future prospects. These technologies were applied to pollen, embryonic axes, shoot tips and somatic embryos.

2 Cryopreservation of *Juglans* Germplasm (Table 1)

2.1 Cryopreservation of Pollen

First, Hall and Farmer (1971), then Farmer and Barnett (1974), froze pollen of *Juglans nigra* in LN. After thawing, they observed a slight reduction in germination rates. Luza and Polito (1985, 1988) stored pollen from eight clones of English walnut (*J. regia*) for 1 year in liquid nitrogen. The authors showed that pollen survival was correlated with its moisture content. Pollen samples with a moisture content higher than 7.5% were killed by freezing, while most pollen samples with a moisture content between 3.2 and 7.5% survived. In the absence of freezing, the germinative capacity of fresh pollen from these eight clones was lost entirely within 5 to 6 days. If pollen was frozen immediately after collection, the percent germination was nil for the cultivars Adams, Chico, and Sharkay, of which the moisture content was higher than 7.5%. The percent germination varied from 18.4 to 65.2%, for the five other cultivars and from 82.4 to 95.6% for the eight controls.

However, when pollen was kept for 24 and 72 h before freezing, even pollen of Adams, Chico, and Sharkey survived after LN storage with good germination rate following 24 h (72.26, 72, and 81% respectively) and 72 h (16.2, 24.2, and

Table 1. Cryopreservation studies on *Juglans* species

Expant	Freezing method	Response	Reference
Pollen	Slow freezing	Germination	Griggs et al. (1971) Hall and Farmer (1971) Farmer and Barnett (1974) Luza and Polito (1985)
Embryonic axes excised from mature seeds	Rapid freezing LN	Shoots	Pence (1990) de Boucaud et al. (1991)
Shoot tips	Slow freezing (−40 °C) then LN	Calli, shoots	Brison et al. (1992)
Somatic embryos	Slow freezing (to −30 °C) then LN	Regrowth and multiplication	Lee (1989) de Boucaud and Brison (1994)

20.2% respectively) of drying. However, in the latter case, the percent germination of the control was not very different. The application of glycerol as a cryoprotectant was not successful for freeze preservation of walnut pollen.

A cooling rate of 1 °C/min combined with rapid thawing maintains a high pollen germination capacity. Thus, pollen storage in LN allows access to walnut pollen germplasm from one season to the next and provides breeders with the possibility of crossing cultivars having different times of flowering. This biotechnology has favoured the extension of walnut breeding programs throughout the world.

2.2 Cryopreservation of Embryonic Axes of *Juglans regia* L. cv. Franquette

2.2.1 Material and Cryopreservation Procedure

Walnut material was provided by the Institut National de la Recherche Agronomique (INRA), Station de Recherches Fruitières (Bordeaux, France). The nuts were harvested in October, washed, and lightly dried. Experiments were carried out with either fresh nuts, or with nuts having undergone a spontaneous dehydration (until 5% moisture content). The procedure was as follows:

1. Nuts sterilized by soaking in sodium hypochlorite (24% chlorine content) for 5 min.
2. Embryonic axes excised.
3. Embryonic axes rapidly sterilized in 3% calcium hypochlorite for 1 min and rinsed three times in sterile deionized water.
4. Preparation of the samples. Excised embryonic axes were divided into three groups:
   - CC, control samples without any treatment,
   - C1, control samples pretreated for 1 h at 0 °C in cryotubes containing 1 ml 5 M 1,2-propanediol supplemented with 20% w/v sucrose,
   - F, frozen samples pretreated for 1 h as above, then frozen in LN and thawed rapidly by incubating in a 45 °C water bath.
5. All embryonic axes (i.e., CC, C1, F) were cultured on a modified version (de Préville 1990) of the Driver and Kuniyuki (1984) walnut medium (DKW) supplemented with benzylaminopurine (BAP 4,4 μml^{-1}). Cultures were incubated in the dark at 26 °C for 15 days under a 14 h photoperiod (Fig. 1a).
6. Survival of embryonic axes was assessed after 2 weeks and again after 2 months of culture. Results were expressed as:
   - % survival: percentage of axes exhibiting green tissues (shoot and callus growth after 2 weeks of culture),
   - % regrowth: percentage of shoots obtained either directly from embryonic axes or indirectly from some green calli after 2 months of culture.

Moisture content was determined in each experiment on three samples each containing 20 embryonic axes, weighed before and after drying to constant weight at 70 °C for 2 days. Moisture content relative to dry weight was determined from the formula: (fresh weight – dry weight)/dry weight.

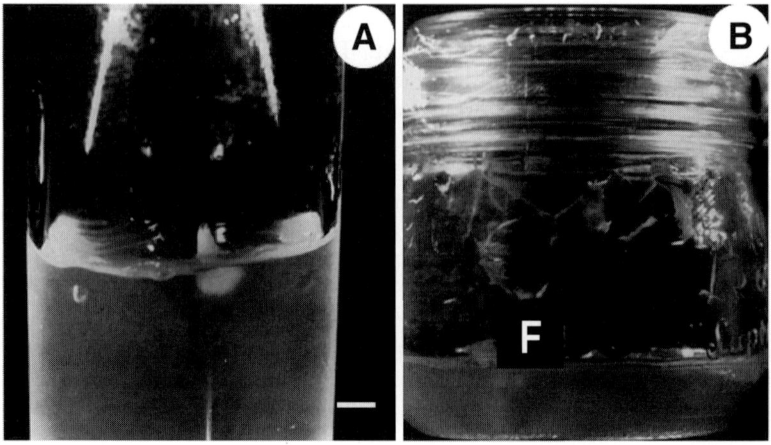

Fig. 1. **a** Embryonic axes at the beginning of culture on DKW medium; *bar* 5 mm. **b** 2-month-old walnut plant from frozen (*F*) embryonic axes. (de Boucaud et al. 1991)

2.2.2 *Results and Discussion*

Effects of Moisture Content on Survival and Regrowth. Results from the different experiments carried out on embryonic axes having different moisture contents, in the absence or presence of cryoprotectant are reported in Table 2. Survival was higher than regrowth. According to Pritchard and Prendergast (1986), this is due

Table 2. Development of embryonic axes of *Juglans regia* with variable moisture content in the absence or presence of cryoprotectant

Average hydration of cultured embryonic axes (relative to dry weight)		+Cryoprotectant[a]			−Cryoprotectant	
		CC	C1	F1	CC	F2
5%[b]	Number of cultured embryonic axes	37	36	68	32	39
	% Survival[d]	86	52	82	87	72
	% Regrowth[e]	56	30	61	56	56
20%[c]	Number of cultured embryonic axes	27	26	62	14	14
	% Survival[d]	92	96	85	86	21
	% Regrowth[e]	91	76	75	79	29

[a]Cryoprotectant treatment: 5 M 1,2-propanediol supplemented with 20% sucrose for 1 h at 0 °C.
[b]Average moisture content of the embryonic axes (8 months after harvesting).
[c]Average moisture content of the embryonic axes (after sterilization and rinsing).
[d]Survival: percent survival (including calli) after 15 days of culture.
[e]Regrowth: percent of normal shoots after 2 months of culture.
CC, control samples without any treatment, C1, control samples pretreated for 1 h at 0 °C in cryotubes containing 1 ml 5 M 1,2-propanediol supplemented with 20% (w/v) sucrose, F1, frozen samples pretreated for 1 h as above, then frozen in LN and thawed rapidly by incubation in a 45 °C water bath. F2, frozen samples without treatment and thawed as F1.

to embryonic axis growth being disturbed, as a result of the dehydration, freezing, and thawing treatments, with the consequence that the live cells remaining produce only calli and not shoots.

The viability of embryonic axes of walnut is reduced by desiccation and aging: survival was 86% and regrowth was 56% at 5% moisture content for CC in comparison with 92 and 91% respectively for 20% moisture content. Simultaneously, however, they became more LN-tolerant (72% survival after freezing without cryoprotectant at 5% moisture content compared to 21% survival at 20% moisture content). Similarly, providing that the nuts were not from a 1-year-old stock, Pence (1990) obtained good survival of *Juglans* embryonic axes after drying overnight under a laminar air-flow and freezing in LN.

Effects of Cryoprotectant on Survival and Regrowth. For dehydrated (i.e., 5% moisture content) and frozen embryonic axes (F), it was not absolute necessary to use cryoprotectant even if its presence increased regrowth slightly 61% (F1) with cryoprotectant compared to 56% (F2) without and appeared to have a beneficial effect on survival 82% (F1) with cryoprotectant, 72% (F2) without. However, cryoprotectant was necessary, in the case of axes hydrated to 20% moisture content, to obtain satisfactory survival and regrowth. In this case, the cryoprotectant allowed 75% (F1) regrowth compared to 29% (F2) without.

1,2-Propanediol has been found to have a very low toxicity and to be a satisfactory cryoprotectant for walnut embryonic axes. 1,2-propanediol, a permeating cryoprotectant of low molecular weight, allows a moderate outflow of water and prevents the increase of salt concentration in the cell to a lethal critical value during freezing. In addition, it has been found to favour vitrification more than other cryoprotectants (Boutron and Mehl 1986). Previous experiments carried out on imbibed *Zea mays* seeds (de Boucaud and Cambecedes 1988) have shown that the cryoprotectant function of 1,2-propanediol is increased with the addition of sucrose.

High concentrations of 1,2-propanediol appear to be toxic for axes with a 5% moisture content (only 30% of regrowth of the C1 control) due to a solute concentration effect within cells of the embryonic axes as a result of dehydration. Toxicity disappears after freezing of the material (61% regrowth). A similar effect was also observed in calli of *Picea abies* and *Picea glauca* (Bercetche et al. 1990) after freezing.

In all cases, freezing induced a delay in growth at the beginning of culture which disappeared after 2 months (Fig. 1b). At this stage, micropropagation of walnut shoots obtained from control and frozen embryonic axes is straight forward using conventional culture methods.

The cryopreservation procedure reported here is easily performed and requires no expensive laboratory apparatus. The use of 1,2-propanediol increases dehydration of axes without a significant influence on survival and regrowth if the treated nuts are not too old. Thus, when embryonic axes are excised in the first 3 months after nut harvesting, regrowth is high, and the multiplication of shoots obtained poses no problem. Cryopreservation of embryonic axes is a reliable method for germplasm storage of many large-seeded species.

2.3 Cryopreservation of Excised Shoot Tips from In Vitro Plantlets

The present research was carried out to establish a method for the cryopreservation and subsequent micropropagation of a *Juglans regia* clone selected for its quality fruits by the INRA Station de Recherches Fruitières, Bordeaux, France.

2.3.1 Material and Methods

2.3.1.1 Plant Material

1. Walnut plantlets were obtained from germinated embryonic axes and micropropagated on modified DKW medium supplemented with 4.4 μM benzylaminopurine (BAP) and 4.9 nM indolbutyric acid (IBA) (de Boucaud et al. 1991).

2. Walnut shoot tips were cultured on a walnut meristem medium (WMM) which was considered suitable to stimulate rapid meristematic growth. This medium contained half-strength Murashige and Skoog medium (1962) (MS/2) with a reduced ammonia concentration, while the micronutrients were increased by one-third. The medium was adjusted to pH 5.8 and gelled with 0.125% Phytagel (Sigma) and 0.3% Difco agar before autoclaving. Sucrose (3% w/v), vitamins (Morel and Wetmore 1951), glycine, cystein HCl, L-glutamine, adenine sulfate, and growth regulators (BAP: 4.4 μM, IBA: 0.49 μM or 4.9 nM) were also added to the medium. Plant material was maintained under a 23 °C day/18 °C night regime with a 14-h photoperiod (39 $\mu mol/m^2$ s).

2.3.1.2 Cryopreservation Procedure

This included the following steps:

1. Hardening. In vitro plants showing active shoot production were maintained under a 23 °C days/4 °C night regime for at least 8 days in order to induce cold hardening (Reed 1990).

2. Preculture. Shoot tips (0.5–1 mm) were excised aseptically from hardened shoots and precultured under hardening conditions for 24 h on walnut meristem medium (WMM) supplemented with 5% dimethylsulfoxide (DMSO) and 0.5% proline.

3. Cryoprotectant treatment. Shoot tips were transferred to 1 ml cryoprotectant in 1.2 ml cryotubes placed on ice (10–12 shoot tips per tube). Cryoprotectant (modified PVS2 of Sakai et al. 1990), containing 10, 12.5, or 15% w/v DMSO, 15% w/v ethylene glycol, 30% w/v glycerol, and 0.2 M or 0.4 M sucrose in MS/2, was gently mixed with the shoot tips using a Pasteur pipette. 0.5 ml was removed from the cryotubes which were maintained on ice for 10 to 30 min.

4. Cooling. Samples were either plunged directly in LN (rapid freezing) or frozen at a cooling rate of 0.5 or 1 °C/min to −40 °C and then immersed in LN (slow freezing).

5. Thawing. Samples were thawed by rapid warming in water at 40 °C. Cryoprotectant was drained and replaced with liquid MS/2 medium containing 1.2 M sucrose (Yamada et al. 1991). Rinsed shoot tips were then placed on meristem medium and maintained for 48 h in darkness, then under low illumination (19 μmol/m^2/s) until regrowth had occurred and thereafter maintained under standard conditions.

2.3.2 *Results and Discussion*

1. 100% of nontreated shoot tips grew rapidly on WMM supplemented with 4.4 μM BAP and 0.49 μm IBA.

2. Cryoprotectant-treated shoot tips exhibited an irregular regrowth ability which depended on the following factors:

   Their previous state: meristems smaller than 0.5 mm which were not protected by two leaflets did not develop into shoots but into callus.
   The time of application and the cryoprotectant concentration: regrowth rates varied from 70% (when shoot tips were exposed for 20 min to a DMSO concentration of 10 to 12.5%) to 36% (when they were exposed for – 30 min to a DMSO concentration of 15%).

3. Frozen-thawed shoot tip survival depended on the following factors:

   The cooling rate: sudden immersion in LN resulted in no survival under the conditions described above. At a cooling rate of 1 °C/min, we obtained only approximatively 10% survival with a 15-min cryoprotectant application containing 10% (w/v) DMSO (Brison et al. 1992). At a cooling rate of 0.5 °C/min, under the same conditions, shoot tip viability increased to 15%.
   The time of application and the cryoprotectant concentrations: improvement of survival was obtained by varying the exposure time of shoot tips to the cryoprotective mixture and by increasing the DMSO concentration. When subjected to a mixture containing 15% DMSO for 20 min, 34% of shoot tips survived freezing. Thus, the viability of meristems was greatly increased, though the number of regenerated plantlets did not increase proportionally. We observed that a high proportion of meristems regrew with a slight red coloration, or regenerated as calli. This phenomenon appears less frequently after treatment with 10% DMSO. Our last experiment, suggests that 30-mins exposure to 15% DMSO is injurious, since the majority of the recovering meristems show callus growth, among both the treated control and frozen-thawed samples.

4. Postculture
   0.49 μM IBA concentration in the culture medium allowed a more rapid meristem recovery. However, this concentration was reduced to 4.9 nM soon after recovery so as to avoid callus formation.
   Nontreated control meristems grew more rapidly. For treated control and frozen meristems growth was delayed by 1 to 2 weeks depending on treatment and plantlet vigour (Fig. 2c).

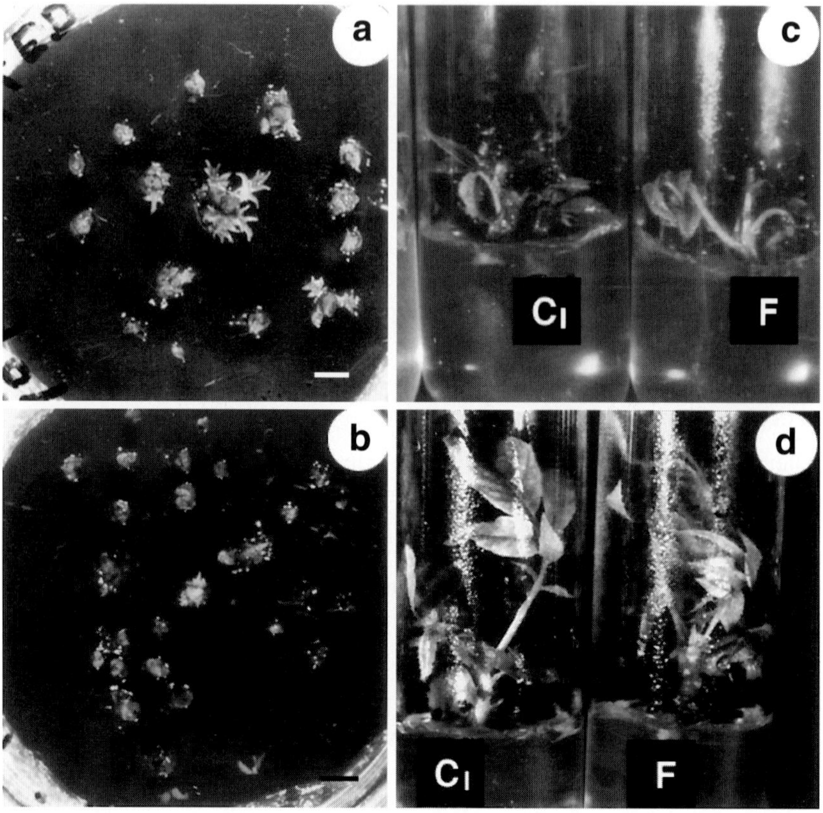

Fig. 2. **a** Cryoprotected and unfrozen (control *Cl*) shoot tips after 45-day culture; *bar* 5 mm. **b** Cryoprotected and frozen shoot tips (*F*) after 45-day culture; *bar* 5 mm. **c** 3-month-old walnut plant from frozen (*F*) and control (*Cl*) shoot tips. **d** 5-month-old walnut plant from frozen and control (*Cl*) shoot tips. *Cl* Control cryoprotected but unfrozen samples, *F* Cryoprotected and frozen samples. Frozen samples which have experienced a slight delay in growth (**c**) attain the same size as the controls after 5 month culture (**d**)

Normal (noncallused) shoots of each treatment produced a large number of buds which were micropropagated under the standard conditions (Fig. 2d).

5. Conclusion and Future Prospects. The results of this study are very encouraging and allow us to expect further improvements as a result of:

A more precise approach to the many factors involved in the cryopreservation of woody plant shoot tips.

Better selection of plant material.

The establishment of clones: within the same experiment we observed a large difference in survival between meristems excised from two different plantlets.

2.4 Cryopreservation of Somatic Embryos

Work on the cryopreservation of walnut somatic embryos was first carried out by Lee (1989). The four cultivars used were obtained from the immature cotyledons of open-pollinated seeds of *J. regia*. They were treated with several cryoprotective solutions, particularly solutions containing 10% (w/v) polyethylene glycol, 8% (w/v) glucose, and 10% (w/v) DMSO or 10% DMSO alone. The somatic embryos were frozen in polypropylene tubes at 1 °C/min to −30 °C, then plunged into LN. Rapid thawing was carried out in a water bath at 40 °C. After rinsing with DKW medium, somatic embryos were subcultured on a DKW basal subculture medium in the absence of growth regulators and solidified with Phytagel at 0.24% (w/v).

Both nonfrozen and frozen pretreated somatic embryos showed a delay in regrowth. This delay was 3 weeks for nonfrozen embryos, and from 3 to 6 weeks for frozen embryos. Survival was relatively low (<40%) compared to nonfrozen pretreated control (90%). Embryo proliferation was variable and appeared to be cultivar-dependent.

2.4.1 Material and Cryopreservation Procedure

2.4.1.1 Plant Material

Somatic embryos of walnut were provided by Cornu and Jay-Allemand, INRA, Station d'Amélioration des Arbres Forestiers Ardon F. 45160 Olivet France. This embryogenic culture line was derived from immature cotyledons of a hybrid nut collected from a *J. nigra* pollinated by *J. regia* (Cornu 1988). This strain was subcultured in the dark every 2 to 3 weeks on a DKW basal medium in the absence of growth regulators and solidified by the addition of Phytagel (Sigma) 0.28% w/v.

2.4.1.2 Cryopreservation Procedure

1. Hardening. Cold hardening of somatic embryos was carried out at least a week at 4 °C night and 21 °C day temperatures in the dark.

2. Preculture. Hardened somatic embryos were precultured on DKW medium supplemented with 5% DMSO and 0.5% proline under hardening conditions.

3. Cryoprotective Treatment. One to two mm isolated somatic embryos were transferred to DKW medium containing increasing concentration of sucrose (from 0.25 to 1 M) for 24 h. Before freezing, they were placed under a laminar air-flow hood for 1.5 h to accelerate sample dehydration.

4. Cooling. Samples were directly plunged in LN (rapid freezing) or frozen at 1 °C/min to −40 °C (slow freezing), then plunged in LN.

5. Thawing and Postculture. Somatic embryos were thawed in a 40 °C water bath for 1 min. Successive rinses in DKW liquid medium containing 1 M, then

0.5 M sucrose were then made. For postculture, the somatic embryos were isolated as carefully as possible in Petri dishes in order to check for further regrowth which was defined as (1) an increase in size, and (2) the capacity to produce new embryos. They had to be freed from the calcium alginate beads 24 h after thawing and subcultured every week.

6. Viability Test. Fluorescein diacetate (FDA) was used to check embryo survival either immediately after thawing or when the delay in regrowth was considered excessive.

7. Moisture Content. At each step of the cryopreservation procedure, the moisture content was determined by desiccation in an oven at 70 °C for 48 h until constant weight was attained. Moisture content was expressed as mg H_2O/mg dry weight.

8. Differential Scanning Calorimetry (DSC). Thermal analysis of somatic embryos was performed with a differential microcalorimeter (DSC-7 Perkin Elmer) using cooling and thawing rates of 50 °C/min. These cooling and thawing rates were chosen to facilitate the registration of the curves, which was not possible by directly plunging in LN followed by rapid thawing in a water bath at 40 °C.

2.4.2 Results and Discussion

1. Somatic embryos used for cryopreservation treatment were initially similar in appearance. They were sampled at the early globular stage (1–2 mm in length), as white as possible, and dehydrated slowly by subculturing with a progressive increase in sucrose concentration. Table 3 shows the moisture content at the different stages of the treatment. The average moisture content of the material on standard culture medium was 7.75 mg H_2O/mg dry weight decreasing to 2.62 mg H_2O/mg dry weight after plasmolysis on 1 M sucrose, and decreasing to 0.36, 0.24, 0.19 mg H_2O/mg dry weight following 1, 2, and 3 h of drying respectively.

2. The analysis of thermal behaviour of somatic embryos carried out by differential scanning calorimetry (DSC), after plasmolysis and 1.5 h dehydration under a laminar air-flow hood, showed no exotherm during cooling and warming. Only

Table 3. Average moisture content of somatic embryos at different stages of pretreatment before freezing. The results are relative to dry weight. Average of three replicates

	Before plasmolysis by 1 M sucrose	After plasmolysis by 1 M sucrose	After drying under laminar air-flow		
			1 h	2 h	3 h
Moisture content (mg H_2O/mg dry weight)	7.75	2.62	0.36	0.24	0.19
Total number of somatic embryos weighted	225	300	320	200	120

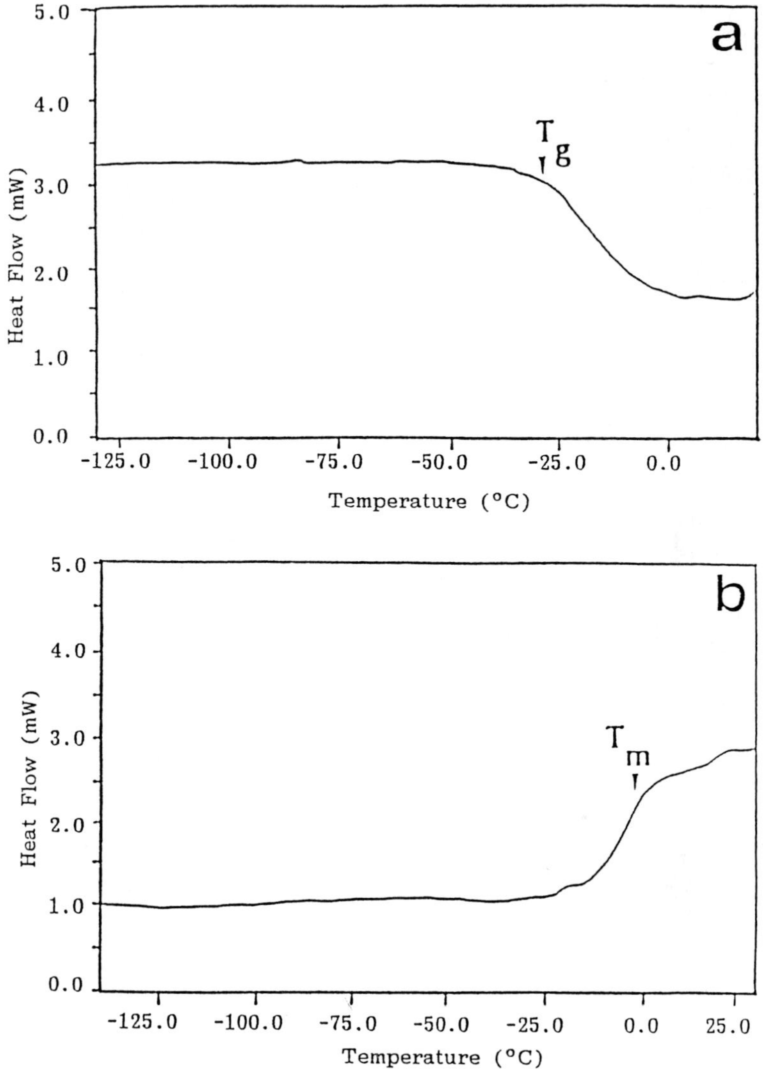

Fig. 3a, b. Differential scanning calorimetry of somatic embryos after 1.5 h drying. Both the cooling (**a**) and warming (**b**) rates were 50 °C/min. T_g, Temperature of the glass transition (−30 °C); T_m temperature of melting (−2 °C)

a glass transition at −25 °C was observed. During this thermal analysis, vitrification of the material occurred at 50 °C/min cooling and warming rate (Fig. 3a, b) and more so by plunging in LN and by rapid warming. It was thought that 1.5 h dehydration was sufficient and that further dehydration could have caused damage to the cytomembranes because we observed that the 3 h dried unfrozen controls had a slower regrowth than those that had been dried for 1 or 2 h. Thus,

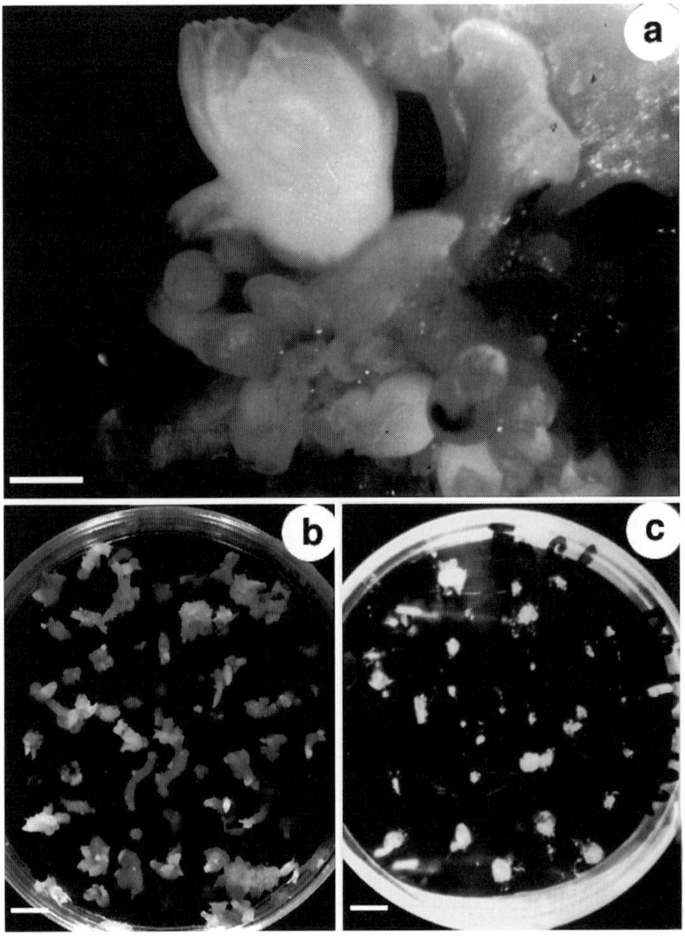

Fig. 4. a Somatic embryos at globular stage *bar* 2 mm, and with differentiated cotyledons. **b** Cryoprotected unfrozen somatic embryos after 4-week culture; *bar* 5 mm. **c** Cryoprotected and frozen somatic embryos after 8-week culture; *bar* 2 mm

progressive plasmolysis with an increasing sucrose concentration up to 1 M, followed by 1.5 h dehydration under a laminar air-flow was adopted as a cryopreservation protocol. All treated samples underwent immediate hydration after rinsing in 1 M, and subsequently in 0.5 M sucrose, and placing on standard DKW medium.

3. The majority (90%) of cryoprotected samples (control C1, Fig. 4b) increased in size, with some delay (2–3 weeks) and then rapidly produced new embryos (Fig. 4c).

4. The frozen-thawed samples subcultured every week did not resume growth before 3–4 weeks. The regrowth appeared firstly as increase in size and secondly

as embryo multiplication occurring after 5–6 weeks. While increase in size was frequent (50–60%), multiplication was highly variable (20–30%). This multiplication may be either weak, producing only four or five yellow white new embryos which did not rapidly develop or very actively, producing many white somatic embryos filling the whole Petri dish. FDA tests were done to check if embryos with tardy regrowth were still alive. Most of the tests were positive.

5. Factors affecting the cryopreservation of somatic embryos.

Hardening. Results of somatic embryos cryopreservation seem to be improved by a week's hardening at 4 °C (16 h)/25 °C (8 h) thermoperiod under dark conditions. However, somatic embryo hardening did not appear to be as efficient as meristem hardening, as emphasized by other workers (Reed 1990; Stushnoff 1987, 1991; Alberdi and Corcuera 1991).

Physiological State and Size of the Embryos. All embryos did not show the same tolerance to dehydration and freezing, since this depended on the physiological state and the size of the embryos. Thus, some embryos (>1–2 mm in size), after either dehydration or thawing, released yellow polyphenols into the culture medium. These compounds were toxic for embryo development. However, polyphenol release was prevented by successive rinses of the frozen-thawed somatic embryos in DKW liquid medium containing 1 M, then 0.5 M sucrose. The early globular stage, was more freezing-tolerant. At this stage, somatic embryos often clumped together forming a solid mass. A treatment involving a progressive increase in sucrose concentration facilitated embryo isolation.

Preculture. The culture of material on a medium containing a low concentration of cryoprotectant has been found beneficial (Bajaj 1983; Uragami et al. 1989; Towill 1991). As for walnut meristems, the use of proline and DMSO for somatic embryos in the preculture medium gave good results (0.5% proline and 5% DMSO yielding the best response).

Sucrose and Dehydration. Somatic embryos placed on high sucrose concentration medium underwent intensive plasmolysis without damage. Their moisture content decreased from 7.75 mg H_2O/mg dry weight to 2.62 mg H_2O/mg dry weight and then after further 2 h drying to 0.24 mg H_2O/mg dry weight. After this treatment 90% of controls C1 can easily resume growth.

Coating embryos in calcium alginate beads before progressive plasmolysis makes them easier to handle, especially after drying under a laminar air-flow hood, and protects them from the injuries of excessive dehydration.

The use of sucrose and dehydration as a cryoprotectant (Engelmann and Dereuddre 1988) associated with alginate coating was suggested by Dereuddre et al. (1991) for cryoconservation of somatic embryos of carrot. Sucrose pretreatment associated with desiccation was envisaged as a way of inducing a quiescent state in somatic embryos, mimicking the behaviour of true seed. The use of dried somatic embryos could constitute another method of germplasm preservation (Lecouteux et al. 1992).

Freezing-Thawing. We used either rapid freezing or slow freezing to −40 °C followed by plunging in LN. Both methods can give good results but we noted that slow freezing favoured a rapid and stronger multiplication than rapid freezing. However, it is also possible to obtain good proliferation by rapid freezing that does not require expensive apparatus.

After freezing, somatic embryos were thawed at 40 °C by rapid warming to avoid crystallization within cells.

3 Summary and Conclusions

During the last decade, considerable progress has been made in the area of cryopreservation of woody plant germplasm. Walnut species are considered to be difficult to culture and cryopreserve because of the high phenolic compound content of their tissues, but successful cryopreservation methods were recently developed. Now, cryostorage of pollen, embryonic axes, meristems, and somatic embryos of walnut appears to be feasible even if regrowth rates of these different germplasm sources remained highly variable.

Protocols for survival, based on empirical trials, had to be adapted to different materials and particularly to their hydration level. Theoretically, survival in cryogenic storage depends on: (1) the potential for tissue to tolerate water removal by dehydration, (2) the glass-forming ability of the tissue, (3) rewarming and recovery of the hydrated state without incurring excessive injury. Furthermore, a previous cold hardening is considered necessary, or even essential, for meristem freezing (Sakai 1985; Niino and Sakai 1992).

Cryopreservation appears as an interesting biotechnology, (Bajaj 1985, 1989, 1990, 1991), which allows (1) the conservation of germplasm such as pollen and nut embryonic axes for enhanced longevity, (2) conservation of rare elite clones which are usually vegetatively propagated, (3) maintenance of pathogen-free stock, (4) conservation of somatic embryos with very high propagation potential which may be considered as a useful substitute for seeds, and (5) international exchange of germplasm which may be carried out from a central bank responsible for the storage, maintenance, and distribution of managed stock.

References

Alberdi M, Corcuera LY (1991) Cold acclimation in plants. Phytochemistry 30(10): 3177–3184

Bajaj YPS (1983) Production of normal seeds from plants regenerated from the meristems of *Arachis hypogaea* and *Cicer arietinum* cryopreserved for 20 months. Euphytica 32: 425–430

Bajaj YPS (1985) Cryopreservation of embryos. In: Kartha KK (ed) Cryopreservation of plant cells and organs. CRC Press, Boca Raton, pp 227–242

Bajaj YPS (1986) Biotechnology of tree improvement for rapid propagation and biomass energy production. In: Bajaj YPS (ed) Biotechnology in agriculture and forestry, vol 1, Trees 1. Springer, Berlin Heidelberg New York, pp 1–23

Bajaj YPS (1989) Induction and cryopreservation of somaclonal variation in wheat and rice. Review of advances in plant biotechnology 1985–1988. CIMMYT, Mexico, pp 195–203

Bajaj YPS (1990) Cryopreservation of germplasm of vegetatively propagated crops. Bull Soc Bot Fr 137 Actual Bot 3: 99–114

Bajaj YPS (1991) Storage and cryopreservation of in vitro cultures. In: Bajaj YPS (ed) Biotechnology in agriculture and forestry, vol 17. High-tech and micropropagation I. Springer, Berlin Heidelberg New York, pp 361–381

Bercetche J, Galerne M, Dereuddre J (1990) Augmentation des capacités de régénération de cals embryogènes de *Picea abies* (L.) Karst après congélation dans l'azote liquide. CR Acad Sci Paris Sér III, 310: 357–363

Boutron P, Mehl P (1986) Nouveaux cryoprotecteurs pour la cryopréservation des cellules par vitrification totale. Bull Soc Bot Fr 133 Actual Bot 3: 27–39

Brison M, Paulus V, de Boucaud MT, Dosba F (1992) Cryopreservation of walnut and plum shoot tips. Cryobiology 29(6): 738

Chalupa V (1981) Clonal propagation of broad-leaved forest trees in vitro. Commun Inst For Czech 12: 255–271

Cornu D (1988) Somatic embryogenesis in tissue cultures of walnut (*Juglans nigra*, *J. major* and hybrids *J. nigra* x *J. regia*). In: Ahuja MR (ed) Somatic cell genetics of woody plants, Kluwer Academic, Dordrecht, pp 45–49

Cornu D (1989) Walnut somatic embryogenesis: physiological and histological aspects. Ann Sci For 46 (Suppl) 133s–135s

Cummins JN, Ashby WC (1969) Aseptic culture of *Juglans nigra* L. stem tissues. For Sci 15: 102–103

de Boucaud MT, Cambecedes (1988) The use of 1,2-propanediol for cryopreservation for recalcitrant seeds: the model case of *Zea mays* imbibed seeds. Cryo Lett 9: 94–101

de Boucaud MT, Brison M, Ledoux C, Germain E, Lutz A (1991 Cryopreservation of embryonic axes of recalcitrant seed: *Juglans regia* L. cv Franquette. Cryo Lett 12: 163–166

de Boucaud MT, Brison M (1994) Cryopreservation of walnut somatic embryos. Cryo Lett 15: 151–160 and 201–202

de Deng M, Cornu D (1992) Maturation and germination of walnut somatic embryos. Plant Cell Organ Tissue Cult 28: 195–202

de Préville D (1990) Contribution à l'étude de la maladie de la ligne noire du Noyer. DEA Biol Santé, option Pathologie microbienne et virale, Station de Recherches Fruitières, INRA, Pont de la Maye, 33140 Villenáve d'Ornon, France, 20 pp

Dereuddre J, Blandin S, Hassen N (1991) Resistance of alginate-coated somatic embryos of carrot (*Daucus carota* L.) to desiccation and freezing in liquid nitrogen. 1. Effects of preculture. Cryo Lett 12: 125–134

Driver JA, Kuniyuki AH (1984) In vitro propagation of paradox walnut rootstock. HortScience 19: 507–509

Engelmann F, Dereuddre J (1988) Cryopreservation of oil palm somatic embryos: importance of the freezing process. Cryo Lett 9: 220–235

Farmer RE, Barnett PE (1974) Low temperature storage of black walnut pollen. Cryobiology 11: 366–376

Germain E (1990) Walnut production and industry in Europe, the middle east and north Africa. In: Proc Expert consultation on the promotion of nut production in Europe and near east regions. Yalova (Istanbul), Turkey 19–22 June 1990. Reur Tech Ser 13, FAO, Rome, pp 119–131

Germain E (1992) Le Noyer. In: Gallais A, Bannerot H (eds) Amélioration des espèces végétales cultivées. INRA, pp 620–632

Griggs WH, Forde HI, Iwakiri BT, Asay RN (1971) Effect of sub-freezing temperature on the viability of Persian walnut pollen. HortScience 6: 235–237

Hall GC, Farmer RE (1971) In vitro germination of black walnut pollen. Can J Bot 49: 799–802

Heile-Sudholt C, Huetteman CA, Preece JE, Van Sambeek JW, Gaffney GR (1986) In vitro embryonic axis and seedling shoot-tip culture of *Juglans nigra*. Plant Cell Tissue Organ Cult 6: 189–197

Lecouteux C, Tessereau H, Florin B, Coutrois D, Petiard V (1992) Conservation d'embryons somatiques de carotte (*Daucus carota* L.) par déshydratation après prétraitement avec du saccharose. CR Acad Sci Paris Sér III, 314: 423–428

Lee BC (1989) Survival and regeneration of somatic embryos of *Juglans regia* L. (English walnut) after being frozen in liquid nitrogen. Res Rep Inst Genet Korea 25: 159–165

Lenartowicz A, Millikan DF (1977) Meristem culture of *Juglans nigra* L. Trans MO Acad Sci 10/11: 279

Leslie C, McGranahan G (1992) Micropropagation of Persian walnut (*Juglans regia* L.). In: Bajaj YPS (ed) Biotechnology in agriculture and forestry, vol 18. High-tech and micropropagation II. Springer, Berlin Heidelberg New York, pp 136–150

Luza JG, Polito VS (1985) In vitro germination and storage of English walnut pollen. Sci Hortic 27: 303–316

Luza JG, Polito VS (1988) Cryopreservation of English walnut (*Juglans regia* L.) pollen. Euphytica 37: 141–148

Meynier V (1985) Mise en culture in vitro de méristèmes de noyers hybrides. CR Acad Sci Paris Sér III, 301: 261–264

Morel G, Wetmore RH (1951) Fern callus tissue culture. Am J Bot 38: 141–143

Murashige T, Skoog F (1962) A revised medium for rapid growth and bioassays with tobacco tissue cultures. Physiol Plant 15: 473–497

Niino T, Sakai A (1992) Cryopreservation of alginate-coated in vitro-grown shoot tips of apple, pear and mulberry. Plant Sci 87: 199–206

Pence VC (1990) Cryostorage of embryo axes of several large seeded temperate tree species. Cryobiology 27: 212–218

Pritchard HW, Prendergast FG (1986) Effects of desiccation and cryopreservation on the in vitro viability of recalcitrant seed species *Araucaria hunsteinii* K. Schum. J Exp Bot 37: 1388–1397

Reed B (1990) Survival of in vitro-grown apical meristems of *Pyrus* following cryopreservation. HortScience 25(1): 111–113

Roberts EH (1973) Predicting the storage life of seeds. Seed Sci Technol 1: 499–514

Rodriguez R (1982a) Callus initiation and root formation from in vitro culture of walnut cotyledons. HortScience 17: 195–196

Rodriguez R (1982b) Stimulation of multiple shoot buds formation in walnut seeds. HortScience 17: 592

Rodriguez R, Sanchez-Tames R (1981) Cultivo de tejidos y diferenciacion en nogal. Rev Fac Cienc Univ Oviedo Ser Biol 22: 21–28

Rodriguez R, Revilla A, Albuerne M, Perez C (1989) Walnut (*Juglans* spp.). In: Bajaj YPS (ed) Biotechnology in agriculture and forestry, vol. 5. Trees II. Springer, Berlin Heidelberg New York, pp 99–126

Sakai A (1985) Cryopreservation of shoot-tips of fruit trees and herbaceous plants. In: Kartha KK (ed) Cryopreservation of plant cells and organs. CRC Press, Boca Raton, pp 135–158

Sakai A, Kobayashi S, Oiyama I (1990) Cryopreservation of nucellar cells of navel orange (*Citrus sinensis* Obs. var. *brasiliensis* Tanaka) by vitrification. Plant Cell Rep 9: 30–33

Somers PW (1982) In vitro propagation of black walnut (*Juglans nigra* L.). Master's Thesis, Southern Illinois University, Carbondale, IL 109 pp

Steffan SJ, Millikan DF (1983) Micropropagation of mature black walnut trees. Annu Rep North Nut Growers Assoc 74: 26–29

Stushnoff C (1987) Cryopreservation of apple genetic resources. Can J Plant Sci 67: 1151–1154

Stushnoff C (1991) Cryopreservation of fruit crop genetic resources. Implications for maintenance and diversity during conservation. HortScience 26(5): 518–522

Towill LE (1988) Genetic consideration for germplasm preservation of clonal materials. HortScience 23(1): 91–97

Towill LE (1991) Cryopreservation. In: Dodds JH (ed) In vitro methods for conservation of plant genetic resources. Chapman and Hall, London, pp 41–70

Tulecke W, McGranahan GH (1985) Somatic embryogenesis and plant regeneration from cotyledons of walnut *Juglans regia* L. Plant Sci 40: 57–63

Tulecke W, McGranahan GH, Leslie CA (1995) Somatic embryogenesis in walnuts (*Juglans* species). In: Bajaj YPS (ed) Biotechnology in agriculture and forestry, vol 30. Somatic embryogenesis and synthetic seed I. Springer, Berlin Heidelberg New York, 370–377

Uragami A, Sakai A, Nagai M, Takahashi TA (1989) Survival of cultured cells and somatic embryos of *Asparagus officinalis* L. cryopreservation by vitrification. Plant Cell Rep 8: 418–421
Yadrov AA (1990) Resources of genetic improvement of walnuts. In: Szentivanyi P (ed) 1st Int Symp walnut production. Budapest, Hungary, 25–29 Sept 1989, Acta Hortic 284: 363–366
Yamada I, Sakai A, Matsumura T, Higuchi S (1991) Cryopreservation of apical meristems of white clover (*Trifolium repens* L.) by vitrification. Plant Sci 78: 81–87

II.7 Cryopreservation of Date Palm, Oil Palm, and Coconut

F. ENGELMANN[1], B. ASSY-BAH[2], S. BAGNIOL[3], D. DUMET[1], and N. MICHAUX-FERRIÈRE[4]

1 Introduction

In tropical and subtropical regions, palms are well represented among cultivated species. Date palm, coconut palm, and oil palm are among the most important crops in the economy of many developing countries around the world. The safe conservation of genetic resources of these species is faced with various problems. Indeed, germplasm collections of these species are conserved under field conditions since seeds of oil palm and coconut are recalcitrant and those of date palm have a germination rate which decreases rapidly even after short storage periods. In vitro culture techniques present interesting solutions to improve the conditions for their propagation and conservation. Indeed, tissue culture systems allow to multiply and store plant germplasm in aseptic conditions, with drastic reduction of space requirements and of labour costs.

The aim of this chapter is to review the literature and to present the most recent work carried out at ORSTOM/Montpellier in collaboration with various institutes for the development of cryopreservation processes for these species.

2 Date Palm (*Phoenix dactylifera* L.)

2.1 Plant Distribution and Importance

Date palm is present on all continents (Branton and Blake 1989): on the mediterranean borders of Europe and Africa, in west and east Africa. In Asia, the most important producing countries are located in the Middle East (Saudi Arabia, Iraq). Plantations also exist in India and China. Date palm was introduced in America in the 16th century and is now found in the USA and South America. Finally, it was introduced in Australia a century ago.

[1]ORSTOM, 911 av. Agropolis, BP 5045, 34032 Montpellier Cédex 1, France. (*Present address*: IPGRI, Via delle Sette Chiese 142, 00145 Rome, Italy)
[2]IDEFOR/DPO, Station Principale de La Mé, 13 BP 989, Abidjan 13, Ivory Coast
[3]GRFP, Conservatoire National Botanique dePorquerolles, 83400 Ile de Porquerolles, France
[4]CIRAD, BIOTROP, BP 5035, 34032 Montpellier Cédex 1, France

Biotechnology in Agriculture and Forestry, Vol. 32
Cryopreservation of Plant Germplasm I (ed. by Y.P.S. Bajaj)

About 5000 named varieties have been identified (Anonymous 1914); most cultivars originated from chance seedlings which were propagated vegetatively through offshoots. Also, a multitude of unnamed varieties occur, since each seedling may be considered a potential cultivar due to the high degree of heterozygosity in date palm (Tisserat et al. 1983b). Some offshoot varieties have been cloned for hundreds of years; isolated cases of offshoots cloned for up to 1300 years are known (Anonymous 1914).

Oases have a double economical importance, firstly for the production of dates, secondly by providing shed which allows associated cultures to grow (fruits, vegetables, cereals). World date production was estimated to 2 883 000 t in 1988 (Anonymous 1989). The main production areas are located in the Middle East (Iran, Iraq, Saudi Arabia) and in North Africa (Algeria, Morocco, Tunisia, Egypt).

2.2 Methods for Storage and Need for Cryopreservation

Date palm germplasm is traditionally conserved in the form of field collections of trees in its cultivation areas. Seeds are not recalcitrant, since they can remain viable for up to 15 years at room temperature. However, germination rate markedly decreases after 6 years of conservation (Munier 1973). Date palm is traditionally vegetatively propagated using sprouts. However, it produces a limited number of sprouts only, around 10 to 20 maximum during the whole life of the palm (Toutain and Rhiss 1973), which considerably slows down its propagation. Moreover, due to the mode of reproduction of this species, seed progenies are highly heterogenous. Therefore, various in vitro techniques including micropropagation through axillary budding and somatic embryogenesis have been developed to multiply genotypes (Tisserat 1987). In vitro collections of germplasm in the form of plantlets have been established, notably in France at GRFP (Groupement de Recherche Français sur le Palmier Dattier).

Date palm is threatened by various pests (acarians) and diseases (fungi, bacteria) (Munier 1973). The most important is a fungal disease induced by *Fusarium oxysporum* Schlect. f. sp. *albedinis* commonly termed bayoud, which occurs in North Africa. Preservation of tolerant or resistant genotypes is therefore essential in the frame of breeding programs against this disease. Maintenance of in vivo collections is costly due to the large size of palms, which induces low planting densities and to their slow growth habit. Thus, at the US Date Palm Repository (Indio, California), the in vivo collection which previously comprised numerous Old World and local varieties was restricted to those female varieties which were considered useful commercially and to five males showing interesting characters (Carpenter 1979). Moreover, plants conserved in natural conditions remain exposed to pests, pathogens, and natural disasters.

Management of large-scale in vitro collections poses numerous practical problems. Moreover, risks of contaminations and of somaclonal variation increase with time. It is therefore essential to develop long-term conservation techniques for the germplasm of numerous potentially useful varieties of date palm.

2.3 Cryopreservation Studies

Only a limited amount of work has been carried out on cryopreservation of date palm, using in vivo (seeds, pollen, apices) or in vitro material (calli) (Table 1). The first experiments were performed in the 1980s with embryogenic calli of different varieties (Deglet Noor, Khadrawy, Medjool) (Tisserat et al. 1981; Finkle et al. 1982; Ulrich et al. 1982) using a classical freezing process. Plantlets could be regenerated from the cryopreserved calli. Pollen grains were successfully cryopreserved and germinated subsequently (Tisserat et al. 1983a). Fruits of the variety Deglet Noor obtained from flowers pollinated with cryopreserved pollen showed normal development (Tisserat et al. 1985). Seeds of the variety Zahidi could be cryopreserved, stored for up to 546 days in LN and thawed with a recovery rate between 85 and 100% (Al-Madeni and Tisserat 1986). Our work on the cryopreservation of date palm apices is discussed here. The successive steps of the cryopreservation process for apices were optimized (Bagniol and Engelmann 1991, 1992) and histological studies were performed after each step of the cryopreservation process and during the first days of recovery (Bagniol et al. 1992).

2.3.1 Materials and Methods

Apices (Fig. 1) were taken from in vitro plantlets of several varieties (Bou Sthammi noir, Zahidi, and Nabut Seif), at the end of a 6-week culture period on a multiplication medium containing 0.1 M sucrose (Ferry et al. 1987). They

Table 1. Summary of work done on cryopreservation of date palm, oil palm, and coconut

Plant species	Explant frozen	Results/remarks	Reference
Date palm	Embryogenic callus	Callus + plantlets	Tisserat et al. (1981), Finkle et al. (1982), Ulrich et al. (1982)
	Pollen grains	Germination + fruits	Tisserat et al. (1985)
	Seeds	85–100% germination	Al-Madeni and Tisserat (1986)
	Apices from in vitro plantlets	11.8–48.2% survival + plants	Bagniol and Engelmann (1991, 1992)
Oil palm	Zygotic embryos	Survival + plants in soil	Grout et al. (1983)
	Somatic embryos	31–55% survival	Engelmann et al. (1985)
		Normal plants in soil	Engelmann (1991b)
		13–100% survival	Dumet et al. (1993)
Coconut	Transverse halves of young embryos	Callus	Bajaj (1984)
	Mature embryos	Callus + 1 plantlet	Chin et al. (1989)
	Immature embryos	0–43% survival	Assy-Bah and Engelmann (1992a)
	Mature embyros	73–93% survival and plant production	Assay-Bah and Engelmann (1992b)

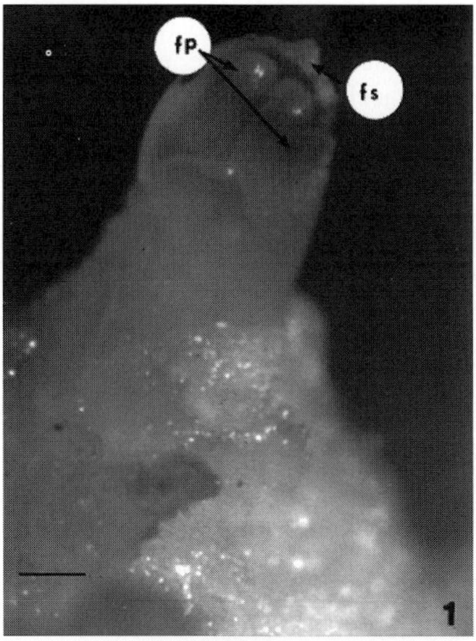

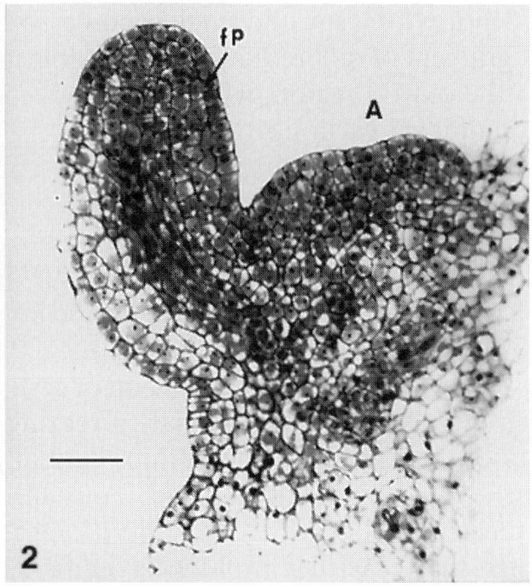

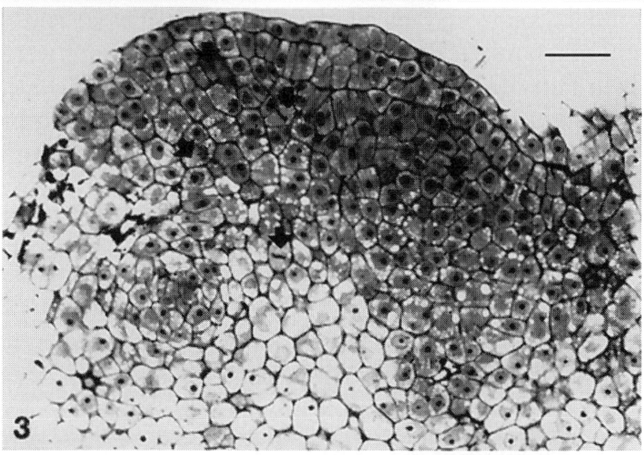

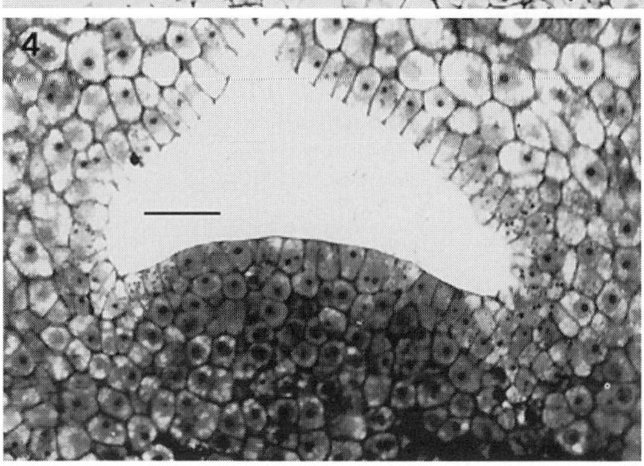

Figs. 1–12. Date palm

Fig. 1. Morphological aspect of an explant. Apical zone with two foliar primordia recently formed (*fp*) surrounded by a foliar sheath (*fs*). *Bar* 0.2 mm. (Bagniol 1992)

Fig. 2. Histological structure of a control sample. *A* Apical dome; *fp* foliar primordium. *Bar* 125 μM. (Bagniol et al. 1992)

Fig. 3. Longitudinal section of an apex after a 24-h culture on standard medium. Structure of the apical dome. *Arrows* indicate figures of mitosis. *Bar* 140 μM. (Bagniol et al. 1992)

Fig. 4. Sample after cryoprotective treatment. Note starch grains (*dark points*) in superficial cells. *Bar* 130 μm. (Bagniol et al. 1992) consisted of the apical dome, foliar primordia, and a basal part (Fig. 2). Cells showed characteristics of meristematic cells: they had a high nucleoplasmic ratio, a dense cytoplasm, and small vacuoles. No polysaccharide reserves were present. A gradient of differentiation was visible in meristems, from the superficial layers to the central region, where cells had a lower nucleoplasmic ratio and a more accentuated vacuolization.

Before pretreatment, meristems could be maintained for 1 to 7 days, immediately after their dissection, on an induction medium, favoring growth of meristems (Ferry et al. 1987). For pregrowth, meristems were placed for 24h on media containing 0.3 or 0.5 M sucrose. For cryoprotective treatment, apices were placed at 0 °C for 2h in liquid sucrose-enriched medium supplemented with 0 to 15% DMSO.

Meristems, suspended in 1 ml of cryoprotective medium, were placed in 2 ml sterile polypropylene cryovials. Freezing was carried out at 0.5 to 5 °C/min to various prefreezing temperatures (end of programmed freezing). Once prefreezing temperature was reached, ampoules were either thawed immediately or immersed in LN, where they were kept for 1 h before being thawed. Meristems were thawed either rapidly by immersion of the cryotubes in a water bath thermostated at 40, 60, or 80 °C, or slowly by placing the cryotubes at room temperature in the air current of a laminar flow cabinet. Thawed meristems were transferred at 24-h intervals in Petri dishes containing 10 ml of solid media with progressively lower sucrose concentrations (0.5, 0.3 M) until standard concentration of 0.1 M was reached. These media were used with or without 1g/l activated charcoal. Cultures were placed either directly under standard lighting conditions (photoperiod of 16h/24, 42 $\mu E/m^2/s$) or in the dark for the first 10 days following thawing and then under attenuated light (7 $\mu E/m^2/s$, same photoperiod).

2.3.2 Results and Discussion

Effect of Culture on Induction Medium Prior to Pregrowth Treatment. In order to determine the survival rate of apices after freezing, it was necessary to submit them after dissection to a culture on induction medium for 1 to 3 days (Table 2). Extension of this culture duration to 5 or 7 days did not allow growth recovery.

Table 2. Effect of the duration of culture on induction medium on the survival rate of control (−LN) and frozen (+LN) meristems. Apices were pregrown for 24h on a medium with 0.5 M sucrose, cryoprotected with 0.5 M sucrose +15% DMSO and frozen to −30 °C at 1 °C min. (Bagniol and Engelmann 1991)

	Duration of culture (days)				
	0	1	3	5	7
−LN	57.1	79.1	96.6	96.6	87.0
+LN	0.0	22.0	15.1	0.0	0.0

Table 3. Effect of sucrose and DMSO concentration during the cryoprotective treatment on the survival rate of control (−LN) and frozen (+LN) meristems. (Bagniol and Engelmann 1991)

Sucrose (M)		DMSO (%) 0	5	10	15
0.3	−LN	62.5	62.9	62.9	61.5
	+LN	0.0	0.0	3.6	7.5
0.5	−LN	66.6	56.6	56.6	57.1
	+LN	0.0	18.9	14.0	17.1

Histological examination of apices after 24h culturing on standard medium revealed that this culture led to recovery of cellular activity (Fig. 3): several mitotic figures as well as numerous divisions were visible. Starch grains were also present in some cells (not shown in Fig. 3). It seems that recovery of cellular activity is necessary for survival. Also, it could have the effect of healing meristems as well as synchronizing explants, thus reducing their physiological heterogeneity.

Effect of Pregrowth and Cryoprotective Treatment. Pregrowth of apices consisted of a 24-h culture on medium containing 0.3 or 0.5 M sucrose. For cryoprotective treatment, meristems were suspended in liquid medium with the same sucrose concentration and various concentrations of DMSO (0 to 15%) were added. For unfrozen control meristems, there was no effect of DMSO concentration, whatever the sucrose level (Table 3). For cryopreserved apices, no survival was obtained without DMSO; however, survival was optimal after a cryoprotective treatment with 0.5 M sucrose and 5 to 15% DMSO. During pregrowth and cryoprotective treatment, synthesis of starch increased in intensity in some of the samples (Fig. 4). Starch grains were located in the superficial zones of the meristem where cells retained a very meristematic appearance. Other samples showed no starch synthesis.

The increased starch synthesis occurring when meristems are placed on a sucrose-enriched medium indicates that, contrary to previous hypotheses (Finkle et al. 1985), sucrose does not act as an osmoticum only but enters cells in large quantities. According to Grout et al. (1988), sugars, dissoluted or attached to the membrane through covalent bounds stabilize the membrane bilayer, thus protecting integrity of cell membranes during dehydration and freezing. Therefore meristems which showed no starch synthesis may be those which were unable to survive after cryopreservation.

Effect of Freezing and Thawing Conditions. Among the various combinations of cooling rates (0.5 to 5 °C/min) and prefreezing temperatures experimented (−20 to −40 °C), survival after cryopreservation was achieved after prefreezing at 1 °C/min down to −30 or −35 °C only and was optimal at −30 °C. Survival of prefrozen meristems decreased progressively in line with the lowering of prefreez-

Table 4. Effect of prefreezing temperature on the survival rate of prefrozen (−LN) and cryopreserved (+LN) meristems. (After Bagniol and Engelmann 1991)

Prefreezing temperature (°C)	−12	−15	−20	−25	−30	−35	−40
−LN	53±15	64±17	37±10	28±10	32±14	17±13	8±12
+LN	0	0	0	0	18± 8	10±12	0

ing temperature (Table 4). No survival of cryopreserved meristems was noted after slow thawing at room temperature. Recovery of meristems was noted after thawing at 40, 60, and 80 °C, but was significantly higher at 40 °C.

Insufficient dehydration during prefreezing down to −25 °C caused death of cryopreserved tissues, whereas from −40 °C onwards excessive dehydration was responsible for death of meristems. Only a very narrow range of temperatures (−30 to −35 °C) ensured survival after freezing in liquid nitrogen. With date palm meristems, rapid thawing in a water bath at 40 °C was preferable to slow rewarming and higher thawing temperatures. Indeed, rapid rewarming makes it possible to avoid recrystallization phenomena which induce lethal damages of the cells (Dereuddre 1985).

Histological examination revealed that after freezing, apices showed cellular heterogeneity (Fig. 5). Some cells conserved their meristematic characters with nuclei sometimes intensely stained (N). This was the case in cellular layers corresponding to the meristem itself, whereas in the underlying zone where cells were more vacuolated, some were damaged, showing broken cell walls. Finally, cells with cytoplasm contracted away from the cell wall were observed in some samples (Fig. 6). Many nuclei displayed a particular appearance, intensely and uniformly stained (Fig. 6). Starch grains were visible inside some cells, more often located in the superficial zone.

Date palm meristems did not withstand freezing in LN as intact structures. It seems that the more meristematic the cells were, the more they resisted. During the steps which precede freezing, gradients may be established both for outflow of water and penetration of cryoprotectants. In the case of date palm, as in that of other plants, the central cells as well as the ones located at the base of the explant are more vacuolated and their dehydration may be more difficult or they may be less tolerant to the necessary level of dehydration. This may explain their incapacity to withstand freezing. Superficial cells which are more meristematic are more likely to reversibly dehydrate since they have no or only a few vacuoles.

Effect of Recovery Conditions. Presence of activated charcoal in the recovery medium improved survival of frozen meristems (Table 5). Their response to lighting conditions depended on the composition of recovery medium, the presence of activated charcoal being the determining factor. Indeed, a minimal survival rate of 19.6% was obtained, whatever the lighting conditions if medium contained activated charcoal. Without charcoal, light totally inhibited recovery of frozen meristems.

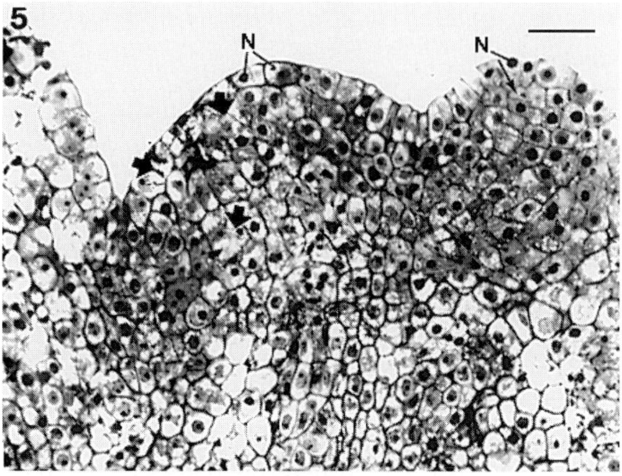

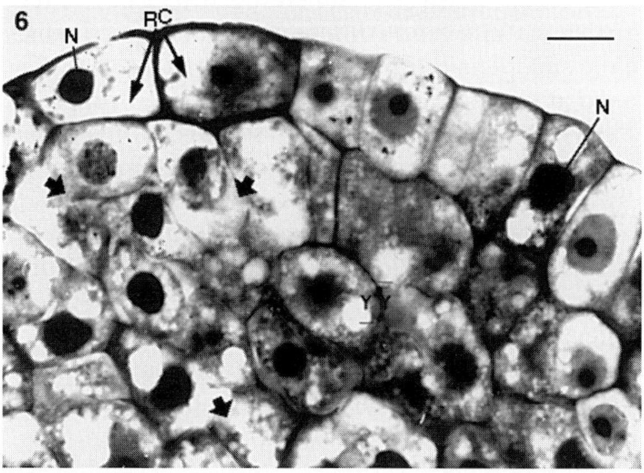

Figs. 5,6. Samples after prefreezing and immersion in liquid nitrogen. *N* Intensely and uniformly stained nuclei; *RC* cells with retracted cytoplasm; *arrows* rupture and deformation of cell walls. *Bar* 125 μm (**Fig. 5**); 25 μm (**Fig. 6**). (Bagniol et al. 1992)

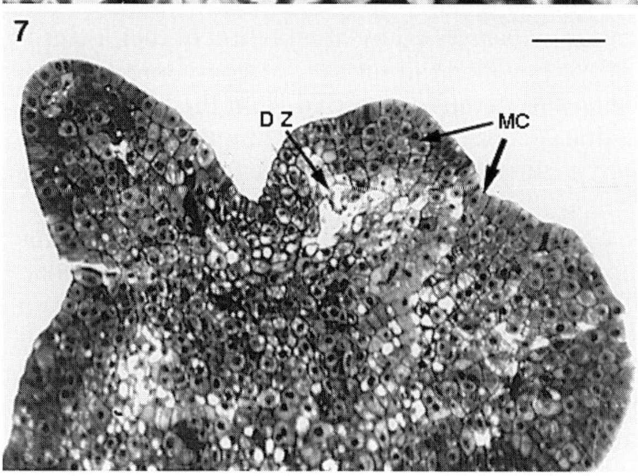

Fig. 7. Zonation in a sample 2 days after thawing showing meristematic cells (*MC*) and disrupted zones (*DZ*). *Bar* 200 μm. (Bagniol et al. 1992)

Table 5. Effect of lighting conditions and activated charcoal on the survival rate ($\pm$S.D.) of cryopreserved meristems after two weeks of culture. After thawing, meristems were exposed directly to a photon dose of 42 $\mu E/m^2/s$ or placed in the dark for 10 days and then exposed to a photon dose of 7 $\mu E/m^2/s$. (Bagniol and Engelmann 1992)

	Activated charcoal (g/l)	
	0	1
42 $\mu E/m^2/s$	0.0	19.6 $\pm$ 5.3
7 $\mu E/m^2/s$	9.1 $\pm$ 0.0	25.9 $\pm$ 11.6

At the histological level, samples showed a cellular heterogeneity 2 days after thawing (Fig. 7): zones with meristematic cells and differentiated or harmed cells. The first mitoses occurred in the superficial layer 4 days after thawing (Fig. 8). However, not all apices observed at this stage showed mitotic figures. From this moment on, thickening of the cell wall could be observed in some cells of various areas. From the 6th day onwards, walls of meristematic or nonmeristematic cells had a sinuous aspect (Fig. 9). Up to the 7th day, the outline of the original meristem remained recognizable. Afterwards, in most cases, the initial apex was no longer distinguishable, hidden by proliferation of meristematic cells of the apical dome and of the foliar primordia (Fig. 10). Intense synthesis of starch occurred inside cells of superficial and underlying zones (Fig. 11). In samples which had shown no morphological sign of regrowth for 6 weeks and were thus considered dead, cells were poorly stained and had a retracted cytoplasm (Fig. 12).

Modifications of recovery conditions had a very positive effect on the growth pattern of cryopreserved meristems, as assessed by observation of their macroscopic evolution. Regrowth of apices which had survived after cryopreservation was visible 7 days after thawing. Proliferation occurred from the basal part of explants and led to the formation of a callus which had a limited development. Each sample which had formed a callus gave rise during the following weeks to a single shoot.

Activated charcoal has an effect of adsorbing growth regulators, thus decreasing the level of hormones available to the cultures (Weatherhead et al. 1978; Ebert and Taylor 1990). After freezing in LN, only groups of cells remain alive but not the whole meristems. When date palm meristems are transferred on a medium without activated charcoal, the quantity of hormones available in the medium may be too high compared with the number of living cells and induces intense callusing. With activated charcoal, the level of free hormones may become sufficiently low to be compatible with the number of living cells and allow regrowth of frozen meristems with limited callus formation. We observed with date palm that growth recovery in the dark improved survival of cryopreserved meristems. Chemiluminescence analyses showed an increase of oxygen singlets in cryopreserved tissues submitted to light immediately after thawing (Benson and Noronha-Dutra 1988). Light has an oxidative effect which induces browning and

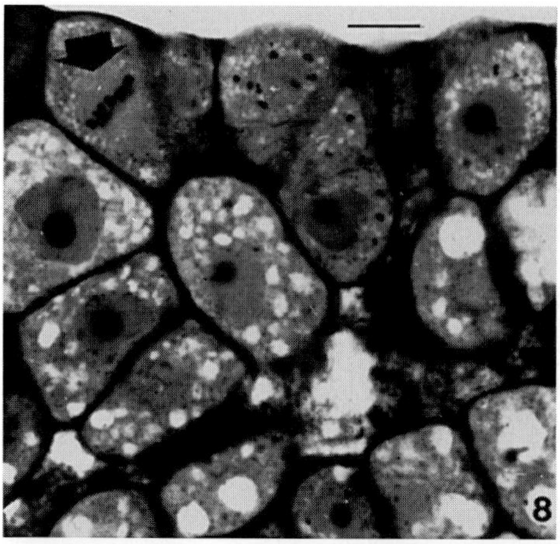

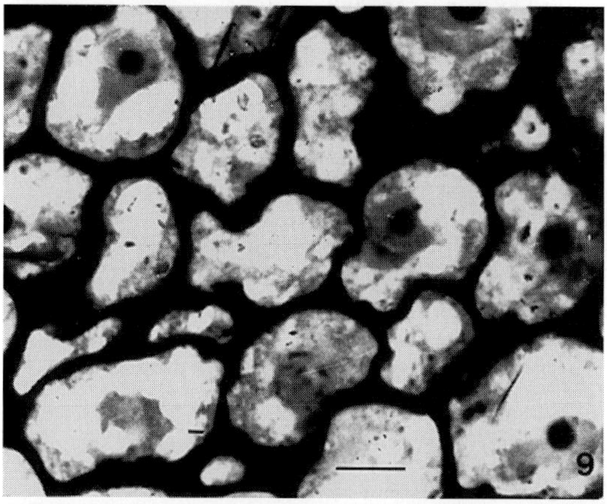

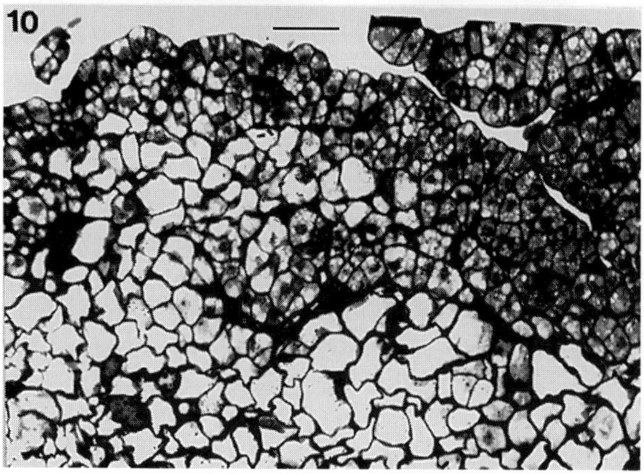

Fig. 8. First mitosis (*arrow*) observed in a cryopreserved sample 4 days after thawing. *Bar* 30 μm. (Bagniol et al. 1992)

Fig. 9. Thickening of cell walls in a cryopreserved sample 6 days after thawing. *Bar* 35 μm. (Bagniol et al. 1992)

Fig. 10. Sample 15 days after thawing. Cell proliferation inside meristem and foliar primordia. *Bar* 150 μm. (Bagniol et al. 1992)

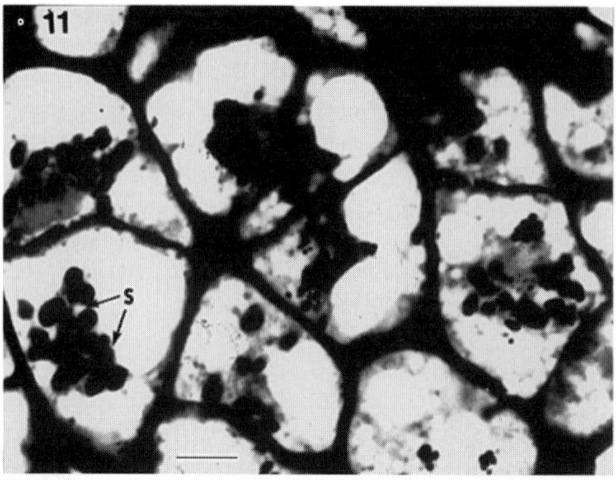

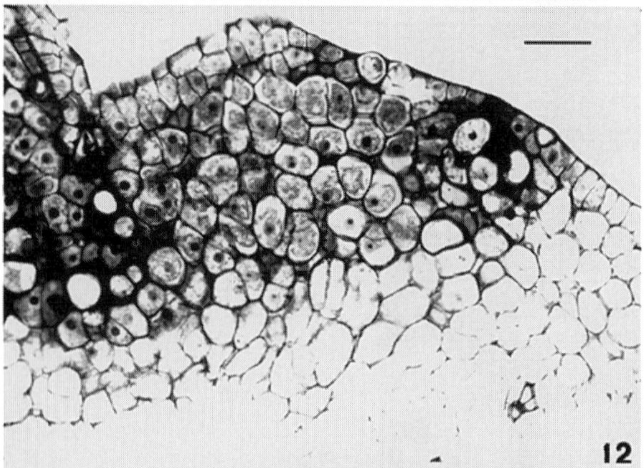

Fig. 11. Richness in starch content (*s*) in the underlying zone of the meristem, 15 days after thawing. *Bar* 40 μm. (Bagniol et al. 1992)

Fig. 12. Sample showing no morphological sign of regrowth 6 weeks after thawing. Note that cells have a retracted cytoplasm. *Bar* 115 μm. (Bagniol et al. 1992)

necroses which can lead to death of cells (Benson et al. 1989). This was observed in the case of date palm meristems. Presence of activated charcoal in recovery medium could also allow adsorption of oxidated substances which are toxic to cells, and thus increase the recovery rate, even under lighted conditions.

2.3.3 Conclusion

The present study allowed a protocol ensuring the cryopreservation of date palm apices, with satisfactory survival rates for two varieties out of the three experimented: 30.3% for Bou Sthammi noir and 48.2% for Nabut Seif, but 11.8% only for Zahidi (Bagniol and Engelmann 1992). However, these values are comparable to those mentioned in the literature dealing with cryopreservation of meristems (Kartha 1985; Bagniol 1992). These results may be related to the behavior of these varieties under in vitro conditions: Zahidi has a relatively low growth rate in comparison with that of the two other varieties. The cryopreservation protocol developed is a classical one, using cryoprotective treatment and programmed freezing. Importance of various parameters such as that of freezing rate could be underlined. The originality of the present work is linked with the observations made concerning the importance of culture conditions prerequisite to cryoprotective treatment and of recovery conditions. The extensive histological study performed also allowed the demonstration of the importance of sucrose (assessed by starch accumulation) during pregrowth treatment in order to ensure high survival rates. More detailed studies on the role of this compound should be performed, since it appears as a key substance for acquisition of tolerance to freezing at the temperature of liquid nitrogen.

3 Oil Palm (*Elaeis guineensis* Jacq.)

3.1 General Account

Oil palm is mainly cultivated in industrial plantations between 7 °S and 7 °N, at an altitude lower than 400 m in areas with high pluviometry. Plantations now exist on all continents, South East Asia, Africa, South America in Brazil and the Amazon basin (Equator, Columbia, Peru), and the Pacific Coast (Costa Rica). The second important species is *Elaeis oleifera* which is cultivated in South America. It is employed for interspecific crosses with *E. guineensis* in order to increase tolerance to pests and pathogens existing in this region.

Oil palm is after soybean, the second source of vegetable edible oil. Annual production was over 11 millions tons in 1991 and is still increasing due to a continuous rise in planted surfaces which reach today 3.5 million ha (Anonymous 1991).

Genetic resources of oil palm are conserved in the form of field collections. Seeds are not truly recalcitrant, but retain a good viability for short periods only, 1–2 years under optimal conditions according to Hartley (1988). In vitro propagation techniques based on somatic embryogenesis have been developed on a large scale in order to clonally propagate elite trees (see Duval et al. 1995 for a review). These elite genotypes are conserved as cultures of somatic embryos under standard conditions. Experiments using controlled atmospheres have also been performed (Engelmann 1990).

3.2 Cryopreservation

Before the research program developed by ORSTOM/CIRAD was initiated, cryopreservation experiments had been carried out with oil palm zygotic embryos only (Grout et al. 1983). Zygotic embryos extracted from seeds were partially desiccated under the laminar flow and successfully frozen in liquid nitrogen.

3.2.1 Methodology/Protocol

Original Protocol. The original protocol set up by ORSTOM/CIRAD (Engelmann et al. 1985) comprised the following steps: (1) production of a particular type of material, finger-shaped embryos by a 2-month culture on a medium supplemented with 0.3 M sucrose (Engelmann and Dereuddre 1988); (2) pretreatment for 7 days on a medium containing 0.75 M sucrose; (3) rapid freezing and rapid thawing; (4) recovery on media with progressively lowered sucrose concentrations and supplemented with 2,4 dichlorophenoxyacetic acid (2,4-D) for 3 weeks; (5) transfer onto standard medium devoid of growth regulators for proliferation recovery.

This process was developed on a large scale, since more than 150 different clones were frozen in five different units in Indonesia, Malaysia, the Ivory Coast, and France (Engelmann 1991b). Embryos could be stored for up to 52 months without any change in recovery rate (Engelmann 1992). Moreover, palms regenerated from cryopreserved material and planted in the field were phenotypically comparable to controls as regards their vegetative and floral development (Durand-Gasselin, pers. comm.). However, routine development of this technique was drastically slowed down by the low and erratic production of finger-shaped embryos which were at that time the only type likely to withstand freezing.

Improved Process. Dramatic improvement was made recently to the original technique by adding a dehydration step after treatment on high sucrose medium, before freezing in liquid nitrogen. Indeed, this allowed the use of standard somatic embryos (Fig. 13). Dehydration could be performed either by placing embryos for 0 to 10 h under the laminar flow, or in air-tight boxes containing silica gel for 0 to 18 h (Dumet et al. 1993).

3.2.2 Results and Discussion

Survival of unfrozen controls was affected by dehydration periods under the laminar flow longer than 4 or 6 h for clones A, D and B, C respectively (Table 6). However, this dehydration was indispensable in most cases to achieve survival after cryopreservation. A 6-h dehydration seemed optimal since it allowed to obtain survival with three out of the four clones tested. No survival was noted with clone B. When dehydration was performed with silica gel, survival of controls decreased from 16-h dehydration onwards (Table 7). With cryopreserved clumps, survival was sometimes higher than that of controls for the longest dehydration periods (16 to 18 h). Dehydration treatment increased the resistance of embryos to freezing in liquid nitrogen. Survival of clone B, which was previously 0%, was maximal after a 16-h dehydration. Survival of clone C was

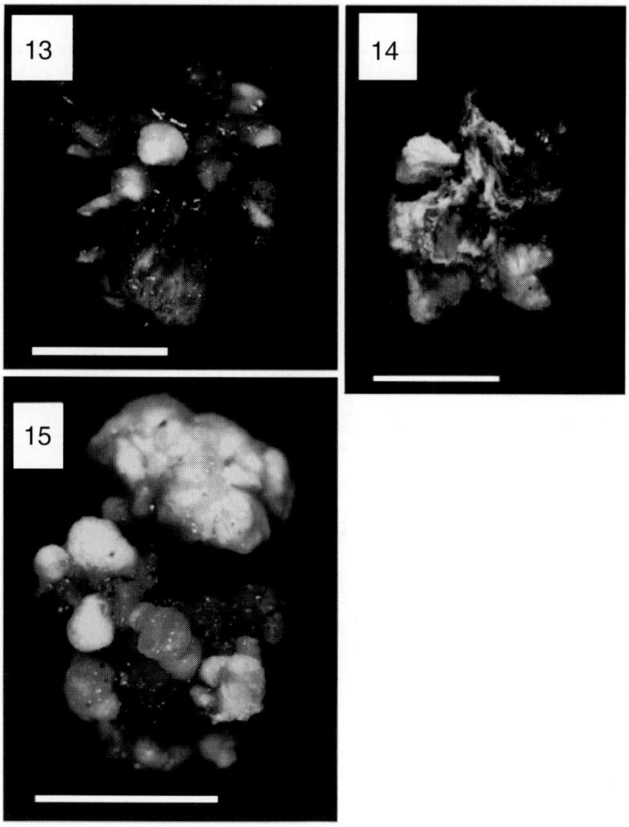

Figs. 13–15. Oil palm

Fig. 13. Clumps of somatic embryos after 1 week pregrowth on a medium containing 0.75 M sucrose. *Bar* 5 mm. (Dumet et al. 1993)

Fig. 14. Clumps of somatic embryos after 1 week pregrowth and 16 h of dehydration with silica gel. *Bar* 5 mm. (Dumet et al. 1993)

Fig. 15. Proliferation recovery of somatic embryos 3 weeks after cryopreservation. *Bar* 1 cm. (Dumet et al. 1993)

improved, since nine clumps out of ten recovered after 18 h of dehydration. However, survival of clone D was not improved but remained high. These pretreatment conditions (16 h of dehydration with silica gel) were applied to seven additional clones. Survival rates of clumps after freezing in liquid nitrogen varied between 13 and 53% (Table 8), depending on the clone. Even though clumps of embryos seemed to be severely damaged by pregrowth and desiccation treatments (Fig. 14) proliferation recovery of clumps surviving after cryopreservation was very rapid and intense (Fig. 15). The dehydration period chosen was not optimal for all clones, since for some of them unfrozen controls had a low survival rate.

Table 6. Survival of unfrozen control (−LN) and cryopreserved (+LN) oil palm somatic embryos of four clones, as a function of the desiccation duration (DD) performed under a laminar flow cabinet. (After Dumet et al. 1993)

	Clone A		Clone B		Clone C		Clone D	
DD (h)	−LN	+LN	−LN	+LN	−LN	+LN	−LN	+LN
0	10/10	0/10	10/10	0/10	10/10	2/10	10/10	0/10
4	10/10	0/10	10/10	0/10	10/10	4/10	10/10	0/10
6	7/10	2/10	10/10	0/10	10/10	7/9	8/10	7/10
8	5/10	2/10	8/10	0/10	9/10	4/9	9/10	3/10
10	6/10	0/10	4/10	0/10	7/10	5/9	7/10	1/9

Table 7. Survival of unfrozen control (−LN) and cryopreserved (+LN) oil palm somatic embryos of three clones, as a function of the desiccation duration (DD) using silica gel. -: not measured. (After Dumet et al. 1993)

	Clone B		Clone C		Clone D	
DD (h)	−LN	+LN	−LN	+LN	−LN	+LN
0	10/10	1/7	10/10	6/10	10/10	6/10
7	10/10	0/10	10/10	7/10	10/10	5/10
16	6/10	10/10	-	7/10	8/10	5/10
18	5/10	8/10	4/10	9/10	4/10	7/10

Table 8. Survival of unfrozen control (−LN) and cryopreserved (+LN) oil palm somatic embryos of seven clones. Pregrowth consisted of a 7-day culture on a medium containing 0.75 M sucrose followed by 16 h of dehydration with silica gel. (After Dumet et al. 1993)

	Clone						
	E	F	G	H	I	J	K
−LN	50	66	13	70	50	20	40
+LN	16	13	16	53	46	37	13

The improvements made to the original cryopreservation technique will facilitate its routine application. Indeed, due to the possibility of using standard embryos, we are no longer limited by the production of finger-like embryos. Moreover, desiccation with silica gel will ensure more reproducible conditions than those obtained with a laminar flow cabinet which can vary, depending notably on air humidity. However, it will be necessary to determine for each clone its optimal dehydration period before its cryopreservation and its long-term conservation. This improved technique is now experimented on a large scale in the different industrial units producing clonal oil palms.

4 Coconut Palm (*Cocos nucifera* L.)

Coconut is widely spread in all tropical regions. Planted areas are estimated to around 10 million ha in three main culture zones: Asia-Oceania, Africa, Central and South America, and the Caribbean. It is a very important crop for many tropical countries. Even though industrial plantations exist, it is mainly cultivated by smallholders. Around 50% of the production is consumed as food and 50% used for oil. The annual world production of oil is estimated to 2.7 million tons, which represents 5.1% of the total vegetable oil (Anonymous 1991).

Coconut is a species with truly recalcitrant seeds which have no dormancy (nuts even germinate on the mother palm). There is no possibility of storing them even for short periods of time. Therefore, germplasm is conserved as field collections in national or regional centers. In vitro culture techniques have been developed for zygotic embryos which facilitate the collection and exchange of germplasm (Assay-Bah et al. 1987, 1989). Also, in vitro medium-term storage techniques are now available (Assay-Bah and Engelmann 1993) which make it possible to preserve zygotic embryos for 1 year under standard conditions with a slight viability loss.

Only a small proportion of the variability that exists within the species is currently available in existing field collections, and many sources of material are still almost unexploited (de Nucé de Lamothe 1992). The situation is all the more serious in that certain accessions are now in danger of dying out, due to the extension of the most widely grown varieties of hybrids and to urbanization. Their introduction and conservation in vitro as embryos is therefore of prior importance. There is also a need for conservation techniques because of problems and constraints linked to occurrence of diseases of coconut in different coconut-growing areas (IBPGR 1992). For example, for material originating from areas where lethal yellowing occurs, parent palms that have not shown disease symptoms should be observed for at least 1 year (the incubation period of the disease) before embryos may be released and exported to a country free of lethal yellowing. In the meantime, excised embryos need to be maintained in vitro, under slow growth or cryopreservation.

In the literature, only two papers report attempts to cryopreserve coconut embryos (Table 1). Bajaj (1984) observed callusing on transversely cut halves of embryos after a freeze-thaw cycle. With whole embryos, Chin et al. (1989) obtained callusing and normal germination of one embryo, 15 months after freezing.

4.1 Freezing Procedure

Experiments were performed with immature and mature embryos (Assy-Bah and Engelmann 1992a,b) of different varieties. Immature embryos (7–8 months after pollination) were placed for 4 h in Petri dishes on standard medium defined by Assy-Bah (1992) supplemented with 600 g/l glucose and 0 to 15% glycerol, sorbitol or polyethyleneglycol (PEG 6000). After rapid freezing and thawing, they were transferred onto standard medium for regrowth.

Table 9. Survival of control (−LN) and cryopreserved (+LN) immature coconut embryos after 7 month in culture. −pgr: no cryoprotective treatment; +pgr: pregrowth with 600 g/l glucose only; *: contaminated. (Assy-Bah and Engelmann 1992a)

			Glycerol (%)				Sorbitol (%)				PEG (%)			
	−prg	+prg	1	5	10	15	2	5	10	15	2	5	10	15
−LN	85	100	95	95	94	73	100	100	94	–	100	78	85	–
+LN	0	0	0	0	25	10	0	0	43	0	0	0	0	–

Mature embryos (11–12 months after pollination) were placed under the laminar flow for 4 h in empty Petri dishes and then transferred onto a culture medium containing 600 g/l glucose and 15% glycerol where they were kept for an additional 11 to 20 h. They were then frozen and thawed rapidly and transferred for regrowth on standard medium (Assay-Bah et al. 1989).

4.2 Results/Discussion

With immature embryos, survival of unfrozen controls after 1 month's culturing was satisfactory for all pregrowth conditions, ranging from 73 to 100% (Table 9). No difference was noted with control embryos not submitted to cryoprotective treatment (-prg). After freezing in liquid nitrogen, survival was obtained in three conditions only: pregrowth with 10 and 15% glycerol (25 and 10% survival respectively) and 10% sorbitol (43% survival). PEG had no cryoprotective effect at the concentrations studied. After 2.5 months culturing, 100% of control embryos germinated but many of them displayed abnormal development. With cryopreserved ones, surviving embryos increased in size but a limited number of them only could give rise to whole plantlets.

Pretreatment consisting of a desiccation which is usually employed with embryos (Engelmann 1991a) was impossible to use since immature embryos were too small for a precise control of their dehydration in air. Therefore, we used a classical pregrowth treatment on semisolid medium with cryoprotectants. As regards cryopreservation, immature embryos initially appeared as interesting material in that they were smaller and less differentiated than mature ones, and may be more likely to resist as a whole. However, the drawback of utilizing such young embryos was that the conditions for their in vitro culture are not yet perfectly mastered. Unlike mature ones, not all immature embryos could develop into whole plantlets. This considerably reduced the interest of using such material for cryopreservation for this crop.

With mature embryos, survival of controls was 100% without pretreatment (Table 10) and only slightly decreased after 19 or 24 h of pretreatment. With frozen embryos, no survival was observed without pretreatment. Survival increased with increasing pretreatment durations for PB 121, Renell Tall (RT) and Indian Tall (IT7) embryos. For 15 h of pretreatment, survival ranged between 7 (RT and IT7) and 35% (PB 121) and reached 86% (RT) to 93% (PB 121) after

Table 10. Survival after 1 month culturing of control (−LN) and cryopreserved (+LN) mature embryos of four varieties of coconut as a function of pretreatment duration. (After Assy-Bah and Engelmann 1992)

		Pretreatment duration (h)				
		0	15	17	19	24
PB 121	−LN	100	100	100	100	87
	+LN	0	35	53	50	93
CRD	−LN	100	67	76	65	78
	+LN	0	61	73	59	56
RT	−LN	100	100	86	86	83
	+LN	0	7	12	38	86
IT7	−LN	100	100	100	82	100
	+LN	0	7	27	24	88

24h pregrowth. In the case of Cameroon Red Dwarf (CRD) embryos, survival was greatest after 17h of pretreatment (73%) and decreased afterwards. The haustorium of frozen embryos browned rapidly and did not develop further, contrarily to unfrozen controls (Fig. 16). Further development of cryopreserved embryos into plantlets was comparable to that of unfrozen controls (Fig. 17) and plantlets with fully expanded leaves could be obtained after 4.5 months culturing (Fig. 18).

The cryopreservation technique established with mature embryos was very efficient since high survival rates and rapid regrowth and development could be obtained with a large number of cryopreserved embryos. Considerable improvements have been made in comparison with previous work on the same material (Bajaj 1984; Chin et al. 1989). This technique is original since pretreatment combines desiccation of embryos in air and culture on medium with high levels of cryoprotectants. Such a combination has been employed up to now only with oil palm somatic embryos (Dumet et al. 1993). Differences observed between varieties concerning the optimal pretreatment period were related with the initial weight of embryos. For CRD, which has the smallest embryos, only 17h were necessary in comparison with 24h with other varieties which have larger embryos. The only difference in regrowth pattern between control and cryopreserved embryos concerned the nondevelopment of haustorium of frozen material. Haustorium is formed of highly vacualoted cells which are completely destroyed during freezing. This is in accordance with previous observations of Chin et al. (1988) on embryos of *Veitchia* and *Howea*.

This process is already very efficient in its present form but still has to be improved in several ways. Pretreatment duration has to be adapted to the different varieties and its conditions standardized.

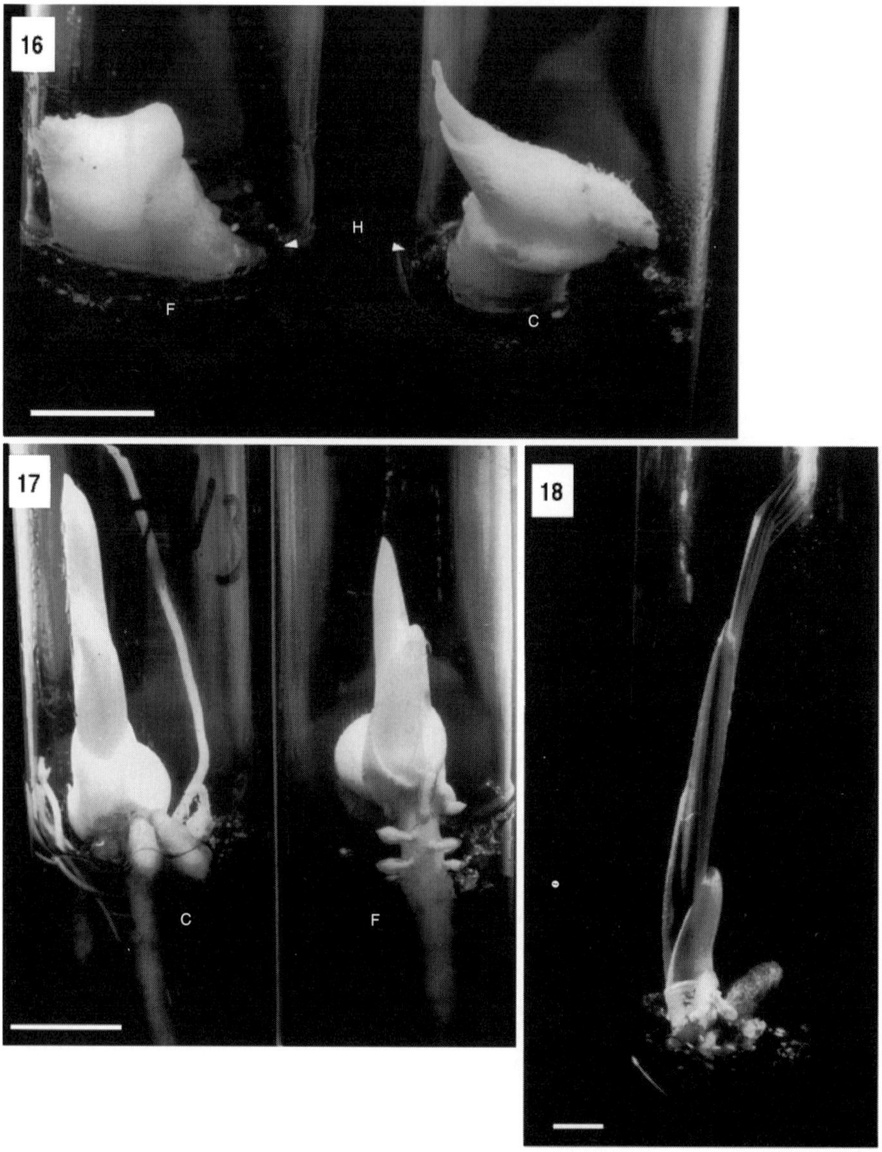

Figs. 16–18. Coconut

Fig. 16. Development of control (*C*) and frozen (*F*) mature embryos after 1 month's culturing. *H* haustorium. *Bar* 1 cm. (Assy-Bah and Engelmann 1992b)

Fig. 17. Development of plantlets originating from control (*C*) and frozen (*F*) mature embryos after 3 months' culturing. *Bar* 1 cm. (Assy-Bah and Engelmann 1992b)

Fig. 18. Development of a plantlet originating from a cryopreserved mature embryo after 4.5 months' culturing. *Bar* 1 cm. (Assy-Bah and Engelmann 1992b)

5 Summary and Conclusions

Significant progress has been made for the cryopreservation of date palm, oil palm, and coconut. The present chapter illustrates the various approaches undertaken for freezing meristems, somatic or zygotic embryos. Date palm apices were frozen using a classical process involving cryoprotection in liquid medium and slow controlled freezing. Survival varied between 11.8 and 48.2%, depending on the variety. Growth recovery was generally slow and transitory callusing was often noted. Indeed, as revealed by the histological study performed, apices did not withstand freezing as a whole but only groups of cells remained alive. However, growth recovery pattern could be improved by placing apices in the dark on a medium supplemented with activated charcoal.

Standard oil palm somatic embryos could withstand rapid freezing after pregrowth on semisolid medium with high sucrose concentration and partial desiccation in air or using silica gel. Survival was generally high and proliferation recovery was very rapid and intense. Results may be improved by modifying the dehydration period which may differ between clones.

Immature coconut embryos could withstand cryopreservation, but only a limited number of them could give rise to plantlets due to their relatively low survival rate and possibly to inadequate in vitro culture conditions. On the contrary, a high percentage (73 to 93%) of mature coconut embryos survived to cryopreservation after pretreatment on semisolid medium with high concentration of glucose and glycerol followed by partial desiccation in air. All surviving cryopreserved embryos of the four varieties tested developed rapidly into whole plantlets.

In conclusion, the experiments described here show that even if the process set up for oil palm somatic embryos is now tested on a large scale and is routinely employed very soon, additional research is still needed for date palm and coconut.

Acknowledments. Research on the development of cryopreservation techniques for oil palm and coconut is performed in the framework of a joint program between ORSTOM (Institut Français de Recherche Scientifique pour le Développement en Coopération), CIRAD/CP (Centre de Coopération Internationale en Recherche Agronomique pour le Développement/Département Cultures Pérennes), and IDEFOR/DPO (Institut des Forêts/Département Plantes Oléagineuses, Ivory Coast). The research project on coconut is supported by IBPGR (International Board for Plant Genetic Resources, Rome).

References

Al-Madeni MA, Tisserat B (1986) Survival of palm seeds under cryogenic conditions. Seed Sci Technol 14: 79–85

Anonymous (1914) Origin of the date palm. J Hered 5: 498–508

Anonymous (1989) FAO. Quartely Bulletin of Statistics, vol 2. FAO, Rome 147 pp

Anonymous (1991) Oil World Annual, ISTA Mielke GmbH, Hamburg, 108 pp

Assay-Bah B (1992) Utilisation de la culture in vitro pour la collecte et la conservation des ressources génétiques du cocotier (*Cocos nucifera* L.). Thèse Univ, Paris 6, 146 pp

Assy-Bah B, Engelmann F (1992a) Cryopreservation of immature embryos of coconut (*Cocos nucifera* L.) Cryo Lett 13: 67–74

Assy-Bah B, Engelmann F (1992b) Cryopreservation of mature embryos of coconut (*Cocos nucifera* L.) and subsequent regeneration of plantlets. Cryo Lett 13: 117–126

Assy-Bah B, Engelmann F (1993) Medium-term conservation of mature embryos of coconut (*Cocos nucifera* L.). Plant Cell Tissue Organ Cult 33: 19–24

Assy-Bah B, Durand-Gasselin T, Pannetier C (1987) Use of zygotic embryo culture to collect germplasm of coconut (*Cocos nucifera* L.). FAO/IBPGR Plant Genet Res Newsl 71: 4–10

Assay-Bah, Durand-Gasselin T, Engelmann F, Pannetier C (1989) Culture in vitro d'embryons zygotiques de cocotier (*Cocos nucifera* L.). Méthode révisée et simplifiée d'obtention de plants de cocotiers transférables aux champs. Oléagineux 44: 515–523

Bagniol S (1992) Cryoconservation d'apex prélevés sur des vitroplants de palmier dattier (*Phoenix dactylifera* L.). Mise au point du protocole et étude histo-cytologique des échantillons au cours de la congélation et de la remise en culture. Thèse Univ, Paris 6, 125 pp

Bagniol S, Engelmann F (1991) Effects of pregrowth and freezing conditions on the resistance of meristems of date palm (*Phoenix dactylifera* L. var Bou Sthammi noir) to low temperatures and to freezing in liquid nitrogen. Cryo Lett 12: 279–286

Bagniol S, Engelmann F (1992) Effect of thawing and recovery conditions on the regrowth of meristems of date palm (*Phoenix dactylifera* L.) after cryopreservation in liquid nitrogen. Cryo Lett 13: 253–260

Bagniol S, Engelmann F, Michaux-Ferrière N (1992) Histo-cytological study of apices from in vitro plantlets of date palm (*Phoenix dactylifera* L.) during a cryopreservation process. Cryo Lett 13: 405–412

Bajaj YPS (1984) Induction of growth in frozen embryos of coconut and ovules of *Citrus*. Curr Sci 53: 1215–1216

Benson EE, Noronha-Dutra AA (1988) Chemiluminescence in cryopreserved plant tissue culture: the possible involvement of singlet oxygen in cryoinjury. Cryo Lett 7: 120–131

Benson EE, Harding K, Smith H (1989) Variation in recovery of cryopreserved shoot tips of *Solanum tuberosum* exposed to different pre- and post-freeze light regimes. Cryo Lett 10: 323–344

Branton RL, Blake J (1989) Date palm (*Phoenix dactylifera* L.). In: Bajaj YPS (ed) Biotechnology in agriculture and forestry, vol. 5. Trees II. Springer, Berlin Heidelberg New York, pp 161–175

Carpenter JB (1979) The national date palm germplasm repository. Date Growers' Inst Rep 54: 29–32

Chin HF, Krishnapillay B, Alang ZC (1988) Cryopreservation of *Veitchia* and *Howea* palm embryos: non-development of the haustorium. Cryo Lett 9: 372–379

Chin HF, Krishnapillay B, Hor YL (1989) A note on the cryopreservation of embryos from young plants of coconuts (*Cocos nucifera* var. Mawa). Pertanika 12: 183–186

Crowe JH, Crowe LM, Carpenter JF, Rudolph AS, Wistrom CA, Spargo BJ, Anchordoguy TJ (1988) Interactions of sugars with membranes. Biochim Biophys Acta 947: 367–384

De Nucé de Lamothe M (1992) Coconut improvement – needs and opportunities. In: IBPGR (ed) Coconut genetic resources. IBPGR, Rome, Int Crop Network Ser 8: 31–34

Dereuddre J (1985) Problèmes posés par la conservation des souches cellulaires et des organes cultivés in vitro. Bull Soc Bot Fr Actual Bot 314: 123

Dumet D, Engelmann F, Chabrillange N, Duval Y (1993) Cryopreservation of oil palm (*Elaeis guineensis* Jacq.) somatic embryos involving a desiccation step. Plant Cell Rep 12: 352–355

Duval Y, Engelmann F, Durand-Gasselin T (1995) Somatic embryogenesis in oil palm (*Elaeis guineensis* Jacq.). In: Bajaj YPS (ed) Biotechnology in agriculture and forestry, vol 30. Somatic embryogenesis and synthetic seed I. Springer, Berlin Heidelberg New York, pp 335–352

Ebert A, Taylor HF (1990) Assessment of the changes of 2,4-dichlorophenoxyacetic acid concentrations in plant tissue culture media in the presence of activated charcoal. Plant Cell Tissue Organ Cult 20: 165–172

Engelmann F (1990) Utilisation d'atmosphères à teneur en oxygène réduite pour la conservation de cultures d'embryons somatiques de palmier à huile (*Elaeis guineensis* Jacq.). CR Acad Sci Paris 310 Sér III: 679–684

Engelmann F (1991a) In vitro conservation of tropical plant germplasm – a review. Euphytica 57: 227–243

Engelmann F (1991b) Le développement actuel de la cryoconservation des embryons somatiques de palmier à huile. In: IIR (ed) Proc XVIIIth Refrigeration Congr, Montréal, 10–17 Aug, vol IV, pp 1676–1680
Engelmann F (1992) Cryopreservation of embryos. In: Dattée Y, Dumas C, Gallais A (eds) Reproductive biology and plant breeding. Springer, Berlin Heidelberg New York, pp 281–290
Engelmann F, Dereuddre J (1988) Effect du milieu de culture sur la production d'embryoïdes destinés à la cryoconservation chez le palmier à huile (*Elaeis guineensis* Jacq.). CR Acad Sci Paris 306 Sér III: 515–520
Engelmann F, Duval Y, Dereuddre J (1985) Survie et prolifération d'embryons somatiques de palmier à huile (*Elaeis guineensis* Jacq.) après congélation dans l'azote liquide. CR Acad Sci Paris 301 Sér III: 111–116
Ferry M, Louvet J, Louvet JM, Monfort S, Toutain G (1987) The specific character of the research into in vitro propagation and mass production of the date palm. Acta Hortic 212: 576
Finkle BJ, Ulrich JM, Tisserat B (1982) Responses of several lines of rice and date palm callus to freezing at $-196\,^{\circ}C$. In: Li PH, Sakai A (eds) Plant cold hardiness and freezing stress. Academic Press, New York, pp 643–660
Finkle BJ, Zavala ME, Ulrich JM (1985) Cryoprotective compounds in the viable freezing of plant tissues. In: Kartha KK (ed) Cryopreservation of plant cells and organs. CRC Press, Boca Raton, Florida, pp 75–113
IBPGR (1992) Coconut genetic resources. Papers of an IBPGR Workshop, Cipanas, Indonesia, 8–11 Oct, 1991. International Crop Network Series No 8. IBPGR, Rome, 78 pp
Grout BWW, Shelton K, Pritchard HW (1983) Orthodox behaviour of oil palm seed and cryopreservation of the excised embryo for genetic conservation. Ann Bot 52: 381–384
Hartley CWS (1988) Germination and the preparation and storage of seed. In: Hartley CWS (ed) The oil palm, 3rd edn. Tropical Agric Ser, Longman, London, pp 304–328
Kartha KK (1985) Meristem culture and germplasm preservation. In: Kartha KK (ed) Cryopreservation of plant cells and organs. CRC Press, Boca Raton, Florida, pp 115–134
Munier P (1973) Le Palmier Dattier. Maisonneuve and Larose, Paris, 221 pp
Tisserat B (1987) Palms. In: Bonga JM, Durzan DJ (eds) Cell and tissue culture in forestry. Martinus Nijhoff, Dordrecht, pp 338–356
Tisserat B, Ulrich JM, Finkle BJ (1981) Cryogenic preservation and regeneration of date palm tissue. Hortic Sci 16: 47–48
Tisserat B, Ulrich JM, Finkle BJ (1983a) Survival of *Phoenix* pollen grains under cryogenic conditions. Crop Sci 23: 254–256
Tisserat BJ, Nelson MD, Ulrich JM, Finkle BJ (1983b) Cryostorage techniques to preserve date palm germplasm. Proc 1st Symp on the Date Palm in Saudi Arabia, Al-Hassa, King Faisal Univ, March 23–25, 1982, pp 108–119
Tisserat B, Gabr MF, Sabour MT (1985) Viability of cryogenically treated date palm pollen. Date Palm J 4: 25–32
Toutain G, Rhiss A (1973) Production du palmier dattier. II. Formation de rejets sur de jeunes palmiers dattiers. Al Awamia 48: 79–89
Ulrich JM, Finkle BJ, Tisserat B (1982) Effects of cryogenic treatment on plantlet production from frozen and unfrozen date palm callus. Plant Physiol 69: 624–627
Weatherhead MA, Burdon J, Henshaw GG (1978) Some effects of activated charcoal as an additive to plant tissue culture media. Z Pflanzenphysiol 89: 141–147

II.8 Cryopreservation of Germplasm of Ornamental Palms (*Veitchia*, *Howea*, and *Calamus* Species)

B. KRISHNAPILLAY, H.F. CHIN, and M. MARZALINA[1]

1 Introduction

The storage of seeds is the customary means of conservation and international exchange of germplasm. However, in a number of tropical species, the seeds are recalcitrant. Recalcitrant seeds typically have higher water content and will not tolerate a reduction in water content below a relatively high level without loss of viability (Roberts and King 1980).

The report by Guppy (1912) that seeds of a number of palm species are damaged by drying has often been cited to indicate that they may not be orthodox. Ellis et al. (1991) argue that it is probable that seeds belonging to the palm species (Arecaceae family) may possibly possess storage characters that are intermediate between orthodox and recalcitrant.

Generally, for palm species, various researchers (Bajaj 1984, 1985; Chin et al. 1988, 1989a; Krishnapillay et al. 1992) have advocated that storage of excised embryos would be a better alternative for genetic conservation than a whole seeds.

In this chapter, the work on the cryopreservation of embryos of three species, *Vietchia merrilli* (Veitchia), *Howea fosteriana* (Kentia), and *Calamus manan* (Rotan manau) to explore the feasibility of this procedure for long-term in vitro conservation of germplasm is discussed.

Veitchia merilli, family Arecaceae, subfamily Arecoideae, is a commonly grown single-stemmed palm which grows to a height of 10 m. The leaves are pinnate, dark green with broad leaflets, and are arched. Inflorescences are produced below the leaves on the stem, and eventually clusters of very attractive bright red fruits develop (Chin and Enoch 1988). The species are grown as avenue plants and have great ornamental value. Other related species include *V. filfera* and *V. spiralis* (Uhl and Dransfield 1987).

Howea fosteriana, also family Arecaceae, subfamily Arecoideae, is a slow-growing elegant ornamental palm that is worldwide in demand as an indoor plant. This species is endemic to the lowlands of Lord Howe Island, off the coast of New South Wales, Australia (Pickard 1983). The leaves are pinnate, rather flat and drooping. Several inflorescences are produced in each leaf axial. Flower and fruit mature slowly, so that several inflorescences in different stages are often

[1]Forest Research Institute of Malaysia, Kepong, 52109 Kuala Lumpur, Malaysia

Biotechnology in Agriculture and Forestry, Vol. 32
Cryopreservation of Plant Germplasm I (ed. by Y.P.S. Bajaj)

present on a single tree. The other related species include *H. belmoreana* (Uhl and Dransfield 1987).

Calamus manan is a spiny, climbing plant, family Arecaceae, subfamily Calamoideae. It is an economically and culturally important no-wood forest product. The canes of this species are most sought after by the furniture industry. The genus *Calamus* is distributed from West Africa to Fiji and from South China to Queensland, Australia (Rao et al. 1991).

All three species are propagated mainly through seeds. It has been reported that the seeds of these species are difficult to store for long periods and to date only a 3-week to 2-month storage period has been possible (Mori et al. 1980; Chin and Krishnapillay 1986, 1987, unpubl.; Chin et al. 1987). Owing to the formidable problems encountered in storage of these seeds, in vitro conservation of germplasm of these species is not being practiced. Present methods of conservation include in situ and ex situ conservation strategies. To avoid the loss of germplasm of these species through natural disaster or land clearing owing to population pressures, there is an urgent need to develop in vitro conservation strategies to complement the present strategies.

2 Cryopreservation

2.1 *Veitchia merrilli* and *Howea fosteriana*

Other than the work carried out by Chin et al. (1988) that is reported below, no work relating to the storage of these species through cryopreservation techniques has been recorded.

2.1.1 Materials and Methods

Veitchia merrilli seeds, ripe and red in color, were harvested from the Agricultural University Malaysia farm while the *Howea fosteriana* seeds, green or yellow in color were obtained from Lord Howe Island, Australia, through the mail.

A random sample of 20 seeds from each species was taken for a viability test prior to the start of the main experiment. The seeds were depericarped and surface sterilized with 20% v/v commercial Clorox for 20 min followed by three rinses with sterile distilled water. The embryos were excised from the seeds aseptically in a laminar flow cabinet with the aid of hand pruning shears and a single-edged razor blade and cultured in vitro. Viability was recorded as shoot emergence.

One hundred depericarped seeds were employed for the experiment. The seeds were thoroughly mixed and divided into two groups to 50. Both the groups were further divided into five subgroups of ten seeds each for the various desiccation periods. The desiccation periods selected were 0, 1, 2, 3, and 4 h. The embryos were excised aseptically as described above and were dried in batches of five (i.e., two replicates), for each of the desiccation times. Desiccation was carried out by placing the isolated embryos on square sheets (5 × 5 cm) of sterile aluminum foil in the sterile air flow of a laminar flow cabinet. At the end of each desiccation period, the moisture level in the embryos was determined and one group of ten embryos was directly cultured in vitro while the other group of ten embryos was placed in sterile polypropylene cryovials and immersed directly in liquid nitrogen (−196 °C) for 48 h. The embryos were then removed from the cryo tank, thawed in a water bath (40 °C) and cultured in vitro without washing. This experiment was repeated twice.

The embryos were cultured on a Murashige and Skoog (1962) inorganic salt formula supplemented with 170 mg/l NaH_2PO_4, 30 g/l sucrose, 0.2 mg/l NAA (naphthalene acetic acid), 0.1 mg/l BAP (benzylaminopurine), 2 g/l activated charcoal and solidified with 7 g/l Difco Bacto Agar. All cultures were maintained at 28 ± 1 °C in a temperature-controlled room with a 12-h photoperiod of 8000–10000 lx intensity provided by fluorescent light. Embryos were recorded as viable when they had swollen, turned green, and showed shoot emergence.

2.1.2 Results

Moisture content in freshly excised embryos was 67.58 ± 0.62 on a fresh weight basis and viability percentage was 90%. Reduction in moisture content of the excised embryos was accompanied by a reduction of viability in vitro (Tables 1, 2). After 4 h of desiccation, when the embryo moisture content was around 10%, the viability was reduced from about 88% to about 69%. Freezing of the embryos with 10% moisture content in liquid nitrogen (LN) gave fairly good recovery (63–64%) after 48 h of storage. Embryos with 12% moisture content or more did not survive the LN treatment (Table 3).

An interesting observation made in both species was that viable embryos that were only desiccated and without LN treatment swelled and turned green after about 7–10 days. The haustorium swelled to a maximum diameter of 3–4 mm and elongated to a maximum length of 20 mm in about 3–4 weeks; shoot emergence around the 5th and 6th week in culture. In the case of the viable

Table 1. Water content of excised embryos of *Veitchia merilli* during desiccation[a]. (Chin et al. 1988)

Time (h)	Water content *Veitchia*	fr.wt. (%) *Howea*
0	67.58 ± 0.62	58.48 ± 1.01
1	19.63 ± 0.68	19.63 ± 0.91
2	12.60 ± 0.49	11.48 ± 0.48
3	10.72 ± 0.44	10.64 ± 0.42
4	9.92 ± 0.22	9.89 ± 0.15

[a]Mean of two experiments with 40 embryos for each determination taken in batches of five.

Table 2. Percentage recovery of excised embryos of *Veitchia* and *Howea* after desiccation and without cryopreservation[a]. (Chin et al. 1988)

Time (h)	Recovery (%) *V. merrilli*	*H. fosteriana*
0	87.50 ± 8.29	92.50 ± 8.29
1	82.50 ± 4.33	85.00 ± 8.66
2	79.46 ± 0.96	81.95 ± 4.74
3	70.00 ± 7.07	71.95 ± 7.83
4	68.61 ± 6.40	70.56 ± 7.60

[a]Mean of two experiments with 20 embryos for each determination. Values shown in the table are means of those embryos which were not contaminated in culture.

cryopreserved embryos, greening was observed around the 2nd and 3rd week. However, the haustorium failed to develop and remained rudimentary. Shoot emergence was observed around the 5th and 6th week in culture but still without any development of the haustorium. Also, it was noted that leaf expansion occurred much earlier in developing cryopreserved embryos as compared to noncryopreserved embryos. Figures 1 and 2 show the development of the haustorium in desiccated, unfrozen embryos and the nondevelopment of the haustorium in desiccated cryopreserved embryos.

2.1.3 Discussion

The data presented in Table 2 indicate that the excised embryos of the two palm species under study are markedly resistant to desiccation. Desiccation to a moisture content of around 10% (Table 3) was found to be suitable for cryopreserva-

Table 3. Survival percentage of excised embryos of *Veitchia* and *Howea* after desiccation and cryopreservation[a]. (Chin et al. 1988)

Time (h)	Survival (%) *V. merrilli*	*H. fosteriana*
0	0.00	0.00
1	0.00	0.00
2	0.00	0.00
3	20.00 ± 7.07	22.50 ± 4.33
4	62.50 ± 4.33	64.17 ± 4.33

[a]Mean of two experiments with 20 embryos for each determination. Values shown in the table are means of those embryos which were not contaminated in culture.

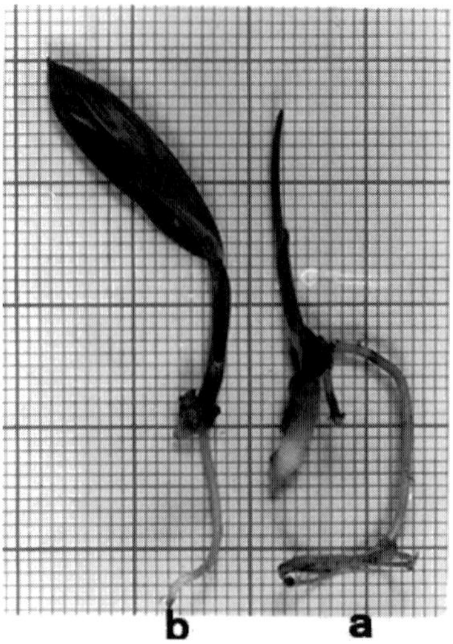

Fig. 1. *Howea* embryos at 3 months in culture. *a* Control; *b* cryopreserved

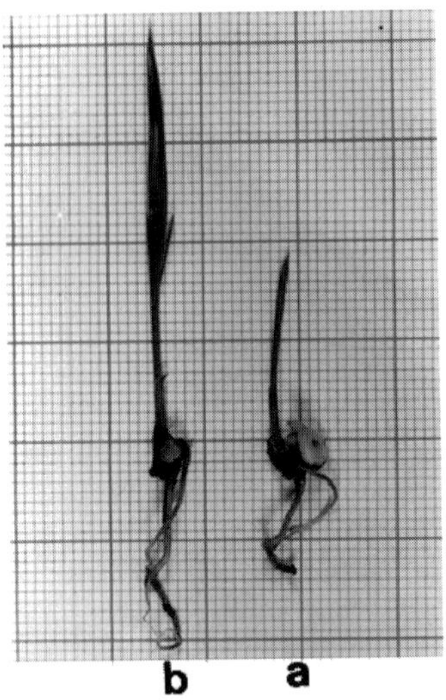

Fig. 2. *Veitchia* embryos at 3 months in culture. *a* Control; *b* cryopreserved

tion studies where the embryos survived and could be revived satisfactorily in culture.

Although the importance of the haustorium in palm seed germination has long been appreciated, the literature shows that few attempts have been made to understand the exact physiological role of this unusual organ (Nagarajan and Pandalai 1963; Keusch 1968; Balasubramaniam et al. 1973; Alang 1982). The structure of the haustorium is adapted to the absorption and transport of nutrients from the endosperm to the developing embryo. The organ is highly metabolic and contains numerous diverse enzymes (Alang 1982). However, the exact mechanism whereby the haustorium controls the breakdown of food reserves from the endosperm has not yet been determined for any palm species.

Rabechault and Cas (1974) in their work on the in vitro culture of oil palm embryos indicated that the haustorium also has a role in the initiation, growth and development of roots and shoots. Other than this work, very little published information is available on the role of the haustorium during the development of embryos in vitro. We have observed that the haustorium develops normally in culture when the embryos are not subjected to cryopreservation. In the cryopreserved embryos, the haustorium fails to develop and remains rudimentary, but the shoot and root appear to develop normally. This observation leads one to speculate that a developed haustorium may not be necessary for normal development of the embryos in in vitro culture, as the required nutrients are available in the medium and can be taken up by the developing embryo by direct absorption. In the whole seed, however, a functional haustorium is necessary to break down the stored food in the endosperm for use by the developing embryo. Al-Madeni and Tisserat (1986) have shown that whole seeds of date palm (*Phoenix dactylifera* var. Zahidi) desiccated to a moisture content of 7.8% could satisfactorily survive cryopreservation for about 1.5 years with no real loss to viability, implying that the haustorium remains functional in this species even after cryopreservation.

It may be possible that in some palm species, cryopreservation of whole seeds at a low moisture content (4–8%) may result in the embryo initials being viable. However, due to the failure of the haustorium to develop within the seed, embryos may fail to develop. Culture of embryos excised from cryopreserved seeds would shed some light on this speculation.

2.2 *Calamus manan* (Rotan manau)

2.2.1 Materials and Methods

Fully ripe seeds, straw yellow in color, were harvested from the forest in Sungei Buloh, Malaysia. A total of 2000 seeds was obtained from one fruiting palm. The seeds were depericarped and the sarcotesta surrounding the seed was removed, using sawdust. The cleaned seeds were left overnight at ambient temperature to surface-dry. Only excised embryos were used in this study. The seeds were first sterilized using a 20% v/v commercial Clorox solution for 20 min followed by three rinses with sterile distilled water. The embryos were excised from the seeds aseptically.

A total of 1640 seeds were used for this study. Each replicate consisted of 20 excised embryos. The desiccation periods selected were 0–8 h, respectively. The embryos were excised as described above and then dried in batches of 20. Nine replicates for each desiccation period were taken. All drying was carried out by placing the isolated embryos on square sheets of sterile aluminum foil (6 × 6 cm) in the sterile air flow of a laminar flow cabinet. At the end of each drying desiccation period, three replicates were taken to determine the moisture content present in the embryos, three replicates were cultured directly in vitro and the other three replicates were placed in sterile polypropylene cryovials, mounted on cryocanes and plunged directly into LN. Moisture content in the embryos were determined using the oven drying method (i.e., 103 °C for 16h). The moisture content was expressed on a wet weight basis. Embryos placed in liquid nitrogen were stored for a period of 1 month before they were tested for viability. When removed from the cryotank the embryos were thawed in a water bath (35–40 °C) and then cultured in vitro without washing.

The culture medium used was a Murashige and Skoog's (1962) inorganic salt formula supplemented with 170 mg/l NaH_2PO_4, 30 g/l sucrose, 0.1 mg/l NAA, 2 m/l BAP, 2 g/l activated charcoal and solidified with 7 g/l Diffco Bacto Agar. The cultures were maintained at 28 ± 2 °C in a temperature-controlled room with a 12-h photoperiod of about 3000–4000 lx intensity provided by fluorescent light. Embryos were recorded as viable if they swelled up, turned green, and if shoots emerged.

2.2.2 Results and Discussion

Freshly excised embryos of rattan manau seeds had a moisture content of 64.41 ± 0.47% of their fresh weight and a percentage viability of 91.67 ± 2.36%.

Drying the embryos in a laminar flow cabinet resulted in a large water loss after 1h, followed by a continued but less drastic loss. A moisture content of around 5% was attainable at around 7h of drying (Table 4). Reduction in water content of the excised embryos was accompanied by a gradual reduction of viability in vitro (Table 4). A drastic loss of viability was observed when moisture content in the embryos fell below 5% (i.e., 8h of drying).

Results of the cryopreservation studies of the embryos are shown in Table 5. After 1 month in LN the stored embryos were tested for survival growth (swelling) and normal development (Table 5). Embryos with more than 10% moisture content (0–4h of drying) did not survive. This is most likely due to damage from freezing of free water present in the embryos. Embryos with 7–8%

Table 4. Moisture content and viability of excised embryos of rotan manau during desiccation. (Krishnapillay et al. 1992)

Drying (h)	Moisture content (%)	Viability (%) in vitro
0	64.409 ± 0.47	91.667 ± 2.36
1	36.531 ± 1.45	90.000 ± 4.08
2	26.531 ± 1.37	88.333 ± 2.36
3	18.278 ± 0.77	86.667 ± 2.36
4	9.817 ± 0.01	86.667 ± 4.71
5	8.326 ± 0.17	83.333 ± 4.71
6	7.276 ± 0.20	76.667 ± 4.71
7	5.421 ± 0.15	53.333 ± 4.71
8	4.425 ± 0.17	13.333 ± 4.71

Table 5. Survival percentage of excised embryos of rotan manau after desiccation and cryopreservation. (Krishnapillay et al. 1992)

Drying (h)	Survival (%) (total)	Swelling of embryos (%)	Normal development (%)
0	0.00 ± 0.0	0.00 ± 0.0	0.00 ± 0.0
1	0.00 ± 0.0	0.00 ± 0.0	0.00 ± 0.0
2	0.00 ± 0.0	0.00 ± 0.0	0.00 ± 0.0
3	0.00 ± 0.0	0.00 ± 0.0	0.00 ± 0.0
4	1.67 ± 2.4	1.67 ± 2.4	0.00 ± 0.0
5	53.33 ± 2.4	43.33 ± 2.4	10.00 ± 0.0
6	63.33 ± 2.4	36.67 ± 2.4	26.67 ± 2.4
7	65.00 ± 4.1	60.00 ± 0.0	5.00 ± 4.1
8	41.67 ± 2.4	41.67 ± 2.4	0.00 ± 0.0

moisture content showed a significant recovery to produce plantlets; 7% moisture content seemed to give optimal results of 27% plantlet formation. Those with moisture content of around 5% (7 h drying) showed good recovery (65%) but only 5% developing plantlets, the bulk of 60% developing into abnormal swelling with no defined growth. This suggests that there is some damage to the embryos in the final stage of dehydration from between 7 to 5% moisture content.

The study indicated that the excised embryos of rotan manau are markedly resistant to desiccation. Drying to a moisture content of 7–8% was suitable for cryopreservation where embryos then survived to be revived satisfactorily in culture. However, the percentage of the recovery embryos that formed normal plantlets was low (i.e., only 40% of those embryos that survived at the 6 h of drying). Grout et al. (1983) and Chin et al. (1989b) carried out similar studies on excised embryos of oil palm (*Elaeis guineensis*) and coconut (*Cocos nucifera*) respectively and they reported successful recovery of desiccated embryos after storage in liquid nitrogen (also see Chap. II.7, this volume).

3 Summary and Conclusions

In all the three palm species studied, i.e., *Veitchia merrilli*, *Howea fosteriana* and *Calamus manan* (Figs. 3, 4) the excised embryos are markedly resistant to desiccation. Excised embryos could be safely desiccated to moisture content below 10% and stored in liquid nitrogen without further treatment.

These results also indicate that while whole seeds of these palm species show recalcitrant behavior, the excised embryos exhibited orthodox characteristics where they could be desiccated to fairly low moisture contents and cryopreserved. Hence, controlled desiccation or cryoprotectant treatment prior to cryopreservation may provide a strategy applicable to the conservation of palm species that are otherwise short-lived and cannot be conserved for long periods using conventional methods of storage.

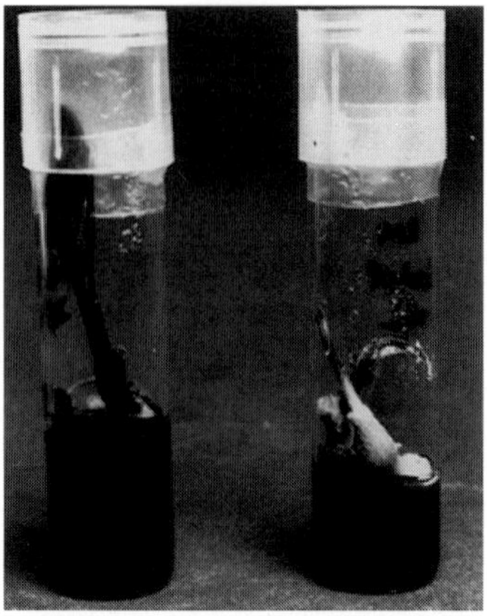

Fig. 3. *C. manan* embryos control *at* 9 weeks in culture

Fig. 4. *C. manan* embryos cryopreserved *at* 6 weeks in culture

References

Alang ZC (1982) Some aspects of the physiology and biochemistry of germination in oil palm (*Elaeis guineensis*). PhD Thesis, CNAA, London

Al-Madeni MA, Tisserat B (1986) Survival of palm seeds under cryogenic conditions. Seed Sci Technol 14: 79–85

Bajaj YPS (1984) Induction of growth in frozen embryos of coconut and ovules of citrus. Curr Sci 53: 1215–1216

Bajaj YPS (1985) Cryopreservation of embryos. In: Kartha KK (ed) Cryopreservation of plant cells and organs. CRC Press, Boca Raton, Florida, pp 227–242

Balasubramaniam K, Atukorala TMS, Wijesundera S, Hoover AA, De Silva MAT (1973) Biochemical changes during germination of the coconut (*Cocos nucifera*). Ann Bot 37: 439–445

Chin HF (1988) Recalcitrant seeds – a status report. Int Board Plant Genet Resour (IBPGR), Rome, 28 pp

Chin HF, Enoch IC (1988) Malaysian trees in colour. Tropical Press Sdn Bhd, Kuala Lumpur, Malaysia, 180 pp

Chin HF, Krishnapillay B, Stanwood PC (1987) Desiccation sensitivity in recalcitrant seeds. Hortic Sci 22(5): 1150

Chin HF, Krishnapillay B, Alang ZC (1988) Cryopreservation of *Veitchia* and *Howea* palm embryos: non-development of the haustorium. CryoLetters 9: 372–379

Chin HF, Krishnapillay B, Stanwood PC (1989a) Seed moisture: recalcitrant v.s. orthodox seeds. In: Stanwood PC, McDonald MB (eds) Seed moisture. CSSA Spec Publ No 14, Madison, Wisconsin, USA, pp 15–22

Chin HF, Krishnapillay B, Hor YL (1989b) A note on the cryopreservation of embryos from young coconuts (*Cocos nucifera* var. Mawa). Pertanika 12(2): 183–186

Ellis RH, Hong TD, Roberts EH, Soetisna U (1991) Seed storage behaviour in *Elaeis guineensis*. Seed Sci Res 1: 99–104

Gupply HB (1912) Studies in seeds and fruits. William and Norgate, London

Grout BWW, Shelton K, Pritchard HW (1983) Orthodox behaviour of oil palm seed and cryopreservation of the excised embryo for genetic conservation. Ann Bot 52: 381–384

Keusch L (1968) Die Mobilisierung des Reservemannans im keimenden Dattelsamen. Planta 78: 321–350

Krishnapillay B, Marzalina M, Aziah MY (1992) Cryopreservation of excised embryos of rattan manau. Proc 3rd Nat Biotechnol Res Confer National Univ Malaysia, pp 325–330

Mori T, Zolfatah AR, Tan CH (1980) Germination and storage of rattan manau (*Calamus manan*) Seeds. Mal For 43(1): 44–45

Murashige T, Skoog F (1962) A revised medium for rapid growth and bioassays with tobacco tissue culture. Physiol Plant 15: 473–497

Nagarajan M, Pandalai KM (1963) Studies on the enzyme activity in the haustorium of the germinating coconut. Indian Coconut J 17(1): 25–34

Pickard J (1983) Vegetation of Lord Howe Island. Cunninghamia 1(2): 1–283

Rao IVR, Aziah MY, Rao AN, Sastry CB (1991) Propagation of bamboo and rattan through tissue culture. IDRC Bamboo and Rattan Res Network Publ, 60pp

Rebechault H, Cas S (1974) Recherches sur la culture in vitro des embryons de palmier à huile (*Elaeis guineensis* Jacq. var. *dura* Becc.). Oleagineux 29: 73–78

Roberts EH, King MW (1980) The characteristics of recalcitrant seeds. In: Chin HF, Roberts EH (eds) Recalcitrant crop seeds. Tropical Press Sdn Bhd, Kuala Lumpur, pp 1–3

Uhl NW, Dransfield J (1987) Genera Palmarum: a classification of palms based on the work of HE Moore Jr., The LH Bailey Hortorium and International Palm Society. Allen Press, Lawrence, Kansas

II.9 Cryopreservation of Germplasm of Rubber (*Hevea brasiliensis*)

M.N. NORMAH[1] and H.F. CHIN[2]

1 Introduction

The genus *Hevea* (family Euphorbiaceae) consists of nine species, namely *H. brasiliensis*, *H. benthamiana*, *H. camporum*, *H. guianensis*, *H. microphylla*, *H. nitida*, *H. pauciflora*, *H. rigidifolia*, and *H. spruceana* (Baldwin 1947). Of these, *H. brasiliensis* (Wild.) Muell.-Arg. is the only commercial species and the primary source of natural rubber. Three other species, *H. benthamiana*, *H. pauciflora*, and *H. spruceana* are used as sources of resistance to South American Leaf Blight (Ho 1979).

Hevea originated from the tropical rainforests of South America within broad regions of the Amazon basin. *H. brasiliensis*, however, occupies most of the area south of the Amazon River (Ho 1979). *Hevea brasiliensis* was introduced in Malaysia in 1877, from the original 22 Wickham seedings received in Singapore within the same year. The gene pool is therefore a narrow one; this has further narrowed down considerably through directional selection. Infusion of new germplasm is necessary for continued progress. With this aim, newly collected and existing germplasm needs to be conserved. *Hevea* germplasm is now conserved in field collections and seeds are stored for short-term storage periods.

2 Storage of Germplasm

The storage of *Hevea* seeds to prolong seed viability is important because seeds are used for the production of seedling root stocks, and as planting materials. Maintenance of seed viability is also important when seedfall does not coincide with scheduled nursery and field planting, when seeds must be transported over long distances from collection site to their destination during germplasm collection, or as in breeding programs where seedling progenies should be planted

[1]Department of Botany, Faculty of Life Sciences, Universiti Kebangsaan Malaysia, 43600 Bangi, Selangor, Malaysia
[2]Department of Agronomy and Horticulture, Faculty of Agriculture, Universiti Pertanian Malaysia, 43400 Serdang, Selangor, Malaysia

Biotechnology in Agriculture and Forestry, Vol. 32
Cryopreservation of Plant Germplasm I (ed. by Y.P.S. Bajaj)

simultaneously. Further, while vegetative propagation is possible and is the accepted practice, it is essential to maintain a viable nucleus gene stock for maximum genetic diversity for future breeding programs.

Hevea seeds are classified as recalcitrant seeds (Roberts 1973), and are normally killed if the moisture content is reduced below a relatively high critical value of 12–31% (Roberts 1973). Even when they are maintained under moist conditions, their longevity is relatively short, typically varying from a few weeks to a few months (King and Roberts 1980). Like many other recalcitrant seeds, *Hevea* seeds are also killed by freezing temperature (Chin et al. 1981).

As recalcitrant seeds are desiccation-sensitive, moist or imbibed storage has been used for rubber. The Rubber Research Institute of Malaya (1966) reported that the best results of storage at ambient temperature were obtained when the seed was mixed with an equal quantity of sawdust (10% moisture content) and spread in shallow layers inside perforated polythene bags. Storage of *Hevea* seeds mixed with moist sawdust in perforated black polythene bags at 7 °C can maintain viability for up to 4 months (Ang 1976). One of the setbacks to moist or imbibed storage of recalcitrant seeds is germination during storage, which often takes place. If loss of such seeds through germination is to be prevented, they must be maintained in a dormant state or they may be stored at lower than optimal temperature for germination (King and Roberts 1980). However, natural dormancy is absent in most *Hevea* seeds except in clones BR.2 and AV.163 ill, which are suspected of exhibiting a certain degree of hard-seededness (Keleny and Van Haaren 1976), and low temperature storage is not suitable for *Hevea* seeds (Normah 1987). An alternative method to prevent germination during storage is by incorporating germination inhibitors into the storage system. Among the artificial germination inhibitors polyethylene glycol (PEG) was found to be the most promising (King and Roberts 1979), as it also does not support mold growth. Treatment with PEG 1500 improved the storage life of *Hevea* seeds by up to 6 months with 25% germination (Sakhibun 1981). The authors' work has shown that *Hevea* seed storage has been prolonged to 12 months with 25% germination when seeds were stored imbibed and treated with 10% PEG 6000 (Normah and Chin 1989a). By observation, 10% PEG, did reduce the occurrence of germination on storage.

Partial drying technique is another method that has been fairly successful in prolonging the storage life of the rubber seeds to 1 year (Normah and Chin 1989b). Using this method, rubber seeds were soaked in 0.3% Benlate, and then drained and surface-dried in air. Seeds were stored in perforated polythene bags at ambient temperatures. Over 50% were viable after a 1-year storage. From this study, it appeared that it is not really necessary to store *Hevea* seeds imbibed, as the dry method of storage also could prolong the storability to the same or longer duration. Furthermore, there was no germination during storage. Ong and Lauw (1963) stored *Hevea* seeds in plastic bags with one or two small holes and maintained viability for up to 4 months at room temperature. These results agreed with those reported by Pereira (1980), who obtained high viability up to six months when *Hevea* seeds were stored at ambient temperature around 27 °C. He suggested that polythene bags were good storage containers as they kept an evenly high moisture. Having treated the seeds with Benlate also significantly reduced mold growth, and thus reduced deterioration of the seeds.

A great deal of interest has been expressed in recent years concerning the application of in vitro techniques to germplasm collection, conservation, and multiplication of species with recalcitrant seeds and those that are vegetatively propagated or that do not produce seed at all. In vitro storage methods are different, depending on the storage duration required. For short- and medium-term storage, the aim is to reduce the growth and increase the intervals between subcultures. This can be achieved by modifying the culture medium, lowering the culture temperature, modifying the gaseous environment in culture vessels, encapsulating of somatic embryos in alginate, and desiccating plant materials such as calli and somatic embryos. However, for long-term storage of germplasm, cryopreservation is employed.

3 Cryopreservation of Embryonic Axes

There was no earlier work on cryopreservation of *Hevea* being reported before the authors' publication (Normah et al. 1986). Nevertheless, it has been reported that materials dehydrated by drying in an oven (Sun 1958) or under vacuum (Withers 1979) exhibit remarkable resistance to cryogenic damage. Grout et al. (1983) reported that embryos of oil palm seeds, which were formerly classified as recalcitrant, can be successfully cryopreserved after desiccation. Bajaj (1985) suggested that in such cases where the seeds are short-lived, germplasm could possibly be conserved through the cryopreservation of excised embryos or their segments. Bajaj (1984) observed that partially desiccated, immature coconut embryos and their segments subjected to sudden freezing showed a lag period of 4 months before continuing growth in cultures. Chin et al. (1988) cryopreserved desiccated embryos of *Veitchia* and *Howea* palm, and observed that the haustorium of the cryopreserved embryos failed to develop. More recently, Assy-Bah and Engelmann (1992a,b) reported successful cryopreservation of coconut immature and mature embryos. For their work, pretreatment consisted of desiccation and preculture in high glucose (600 g/l) and glycerol (15%) medium for mature embryos and pregrowth on semisolid medium containing 600 g/l glucose and 10 or 15% glycerol or sorbitol for immature embryos. Marin et al. (1990) reported successful cryopreservation of cassava embryonic axes by slow cooling (2 °C/min) and thawing (−70 °C for 1 h and −15 °C before reaching room temperature). Thus, in our study, attempts were made to cryopreserve *Hevea* embryonic axes by manipulating the moisture content, concentration of cryoprotectants, and the cooling and thawing rates. Another vital prerequisite to in vitro storage of embryonic axes is the elucidation of a reliable method for normal development of embryonic axes into seedlings under in vitro condition and for the acclimatization and planting out of in vitro seedlings in the glasshouse and later in the field.

3.1 Embryo Culture of *Hevea*

Embryo culture of *Hevea* was first studied by Muzik (1956). It was found that mature embryos failed to develop into seedlings when the cotyledons were completely removed. At least one-third of a cotyledon was essential for the embryo to develop into a healthy plant while those embryos with two-thirds of a cotyledon attached developed into complete seedlings as rapidly as the embryos with full cotyledons. Paranjothy and Ghandimathi (1975), however, reported the ability of embryonic axes (the decotyledonized embryos) to develop into complete seedlings when cultured in MS liquid medium (Murashige and Skoog 1962). The growth of the seedlings was, however, relatively poor. Chen (1984) observed that it was more difficult for the embryos without cotyledons to germinate than those with a small part of cotyledon attached.

In the authors' study, the culture procedure was as follows: The seeds of *Hevea brasiliensis* (Clone PB260 obtained from Prang Besar Research Station in Prang Besar, Malaysia), were surface sterilized in 20–30% Chlorox solution containing a few drops of Teepol for 30 min, and rinsed with sterile distilled water. The seeds were then aseptically cracked open, using a hammer which had been swabbed with ethanol. The excision of the embryonic axes was done by removing the endosperm and cotyledons completely under aseptic conditions. The axes were transferred temporarily to sterile Petri dishes until all axes were excised and ready for use. They were then cultured individually on MS nutrient agar medium supplemented with 4 g/l activated charcoal and on basal MS medium as the control. Cultures were incubated in light (3300 lx) under a 12-h photoperiod.

It was observed that embryonic axes (with no cotyledon attached) swelled and expanded after 2 days and formed seedlings within 7 to 10 days when cultured on basal MS medium supplemented with 4 g/l activated charcoal. On the other hand, on basal medium without charcoal, they swelled but expanded less, and remained green without any further development even after 1 month of culture (Normah et al. 1988). Charcoal was therefore used to supplement the medium for all further experiments.

3.2 Methodology for Cryopreservation

Since *Hevea* seeds are known to be sensitive to desiccation, in the initial studies, no attempt was made to desiccate the embryonic axes before cryopreservation. Instead, the application of cryoprotectants was considered first. Embryonic axes were excised under aseptic conditions. In the first experiment, the cryoprotectants applied were 5% dimethylsulfoxide (DMSO), 5% glycerol, 5% sucrose, 5% DMSO + 5% glycerol, 5% DMSO + 5% sucrose, 5% sucrose + 5% glycerol, and 5% DMSO + 5% glycerol + 5% sucrose. After the excised embryonic axes were placed in cryovials (five axes to one vial), 1 ml of sterile liquid basal MS medium was added to the axes. One ml of chilled cryoprotectants (prepared at twice the final concentration in culture medium and sterilized by filtration) was added under chilled conditions into each cryovial by slow addition over 40 min. The embryonic axes with the cryoprotectants were left for 40 min in ice before beginning the cryopreservation procedure. For the control, no cryoprotectant was applied. The cooling rates employed were step-wise cooling (5 °C for 30 min, −15 °C for 30 min, −30 °C for 30 min, then in liquid nitrogen phase for 30 min, and finally in liquid nitrogen) and direct immersion into LN.

The embryonic axes were stored in LN for 16 h, after which thawing was carried out either by the slow thawing or by the rapid thawing in warm water bath. Slow thawing was carried out by suspending the cryovials in the vapor phase of LN for 10 min, followed by 15 min at room temperature. Rapid thawing was carried out by immersing the cryovials directly in a water bath at 37 ± 2 °C for 3–5 min.

Each embryonic axis was then cultured individually on a basal MS medium supplemented with 4 g/l activated charcoal. Cultures were incubated in light (3300 lx) under a 12-h photoperiod. The axes were recorded as viable when they expanded and turned green.

For the second experiment, the same cryoprotectants as described above were employed. However, after the application of cryoprotectants, the axes were desiccated in a laminar flow hood for 2 h. After desiccation, the axes were enclosed in tightly folded, sterile aluminum foil envelopes, cooled, stored, and thawed as described above. After thawing, each individual embryonic axis was cultured and kept in conditions as above.

From the second experiment, it was found that the desiccated axes without cryoprotectants were viable after storage in LN even though no normal seedling was observed. Thus, the desiccation of *Hevea* embryonic axes was investigated further.

In the third experiment, embryonic axes were excised and desiccated for 0, 1, 2, 3, 4, or 5 h as described earlier. Four replicates of ten embryonic axes were used for moisture content determination. Another three replicates of ten axes of each treatment were enclosed in tightly folded sterile aluminum foil envelopes, cooled, stored, and thawed as in the second experiment. After thawing, each individual embryonic axis was then cultured on a modified MS medium with a double dose of microelements and supplemented with 0.6–0.7 μM kinetin, 1 μM naphthaleneacetic acid, 1.4 μM gibberellic acid, and 4 g/l activated charcoal. Cultures were incubated as in the first two experiments. Observations were made on the growth of the embryonic axes. The axes were recorded as viable when they expanded and turned green, and as having survived cryopreservation when a fully developed seedling (with normal shoot and root) was obtained. A similar number of embryonic axes were cultured immediately following each desiccation time.

To determine the moisture content after desiccation, the axes were weighed and then dried in an oven at 105 °C for 16 h. They were then reweighed after cooling in desiccators. Moisture content was expressed as percentage of fresh weight.

3.3 Results

In the first experiment, none of the cryoprotectant treatments used enabled embryonic axes to survive storage in LN. In the next experiment no complete seedling was obtained. The results were recorded as viability of the axes (axes swollen and green, or axes with roots). The percentage viability of axes after cryopreservation are shown in Table 1. The embryonic axes which had been desiccated but not cryoprotected showed 25–50% viability after cryopreservation depending on the cooling and thawing procedure. Thus, desiccation of the axes without cryoprotectants was investigated further in the third experiment. Part of this experiment was reported in Normah et al. (1986).

The results show that freshly excised embryonic axes had a moisture content of approximately 55%. The moisture content of axes decreased rapidly within the first 2 h of desiccation under the condition employed, but the decrease was more gradual thereafter. Embryonic axes with no desiccation showed 100% viability and survival when cultured in vitro. However, with increasing duration of desiccation (1–5 h), there was a decrease in viability down to 50% at 5 h.

Following cryopreservation, none of the axes with moisture content of 55 and 35% (0 and 1 h desiccation) were viable (Table 2). The highest percentage viability (97–99%) was shown by axes that were desiccated for 2, 3, and 4 h (moisture content 15–19%), cooled by the stepwise method, and thawed rapidly.

Percentage survival of *Hevea* embryonic axes after cryopreservation is shown in Table 3.

The highest percentage survival (69–71%) was obtained when embryonic axes were desiccated for 2 and 3 h, cooled by the stepwise method or by direct immersion respectively, and thawed rapidly. Embryonic axes with desiccation of 5 h (moisture content 13.69%), cooled by the stepwise method and thawed slowly also showed a high percentage survival (57%).

Table 1. Percentage viability after cryopreservation of cryoprotected and desiccated *Hevea* embryonic axes. (Normah 1987)

	Cooling			
	Stepwise		Direct immersion	
	Thawing		*Thawing*	
Cryoprotectants (%)	*Slow*	*Rapid*	*Slow*	*Rapid*
DMSO 5	30	53	70	24
DMSO 5 + sucrose 5	64	90	69	50
DMSO 5 + glycerol 5	46	83	40	30
Sucrose 5	48	90	13	20
Glycerol 5	20	75	10	30
Sucrose 5 + glycerol 5	25	51	11	9
Sucrose 5 + DMSO 5 + glycerol 5	19	51	10	32
No cryoprotectants	50	25	39	39

Table 2. Percentage viability of *Hevea* embryonic axes after desiccation and cryopreservation. (Normah 1987)

		Cooling			
		Stepwise		Direct immersion	
Desiccation (h)	Moisture content (%)	*Thawing* *Slow*	*Rapid*	*Thawing* *Slow*	*Rapid*
0	54.51	0^{a}	0^{a}	0^{a}	0^{a}
1	35.21	0^{a}	0^{a}	0^{a}	0^{a}
2	19.17	64^{cd}	97^{fg}	23^{b}	50^{c}
3	16.23	85^{e}	99^{g}	65^{cd}	87^{ef}
4	15.28	78^{de}	97^{fg}	65^{cd}	80^{de}
5	13.69	86^{e}	73^{de}	30^{b}	80^{de}

Values having the same superscript are not significantly different at $p = 0.05$ based on Duncan Multiple Range Test.

Table 3. Percentage survival of *Hevea* embryonic axes after desiccation and cryopreservation. (Normah 1987)

	Cooling			
	Stepwise		Direct immersion	
Desiccation (h)	*Thawing* *Slow*	*Rapid*	*Thawing* *Slow*	*Rapid*
0	0^{a}	0^{a}	0^{a}	0^{a}
1	0^{a}	0^{a}	0^{a}	0^{a}
2	50^{ef}	71^{h}	5^{b}	20^{c}
3	29^{cd}	47^{ef}	44^{def}	69^{gh}
4	52^{fg}	45^{def}	52^{f}	25^{c}
5	57^{fgh}	30^{cd}	20^{c}	32^{cde}

Values having the same superscript are not significantly different at $p = 0.05$ based on Duncan Multiple Range Test.

Treatments which resulted in high percentage viability of axes (Table 2) did not necessarily result in high percentage survival (Table 3). Many of the axes which were viable developed abnormally, thus resulting in low percentage survival.

Figures 1A, B, and C show the development of the embryonic axes after cryopreservation. Viable axes swelled and turned green within 7–10 days of culture. Shoot and root emergence was observed between 2 and 3 weeks in culture. The first trifoliate leaves formed within 3–4 weeks. The hypocotyl of the control seedling elongated and swelled more than the hypocotyl from the cryopreserved axis. Compared to the control, the development of the root and shoot of the cryopreserved axis was slower by 2 weeks. However, after the first trifoliate leaves formed, both types of seedlings were very similar in appearance.

Fig. 1A–C. Growth of cryopreserved *Hevea* embryonic axis. **A** After 1 week of culture. **B** After 3 weeks of culture. **C** After 5 weeks of culture. (Normah 1987)

Thus, the cryopreserved embryonic axes took a longer time to develop root and shoot, but once these had developed, normal growth continued.

3.4 Discussion

The non-desiccated *Hevea* embryonic axes could not survive storage in LN even when they were treated with cryoprotectants and subjected to various cooling and thawing rates. Desiccation after cryoprotectant treatments gave promising result even though no normal seedling was observed. Five percent glycerol, 5% sucrose, and a combination of 5% DMSO + 5% glycerol and 5% DMSO + 5% sucrose were found to be the best in this investigation.

It has been reported that zygotic embryos of citrus (Mumford and Grout 1978) and oil palm (Grout et al. 1983) withstood freezing after being subjected to partial desiccation. Desiccation followed by rapid freezing has been a successful technique for zygotic embryos of *Musa* (Abdelnour-Esquivel et al. 1992a), *Coffea arabica* and *Coffea canephora* (Abdelnour-Esquivel et al. 1992b), and *C. liberica* (Normah and Vengadasalam 1992). Pence (1991) suggested that cryostorage should be possible for a number of difficult-to-store temperate tree seeds and later successfully cryopreserved immature zygotic embryos of *Theobroma cacao* L. (Pence 1991). The authors' study is believed to be the first to demonstrate the potential of desiccation and cryopreservation of *Hevea* embryonic axes (Normah et al. 1986). The study suggests the existence of critical optimum moisture content of the axes in order to survive cryopreservation.

Though the *Hevea* embryonic axes used in this study were desiccated, moisture content of the axes was still relatively high (14–19%) compared to that used in the cryopreservation of orthodox seeds such as *Vigna* (6.7–10.5%) (Normah and Vengadasalam 1992) and *Musa* (9–14%) (Abdelnour-Esquivel et al. 1992a). The critical moisture content of rubber embryonic axis for cryopreservation is quite similar to those of jack fruit (14–15%) (Krishnapillay 1989) and *Coffea liberica* (15–20%) (Normah and Vengadasalam 1992).

For *Hevea* embryonic axes at about 19% moisture content, the best cooling was by the stepwise method and thawing at a rapid rate. It seems that rapid thawing is most suitable for plant cells which have been frozen slowly. At a moisture content of 16%, direct immersion into LN and rapid thawing resulted in the best survival. At lower moisture content, stepwise cooling and slow thawing appeared to enhance survival, and this agrees with results on cassava embryos (Marin et al. 1990).

In a study by Normah (1987) it was also shown that adding charcoal to the medium improved the growth and development of *Hevea* embryonic axes. So far, only Toraun and Saryatmana (1977) have obtained seedlings from *Hevea* embryonic axes cultured on basal MS medium. Their success could not be repeated in other studies, which found that it was more difficult for the embryos without cotyledons to grow than those with a small number of cotyledons attached (Chen 1984). However, in the present study, the embryonic axes that were cultured on MS medium supplemented with charcoal grew and developed into normal seedlings within 7 to 10 days, while the axes on the medium without charcoal remained green without any further development. Fridborg et al. (1978) suggested that the effects of activated charcoal are due to the removal of substances from the medium which promote unorganized growth and inhibit embryogenesis, root formation, and elongation.

4 Summary and Conclusions

It is possible to lower the moisture content of *Hevea* embryonic axes (classified as recalcitrant seeds) without deleterious effect. Axes desiccated to 14–19% moisture content survived storage in liquid nitrogen (−196 °C). The precise cooling and thawing method, however, appeared to vary with specific moisture content. Nevertheless, direct immersion or stepwise cooling followed by rapid thawing in a warm water bath (37 ± 2 °C) appeared to be suitable regimes. No cryoprotectant was necessary. The growth and development of cryopreserved axes in culture appeared normal even though the growth was somewhat slower at the beginning of culture.

References

Abdelnour-Esquivel A, Mora A, Villalabos V (1992a) Cryopreservation of zygotic embryos of *Musa acuminata* (AA) and *M. balbisiana* (BB). Cryo Lett 13: 159–164

Abdelnour-Esquivel A, Villalabos V, Engelmann F (1992b) Cryopreservation of zygotic embryos of *Coffea* spp. Cryo Lett 13: 297–302

Ang BB (1976) Problems of rubber seed storage. In: Chin HF, Enoch IC, Raja Harun RM (eds) Seed technology in the tropics. Univ Pertanian, Malaysia, pp 117–122

Assy-Bah B, Engelmann F (1992a) Cryopreservation of immature embryos of coconut (*Cocos nucifera* L.). Cryo Lett 13: 67–74

Assy-Bah B, Engelmann F (1992b) Cryopreservation of mature embryos of coconut (*Cocos nucifera* L.) and subsequent regeneration of plantlets. Cryo Lett 13: 117–126

Bajaj YPS (1984) Induction of growth in frozen embryos of coconut and ovules of citrus. Curr Sci 53: 1215–1216

Bajaj YPS (1985) Cryopreservation of embryos. In: Kartha KK (ed) Cryopreservation of plant cells and organs. CRC Press, Boca Raton, pp 227–242

Baldwin JR JT (1947) *Hevea*: a first interpretation. J Hered 38: 54

Chen Z (1984) Rubber (*Hevea*). In: Sharp WR, Evans DA, Ammirato PV, Yamada Y (eds) Handbook of plant cell culture, vol 2 Crop species. Macmillan, New York, pp 546–571

Chin HF, Maheran A, Ang BB, Samsidar H (1981) The effect of moisture and temperature on the ultrastructure and viability of seeds of *Hevea brasiliensis*. Seed Sci Technol 9: 411–422

Chin HF, Krishnapillay B, Alang ZC (1988) Cryopreservation of *Veitchia* and *Howea* palm embryos. Cryo Lett 9: 372–379

Fridborg G, Pederson M, Landstrom L, Eriksson T (1978) The effect of activated charcoal on tissue cultures: absorption of metabolites inhibiting morphogenesis. Physiol Plant 43: 104–106

Grout BWW, Shelton K, Pritchard HW (1983) Orthodox behavior of oil palm seed and cryopreservation of the excised embryo for genetic conservation. Ann Bot 52: 381–384

Ho CY (1979) Conservation and utilization of *Hevea* germplasm. Malays. Appl Biol 8(1): 1–14

Keleny GP, Van Haaren AJH (1976) Progress report on rubber seed investigation. Papua New Guinea Agric J 19(2): 72–87

King MW, Roberts EH (1979) The storage of recalcitrant seeds – achievement and possible approaches. Dep Agric Hortic Univ Reading, UK, pp 1–45

King MW, Roberts EH (1980) Maintenance of recalcitrant seeds in storage. In: Chin HF, Roberts EH (eds) Recalcitrant crop seeds. Tropical Press, Sdn Bhd, Kuala Lumpur, pp 53–89

Krishnapillay B (1989) Towards the development of a protocol for cryopreservation of embryos of a recalcitrant seed (*Artocarpus heterophyllus* Lam.). PhD Thesis, Univ Pertanian Malaysia, Malaysia

Marin ML, Mafla G, Roca WM, Withers LA (1990) Cryopreservation of cassava zygotic embryos and whole seeds in liquid nitrogen. Cryo Lett 11: 257–264

Mumford PM, Grout BWW (1978) Desiccation and low temperature (–196 °C) tolerance of *Citrus limon* seed. Seed Sci Technol 7: 407–410

Murashige T, Skoog F (1962) A revised medium for rapid growth and bioassays with tobacco tissue cultures. Physiol Plant 15: 473–497

Muzik TJ (1956) Studies on the development of the embryo and seed of *Hevea brasiliensis* in culture. Lloydia 19: 86–91

Normah MN (1987) Effects of temperature on rubber (*Hevea brasiliensis* Muell.-Arg.) seed storage. PhD Thesis, Univ Pertanian Malaysia, Malaysia

Normah MN, Chin HF (1989a) Effects of temperature on storage of *Hevea* seeds treated with polyethylene glycol 6000. Malays Appl Biol 18(2): 133–138

Normah MN, Chin HF (1989b) Recalcitrant seed storage by partial desiccation technique. Acta Hortic 253: 258–259

Normah MN, Vengadasalam M (1992) Effects of moisture content on cryopreservation of *Coffea* and *Vigna* seeds and embryos. Cryo Lett 13: 199–208

Normah MN, Chin HF, Hor YL (1986) Desiccation and cryopreservation of embryonic axes of *Hevea brasiliensis* Muell. Arg Pertanika 9(3): 299–303

Normah MN, Chin HF, Hor YL (1988) Conservation of *Hevea* by cryopreserved embryo culture technique. In: Umaly R, Umboh I, Halos S, Normah MN (eds) The application of tissue culture techniques in economically important tropical trees. SEAMEO – BIOTROP, Bogor, BIOTROP Spec Publ 35: 127–131

Ong TP, Lauw IK (1963) Results of storage tests with seeds of *Hevea brasiliensis.* Menara Perkebunan 32: 183–192

Paranjothy K, Ghandimathi H (1975) Morphogenesis in callus cultures of *Hevea brasiliensis* Muell Arg. Proc Natl Plant Tissue Culture Symp, Kuala Lumpur, pp 19–25

Pence VC (1991) Cryopreservation of immature embryos of *Theobroma cacao.* Plant Cell Rep 10: 144–147

Pereira JP (1980) Conservacao da viabilidade do Poder germinativo da semente de seringueira. Pesqui Agropecu Bras Brasilia 15(2): 237–244

Roberts EH (1973) Predicting the storage life of seeds. Seed Sci Technol 1: 499–514

Rubber Research Institute of Malaya (1966) Annu Rep RRIM 1966, p 27

Sakhibun MH (1981) Storage and viability of *Hevea* seeds. M Agric Sci Thesis, Univ Pertanian Malaysia, Malaysia

Sun CN (1958) The survival of excised pea seedlings after drying and freezing in liquid nitrogen. Bot Gaz 119: 234–236

Toraun NL, Saryatmana N (1977) Tissue culture of *Hevea brasiliensis* Muell. Arg Menara Perkebunan 45: 17–21

Withers LA (1979) Freeze preservation of somatic embryos and clonal plantlets of carrot (*Daucus carota* L.). Plant Physiol 63: 460–467

II.10 Cryopreservation of Germplasm of *Larix* and *Picea* Species

P.J. CHAREST and K. KLIMASZEWSKA[1]

1 Introduction

Both *Larix* and *Picea* are widely distributed conifers in temperate climates and are economically valuable to the pulp and paper and lumber industries. These genera include species that are biologically interesting because of their various growth habits and their ability to produce interspecific hybrids (Boyle et al. 1989; Burns and Honkala 1990). The distribution and biology of these species have been dealt with by Burns and Honkala (1990). *Larix* and *Picea* include species that were successfully propagated through organogenesis and somatic embryogenesis (Diner et al. 1986; Klimaszewska 1989; von Aderkas et al. 1990; Attree and Fowke 1991; Bonga and Pond 1991; Tautorus et al. 1991; Bonga and von Aderkas 1992; Lelu et al. 1993). The germplasms of these species have traditionally been stored at 4 °C in the form of seeds. These trees are not threatened with extinction per se, but germplasms from certain provenances need to be preserved.

Because of the intensive use of *Picea* and *Larix* in tree improvement programmes, somatic embryogenesis might become a critical tool to accelerate the tree breeding cycle (Cheliak and Rogers 1990; Charest et al. 1992). Somatic embryogenesis is the preferred method of conifer micropropagation, over organogenesis and rooted cuttings. The problems associated with organogenesis include high somaclonal variation and difficulty in producing large numbers of propagules. With rooted cuttings, the problems are the rooting recalcitrance of older tree branches and their potential plagiotropic growth. In contrast, somatic embryogenesis is amenable to automation and upscaling and there is no evidence yet of somaclonal variation induction (Eastman et al. 1991; Leggo, Rutlegde, and Sutton, unpubl.). The regenerated plants have displayed normal growth. The oldest somatic embryo-derived tree plantation (*Larix* × *eurolepis*) was 7 years old in 1993 and is located at the Petawawa National Forestry Institute (Chalk River, Ontario, Canada). These trees have a normal phenotype when compared to seedling-derived control trees.

Through somatic embryogenesis, an unlimited number of propagules could theoretically be produced. However, so far somatic embryogenesis has been possible only from juvenile tissues of conifers, and testing of the regenerated trees

[1]Molecular Genetics and Tissue Culture Group, Petawawa National Forestry Institute, Chalk River, Ontario, Canada K0J 1J0

Biotechnology in Agriculture and Forestry, Vol. 32
Cryopreservation of Plant Germplasm I (ed. by Y.P.S. Bajaj)

is required before their large-scale use. Consequently, tissue culture lines need to be maintained by regular subculturing, slow growth, or cryopreservation methods until their elite characteristics are proven. For long-term germplasm preservation, cryopreservation is the ideal method that ensures the safety and stability of the tissues preserved. (see Bajaj 1991).

2 Cryopreservation of *Larix* and *Picea* Embryonal Masses

2.1 Review of the Literature

The cryopreservation of *Larix* and *Picea* tissue cultures has been attempted by a number of workers (Table 1); however, much of what has been done relates to the use of embryonal masses for the freezing process (also see Cyr et al. 1994). One study used cotyledons of young seedlings of *Picea glauca* that were cryopreserved successfully, and organogenesis was performed after thawing. With all species and tissues tested, it was possible to recover the tissues after thawing; however, there was no quantification of the success rate. Cryopreservation has also been achieved with embryonal masses of other conifers such as *Pinus taeda* (Gupta et al. 1987) and *Pinus caribaea* (Laine et al. 1992). All the published protocols are very similar and include four steps: (1) choice of tissue to be frozen, (2) pretreatment of the cells with high osmoticum and dimethylsulfoxide (DMSO), (3) gradual freezing to $-40\,^{\circ}C$ and rapid freezing to $-196\,^{\circ}C$, and (4) fast thawing and regrowth of the tissue.

2.2 Freezing Embryonal Masses of *Larix* and *Picea*

2.2.1 Tissue Culture Material

The embryogenic cell lines we used were derived from mature zygotic embryos of *Picea mariana*, *Picea rubens*, and *Picea glauca* (Tremblay 1990; Cheliak and Klimaszewska 1991; Tremblay and Tremblay 1991a, b) and from immature zygotic embryos of *Larix decidua*, *Larix leptolepis*, and their hybrids (*Larix* × *leptoeuropaea* and *Larix* × *eurolepis*) (Klimaszewska 1989; von Aderkas et al. 1990).

Step 1. Choice of material to be frozen
The technique works best with small groups of cells and with actively growing tissues such as cell suspensions. The protocol established at the Petawawa National Forestry Institute is highly effective for *Picea* and *Larix* embryonal masses (Fig. 1) and recovery of all the lines used has been achieved after thawing. Trees regenerated from cryopreserved embryonal cell masses of *Larix* have been growing for 6 years (1993) at the Petawawa National Forestry Institute and are no different from control trees.

Table 1. Cryopreservation of *Picea* and *Larix*

Species	Pre-treatment	Freezing	Regrowth	Reference
Picea abies	0.5 M sucrose + tissue culture medium for 18 h ↓ 0.5 M sucose + tissue culture medium + 0.5% DMSO added stepwise over 30 min at 0 °C ↓ 2 h under the same conditions	0 to −40 °C rate = 0.5 °C/min ↓ −196 °C	Fast thawed at +40 °C ↓ Put one tissue culture medium with daily transfer and sucrose concentration from 0.5 M to 0.09 M over a period of 5 days	Galerne and Dereuddre (1987)
Picea abies	10% PEG + 8% glucose + 10% DMSO added stepwise over 30 min plus another 30 min in final concentrations on ice	0 to −30 °C rate = −1 °C/min ↓ −196 °C	Fast thawed ↓ Directly on tissue culture media	Gupta et al. (1987)
Picea glauca	0.2 M sorbitol + tissue culture medium for 24 h ↓ 0.4 M sorbitol + tissue culture medium for 24 h ↓ 0–5% DMSO added stepwise over 30 min on ice	0 to −35 °C rate = −30 °C/min ↓ −196 °C	Fast thawed at 26 °C ↓ Change of tissue culture medium at 1 and 18 h	Kartha et al. (1988)
Picea glauca Cotyledons from mature embryos	Placed in tissue culture medium + 2% sucrose ↓ 0–0.4 M sorbitol stepwise over 3 days at 4 °C ↓ 5% DMSO added with another 15 min at 0 °C	0 to −40 °C rate = −0.5 °C/min ↓ −196 °C	Not specified	Toïvonen and Kartha (1989)
Picea mariana *Larix × eurolepis*	0.4 M sorbitol + tissue culture medium for 24 h ↓ 0–10% DMSO added stepwise over 30 min at 0 °C with another 30 min in final concentrations at 0 °C	0 to −40 °C rate = −0.30 °C/min ↓ −196 °C	Fast thawed at 37–42 °C ↓ Put on tissue culture medium for 24 h and transferred to fresh one	Klimaszewska et al. (1992)

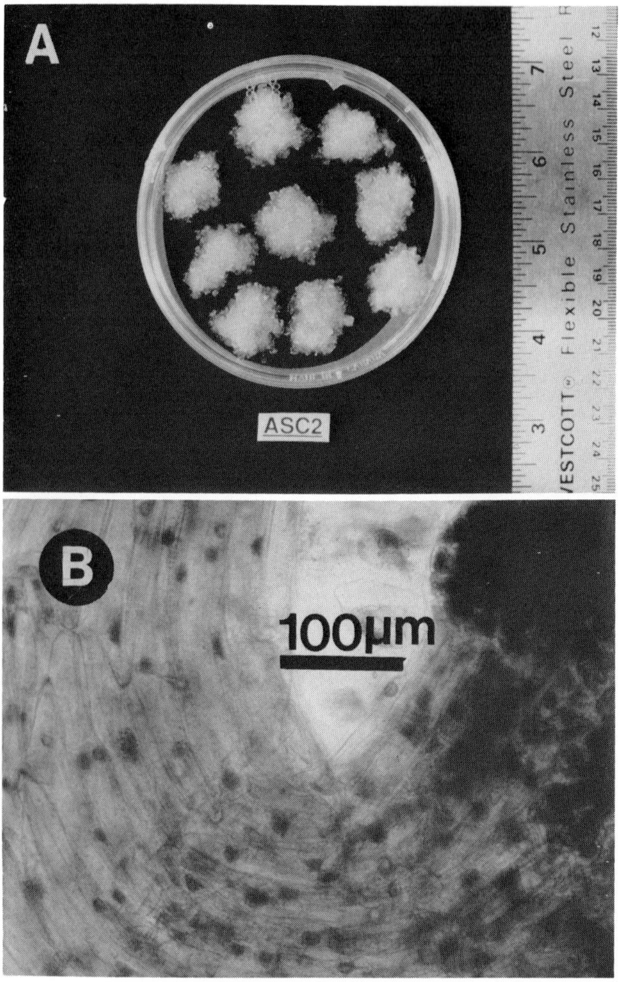

Fig. 1A, B. Embryonal mass (**A**) and embryoids of *Picea glauca* (**B**)

Step 2. Pretreatment of the tissues

1. Embryonal masses grown on solid tissue culture media are resuspended in liquid medium by shaking vigorously or by using a spatula to break up cell clusters.
2. Cells are collected by filtering the suspension through a nylon mesh of 73 µm pore size.
3. Cell aliquots of 1 to 2 g (fresh weight) are transferred to 125-ml Erlenmeyer flasks each containing 7 ml of liquid medium supplemented with 0.4 M sorbitol.
4. The flasks are placed on a rotary shaker at 100–115 rpm for 20–24 h.
5. Flasks are placed on ice and 3 ml of a cold solution of DMSO (33.3%; see

tissue culture media section below) in 0.4 M sorbitol medium are added stepwise over a period of 30 min. The final concentration will be 10% DMSO.
6. The mixture is shaken gently and left on ice for 30 min.

Step 3. Freezing the tissues
7. The flasks are shaken gently and 1-ml aliquots of the suspended cells are dispensed into 1.2-ml cryogenic vials.
8. The vials are placed inside a programmable freezer precooled to 0 °C. The programme used is as follows:
   a) The temperature of the sample is held at 0 °C for 10 min.
   b) The sample is cooled to −40 °C at the rate of 0.33 °C per min.
   c) The temperature is held at −40 °C for 10 min.
9. The vials are removed from the programmable freezer chamber and immediately immersed in liquid nitrogen (−196 °C).

A typical freezing profile is given in Fig. 2, where the temperature of the sample was monitored.

Step 4. Storage of the tissue
The cryopreserved cells may be stored directly in LN or in the vapour phase of LN. The temperature of the gaseous phase will range from −196 °C at the surface of the liquid nitrogen to −140 °C at the top of most freezers equipped for this type of storage. The gaseous phase is preferable to liquid phase because, in

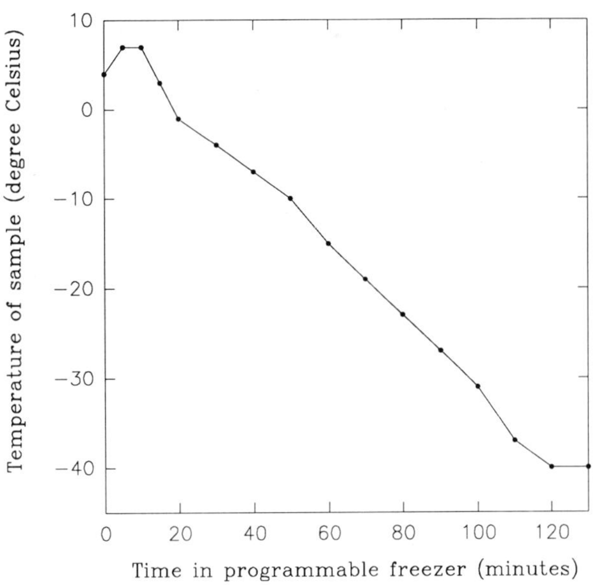

Fig. 2. Freezing profile of embryonal mass in cryoprotectant solution in the programmable freezer. Shown is the sample temperature. Initial temperature +5 °C, final temperature −40 °C

liquid storage, the nitrogen can seep into the cryovials and explode upon thawing (Stanwood 1984). The cryopreserved cells should not be warmed above −130 °C at any time during storage; this will cause the formation of intracellular ice crystals and damage the cells.

Step 5. Thawing and regrowth
1. The vials are removed from the freezer and held in LN.
2. The vials are then placed in a 37 to 40 °C water bath for quick thawing and removed as soon as the ice pellet has melted (about 2 min).
3. They are left at room temperature for 15 min.
4. The contents of the vials are poured onto a sterile filter paper (Whatman No. 2) placed on the surface of a gelled tissue culture medium in a Petri dish.
5. The cells are grown in the dark at 25 °C for 24 h and the filter paper with cells transferred to fresh medium to remove the DMSO and excess liquid.
6. The cells are then subcultured at the regular 2-week intervals and are ready for experimentation.

2.2.2 Tissue Culture Media

The tissue culture medium for *Picea* is half-strength LM (Cheliak and Klimaszewska 1991) and for *Larix*, it is MSG (Klimaszewska 1989). The concentrated DMSO solution is prepared by gradually adding 1 ml filter sterilized DMSO to 2 ml of the tissue culture liquid medium (with 0.4 M sorbitol) on ice. This is added to the 7 ml of cell suspension already in the Erlenmeyer flasks. The solution has to be made fresh and on ice.

2.2.3 Equipment

Two main pieces of equipment are required; a programmable freezer for gradual cooling of the tissues from 0 to −40 °C and a liquid nitrogen freezer for storage. The best programmable freezers are computer-controlled, and allow the maximum flexibility in programming and operating. Ideally, the liquid nitrogen freezer should be equipped with automatic LN supply and an alarm system for LN level. It should have the potential for gazeous or liquid phase storage. A schematic of a typical freezer is given in Fig. 3. Alternatively, electrical freezers that can hold −140 °C are available but they are less reliable.

2.3 Results

To establish an efficient cryopreservation procedure for operational use, the methodology developed must be applicable to a wide range of species, has to be easy to perform, and give consistent results. The protocol presented here meets all these criteria. It has been used successfully with nine species of *Larix* and *Picea* (Table 2) from a wide range of genotypes (326 lines). In addition, the protocol can be used both with embryonal cells in suspension or with embryonal masses

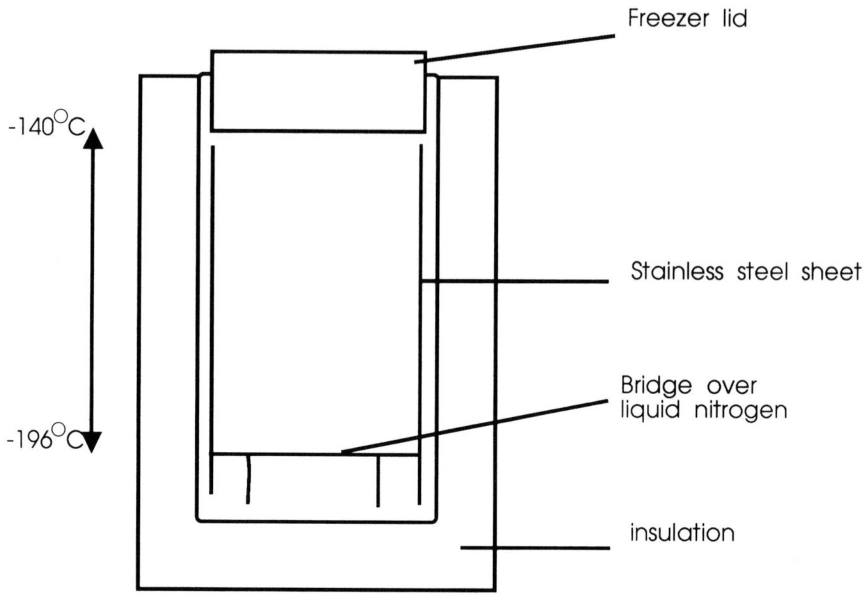

Fig. 3. Schematic of a liquid nitrogen freezer using vapour phase. The temperature ranges from −196 °C in liquid nitrogen to −140 °C at the top of the freezer

Table 2. Cryopreservation of *Larix* and *Picea* embryogenic lines at the Petawawa National Forestry Institute

Species	No. of lines in cryopreservation[a]
Larix decidua	3
L. decidua (haploid)	4
L. leptolepis	3
Larix leptolepis (haploid)	1
Larix × *eurolepis*	9
Larix × *leptoeuropaea*	23
L. occidentalis	3
Picea glauca	7
P. glauca (transgenic)	7
P. glauca engelmannii complex	16
P. mariana	254
P. mariana (transgenic)	1
P. rubens	2
Total of species and lines	9 species and 326 lines

[a]No. of lines cryopreserved successfully as of January 1993.

grown on solid medium. A number of factors have been studied in more detail to optimize the procedure, such as the effect of sorbitol and DMSO concentrations (Klimaszewska et al. 1992), the growth cycle of the cultures, and the temperature of storage. In addition, preliminary data with a slow growth method are presented.

2.3.1 *Effect of Sorbitol and DMSO Concentrations*

It was found that for both *Larix × eurolepis* and *Picea mariana*, the optimal concentrations were 0.4 M sorbitol and 5% DMSO (Klimaszewska et al. 1992; see also Fig. 4A, B). These concentrations were used with the species listed in Table 2.

2.3.2 *Effect of the Tissue Culture Growth Cycle*

To examine the effect of the culture cycle, embryonal cell masses of *Picea glauca*, *Picea mariana*, and *Larix × eurolepis* were subjected to the cryopreservation procedure at different times in their growth cycle. The cycle had a pronounced effect on most of the lines tested (Fig. 5a, b) and, as a general trend, the earlier it was in the growth cycle, the better the rate of regrowth after thawing.

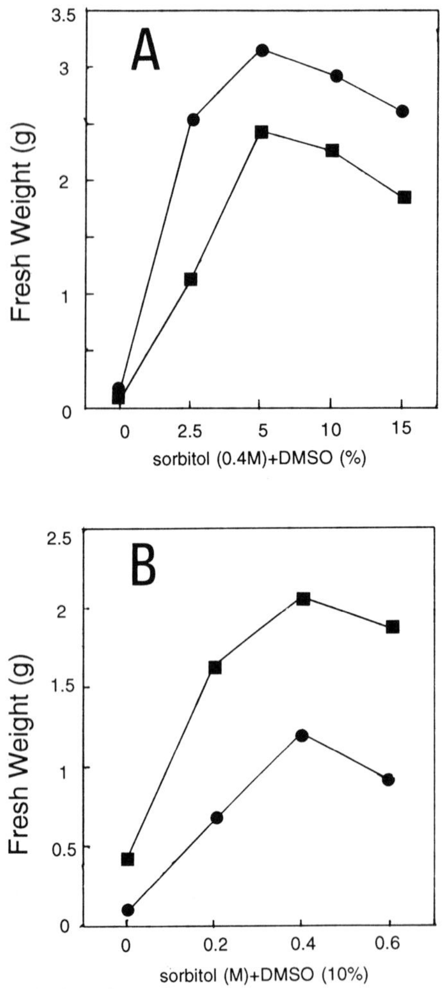

Fig. 4A, B. Effect of sorbitol and DMSO on postfreezing growth of *Larix × eurolepis* (■) and *Picea mariana* (●). **A** After treatment with 0.4 M sorbitol and various concentrations of DMSO. **B** After treatment with 10% DMSO and various concentrations of sorbitol. All curves are the average of three experiments except for *P. mariana* in **A**, where only one experiment is shown. (Klimaszewska et al. 1992, by permission of Oxford University Press)

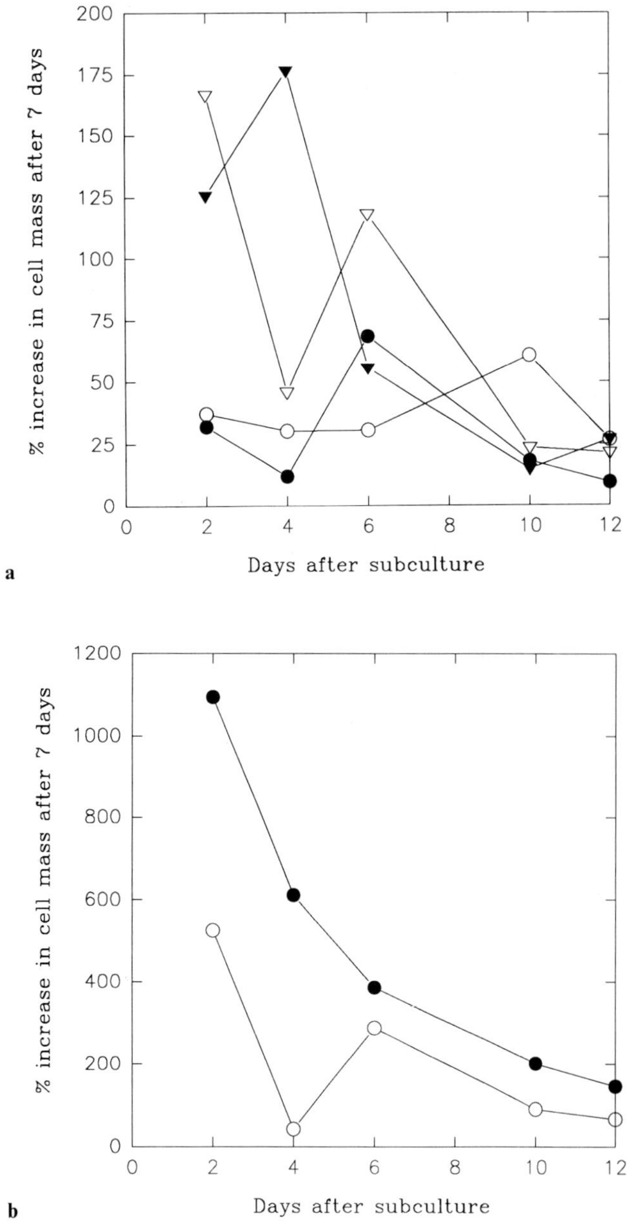

Fig. 5a, b. Effect of time of subculture on post-freezing growth of *Larix* and *Picea*. **a** *Larix* × *eurolepis* lines 1 × d23 (▼), *Picea glauca* line ASC2 (●), and line W14 (▽), and *Picea mariana* line 2 (○). **b** *Larix* × *eurolepis* line LB (●) and *Picea mariana* line R4F14 (○). Average of two experiments

2.3.3 *Effect of Temperature During Storage*

The temperatures tested were −196, −140, −100, and −80 °C. The temperatures below −140 °C had a negative effect on tissue regrowth (Fig. 6). No survival of the tissues after thawing was observed when stored at −80 °C and a reduction of tissue survival was observed when stored at −100 °C even though the storage time was from 2 to 5 days.

2.3.4 *Comparison with Slow-Growth Method*

A slow-growth method was recently developed for embryonal cell masses of *Picea glauca* (Joy et al. 1991) where an 80% survival of the stored cultures was observed after 1 year. This method was tested with *Picea mariana* and *Larix* × *eurolepis* embryonal masses to compare its effectiveness with cryopreservation. However, the method did not result in high frequency survival of the tissues tested. After 14 weeks in storage, no regrowth was observed with *Picea mariana* and, after 22 weeks, 10% survival was observed with *Larix* × *eurolepis.*

2.3.5 *Effect of Cryopreservation on Embryonal Cells*

In both *Larix* × *eurolepis* and *Picea mariana*, the cryopreservation procedure killed most of the suspensor cells and not the meristematic region of the somatic

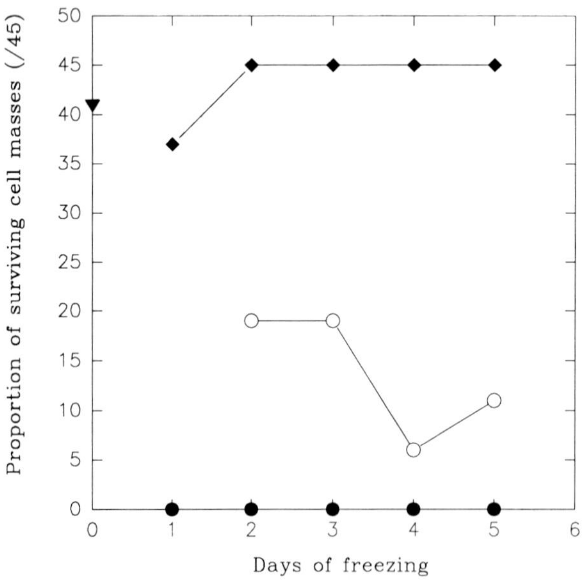

Fig. 6. Effect of temperature of freezing on survival after freezing *Larix* × *eurolepis* embryonal masses. Temperatures of storage were −196 °C (◆), −100 °C (○), and −80 °C (●). (▼) is the growth of the tissues treated with cryoprotectants but not frozen

embryos, as demonstrated by fluorescein diacetate viability staining (Klimaszewska et al. 1992).

2.4 Discussion

For *Picea* and *Larix* species, seeds can be used effectively as a means to preserve germplasm; however, to accelerate the tree breeding cycle, tissue culture is used to produce clonal trees from improved individuals. The method most suitable for commercial reforestation is somatic embryogenesis. Four types of tissues used in this process could be employed for cryopreservation: the original explants (immature and mature zygotic embryos, cotyledons, or needles from young seedlings) from which the embryogenic culture will be initiated, the embryogenic cultures consisting of embryonal masses, the maturing somatic embryos, and the germinating somatic seedlings. The embryonal masses are the most suitable for cryopreservation because they are actively growing structures composed of a few cells that can be suspended uniformly in liquid. Our results showed that the meristematic regions of the embryoids survived freezing while the suspensor cells (highly vacuolated) did not.

The mature embryos and seedling parts probably require a different protocol. Such protocols exist for somatic embryos of carrot and coffee (Lecouteux et al. 1991; Tessereau et al. 1991), where desiccated somatic embryos were used for cryopreservation. Germinated somatic seedlings would require a special cryopreservation protocol (vitrification) that is suitable for heterogeneous structures. The tree seedlings in this protocol would be frozen at rates between −1800 and −4800 °C/min in high concentrations of cryoprotectants, which would lead to intracellular water glass formation instead of ice (Towill 1990; Sakai et al. 1990). The cryopreservation of original explants before induction of embryonal masses could be useful, but is not really practical because only a portion of the explants are capable of undergoing somatic embryogenesis.

Other methods, such as slow growth, could be used, but further optimization would be required for these two genera. Cryopreservation offers longer-term storage of the frozen tissues; however, the length of time that the tissues can be preserved is limited because background natural radiation may cause DNA damage. With cryopreserved animal cells, it was calculated that it would be 30 000 years before there was detrimental radiation damage (Ashwood-Smith and Friedmann 1979).

In our experience, the choice of the embryogenic line and its growth cycle are the two factors that are extremely important but yet are often underestimated. It is more than likely that changes in sorbitol and DMSO concentrations can be made to optimize the freezing protocol because the tissue culture medium and genotype of the cell lines used will influence the response of the tissues frozen with these cryoprotectants. The temperature of storage should be between −140 and −196 °C and the vapour phase presents less risk of cryovial explosion than the liquid phase. Shipments of frozen lines should be done in liquid nitrogen and, if that is not possible, growing tissues should be sent and frozen on arrival at the destination.

3 Summary and Conclusions

An efficient protocol for the cryopreservation of *Larix* and *Picea* embryogenic masses has been developed. The process is straightforward and includes (1) choice of the tissue, (2) pretreatment of the cells, (3) freezing, and (4) thawing. In particular, the effects of sorbitol, DMSO, tissue culture growth cycle, and temperature of storage have been evaluated. High success rates have been achieved with the nine species tested at PNFI. Cryopreservation protocols should be developed for other tissues such as excised immature and mature zygotic embryos, mature somatic embryos, and cotyledons. This would not only extend the availability of explants required to induce somatic embryogenesis, but also provide the potential to accelerate the production of somatic seedlings by storing tissues that are closer to the plantlet stage. This will contribute to a better integration of the technology into tree improvement programs.

Acknowledgments. The authors would like to thank K. Laplante, J. Doubt, C. Wood, and I. Hay for technical help and A. Yapa, G. Murray, D. Lachance, and C. Jones for critical reviews of the manuscript.

References

Ashwood-Smith MJ, Friedmann JB (1979) Lethal and chromosomal effects of freezing, thawing, storage time and X-irradiation on mammalian cells preserved at −196 °C in dimethylsulfoxide. Cryobiology 16: 132–140

Attree SM, Fowke LC (1991) Micropropagation through somatic embryogenesis in conifers. In: Bajaj YPS (ed) Biotechnology in agriculture and forestry, vol 17. High-tech and micropropagation I. Springer, Berlin Heidelberg New York Tokyo, pp 53–70

Bajaj YPS (1991) Storage and cryopreservation of in vitro cultures. In: Bajaj YPS (ed) Biotechnology in agriculture and forestry, vol 17. High-tech and micropropagation I. Springer, Berlin Heidelberg New York Tokyo, pp 361–381

Bercetche J, Galerne M, Dereuddre J (1990) Augmentation des capacités de régération de cals embryogènes de *Picea abies* (L.) Karst après congelation dans l'azote liquide. CR Acad Sci Paris Ser III 310: 357–363

Bonga JM, Pond SE (1991) Adventitious shoot formation in cultures of 30-year-old *Larix decidua*, *L. leptolepis*, *L. eurolepis* and *L. laricina* trees. Plant Cell Tissue Org Cult 26: 45–51

Bonga JM, von Aderkas P (1992) In vitro culture of trees, Kluwer, Boston, 236 pp

Boyle TJB, Nieman TC, Magnussen S, Veen J (1989) Species, provenance and progeny tests of the genus *Larix* by the Petawawa National Forestry Institute. PNFI Inf Rep PI-X-94, Chalk River, Ont, Canada, 70 pp

Burns RM, Honkala BH (1990) Silvics of North America. Vol 1: Conifers. USDA For Serv, Washington DC, Agric Handb 654: 675

Charest PJ, DeVerno LL, Klimaszewska KK, Lelu MA, Ward C (1992) Advanced biotechnology in the genus *Larix*: potential integration into tree improvement programs. Proc Symp Ecology and Management of *Larix* Forests: a look ahead. Oct 1992, Whitefish, Montana. Tech Rep INT-000. US Dept Agric, Forest Service, Intermountain Research Station

Cheliak WM, Klimaszewska K (1991) Genetic variation in somatic embryogenic response in open-pollinated families of black spruce. Theor Appl Genet 82: 185–190

Cheliak WM, Rogers DL (1990) Integrating biotechnology into tree improvement programmes. Can J For Res 20: 452–463

Cyr DR, Lazaroff WR, Grimes SMA, Quian G, Benthune TD, Dunstan DI, Roberts DR (1994) Cryopreservation of interior spruce (*Picea glauca-engelmanni* complex) embryogenic cultures. Plant Cell Rep 13: 574–577

Diner AM, Strickler A, Karnosky AF (1986) Initiation, elongation and remultiplication of *Larix decidua* micropropagules. NZJ For Sci 16: 306–318

Eastman PAK, Webster FB, Pitel JA, Roberts DR (1991) Evaluation of somaclonal variation during somatic embryogenesis of interior spruce (*Picea glauca engelmannii* complex) using culture morphology and isozyme analysis. Plant Cell Rep 10: 425–430

Galerne M, Dereuddre J (1987) Survie de cals embryogènes d'epicea après congelation à −196 °C. Ann Afocel 7–32

Gupta PK, Durzan DJ, Finkle BJ (1987) Somatic polyembryogenesis in embryogenic cell masses of *Picea abies* (Norway spruce) and *Pinus taeda* (loblolly pine) after thawing from liquid nitrogen. Can J For Res 17: 1130–1134

Joy RW, Kumar PP, Thorpe TA (1991) Long-term storage of somatic embryogenic white spruce tissue at ambient temperature. Plant Cell Tissue Organ Cult 25: 53–60

Kartha KK, Fowke LC, Leung NL, Caswell KL, Hakman I (1988) Induction of somatic embryos and plantlets from cryopreserved cell cultures of white spruce (*Picea glauca*). J Plant Physiol 132: 529–539

Klimaszewska K (1989). Plantlet development from immature zygotic embryos of hybrid larch through somatic embryogenesis. Plant Sci 63: 95–103

Klimaszewska K, Ward C, Cheliak WM (1992) Cryopreservation and plant regeneration from embryogenic cultures of larch (*Larix × eurolepis*) and black spruce (*Picea mariana*). J Exp Bot 43: 73–79

Laine E, Bade P, David A (1992) Recovery of plants from cryopreserved embryogenic cell suspensions of *Pinus caribaea*. Plant Cell Rep 11: 295–298

Lecouteux G, Florin B, Tessereau H, Bollon H, Petiard V (1991) Cryopreservation of carrot somatic embryos using a simplified freezing process. Cryo Lett 12: 319–328

Lelu MA, Klimaszenska KK, Jones C, Ward C, von Aderkas P, Charest PJ (1993) A laboratory guide to somatic embryogenesis in spruce and larch. PNFI Inf Rep PI-X-III, Chalk River, Ont, Canada, 57pp

Sakai A, Kobayashi S, Oiyama I (1990) Cryopreservation of nucellar cells of navel orange (*Citrus sinensis* Osb. var *brasiliensis* Tanaka) by vitrification. Plant Cell Rep 9: 30–33

Stanwood PC (1984) Cryopreservation of seeds. Rep 2nd Meet IBPGR Advisory Comm on seed storage. IBPGR Secretariat, Rome, Italy, 27pp

Tautorus TE, Fowke LC, Dunstan DI (1991) Somatic embryogenesis in conifers. Can J Bot 69: 1873–1899

Tessereau H, Lecouteux C, Florin B, Schlienger C, Petiard V (1991) Use of a simplified freezing process and dehydration for the storage of embryogenic cell lines and somatic embryos. Rev Cytol Biol Veg Bot 14: 297–310

Toïvonen PMA, Kartha KK (1989) Cryopreservation of cotyledons of nongerminated white spruce [*Picea glauca* (Moench) Voss] embryos and subsequent plant regeneration. J Plant Physiol 134: 766–768

Towill LE (1990) Cryopreservation of isolated mint shoot tips by vitrification. Plant Cell Rep 9: 178–180

Tremblay FM (1990) Somatic embryogenesis and plantlet regeneration from embryos isolated from stored seeds of *Picea glauca*. Can J Bot 68: 236–242

Tremblay L, Tremblay FM (1991a) Carbohydrate requirements for the development of black spruce (*Picea mariana* (Mill.) B.S.P.) and red spruce (*P. rubens* Sarg.) somatic embryos. Plant Cell Tissue Organ Cult 27: 95–103

Tremblay L, Tremblay FM (1991b) Effects of gelling agents, ammonium nitrate, and light on the development of *Picea mariana* (Mill) B.S.P. (black spruce) and *P. rubens* Sarg. (red spruce) somatic embryos. Plant Sci 77: 233–242

Von Aderkas P, Klimaszewska K, Bonga JM (1990) Diploid and haploid embryogenesis in *Larix leptolepis, L. decidua*, and their reciprocal hybrids. Can J For Res 20: 9–14

II.11 Cryopreservation of Scots Pine (*Pinus sylvestris* L.)

A. Hohtola[1]

1 General Account

Scots pine (*Pinus sylvestris* L.) is widely distributed in the northern hemisphere from Spain to Mid-Siberia, and together with Norway spruce, it belongs to the most widely grown trees in northern Europe. Scots pine is an important source of timber in northern Eurasia, and is also used in the paper industry. In the northern United States, it is used mainly for Christmas trees and landscape planting.

Regeneration of Scots pine is routinely done by seed. Vegetative methods have been developed but are not yet suitable for effective propagation. Good crops of seed occur in Lapland only once in 10–15 years (Lähde 1975). The seed required for forestation has to be stored for considerable periods of time. Gymnosperm seeds generally have a short life span, but it is possible to store seed in cold storage for many years without a significant decrease in germinability (Allen 1957; Huss 1967; Kozlovski 1971). The storage conditions are critical; during storage, at 0–5 °C, the most suitable water content for pine seeds is 4.5–8%. The seeds retain their germination capacity best at temperatures below 0 °C and then the moisture content should be near 4.5% (Ryynänen 1980). Long-term storage of pine seeds at different storage temperatures was tested by Huss (1967). The seeds retained their initial germinability for 3 years at +5 °C, but for 20 years at a storage temperature of −5 °C. Germinability of pine seed after storage at −15 °C was comparable to that at −5 °C, but lower storage temperature resulted in a greater mass of seedlings after 10 days' germination. A short-term storage of Scots pine seeds in liquid nitrogen (LN) was tested by Jörgensen (1990). Percentage of germination after freezing was found to be 35.

In dry seed, changes caused by aging are difficult to see, but during imbibition these effects become evident. Inhibition of development of the mitochondria and plastids is the first indication of senescence (Simola 1974). The nucleus seems to be morphologically the most resistant part of the cell, but the plasmalemma was more rapidly damaged compared to the membrane surrounding the spherosomes. There are no notable differences in the stages of deterioration of different parts of the seeds.

The seeds of many conifers require a period in moist, cold storage (stratification) before germination. Frost-sensitive seeds of Scots pine have shown a 75% loss of viability at relatively high subzero temperature (−5 °C) and mortality

[1]Department of Botany, University of Oulu, 90570 Oulu, Finland

Biotechnology in Agriculture and Forestry, Vol. 32
Cryopreservation of Plant Germplasm I (ed. by Y.P.S. Bajaj)

increased further at lower temperatures after imbibition (Keefe and Moore 1982). On the other hand, the mortality of seeds which had been surface-dried after imbibition was only 20%. In the frost-damaged seeds of Scots pine the greatest changes were seen in the morphology of testa.

It has also been found that a proportion of healthy seeds of various conifer species do not germinate during certain periods of the year. This phenomenon is supposed to be due to seasonal periodicity or biorhythms (Barnett and Mamonov 1989). Periodicity seems to be related to the age of seeds, seed storage conditions, and tree species. Germination showed no periodicity in Scots pine and Norway spruce seeds during the first year of storage, but germination of Siberian larch seeds decreased in March and September. After 1 year of storage, Scots pine seed germination also showed periodicity; germination decreased during the summer (June and July) and winter (December and January) (Barnett and Mamonov 1989; Hohtola, unpubl.).

Conservation of mature pollen of *Pinus sylvestris* in LN was tested by Jörgensen (1990) and Lanteri et al. (1993). Cryoprotectants and precooling were used to prevent cryoshock. Germination of pollen before and 8–24 h after deep freezing was 90%. (Jörgensen 1990). Germination after 24 months was 50% (Lanteri et al. 1993). Pollen of *Pinus brutia*, *P. canariensis*, *P. halepensis*, and *P. pinaster* has been stored at −20 °C for 1 year (Braggio et al. 1990). Germination decreased in *Pinus canariensis* and *P. pinaster*, while in *P. brutia* and *P. halepensis* it increased at low temperature. At room temperature the viability of pollen lowers and disappears within few days. Thus the cryopreservation has to be considered as an important method for the maintenance of the germplasm of these trees.

Many *Pinus* species have been propagated by in vitro from juvenile tissues using excised embryos and young seedlings as explants (Horgan and Aitken 1981; Patel and Thorpe 1984; Mudge 1986; Bajaj 1991; Jang and Tainter 1991; Martinez Pulido et al. 1992). However, there are only a few reports of propagation using tissues from mature *Pinus* species (Bonga 1981; Gupta and Durzan 1985; Abdullah et al. 1987; Lin et al. 1991). A method for vegetative propagation of selected phenotypes from adult trees would be desirable for various forest breeding and reforestation programs. Among *Pinus* species, the Scots pine appears to be especially difficult to handle in cultures, and a method for routine use has not been developed up to now. Some progress has been achieved by using young (Bornman and Jansson 1980; Toribio and Pardos 1981; Skripachenko 1982; Jain et al. 1988; Žel et al. 1988; Chalupa 1989) or adult material (Hohtola 1988). Various aspects of the in vitro culture of *Pinus sylvestris* have been reviewed by Toribio and Pardos (1989).

2 Cryopreservation

2.1 General

Cryopreservation and subsequent culture of plant tissues (Sakai 1985) and even regeneration from frozen protoplasts (Bajaj 1988) has been performed on a number of herbaceous plants. Among trees, successful cryopreservation of dormant buds has been reported (Sakai and Nishiyama 1978; Katano et al. 1983; Tyler et al. 1988; Yakuwa and Oka 1988). A method has also been developed for cryopreservation of embryogenic cultures of *Picea abies* (Bercetche et al. 1990), hybrid larch (Klimaszewska et al. 1992), and embryogenic cell suspensions of *Pinus caribaea* (Lainé et al. 1992). In most cases, freeze preservation of plant material has been achieved in liquid nitrogen, but storage of frozen plant organs at temperatures above that of liquid nitrogen has also succeeded and has been reported by Sakai and Nishiyama (1978), who were able to maintain dormant, prefrozen apple buds at −70 °C for 1 year. Freeze preservation of Scots pine buds at the same temperature was reported by Kuoksa and Hohtola (1991).

The works of Sakai (1973, 1979) showed that in the most cold-hardy coniferous genera such as *Picea*, *Tsuga*, *Pseudotsuga*, *Cedrus*, *Keteleeria*, and *Pinus*, winter survival of the buds appears to be related to the capability for supercooling. Highly hardy twigs were also resistant to rapid cooling. According to Sakai's (1982) results, most of the water may be gradually removed from the primordial shoots during the freezing of whole buds. Extra-organ freezing of this kind could reasonably explain why shoot primordia of very hardy conifers can survive winter temperatures of −60 °C or below in Alaska and Siberia.

Scots pine belongs also to the very hardy conifers; the shoot primordia tolerate slow freezing down to −70 °C or immersion in liquid nitrogen after prefreezing to −40 °C (Sakai 1983). In woody plants, the degree of success varies with the physiological state of the buds. Buds in the cold-hardened stage do not need supplementary acclimation, but for reactivated material it is necessary. Experiments with suspension cultures of *Pinus sylvestris* suggest that cold acclimation is also dependent on the organization of the cells in tissues (Hellergren 1983).

Scots pine is economically the most important tree in Northern countries but, so far, routine methods for its micropropagation have not been developed. The present chapter is part of a project aimed at finding a method to propagate mature Scots pine trees in vitro using buds as a starting material. The best viability of explants in cultures has been found in buds collected in January and April (Hohtola 1988), suggesting that the most reactive explants can be collected only during a limited period each year. Autumn buds grow very poorly in cultures, and thus it was necessary to find a method for storing bud material for autumn experiments. Cryopreservation of buds enables long-term and repeatable experiments to be carried out using material from a certain season. The method has to be rapid enough to make it possible to handle large numbers of buds, and also gentle enough to keep the buds in the condition they have at the time of collection. Cryopreservation of Scots pine buds without cryoprotectants was deemed a suitable method for this purpose.

2.2 Methodology

Twigs of Scots pine (*Pinus sylvestris* L.) were collected from 8 to 30-year-old pines in a natural stand in the district of Oulu (65° N; 25°30′ E) from March to

May 1990. (Preliminary experiments were done during the same period in 1989.) The twigs were collected from the upper part of the tree, from 5 trees on each occasion, giving 350 buds altogether. Since 1990, cryopreservation of pine buds has been done routinely for long-term storage of buds from April for later use in autumn.

The material was placed in a cold room at 0 °C, where the 1–1.5-cm-long buds were cut off and put into 4-ml plastic ampules (15 buds/ampule). Each ampule was wrapped in aluminum foil and placed in an ice bath. Controlled cooling and freezing was achieved in a freezer (Union Carbide, M180/B2, UK) precooled to 0 °C prior to the introduction of the ampules. The ampules were arranged in the freezing chamber equidistantly around the one containing the temperature probe. The samples were cooled at 1 °C/min until the terminal point of freezing, −39 °C, was reached. They were then kept at this temperature for 30 min and subsequently immersed in LN for 10 min. Finally the ampules were transferred to a deep freezer at −80 °C for 1-, 4-, 17-, or 27-week-long storage periods. A part of the samples was stored for 2.5 years. The supercooling point for the buds was between −6 and −10 °C. The buds then warmed up to −2 °C, which is their freezing point. At that point LN was slowly added to the freezing chamber for 35 s to bring the temperature back to −6/−10 °C, after which the freezing procedure continued normally at 1 °C/min.

After storage in the freezer, the buds were thawed rapidly for 2 min in a water bath at +37 °C, and the ampules were then kept at room temperature (+22 °C). The viability of the buds after cryopreservation was determined by staining with fluorescein diacetate (FDA, Kartha et al. 1988). Their viability was also tested by culturing shoot tips on 1/2 MS medium for at least 2 weeks. For tissue culture, the buds were surface sterilized in 3% Na-hypochlorite for 15 min and rinsed four times in sterile distilled water. After peeling, 3–4-mm pieces including apical meristems were cut from the upper parts of the buds and transferred to a modified MS medium (Murashige and Skoog 1962; Hohtola 1988), (0.2 mg/l BAP and 2,4-D, 25 g/l sucrose) for viability testing. The explants were grown under a 16-h light period (white fluorescent Osram 18 W tubes, at 1.8 W/m^2, 400–700 nm) at +26 °C and a 8-h dark period at +23 °C. Blackening of injured tissues was used as the first criterion for rating injury. Viability of cultures was estimated visually after 2 to 6 weeks according to Hohtola (1988). The fresh mass of the samples was measured after that. Control samples were dissected for culture immediately after collection.

The water content of the buds was estimated from a sample of the explants (+105 °C, overnight). Tukey's test was used for the statistical analysis.

2.3 Results

The survival of the frozen buds was satisfactory during the first and second weeks in April, and the material collected in March gave similar results as compared with the controls. Since April is the most interesting month with regard to regeneration, most of the material tested was collected at that time. The material collected during the last 2 weeks of the month survived poorly, more than

Table 1. Collection times, storage times and survival of explants after 2 weeks growth in tissue culture. The viability of explants (diameter 2 mm) was defined by estimating the color (green = still living, black = dead) and by measuring the diameter of the tissues once a week. Samples with diameter under 3 mm after the culture period were regarded as nonviable; n = 15–20. (Data from Kuoksa and Hohtola 1991, modified)

Storage time in −80 °C	Collection time 1990	Survival after 2 weeks (%)
Control	April 4	100
1 week		93
4 weeks		100
17 weeks		90
2.5 years		80
Control	April 11	100
1 week		100
4 weeks		100
17 weeks		98
27 weeks		94
Control	April 18	100
1 week		27
4 weeks		20
Control	April 25	100
1 week		13
4 weeks		7

73–93% dying during the cryopreservation process (Table 1). Both FDA staining and the viability of explants in tissue culture gave similar results. The viability did not decrease markedly during cryopreservation; after 2.5 year it was still 80%. The buds did not survive direct freezing to −25, −40, −80, or −196 °C. Frost damage was easy to observe; dead tissues turned black after only 2 days on a medium.

The growth of the explants in tissue culture showed the same pattern as in the controls, the increase in fresh mass being 20-fold after a growth period of 6 weeks. The tissues consisted of undifferentiated callus cells. There were no noticeable differences between the trees from which the samples were collected in terms of survival after freezing. The degree of contamination, 38%, was similar to that found earlier (Hohtola 1988).

Different rates of freezing were tested (0.55, 0.6, 0.7, 0.9, 1.0, 1.1, 2.3 °C per min and direct immersion in LN and freezers at −25, −40, −80 °C). The best result was achieved when the cooling rate was 1 °C/min up to a terminal freezing temperature of −39 °C, after which the buds were immersed in liquid nitrogen for 10 min before transferring to a deep freezer at −80 °C. The temperature at the initiation of freezing of the buds, reflecting the degree of supercooling of the samples, varied from −6 to −10 °C. A typical freezing curve for a bud is shown

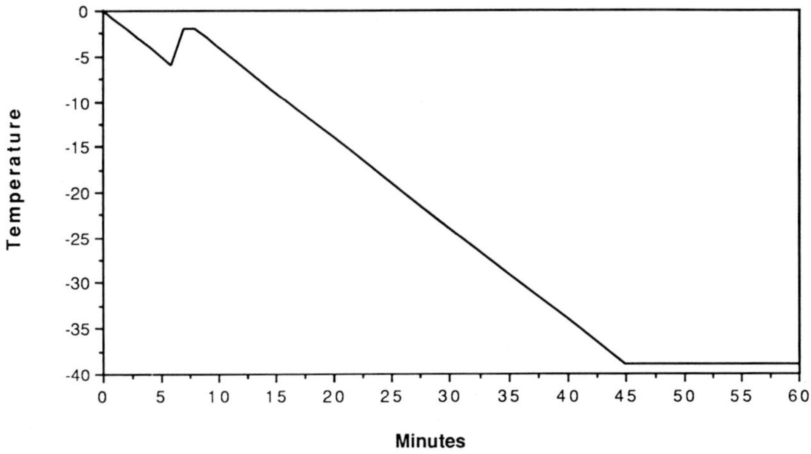

Fig. 1. Typical freezing curve for Scots pine buds showing occurrence of the supercooling point after 6 min (unpubl.)

Table 2. Changes in the water content of buds (means ±SD) from March to May. Differences are significant in Tukeys's test ($p = 0.05$) where the letters differ between the groups; n = 10. (Kuoksa and Hohtola 1991.)

Date 1990	Amount of water in buds (%)	Tukey's test
March 14	48.0 ±2.95	D
March 28	47.4 ±1.76	D
April 11	47.2 ±1.04	D
April 18	50.6 ±1.56	C
April 25	52.1 ±2.20	C
May 9	58.3 ±1.61	B
May 23	63.4 ±1.11	A

in Fig. 1. Upon thawing, the buds reached the room temperature within 5 min. Pretreatments with 5–10% DMSO, 10–40% glycerol, and different concentrations (50–600 g/l) of sucrose and glucose were tried, but the results were not satisfactory.

The amount of water in the buds increased markedly after the middle of April (Table 2), and successful freezing of the buds became very difficult after that (Table 1). Cryopreserved buds and calli derived from cryopreserved buds were found to be suitable also in gene transfer experiments by the particle acceleration method (Aronen et al. 1994).

2.4 Discussion

It has been shown that it is not the low temperature as such but the formation of ice crystals and dehydration that kills the tissue. Thus the ability of cells to endure dehydration and intracellular ice formation will influence the result (Aronsson and Eliasson 1970). In winter buds of *Abies balsamea*, kept at −20 °C for 30 days, the primordial shoots remained supercooled and after dehydration survived subsequent slow cooling to as low as −60 °C, and even immersion in liquid nitrogen from −30 °C (Sakai 1979). Cooling tolerance to ultralow temperature resulted both in supercooling and in subsequent dehydration by extraorgan freezing. Most of the water in the primordial shoot freezes out through the crown (medullary bridge) and consequently it is the crown that plays the major role in freezing avoidance in the primordial shoots of conifer buds (Sakai 1978, 1979). Tyler and Stushnoff (1988) used dormant vegetative apple buds in their cryo-experiments, dehydrating them by keeping them at −4 °C for various lengths of time and then cooling them slowly to −30 °C. The samples were kept at −30 °C for 1 h before immersion in LN. Prefrozen vegetative buds of *Morus bombycis* (Yakuwa and Oka 1988) and hardy shoots of *Malus domestica* (Sakai and Nishiyama 1978) were stored in LN and regenerated successfully after thawing.

The survival of plants, organs, and cells in the deep freezer or in LN is based on the same factors as the frost hardiness in natural stands. The freeze resistance of less hardy plants can be artificially enhanced by various pretreatments. The commonly used cryopreservation methods necessitate either preconditioning with cryoprotectants such as DMSO, glycerol, ethylene glycol, etc. or preconditioning of the tissues at a low temperature (Katano et al. 1983; Kartha et al. 1988; Tyler et al. 1988). On the other hand, it has been found that high levels of cryoprotectant mixtures may cause toxicity and therefore they have to be added with care (Withers 1991). As the present investigation shows, cryoprotectants are not always necessary for cryopreservation.

According to our concept, the buds are no longer in their rest period in April, but most probably certain elements of acquired natural hardiness still persist (Kupila-Ahvenniemi and Hohtola 1980; Kupila-Ahvenniemi et al. 1988). Their metabolic activity changes around the middle of April, and their water content increases (Table 2). The first indication of an increase in the amount of removable water is obtained in March and a more pronounced rise is observed in April (Kupila-Ahvenniemi et al. 1978, 1988). Slow changes in cell structure take place even in midwinter, proving that the dormant phase is not entirely a period of complete rest and minimal metabolism (Kupila-Ahvenniemi et al. 1988). Cell divisions in buds start in April, preceded by a few weeks of marked cellular activity (Kupila-Ahvenniemi et al. 1978; Kupila-Ahvenniemi and Hohtola 1980).

In April, the best survival of buds during 4 weeks of storage at −80 °C was achieved from the beginning to the middle of the month (100% survival), the viability of the corresponding control (unfrozen) buds at the beginning of April being about the same as found earlier by Hohtola (1988). The survival of the frozen buds decreased markedly (down to 7–27%) towards the end of April. The success of the freeze preservation of apple buds was similarly dependent on the month during which the experiments were carried out, and in part on the quantity of liquid water at the temperature to which the tissue was cooled prior to immersion in LN (Tyler and Stushnoff 1988). It was also noticed by Sakai (1965) that prefreezing temperatures below −15 °C were sufficient to enable willow twigs to withstand liquid nitrogen in winter, while in early spring and in early winter prefreezing temperatures below −30 °C were required.

It has been found also by other researchers that the degree of success varies with the amount of water (von Fircks and Cristersson 1982). For these reasons, freeze preservation of vegetative buds is unsuccessful in May (Table 2). Water content decreases in woody stems and needles of Scots pine with cold acclimation (Smit-Spinks et al. 1984). Differences in water content and also in age have a significant effect on the freezing profiles of short shoots of *Pinus sylvestris* (von Fircks and Cristersson 1982); the freezing curves of 1-year-old short shoots usually showed two subcooling temperatures, while curves from the current year's short shoots showed only one.

The meristems of Scots pine seedlings are sensitive to frost damage at the beginning of the growing period; at −7 °C increased numbers of dead seedlings were found (Holopainen 1990). On the other hand, acclimated mature Scots pine tree has survived freezing to −70 °C or below (Sakai 1983). Very hardy buds survive extremely low temperatures by tolerating extracellular freezing (Aronsson and Eliasson 1970; Sakai 1978). There appear to be substantial changes in sugar (Aronsson et al. 1976) and fatty acid composition in the plasma membrane following cold acclimation (Hellergren et al. 1984), and also the protein metabolism plays an important role in the cold acclimation (Kupila-Ahvenniemi et al. 1988). Fast freezing, which is almost always lethal to unacclimated plants, results in high electrolyte leakage immediately after thawing (Sutinen et al. 1992). The best results with Scots pine buds were achieved when cooling rate was 1 °C/min up to terminal freezing temperature. Even 0.1 °C differences in cooling velocity seemed to be critical.

Cryopreserved shoots can suffer severe structural damages as a result of freezing and thawing. Recovery growth of damaged shoots on culture medium containing growth regulators may permit regeneration from a few surviving cells (Withers 1991). In the Scots pine this seems to be the case as the transient expression of the GUS gene could be found in 64% of callus samples started from cryopreserved shoot tips, while the percentage of expressing cryopreserved buds was 5 (Aronen et al. 1994). On the other hand, the whole bud is not needed for a successful micropropagation. Lainé et al. (1992) found that only the rapidly growing suspension cultures of *Pinus caribaea* could be recovered successfully after the freezing process. Post-thaw growth was obtained after a short lag phase.

Micropropagation of adult Scots pine trees has not succeeded in spite of many attempts (Hohtola 1988). In our laboratory, cultures are established throughout the year. Different methods for in vitro propagation are used, and callus is produced for various physiological and molecular biological studies. In the autumn there is often a lack of callus material, since the autumn buds do not grow well. So there was a need for short-term, 4- to 6-month cryopreservation. The present results suggest that cryopreservation of Scots pine buds do possible without any pre-treatment. The method is rapid, and allows processing of 600–1000 buds per day. Buds can be stored conveniently in a deep freezer without the extra work of maintaining the supply of liquid nitrogen.

3 Summary and Conclusions

A procedure has been developed for cryopreservation of buds of the Scots pine (*Pinus sylvestris* L). Instead of liquid nitrogen, cold storage in −80 °C was used. The partly dormant material was obtained directly from a natural stand in Northern Finland and no prefreezing or cryoprotectants for preconditioning wrer used. Cooling velocity was 1 °C/min up to a terminal freezing temperature of −39 °C, after which the buds were immersed in liquid nitrogen at −196 °C for 10 min. The material was then transferred to a deep freezer at −80 °C and stored for up to 2.5 years. The viability of the buds was still 80% after 2.5 years storage in a deep freezer. After rapid thawing, the buds were sterilized, and their viability was tested by FDA staining and by culturing apical meristems on 1/2 MS medium for at least 2 weeks. All the freezing experiments were performed during March and April. The best survival of buds (90–100%) was achieved at the beginning of April, after which a pronounced decline in survival occurred, obviously due to a rise in the water content of the buds. The degree of success varies with the physiological state of the buds.

References

Abdullah AA, Yeoman MM, Grace J (1987) Micropropagation of mature Calabrian pine (*Pinus brutia* Ten.) from fascicular buds. Tree Physiol 3: 123–136

Allen GS (1957) Storage behaviour of conifer seeds in sealed containers held at 0, 3 °C and room temperature. J For 55: 278–281

Aronen T, Häggman H, Hohtola A (1994) Transient β-glucuronidase expression in Scots pine tissues derived from mature trees. Can J For Res 24: 2006-2011

Aronsson A, Eliasson L (1970) Frost hardiness in Scots pine (*Pinus silvestris* L.). I. Conditions for test on hardy plant tissues and for evaluation of injuries by conductivity measurements. Stud For Suec 77: 1–30

Aronsson A, Ingestad T, Lööf L-G (1976) Carbohydrate metabolism and frost hardiness in pine and spruce seedlings grown at different photoperiods and thermoperiods. Physiol Plant 36: 127–132

Bajaj YPS (1988) Regeneration of plants from frozen (−196 °C) protoplasts of *Atropa belladonna* L., *Datura innoxia* Mill. and *Nicotiana tabacum* L. Indian J Exp Biol 26: 289–292

Bajaj YPS (ed) (1991) Biotechnology in agriculture and forestry, vol 16. Trees III. Springer, Berlin Heidelberg New York

Barnett JP, Mamonov NI (1989) Biorhythms in conifer seed germination during extended storage. Ann Sci For 46: 85–88

Bercetche J, Galerne M, Dereuddre J (1990) Efficient regeneration of plantlets from embryogenic callus of *Picea abies* (L.) Karst after freezing in liquid nitrogen. CR Acad Sci Paris 310: 357–363

Bonga JM (1981) Organogenesis in vitro of tissues from mature conifers. In Vitro 17: 511–518

Bornman CH, Jansson E (1980) Organogenesis in cultured *Pinus sylvestris* tissue. Z Pflanzenphysiol 96: 1–6

Braggio GM, Cornara L, Dellacha E (1990) The cryopreservation effect on pollen viability. Boll Soc Ital Biol Sper 66: 821–828

Chalupa V (1989) Micropropagation of *Larix decidua* Mill. and *Pinus sylvestris* L. Biol Plant 31: 400–407

Gupta PK, Durzan DJ (1985) Shoot multiplication from mature trees of Douglas-fir (*Pseudotsuga menziesii*) and sugar pine (*Pinus lambertiana*). Plant Cell Rep 4: 177–179

Hellergren J (1983) Cold acclimation of suspension cultures of *Pinus sylvestris* in response to light and temperature treatments. Plant Physiol 72: 992–995

Hellergren J, Lundborg T, Widell S (1984) Cold acclimation of *Pinus sylvestris*: phospholipids in purified plasma membranes from needles of pine. Physiol Plant 62: 162–166

Hohtola A (1988) Seasonal changes in explant viability and contamination of tissue cultures from mature Scots pine. Plant Cell Tissue Organ Cult 15: 211–222

Holopainen JK (1990) The relationship between multiple leaders and mechanical and frost damage to the apical meristem of Scots pine seedlings. Can J For Res 20: 280–284

Horgan K, Aitken J (1981) Reliable plantlet formation from embryos and seedling shoot tips of radiata pine. Physiol Plant 53: 170–175

Huss E (1967) Long-term storage of conifer seed. Stud For Suec 46: 1–59

Jain SM, Newton RJ, Soltes EJ (1988) Induction of adventitious buds and plantlet regeneration in *Pinus sylvestris* L. Curr Sci 57: 677–679

Jang JC, Tainter (1991) Micropropagation of shortleaf, Virginia and loblolly x shortleaf pine hybrids via organogenesis. Plant Cell Tissue Organ Cult 25: 61–67

Jörgensen J (1990) Conservation of valuable gene resources by cryopreservation in some forest tree species. J Plant Physiol 136: 373–376

Kartha KK, Fowke LC, Leung NL, Caswell KL, Hakman I (1988) Induction of somatic embryos and plantlets from cryopreserved cell cultures of white spruce (*Picea glauca*). J Plant Physiol 132: 529–539

Katano M, Ishihara A, Sakai A (1983) Survival of dormant apple shoot tips after immersion in liquid nitrogen. Hortic Sci 18: 707–708

Keefe PD, Moore KG (1982) Frost damage during stratification: mechanism and protection in *Pinus sylvestris* seed. Seed Sci Technol 10: 485–494

Klimaszewska K, Ward C, Cheliak WM (1992) Cryopreservation and plant regeneration from embryogenic cultures of larch (*Larix* × *eurolepis*) and black spruce (*Picea mariana*). J Exp Bot 43: 73–79

Kozlovski TT (1971) Growth and development of trees, vol 1. Academic Press, New York

Kuoksa T, Hohtola A (1991) Freeze-preservation of buds from Scots pine trees. Plant Cell Tissue Organ Cult 27: 89–93

Kupila-Ahvenniemi S, Hohtola A (1980) Nucleoli of the overwintering strobilal cells of Scots pine. Protoplasma 105: 107–119

Kupila-Ahvenniemi S, Pihakaski S, Pihakaski K (1978) Wintertime changes in the ultrastructure and metabolism of the microsporangiate strobili of the Scots pine. Planta 144: 19–29

Kupila-Ahvenniemi S, Heikkilä S, Kuusela H (1988) Poly(A)+ RNA and the in vitro protein products of bud tissues in the Scots pine. Molecular genetics of forest trees. Proc Frans Kempe Symp in Umeå, June 14–16, 1988, Sveriges Lantbruksuniv Rep 8: 187–200

Lähde E (1975) The size of pine cones, seed crop and seed quality in Northern Finland. Commun Inst For Fenn 86: 1–23

Lainé E, Bade P, David A (1992) Recovery of plants from cryopreserved embryogenic cell suspensions of *Pinus caribaea.* Plant Cell Rep 11: 295–298

Lanteri S, Belletti P, Lotito S (1993) Storage of pollen of Norway spruce and different pine species. Silvae Genetica 42: 104–109

Lin Y, Wagner MR, Heidmann LJ (1991) In vitro formation of axillary buds by immature shoots of Ponderosa pine. Plant Cell Tissue Organ Cult 26: 161–166

Martinez Pulido C, Harry IS, Thorpe TA (1992) Optimization of bud induction in cotyledonary explants of *Pinus canariensis.* Plant Cell Tissue Organ Cult 29: 247–255

Mudge KW (1986) Micropropagation of Mugo pine from embryonic and seedling explants. Hortic Sci 21: 298–299

Murashige T, Skoog F (1962) A revised medium for rapid growth and bioassays with tobacco tissue cultures. Physiol Plant 15: 473–497

Patel KR, Thorpe TA (1984) In vitro differentiation of plantlets from embryonic explants of lodgepole pine (*Pinus contorta* Dougl, ex Loud). Plant Cell Tissue Organ Cult 3: 131–142

Ryynänen L (1980) Storage of Scots pine seed and seed ageing. Folia For 428: 1–11

Sakai A (1965) Survival of plant tissue at super-low temperatures. III. Relation between effective prefreezing temperatures and the degree of frost hardiness. Plant Physiol 40: 882–887

Sakai A (1973) Characteristics of winter hardiness in extremely hardy twigs of woody plants. Plant Cell Physiol 14: 1–9

Sakai A (1978) Low temperature exotherms of winter buds of hardy conifers. Plant Cell Physiol 19: 1439–1446

Sakai A (1979) Freezing avoidance mechanism of primordial shoots of conifer buds. Plant Cell Physiol 20: 1381–1390

Sakai A (1982) Freezing tolerance of shoot and flower primordia of coniferous buds by extraorgan freezing. Plant Cell Physiol 23: 1219–1227

Sakai A (1983) Comparative study on freezing resistance of conifers with special reference to cold adaptation and its evolutive aspects. Can J Bot 61: 2323–2332

Sakai A (1985) Cryopreservation of shoot tips of fruit trees and herbaceous plants. In: Kartha KK (ed) Cryopreservation of plant cells and organs. CRC Press, Boca Raton, pp 135–158

Sakai A, Nishiyama Y (1978) Cryopreservation of winter vegetative buds of hardy fruit trees in liquid nitrogen. Hortic Sci 13: 225–227

Simola, LK (1974) Ultrastructural changes in the seeds of *Pinus sylvestris* L. during senescence. Stud For Suec 119: 1–22

Skripachenko VV (1982) In vitro cultivation of tissues from seedlings of three species. Sov Physiol 29: 162–167

Smit-Spinks B, Swanson BT, Markhart AH III (1984) Changes in water relations, water flux, and root exudate abscisic acid with cold acclimation of *Pinus sylvestris* L. Aust J Plant Physiol 11: 431–441

Sutinen M-L, Palta JP, Reich PB (1992) Seasonal differences in freezing stress of needles of *Pinus nigra* and *Pinus resinosa*: evaluation of the electrolyte leakage method. Tree Physiol 11: 241–254

Toribio M, Pardos JA (1981) In vitro organogenesis of *Pinus sylvestris* L. tissues. Afocel 14: 143–148

Toribio M, Pardos JA (1989) Scots Pine (*Pinus sylvestris* L.). In: Bajaj YPS (ed) Biotechnology in agriculture and forestry, vol 5. Trees II. Springer, Berlin Heidelberg New York, pp 479–506

Tyler N, Stushnoff C (1988) Dehydration of dormant apple buds at different stages of cold acclimation to induce cryopreservability in different cultivars. Can J Plant Sci 68: 1169–1176

Tyler N, Stushnoff C, Gusta LV (1988) Freezing of water in dormant vegetative apple buds in relation to cryopreservation. Plant Physiol 87: 201–205

Von Fircks HA, Christersson L (1982) Artificial ice crystal formation in excised short shoots of *Pinus silvestris* L. Physiol Plant 54: 386–392

Withers LA (1991) In vitro conservation. Biol J Linn Soc 43: 31–42

Žel J, Gogala N, Camloh M (1988) Micropropagation of *Pinus sylvestris*. Plant Cell Tissue Organ Cult 14: 169–175

Yakuwa H, Oka S (1988) Plant regeneration through meristem culture from vegetative buds of mulberry (*Morus bombycis* Koidz.) stored in liquid nitrogen. Ann Bot 62: 79–82

Section III
Cereals and Grasses

III.1 Cryopreservation of Germplasm of Cereals (Wheat, Rice, and Maize)

Y.P.S. Bajaj[1]

1 Introduction

Cereals are the main source of staple carbohydrate food, and are grown in most countries (Table 1). They cover an area of about 699 588 000 ha with a production of 1 952 224 000 metric tonnes (FAO 1992). Their germplasm is preserved through seed, however, with the ever enthusiastic search for novel plants and the rapid increase in the number of cultivars, it is not only becoming difficult, but at times impossible to preserve or maintain the old genetic stocks. Therefore, some of the germplasm which may not be of much use at present, but might be needed in future, is ignored and lost. For example, in wheat, with the introduction of exotic cultivars and their cultivation over large areas, the native genotypes are often ignored and sometimes lost. This has been one of the aftereffects of the green revolution. Likewise, since World War II, 95% of the wheat varieties in Greece have been abandoned (Plucknett et al. 1983), some of them obviously lost, leading to the depletion of genetic variability. In maize, the susceptibility to *Helminthosporium maydis* cost US farmers hundreds of millions of dollars in lost production in 1970 (Pluncknett et al. 1983). This attracted international attention, and now germplasm banks are supported in many countries (Tables 2, 3, 4). The CIMMYT germplasm bank has perhaps the largest maize seed collection: The relevance of the establishment of germplasm banks is especially evident when material is preserved that has vanished elsewhere. For instance, *Oryza perennis* from Taiwan, which is resistant to ragged stunt virus, is now extinct there; however, it was previously stored at IRRI. Likewise in Kampuchea, many unique rice cultivars were lost as far back as 1970, but some have been preserved at IRRI (Plucknett et al. 1983). Thus, there is a need to develop and strengthen germplasm banks, and new technologies have to be applied for long-term and stable storage.

2 Need for Cryopreservation

Seed. Cereal seed is orthodox (nonrecalcitrant), can be refrigerated, and is being stored in various international centers (Tables 2, 3, 4). The CIMMYT germplasm bank has perhaps the largest seed collection of maize and wheat (Tables 2, 4). It

[1]Former Professor of Tissue Culture, Punjab Agricultural University, Ludhiana (Punjab), India. (*Present address*: A-137 New Friends Colony, New Delhi 110 065, India)

Biotechnology in Agriculture and Forestry, Vol. 32
Cryopreservation of Plant Germplasm I (ed. by Y.P.S. Bajaj)

Table 1. The area under cultivation and production of wheat, rice, and maize in various countries (FAO 1992)

Region/country	Wheat Area (1000 ha)	Wheat Production (1000 metric t)	Rice Area (1000 ha)	Rice Production (1000 metric t)	Maize Area (1000 ha)	Maize Production (1000 metric t)
World	220 007	563 649	147 168	525 475	132 266	526 410
Africa	8082	13 308	6930	14 011	2052	23 963
North-central America	40 017	100 415	1842	10 109	39 257	262 479
Mexico	916	3602	84	361	7348	14 997
USA	25 255	66 920	1267	8123	29 195	240 774
South America	7246	14 589	6272	16 343	18 255	46 106
Argentina	4230	9400	148	753	2321	10 699
Brazil	1995	2839	4639	9961	13 429	30 619
Asia	84 384	214 742	130 974	479 588	39 539	132 301
China	30 651	101 003	32 399	188 150	21 085	95 340
India	22 980	55 084	42 000	109 511	5950	9 740
Europe	25 398	115 498	404	2257	11 621	52 824
France	5020	32 600	24	124	1835	14 613
Former USSR	45 737	89 925	605	2006	2999	7362

Table 2. Wheat (*Triticum* species) accessions in gene banks. (Plucknett et al. 1983)

Accessions	Term of storage	Institution	Location
63 000	Medium, long	VIR[a]	Leningrad, Former USSR
50 000	Medium	CIMMYT	El Batán, Mexico
46 186	Long	NSSL[a]	Fort Collins, USA
36 710	Medium	USDA National Small Grains Collection	Beltsville, USA
31 000	Short	ARO	Bet Dagan, Israel
26 000	Medium, long	Instituto del Germoplasma[a]	Bari, Italy
20 200	Long	New South Wales Department of Agriculture	Tamworth, Australia
20 000	Short[b]	CGI	Beijing, China
17 000	Short	ICARDA	Aleppo, Syria
13 600	Short[b]	IPIGR	Plovdiv, Bulgaria
10 000	Medium, long	Zentralinstitut für Genetik und Kulturpflanzenforschung	Gatersleben, Former DRG

[a] IBPGR – designated base collections.
[b] Long-term facility being constructed.
ARO, Agricultural Research Organization; CIMMYT, Centro Internacional de Mejoramiento de Maiz y Trigo; CGI, Crop Germplasm Institute; ICARDA, International Centre for Agricultural Research in the Dry Areas; IPIGR, Institute of Plant Introduction and Genetic Research; NSSL, National Seed Storage Laboratory; USDA, National Small Grains Collection; VIR, All-Union Institute of Plant Industry.

contains a wealth of genetic diversity from over 50 countries (CIMMYT 1986; Taba 1990); the term of storage ranges from short to medium. Likewise, IRRI is responsible for the storage of thousands of rice accessions (Table 3).

Although cereal seed is currently being stored without much difficulty, the possibility of various types of changes, including physiological deterioration and genetic mutations, cannot be ruled out. There is also the risk of contamination by various pathogens, resulting in complete loss or deterioration of seed.

According to Stanwood (1985), "large-scale mechanical refrigeration systems, which hold seed at temperatures down to $-20\,^{\circ}C$ have greatly increased the storage life of a seed sample. However, deterioration and loss of viability can still occur with increased time in storage. Cryopreservation techniques provide the potential for indefinite preservation by reducing metabolism to such a low level that all biochemical processes are significantly reduced and biological deterioration virtually stopped." The work on the cryopreservation of seeds of wheat, rice and maize has shown that 100% germination of retrieved seed can be achieved, thus the possibility of selection pressure is ruled out.

In Vitro Induced Genetic Variability (Table 5). The author, during his visits to the International Rice Research Institute (IRRI) at Los Baños, Philippines, in 1979, and to the Centro Internacional de Mejoramiento de Maiz y Trigo (CIMMYT) at Mexico in 1988, had the opportunity to observe their enormous efforts in developing methods and facilities for the storage of seed of rice, maize, and wheat. However, in view of the recent progress in the induction of genetic

Table 3. Rice (*Oryza* species) accessions in gene banks. (Plucknett et al. 1983)

Species	Accessions	Term of storage	Institution	Location
Common rice (*O. sativa*)	60000	Medium, long	IRRI[a]	Los Baños, Philippines
	30000	Medium	IARI	New Delhi, India
	18065	Long	NSSL[a]	Fort Collins. USA
	18000	Medium, long	NIAS[a]	Tsukuba, Japan
	15249	Medium	CRRI	Cuttack, India
	11230	Short	Agricultural Research Center	Beltsville, USA
	8226	Medium	WARDA	Monrovia, Liberia
	6000	Medium	Central Research Institute for Agriculture	Bogor, Indonesia
	5100	Medium	Bangladesh Rice Research Institute	Dacca, Bangladesh
	4600	Medium, long	Koitotron Seed Bank	Penang, Malaysia
	4227	Medium, long	Agricultural Experiment Station	Suweon, Korea
	4000	Short	Bangkhen Rice Station	Bangkok, Thailand
	3765	Medium[b]	IITA[b]	Ibadan, Nigeria
	3500	Medium, long	VIR	Leningrad, former USSR
	3200	Short	INTA	Cordoba, Argentina
African rice (*O. glaberrima*)	2575	Medium, long	IRRI[a]	Los Baños, Philippines
	1515	Medium	IITA[a]	Ibadan, Nigeria

[a] IBPGR-designated base collections.
[b] Long-term facility about to be operational.
CRRI, Central Rice Research Institute; IARI, Indian Agricultural Research Institute; IITA, International Institute of Tropical Agriculture; INTA, Instituto Nacional de Tecnologia Agropecuaria; IRRI, International Rice Research Institute; NIAS, National Institute of Agricultural Sciences; NSSL, National Seed Storage Laboratory; VIR, All-Union Institute of Plant Industry; WARDA, West Africa Rice Development Association.

variability in these crops through various types of somaclones, genetic transformation, protoplast fusion, and somatic hybridization (Table 5), there is an urgent need to develop methods for their conservation. Cell cultures on repeated subculturing undergo genetic erosions and molecular changes, i.e., polyploidy, aneuploidy, gene amplification, translocation, chromosome elimination, mutation, etc., thus leading to the regeneration of variant plants (see Bajaj 1990b). Most of these variations are negative and are therefore discarded. Nevertheless, sometimes interesting cell lines or high-yielding and disease-resistant somaclones can be isolated. If we wish to maintain their genetic stability and to store the cell lines, then cryopreservation is the method of choice.

Table 4. Corn (*Zea mays*) accessions in gene banks. (Plucknett et al. 1983)

Accessions	Term of storage	Institution	Location
15084	Medium, long	VIR[a]	Leningrad, former USSR
15000	Medium	Institute of Maize Research	Belgrade, Yugoslavia
14000	Short, medium	CIMMYT	El Batán, Mexico
11000	Short, medium	INIA	Mexico City, Mexico
7619	Long	NSSL[a]	Fort Collins, USA
5006	Short	ICA	Colombia
3444	Medium	Universidad Nacional Agraria La Molina	Lima, Peru
3200	Short	Research Institute for Cereals and Technical Plants	Fundulea, Romania
3000	Medium, long	INTA	Pergamino, Argentina

[a] IBPGR-designated collections.
CIMMYT, Centro Internacional de Mejoramiento de Maiz y Trigo; ICA, Instituto Colombiano Agropecuário; INIA, Instituto Nacional de Investigaciónes Agricolas; INTA, Instituto Nacional de Technologia Agropecuária; NSSL, National Seed Storage Laboratory; VIR, All-Union Institute of Plant Industry.

Table 5. Some references on the in vitro induction of variability in wheat, rice, and maize

Crop	Mode of induction of variability	Reference
Wheat	Somaclonal variation	Bajaj (1986a); Metakovsky and Sozinov (1990)
	Genetic transformation	Lee et al. (1990), Vasil et al. (1992), Mooney and Goodwin (1994)
	Protoplast fusion/somatic hybridization	Bajaj (1983b); Vasil et al. (1988); Wang et al. (1993)
Rice	Somaclonal variation	Bajaj and Bidani (1980); Sun and Zheng (1990); Ling (1991)
	Genetic transformation	Hasezawa et al. (1989); Raineri et al. (1990); Uchimiya and Toriyama (1991); Hayakawa et al. (1992)
	Somatic hybrids/cybrids	Terada et al. (1987); Toriyama and Hinata (1988); Kyozuka et al. (1989)
Maize	Somaclonal variation	Earle and Kuehnle (1990); Somers and Anderson (1994); Lazanyi (1994)
	Genetic transformation	Rhodes et al. (1989); Koziel et al. (1993); Golovkin et al. (1993) Dupuis and Pace (1993)
	Somatic hybridization	Wang et al. (1993)

3 Cryopreservation Studies

Extensive studies have been conducted on cryopreservation of wheat (Bajaj 1990a), rice (Bajaj and Sala 1991), and maize (Zhang et al. 1994). Some work has also been reported on barley and rye protoplasts (Takeuchi et al. 1982; Langis and Steponkus 1990). The work on wheat, rice, and maize is summarized here.

3.1 Cryopreservation of Wheat (Tables 6, 7)

3.1.1 Seed

After preservation in liquid nitrogen (LN), wheat seed showed a viability varying from 83–100% (Thiselton-Dryer 1928; Lipman 1936; Stanwood 1980; Stanwood and Bass 1981). In the author's work (Bajaj 1980a; Table 7), mature and dry seed

Table 6. Summary of cryopreservation studies on wheat

Material	Response	Reference
Seed	94.5 germination, normal plants	Bajaj (1980a)
Zygotic embryo	41.8–70% callused/developed plants	Bajaj (1980a, 1984)
	Plants matured, set seed capable of normal germination	Bajaj (1986a)
Pollen embryos/androgenic anther segments	5–19% survival, callus, shoots	Bajaj (1983a, 1984)
Callus cells	11% survival (26% of control) Plant regeneration	Bajaj (1980a)
Cell suspension	2.2–59.4% survival, plants	Chen et al. (1985)
Callus	Freeze-tolerant plants	Kendall et al. (1990)
Somatic hybrid protoplasts (wheat × pea)	12.2–17.9% survival	Bajaj (1983b)

Table 7. Growth response of the seeds, excised embryos, and callus cultures of wheat (*Triticum aestivum* L. cv. Kalyansona) subjected to freezing in liquid nitrogen.[a] (Bajaj 1980)a)

Material	Control	Frozen	
		Growth response (survival)	Growth (percentage of control)
1. Seeds	92% germination	87% germination	94.5
2. Embryos (with a portion of endosperm)	90% grew	63% embryos callus/ develop plants	70
3. Mesocotyl-derived callus	41% cells showed protoplast streaming/ elongation/division	11% cells showed signs of growth	26.8

[a]The seeds were germinated on moist filter paper in a Petri dish, whereas the embryos and calli were cultured on basal MS medium (Murashige and Skoog 1962), and MS + 2,4–D(2 mg/l), respectively.

(about 5% moisture) of *Triticum aestivum* cv. Kalyansona showed viability of up to 94.5% (% of control). On a 2,4-D medium they proliferated to form callus, but on a simple mineral medium they germinated to form normal plants, which in turn produced normal seeds.

3.1.2 Embryos

The significance of the cryopreservation of plant embryos has been discussed (Bajaj 1985), and its potential is manifold. For instance, it could be of use in the case of rare or damaged seeds or seeds which show a low germination rate. Likewise, in incompatible crosses, in which hybrid embryos generally abort at an early stage, the young embryos can be cryopreserved and cultured when the need arises. Moreover, cryopreservation can be extended to somatic and pollen embryos which can be obtained in large numbers.

Zygotic Embryos. The revival of immature zygotic embryos (from spikes 15–19 days afer pollination) varied from 41.8–48.7% (Bajaj 1984). The survival of frozen embryos was judged by their ability to (1) increase in size, (2) turn greenish, (3) proliferate to form callus, and (4) develop root/shoots/plantlets. The control embryos started to elongate or initiated proliferation within 3 days; the frozen embryos underwent a lag period of about 2 weeks, after which they showed signs of growth. These embryos then elongated and proliferated to form a mass of callus (Fig. 1C), and developed shoots and plantlets (Fig. 1D). The callus on transfer to MS + IAA (1 mg/l) + BA (5 mg/l) + CH (500 mg/l) differentiated to form multiple shoots. The callus was loose, friable, and in shake cultures occasionally underwent somatic embryogenesis. The zygotic embryos and the callus-derived plantlets developed further and matured to produce normal plants and spikes (Fig. 2).

Pollen Embryos. Segments of 5–6-week-old cultures of anthers undergoing androgenesis were treated with a mixture of cryoprotectants (7% DMSO, 5% glycerol, and 5% sucrose). The segments from 50 anthers were wrapped in aluminum foil or were put in an ampule containing 1 ml of the cryoprotectant solution. From some anthers a suspension of pollen embryos was also prepared. The cultures were then frozen by two methods: (1) subjection to the vapors of LN and then gradual immersion or (2) rapid freezing by quick immersion in LN (Bajaj 1984).

The retrieved anthers underwent a lag period of 2–4 weeks, thereafter they proliferated to form callus. The callus when transferred to hormone-free medium or to a medium supplemented with IAA + BA + CH differentiated into shoots/ plantlets. Plants could be regenerated from cultures cryopreserved for 1 year (Bajaj 1983a).

A longitudinal slit, or gentle teasing of the androgenic anthers with needles, helped to expose pollen embryos to uniform treatment with the cryoprotectants, resulting in higher viability and a better growth response in comparison to the segments. The pattern of morphogenesis was disturbed in some cultures, as malformed and stunted shoots and abnormal-appearing plantlets were formed.

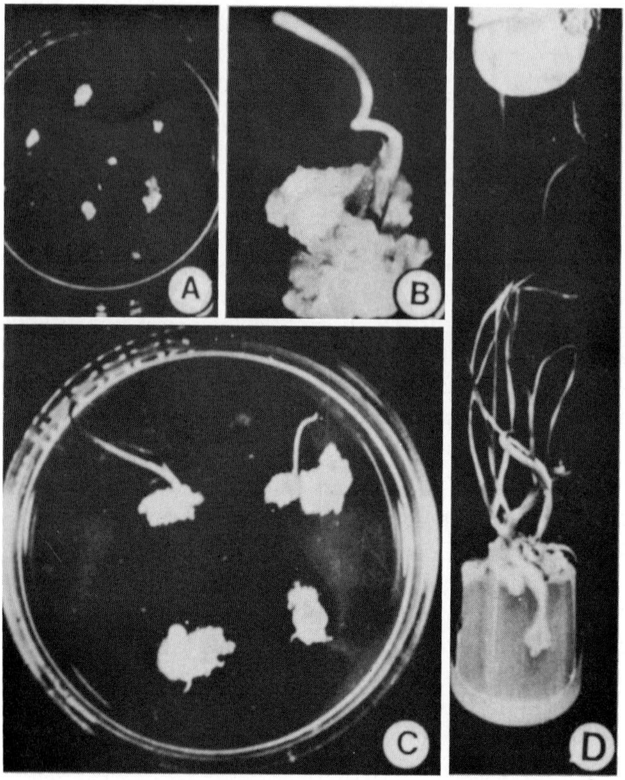

Fig. 1A–D. Cryopreservation of pollen embryos and zygotic embryos of wheat. **A** Callus masses obtained from frozen segments of 5–6-week old androgenic anthers. **B** Same, showing the regeneration of a shoot. **C** Callus and shoot formation from immature zygotic embryos frozen at $-196\,^{\circ}$C, and cultured for 6 weeks on MS + 2,4-D (0.5 mg/l). **D** Regeneration of multiple shoots from retrieved embryos 10 weeks after culture. (Bajaj 1984)

The low survival of anther segments could be partly attributed to the injury caused during cutting as well as to cryodamage.

3.1.3 Callus/Cell Suspension/Protoplasts

The main significance of the cryopreservation of cell cultures lies in the maintenance of the stability of cell lines and avoidance of the need for periodic subculturing. Since in vitro induced somaclones in wheat (Table 5) are now assuming greater importance, it is desirable to employ the cryopreservation method for their conservation.

The cell cultures of cv. Chinese Spring showed a viability of only 26% (Bajaj 1980a). The highest degree of viability was obtained with young and highly cytoplasmic, small, thin-walled, and nonvacuolated cells. From such frozen cells, plants were regenerated (Fig. 3; Bajaj 1990a). Complete plants were also

Fig. 2A–D. **A** Frozen embryo-derived plants transferred to the field being examined by the author. **B** Same, at maturity, showing normal spikes. **C** Uniformity of seeds obtained from plants in **B**. **D** Uniformity of plants obtained as a result of germination of seeds from **C**. (Bajaj 1986b)

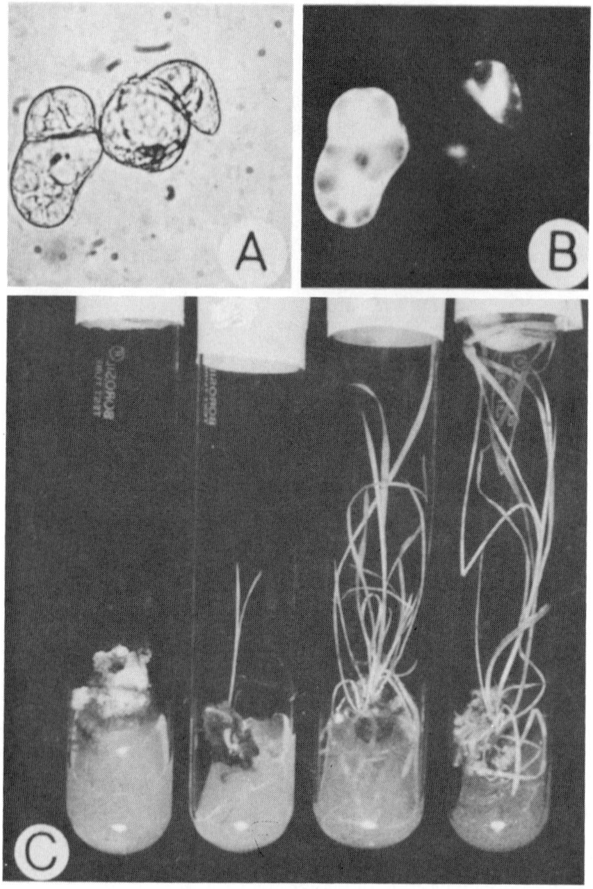

Fig. 3A–C. **A, B** Twin photographs of frozen callus cells of wheat under tungsten (**A**) and ultraviolet light (**B**); note the partial survival of the cell (stained with fluorescein diacetate) in **B**. **C** Various stages in the regeneration of plants from wheat callus. (Bajaj 1980a, 1990a)

obtained from cell cultures of cv. Norstar; cold hardening and ABA treatment increased the freezing resistance (Chen et al. 1985): Later, freeze-tolerant plants were regenerated (Kendall et al. 1990).

Since entire plants can be obtained from frozen protoplasts of some plant species (Bajaj 1988), this work, if extended to cereals, would yield useful information. In addition to the various uses of cell cryopreservation, as already mentioned, isolated protoplasts offer the following advantages:

1. Protoplasts offer a system in which the effect of freezing on the plasma membrane can be studied directly in situ.
2. Freeze preservation of protoplasts reduces the number of steps involved in cloning mutant cells (Weber et al. 1983).
3. The cryomicroscopy of protoplasts allows the direct analysis of intracellular ice formation, and it also enables the precise control of temperature and

cooling rate. With videoimage analysis the temperature at which intracellular nucleation occurs is observed directly (Steponkus and Dowgert 1981; Dowgert and Steponkus 1983).

Somatic hybrid protoplasts obtained as a result of fusion between callus cell protoplasts of wheat × mesophyll protoplasts of pea, and frozen in LN in the presence of cryoprotectants could be revived (Bajaj 1983b). In the control (unfrozen) cultures, the hybrid protoplasts showed mixing of the cytoplasm, yellowing and redistribution, and an overall increase in size in 5–7 days. However, the retrieved (frozen) fused protoplasts were quiescent for 2 weeks in culture. Thereafter, changes in their morphology were observed. There was mixing of the cytoplasm, and visible cyclosis in the cytoplasm. With the passage of time, chloroplasts became scanty and yellowish. The survival of the fused protoplasts was rather low, i.e, between 12.2 and 17.9%. Budding of the protoplasts was a common phenomenon. The heterokaryon showed two nuclei of different size, and occasionally fusion of the nuclei was also observed in the somatic hybrid cells. Only in one instance did the retrieved hybrid cell undergo repeated division (Bajaj 1983b).

3.2 Cryopreservation of Rice (Tables 8, 9)

3.2.1 Seeds, Embryos, and Endosperm

Seeds. The work on rice seed frozen in LN by various workers (Sakai and Noshiro 1975; Gresshoff and Gartner 1977; Ishikawa and Sakai 1978; Stanwood

Table 8. Cryopreservation studies on rice

Material	Response	Reference
Seed	98% germination, normal plants	Bajaj (1981)
Dehusked seed	87% germination, normal plants	Bajaj (1981)
Zygotic embryos	52.5–83% callused, plant regeneration	Bajaj (1981, 1984)
Endosperm	11% callused, plant regeneration	Bajaj (1981)
Microspores	Survival at −196 °C	Coulibaly and Demarly (1979)
Pollen embryo/androgenic anther	Anthers callused, plant regeneration	Bajaj (1980b, 1984)
Callus/cell suspension	60–65% survival	Sala et al. (1979)
	70–100% survival	Cella et al. (1982)
	21–42% survival	Bajaj (1982)
Callus cells	Removal of cryoprotectant favors cell survival	Finkle and Ulrich (1982)
Callus cells	Ammonium ions inhibitory for cell recovery	Kuriyama et al. (1990)
Protoplasts from frozen cells	Plant regeneration	Meijer et al. (1991)
Somatic hybrid protoplasts		
Rice + pea	10.8–13.9% survival	Bajaj (1983b)
Rice + sorghum	10.3–13.8% survival	Bajaj (1983c)

Table 9. Effect of freezing (−196 °C) on the seeds, excised embryos, and endosperm of rice.[a] (Bajaj (1981)

	Control	Frozen	
		Growth response (survival)	Growth (% of control)
1. Seeds	98% germination	96% germination	98
2. Dehusked seeds	94% germination	82% germination	87
3. Excised embryo with a portion of endosperm	86% grew	71% embryos callused and developed shoots	83
4. Segments of mature endosperm	16% callused	11% proliferated to form callus	68

[a] All materials were preserved for 3 weeks in liquid nitrogen, then the seeds were germinated on moist filter paper in a Petri dish, whereas dehusked seeds, excised endosperm, and the embryos were cultured on MS+2,4-D 2 mg/l. Data based on 350 seeds, 92 dehusked seeds, and 360 cultures of embryos and endosperm.

1980; Stanwood and Bass 1981) has shown that germination in the retrieved seeds varied between 80–100%. In the author's work (Bajaj 1981; Table 9), the seeds of cv. B-370 stored in LN for 3 weeks yielded a viability of 87–98% (% of control). The capacity to withstand freezing was inversely related to the water content. The 2-year-old seed (8–11% water content) showed almost complete revival, while immature and freshly harvested seed (35% water content) survived only partially.

Embryos. The excised immature embryos were frozen in the presence of 10% DMSO and 4% sucrose, and then cultured on MS+2,4-D (Bajaj 1984). Their revival varied between 52.5–55.4%. The control embryos started to elongate or initiated proliferation within 3 days; the frozen embryos underwent a lag period of about 2 weeks, after which they showed signs of growth. These embryos then elongated and proliferated to form callus, which upon transfer to MS+IAA+CH medium underwent differentiation to form shoots. The zygotic embryos and the callus-derived plants developed further and matured (Fig. 4). The seeds obtained from such plants in turn showed normal germination and generated plants with no apparent abnormalities.

Endosperm. In culture, the excised endosperm of rice, from both immature and mature seed, can differentiate into entire triploid plants (Bajaj et al. 1980). This morphogenetic potential was retained in cultures that were frozen in LN (Bajaj 1981). The frozen seeds were surface sterilized with chlorine water, dehusked, and cut into two. The endosperm was taken out and cultured on MS medium supplemented with 2,4-D (2 mg/l) in the dark at 23–26 °C. The viability and growth responses are shown in Table 9. Compared with the controls, the initiation of proliferation in the retrieved endosperm was delayed by 1 week. In some cases, callus formation was localized, and a mass of callus was formed in 5–9 weeks. When transferred to MS+kinetin (2 mg/l)+IAA (4 mg/l), the callus underwent morphogenesis to form plantlets in about 3% of the cultures.

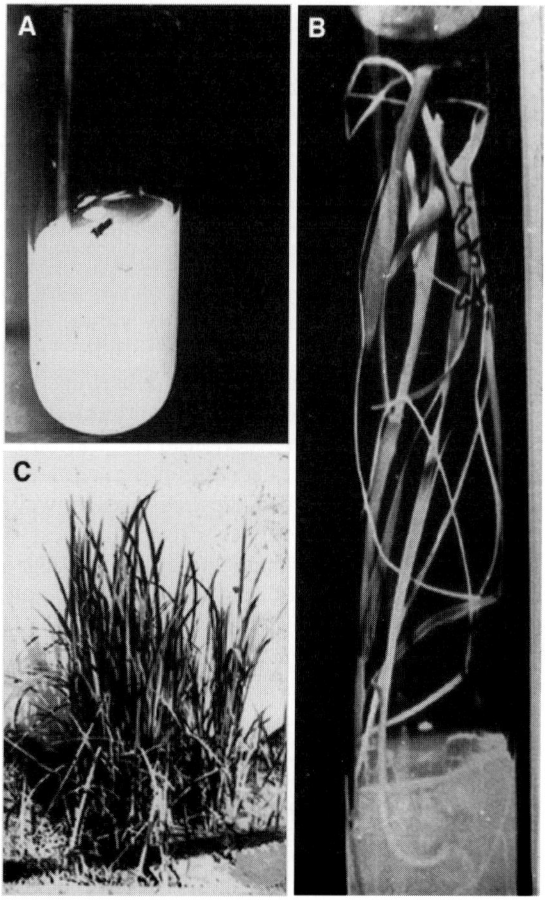

Fig. 4A–C. Regeneration of plants from frozen embryos of rice. **A** An embryo (excised from a frozen seed) 3 weeks after culture; note the formation of root. **B** A normal-looking plant obtained from a frozen embryo. **C** Transfer of test tube-plants to soil. (Bajaj 1981)

3.2.2 Anthers/Pollen Embryos

The method used for the cryopreservation of rice anthers was the same as that already described for wheat. Initially, most of the retrieved anthers became soft and spongy, turned brown, and appeared dead. However, of the 927 anthers cultured, only 5 showed visual signs of growth, i.e., localized callusing. The callus was slow-growing and exhibited various modes of differentiation (Bajaj 1980b, 1984), i.e., only roots, only shoots, or complete plants. One plant obtained was albino, which is a common phenomenon in cereal anther culture, rather than an effect of freezing.

The segments of the androgenic anthers showed a viability of 6%, whereas 21% of the pollen embryos survived freezing. The retrieved cultures underwent a lag period of 2–4 weeks, thereafter proliferation was initiated to form callus. The callus, upon transfer to a hormone-free medium, or to a medium supplemented with IAA + BA + CH, differentiated into shoots/plantlets.

3.2.3 Callus, Cell Suspension, Protoplasts

The rice callus obtained from various explants, i.e., germinating seed (Sala et al. 1979), androgenic anthers (Bajaj 1980b), and endosperm (Bajaj 1981), has been subjected to freezing with various degrees of success. Cell survival varied from 70–100% (Cella et al. 1982). Detailed studies on rice have been discussed earlier (Bajaj 1982; Bajaj and Sala 1991), the following is a summary of the same.

Sala et al. (1979) used cell suspensions in the middle of the exponential phase of growth subcultured at 2-week intervals. The optimum concentration was (5%) DMSO or (15–20%) glycerol. Best results were obtained when very slow cooling rates were used, while the viability values obtained when cells were directly immersed in LN did not substantially differ from those obtained with intermediate cooling rates. In a later study (Cella et al. 1982), rice cells returned to in vitro culture after preservation in LN were characterized by a number of physiological alterations, including: reduction in respiration rate and glucose uptake, loss of intracellular potassium, decrease in the cellular level of ATP, etc. Plant regeneration was not attempted in these cultures, however, it should be recalled that the calli derived from retrieved, excised anthers and the endosperm (Table 8) retained their morphogenetic potential and regenerated plants.

Fused protoplasts of rice × pea, and rice × sorghum (Bajaj 1983b, c) survived freezing; their viability ranged from 10.3–13.9%. Budding of the protoplasts was a common phenomenon. The heterokaryon showed two nuclei of different sizes; fusion of the nuclei was also observed in the somatic hybrid cells.

3.3 Cryopreservation of Maize (Table 10)

Cryopreservation studies on maize have been recently reviewed (Zhang et al. 1994), the following is a summary of the same.

3.3.1 Seeds and Embryos

At CIMMYT maize germplasm is successfully conserved by storing seed at 1 to −15 °C (CIMMYT 1986; Taba 1990). The storage conditions ensure that seed remain viable for 20–25 years before having to be regenerated for seed renewal. de Boucaud and Cambecedes (1988) stored hydrated seeds after rapid freezing in LN. Of all the different cryoprotectants tested (i.e., 1,2-propanediol, DMSO, sucrose, glycerol), 1,2-propanediol (5 M) added with sucrose (20%) was best for damage-free storage. Later, immature zygotic embryos of maize were also cryopreserved (Delvallée et al. 1989).

3.3.2 Embryogenic Callus and Protoplasts

The highly regenerable embryogenic calli initiated from anther culture of genotype Pai-Tang-Bai were used for cryopreservation (Sun and Jian 1989). The

Table 10. Summary of cryopreservation studies on maize (see also Zhang et al. 1994)

Material	Response to freezing	Reference
Seed	Hydrated seed treated with 1,2-propanediol remained viable in LN	de Boucaud and Cambecedes (1988)
Immature embryo	Remained viable	Delvallée et al. (1989)
Callus	Proliferation, plant regeneration	Jian et al. (1987); Sun and Jian (1989)
Callus cells	Proline effective cryoprotectant	Withers and King (1979)
Protoplasts	Cryoprotectant DMSO + glucose, plant regeneration	Zhang et al. (1990)
Protoplasts from frozen cells	Plant regeneration	Shillito et al. (1989)
Pollen	Retained fertilizing ability after storage in LN for one year	Nath and Anderson (1975); Barnabas and Rajki (1976, 1981); Filippova (1985)

results show that higher survival was obtained from embryogenic callus subcultured for 12–15 days, treated with a mixture of DMSO (5%) + glycerine (10%) + sucrose (8%), cooled at the rate of 1 °C/min (from 0 to −40 °C), followed by storage in LN, and thawed at 40 °C. Earlier, proline was also observed to be an effective cryoprotectant (Withers and King 1979). Plantlets were regenerated from retrieved calli.

Protoplasts isolated from light yellow embryogenic calli of an F_1 hybrid (Hiao-pa-thang × Shuai-bai) were used for freezing studies (Zhang et al. 1990). The protoplasts treated with a mixture of 5% DMSO + 0.55 M glucose were frozen at the rate of 1 °C/min to −40 °C, and then immersed in LN. The survival of protoplasts varied between 30–40%, and in culture the first division occurred within 2 days. The recovery phase reported for protoplasts of *Atropa, Datura,* and *Nicotiana* protoplasts (Bajaj 1988) was not observed here. Plantlets were eventually regenerated from retrieved calli (Zhang et al. 1990).

3.3.3 Pollen

The pollen of cereals are generally short-lived. Maize pollen in particular remain viable only from a few hours to 10 days, depending on the storage conditions and the genotype (Sartoris 1942; Pfahler and Linskens 1972). Freeze-dried pollen retained their viability for 2 weeks (Poncova 1959), however, LN storage offers the possibility of long-term conservation (Nath and Anderson 1975; Barnabas and Rajki 1976; Filippova 1985). Nath and Anderson (1975) observed that the viability was extended to 180 days by sudden freezing in LN. Barnabas and Rajki (1976; 1981) demonstrated that the water content of maize pollen is critical for ultra-low temperature storage, and that the critical water content during storage was about 30% of its original moisture. Maize pollen have now been claimed to be successfully stored for 10 years in LN (Barnabas 1994). The single most important reason for the interest in the preservation of pollen is its use in plant breeding as and when needed. Thus, the establishment of pollen banks assumes great importance (Bajaj 1987).

4 Summary

Though cereal seed is orthodox and poses no problems for the conservation of germplasm, the possibility of various changes, including physiological deterioration, genetic mutations, and contamination by pathogens, cannot be ruled out. In this regard, storage of seed in liquid nitrogen would avoid some of these problems.

Moreover, recently, in vitro culture studies have resulted in the creation of genetic variability in wheat, rice, and maize, i.e., somaclonal variants of various nature, genetic transformants, somatic hybrids and cybrids, mutations, etc. At present, no method is available by which genetic stability of cell cultures can be maintained. Cryopreservation of cultures would ensure their long-term conservation. Since complete plants can be regenerated from cell cultures of these three crops when subjected to freezing, cryopreservation has a special significance for the conservation of somaclones having desired agronomic traits.

References

Bajaj YPS (1980a) Freeze-preservation of plant cells – a novel approach to the conservation of germplasm. In: Gupta AK (ed) Genetics and wheat improvement. Oxford & IBH, New Delhi, pp 141–149

Bajaj YPS (1980b) Induction of androgenesis in rice anthers frozen at −196 °C. Cereal Res Commun 8: 365–369

Bajaj YPS (1981) Growth and morphogenesis in frozen (−196 °C) endosperm and embryos of rice. Curr Sci 50: 947–948

Bajaj YPS (1982) Induction and cryopreservation of genetic variability in rice. Int Rice Tissue Culture Planning Conf 1980. IRRI, Los Baños, pp 99–111

Bajaj YPS (1983a) Regeneration of plants from pollen-embryos of *Arachis, Brassica* and *Triticum* spp. cryopreserved for one year. Curr Sci 52: 484–486

Bajaj YPS (1983b) Survival of somatic hybrid protoplasts of wheat × pea and rice × pea subjected to −196 °C. Indian J Exp Biol 21: 120–122

Bajaj YPS (1983c) Cryopreservation of germplasm of cereals—progress and prospects. In: Sakamoto S (ed) Proc 6th Int Wheat Genet Symp, Kyoto Univ, Kyoto, pp 565–574

Bajaj YPS (1984) The regeneration of plants from frozen pollen embryos and zygotic embryos of wheat and rice. Theor Appl Genet 67: 525–528

Bajaj YPS (1985) Cryopreservation of embryos. In: Kartha KK (ed) Cryopreservation of plant cells and organs. CRC Press, Boca Raton, pp 227–242

Bajaj YPS (1986a) In vitro regeneration of diverse plants and the cryopreservation of germplasm in wheat (*Triticum aestivum* L.). Cereal Res Commun 14: 305–311

Bajaj YPS (1986b) In vitro preservation of genetic resources – techniques and problems. Int Symp Nuclear techniques and in vitro culture for plant improvement, 1985, IAEA, Vienna, pp 43–57

Bajaj YPS (1987) Cryopreservation of pollen and pollen embryos, and the establishment of pollen banks. Int Rev Cytol 107: 397–420

Bajaj YPS (1988) Regeneration of plants from frozen (−196 °C) protoplasts of *Atropa belladonna* L., *Datura innoxia* Mill and *Nicotiana tabacum* L. Indian J Exp Biol 26: 289–292
Bajaj YPS (1990a) Cryopreservation of germplasm of wheat. In: Bajaj YPS (ed) Biotechnology in agriculture and forestry, vol 13. Wheat. Springer, Berlin Heidelberg New York, pp 669–681
Bajaj YPS (ed) (1990b) Biotechnology in agriculture and forestry, vol 11. Somaclonal variation in crop improvement I. Springer, Berlin Heidelberg New York
Bajaj YPS, Bidani M (1980) Differentiation of genetically variable plants from embryo-derived callus of rice. Phytomorphology 30: 290–294
Bajaj YPS, Sala F (1991) Cryopreservation of germplasm of rice. In: Bajaj YPS (ed) Biotechnology in agriculture and forestry, vol 14. Rice. Springer, Berlin Heidelberg New York, pp 553–571
Bajaj YPS, Saini SS, Bidani M (1980) Production of triploid plants from the immature and mature endosperm cultures of rice. Theor Appl Genet 58: 17–18
Barnabas B (1994) Preservation of maize pollen. In: Bajaj YPS (ed) Biotechnology in agriculture and forestry, vol 25. Maize. Springer, Berlin Heidelberg New York, pp 607–618
Barnabas B, Rajki E (1976) Storage of maize (*Zea mays* L.) pollen in liquid nitrogen at −196 °C. Euphytica 25: 747–752
Barnabas B, Rajki E (1981) Fertility of deep-frozen maize (*Zea mays* L.) pollen. Ann Bot 48: 861–864
Cella R, Colombo R, Galli MG, Nielsen E, Rollo F, Sala F (1982) Freeze-preservation of rice cells: a physiological study of freeze-thawed cells. Physiol Plant 55: 279–284
Chen THH, Kartha KK, Gusta LV (1985) Cryopreservation of wheat suspension culture and the regenerable callus. Plant Cell Tissue Organ Cult 4: 101–109
CIMMYT (1986) Seed conservation and distribution: the dual role of the CIMMYT maize germplasm bank. CIMMYT, Mexico, DF, pp 1–18
Coulibaly Y, Demarly Y (1979) Sur les conditions de développement des microspores de *Nicotiana tabacum* et d'*Oryza sativa* soumises à la température de l'azote liquide (−196 °C). CR Acad Sci Paris Ser D 286: 1065–1068
de Boucaud M Th, Cambecedes J (1988) The use of 1,2-propanediol for cryopreservation of recalcitrant seeds: the model case of *Zea mays* imbibed seeds. Cryo-Letters 9: 94–101
Devallée I, Guillaud J, Beckert M, Dumas C (1989) Cryopreservation of immature maize embryos after freeze-hardening in the ear and in vitro. Plant Sci 60: 129
Dowgert MF, Steponkus PL (1983) Effect of cold acclimation on intracellular ice formation in isolated protoplasts. Plant Physiol 72: 978
Dupuis I, Pace GM (1993) Gene transfer in maize male reproductive structures by particle bombardment of tassel primordia. Plant Cell Rep 12: 607–611
Earle ED, Kuehnle AR (1990) Somaclonal variation in maize. In: Bajaj YPS (ed) Biotechnology in agriculture and forestry, vol 11. Somaclonal variation in crop improvement I. Springer, Berlin Heidelberg New York, pp 326–351
FAO (1992) Production yearbook, vol 46. FAO, Rome
Filippova TV (1985) Possibility for long-term storage of maize pollen at −196 °C temperature. Nauka Novosibirsk 64–68
Finkle BJ, Ulrich JM (1982) Cryoprotectant removal as a factor in the survival of frozen rice and sugarcane cells. Cryobiology 19: 329–335
Golovkin MV, Abraham M, Morocz S, Bottka S, Feher A, Dudits D (1993) Production of transgenic maize plants by direct DNA uptake into embryogenic protoplasts. Plant Sci 90: 41–42
Gresshoff PM, Gartner E (1977) Cryopreservation of *Arabidopsis thaliana* and other seeds by storage in liquid nitrogen. Arabidopsis Inf Serv 14: 12
Hasezawa S, Baba A, Syono K (1989) Protoplast culture and transformation studies on rice. In: Bajaj YPS (ed) Biotechnology in agriculture and forestry, vol 9. Plant protoplasts and genetic engineering II. Springer, Berlin Heidelberg New York, pp 107–121
Hayakawa T, Zhu Y, Itch K, Kimura Y, Izawa T, Shimamoto K, Toriyama S (1992) Genetically engineered rice resistant to rice stripe virus, and insect transmitted virus. Proc Natl Acad Sci USA 89: 9865–9869
Ishikawa M, Sakai A (1978) Freezing avoidance in rice and wheat seeds in relation to water content. Low Temp Sci B36: 39

Jian LC, Sun DL, Sun LH (1987) A study on some factors in cryopreservation of sugarcane calli. Acta Bot Sin 29: 123–131

Kendall EJ, Qureshi JA, Kartha KK, Leung N, Chevrier N, Caswell K, Chen THH (1990) Regeneration of freezing-tolerant spring wheat (*Triticum aestivum* L.) plants from cryoselected callus. Plant Physiol 94: 1756–1762

Koziel MG, Beland GL, Bowman C, Carozzi NB, Crenshaw R, Crossland L, Dawson J, Desai N, Hill M, Kadwell S, Launis K, Lewis K, Maddox D, PcPherson K, Meghji MR, Merlin E, Rhodes R, Warren GW, Wright M, Evola SV (1993) Field performance of elite transgenic maize plants expressing an insecticidal protein derived from *Bacillus thuringiensis*. Biotechnology 11: 194–200

Kuriyama A, Watanabe K, Ueno S, Mitsuda H (1990) Inhibitory effect of ammonium ion on recovery of cryopreserved rice cells. Plant Sci 64: 231–235

Kyozuka J, Kaneda T, Shimamoto K (1989) Production of cytoplasmic male sterile rice (*Oryza sativa* L.) by cell fusion. Biotechnology 7: 1171–1174

Langis P, Steponkus PL (1990) Cryopreservation of rye protoplasts by vitrification. Plant Physiol 92: 666–671

Lazanyi J (1994) Somaclonal variation in the progeny of CH1-31 maize inbred line derived from tissue culture. In: Bajaj YPS (ed) Biotechnology in agriculture and forestry, vol 25. Maize. Springer, Berlin Heidelberg New York, pp 265–275

Lee BT, Murdoch K, Topping J, Kreis M, Jones MGK (1990) Transient gene expression in wheat (*Triticum aestivum*) protoplasts. In: Bajaj YPS (ed) Biotechnology in agriculture and forestry, vol 13. Wheat. Springer, Berlin Heidelberg New York, pp 659–668

Ling DH (1991) Male sterile mutants from rice somaclones. In: Bajaj YPS (ed) Biotechnology in agriculture and forestry, vol 14. Rice. Springer, Berlin Heidelberg New York, pp 347–367

Lipman CB (1936) Normal viability of seeds and bacterial spores after exposure to temperature near the absolute zero. Plant Physiol 11: 201

Meijer EGM, Van Iren F, Schrijnmakers E, Hensgens LAM, Van Zigderreld, Schilperoort RA (1991) Retention of the capacity to produce plants from protoplasts in cryopreserved cell lines of rice (*Oryza sativa* L.). Plant Cell Rep 10: 171–174

Metakovsky EV, Sozinov AA (1990) Genetics of gliadin proteins and the problems of interpreting results obtained with somaclonal variation in wheat. In: Bajaj YPS (ed) Biotechnology in agriculture and forestry, vol 13. Wheat. Springer, Berlin Heidelberg New York, pp 526–537

Mooney PA, Goodwin PB (1994) Genetic transformation in *Triticum* species (wheat). In: Bajaj YPS (ed) Biotechnology in agriculture and forestry, vol 29. Plant protoplasts and genetic engineering V. Springer, Berlin Heidelberg New York, pp 383–400

Murashige T, Skoog F (1962) A revised medium for rapid growth and bioassays with tobacco tissue cultures. Physiol Plant 15: 473–497

Nath J, Anderson JO (1975) Effect of freezing and freeze-drying on the viability and storage of *Lilium longiflorum* L. and *Zea mays* L. pollen. Cryobiology 12: 81–88

Pfahler PL, Linskens HF (1972) In vitro germination and pollen tube growth of maize (*Zea mays* L.) pollen. VI. Combined effects of storage and the alleles at the waxy (wx), sugary (su1) and shrunken (sh2) loci. Theor Appl Genet 42: 136–140

Plucknett DL, Smith NJH, Williams JT, Murthi Anishetty N (1983) Crop germplasm conservation and developing countries. Science 220: 163–169

Poncova J (1959) Experiments on extending the viability of maize pollen by preservation. Ceskoslov Akad Zemedel Ved Sborn 32: 1637–1650

Raineri DM, Bottino P, Gordon MP, Nester EW (1990) *Agrobacterium*-mediated transformation of rice (*Oryza sativa* L.). Biotechnology 8: 33–38

Rhodes CA, Pierce DA, Mettler IJ (1989) Genetically transformed maize (*Zea mays* L.) plants. In: Bajaj YPS (ed) Biotechnology in agriculture and forestry, vol 9. Plant protoplasts and genetic engineering II. Springer Berlin Heidelberg New York, pp 99–106

Sakai A, Noshiro M (1975) Some factors contributing to the survival of crop seeds cooled to the temperature of liquid nitrogen. In: Frankel O, Hawkes JG (eds) Crop genetics resources for today and tomorrow. Univ Press, Cambridge, p 317

Sala F, Cella R, Rollo F (1979) Freeze-preservation of rice cells grown in suspension culture. Physiol Plant 45: 170–176

Sartoris GB (1942) Longevity of sugarcane and corn pollen: a method for distance shipment of sugarcane pollen by airplane. Am J Bot 42: 395–400

Shillito RD, Carswell GK, Johnson CM, DiMaio JJ, Harms CT (1989) Regeneration of fertile plants from protoplasts of elite inbred maize. Biotechnology 7: 581–587

Somers DA, Anderson PC (1994) In vitro selection for herbicide tolerance in maize. In: Bajaj YPS (ed) Biotechnology in agriculture and forestry, vol 25. Maize. Springer, Berlin Heidelberg New York, pp 293–313

Stanwood PC (1980) Tolerance of crop seeds to cooling and storage in liquid nitrogen (−196 °C). J Seed Technol 5(1): 26

Stanwood PC (1985) Cryopreservation of seed germplasm for genetic conservation. In: Kartha KK (ed) Cryopreservation of plant cells and organs. CRC, Boca Raton, pp 199–226

Stanwood PC, Bass LN (1981) Seed germplasm preservation using liquid nitrogen. Seed Sci Technol 5: 699

Steponkus PL, Dowgert MF (1981) Phenomenology of intracellular ice nucleation in isolated protoplasts. Plant Physiol 67: s-58

Sun LH, Jian LC (1989) The cryopreservation of maize (*Zea mays* L.) calli. Chinese Bull Bot 6: 28–30

Sun Z-X, Zheng K-L (1990) Somaclonal variation in rice. In: Bajaj YPS (ed) Biotechnology in agriculture and forestry, vol 11. Somaclonal variation in crop improvement I. Springer, Berlin Heidelberg New York, pp 288–325

Taba S (1990) Germplasm bank, conservation and use of CIMMYT maize genetic resources at CIMMYT. Maydica 35: 183–186

Takeuchi M, Matsushima H, Sugawara Y (1982) Totipotency and viability of protoplasts after long-term freeze preservation. In: Fujiwara A (ed) Plant tissue culture 1982. Maruzen, Tokyo, pp 797–798

Terada R, Kyozuka J, Nishibayashi S, Shimamoto K (1987) Plantlet regeneration from somatic hybrids of rice (*Oryza sativa* L.) and barnyard grass (*Echinochloa oryzicola* Vasing). Mol Gen Genet 210: 39–43

Thiselton-Dryer W (1928) On the influence of the temperature of liquid hydrogen on the germinative power of seeds. R Soc Proc 62: 161

Toriyama K, Hinata K (1988) Diploid somatic hybrid plants regenerated from rice cultivars. Theor Appl Genet 76: 665–668

Uchimiya H, Toriyama K (1991) Transformation in rice. In: Bajaj YPS (ed) Biotechnology in agriculture and forestry, vol 14. Rice. Springer, Berlin Heidelberg New York, pp 415–421

Vasil V, Ferl RJ, Vasil IK (1988) Somatic hybridization in the Gramineae: *Triticum monococcum* L. (Einkorn) + *Pennisetum americanum* (L.) Schum (pearl millet). J Plant Physiol 132: 160–163

Vasil V, Castillo AM, Fromm ME, Vasil IK (1992) Herbicide resistant fertile transgenic wheat plants obtained by microprojectile bombardment of regenerable embryogenic callus. Biotechnology 10: 667–674

Wang TB, Niizeki M, Harada T, Ishikawa R, Qian YQ, Saito K (1993) Establishment of somatic hybrid cell lines between *Zea mays* L. (maize) and *Triticum* sect. *tritigia* Mackey (*tritigia*). Theor Appl Genet 86: 371–376

Weber G, Roth EJ, Schweiger HG (1983) Storage of cell suspensions and protoplasts of *Glycine max* (L.) Merr., *Brassica napus* (L.), *Datura innoxia* (Mill.), and *Daucus carota* (L.) by freezing. Z Pflanzenphysiol 109: 29–39

Withers LA, King PJ (1979) Proline – a novel cryoprotectant for the freeze-preservation of cultured cells of *Zea mays* L. Plant Physiol 64: 675–678

Zhang SB, Jian LC, Kuo CS, Qu GP, Qian YQ (1990) Bud and root differentiation of maize (*Zea mays* L.) protoplasts after cryopreservation. Acta Biol Exp Sin 23: 117–121

Zhang SB, Jian LC, Qian YQ, Bajaj YPS (1994) Cryopreservation of germplasm of maize. In: Bajaj YPS (ed) Biotechnology in agriculture and forestry, vol 25. Maize. Springer, Berlin Heidelberg New York, pp 619–626

III.2 Cryopreservation of Foxtail Millet (*Setaria italica* L.)

T.G. Lu and C.S. Sun[1]

1 Introduction

There are about 125 species of *Setaria* (Gramineae), but only few of them are important for food production. One of the most important species, foxtail millet (*Setaria italica* L., chromosome number: $2n = 18$), is cultivated in both the old and New World for hay, pasture, and as a cereal. It is an important grass in India, China, and Japan, and is grown in North and South America, Australia, and parts of Africa as a minor cereal. It can be grown in regions of low rainfall and can tolerate a wide range of soil conditions. *Setaria italica* is an annual plant with stems that branch little. The stalk is filled with loose tissue. Several *Setaria* species other than S. *italica* are harvested locally: for example, S. *phacelata* areas in the savannah region of Africa, S. *pallidifusca* in the Upper Volta, S. *verticillata* around the oasis of Bilma (de Wet et al. 1979). Other species are harvested by Aborigines of Australia and in times of scarcity in India. (Blumer 1964).

Setaria constitutes only a small part of cereals produced in the world; its germplasm is customarily stored through seed at several national and international institutes (Table 1). Farmers find it convenient and economical to conserve *Setaria* seed at room temperature. Seeds with good qualities are harvested from plants and dried to a relative low water content (10 to 13%). The ability of seed germination can be preserved for 3 to 5 years. Another method for storage of *Setaria* seeds is the germplasm bank; this can be divided into three types according to its storage period, short, medium, and long-term storage (Table 2). Therefore, in this regard cryopreservation (−196 °C) is of limited application compared to, for instance, in wheat and rice (Bajaj 1983, 1990; Bajaj and Sala 1991). However, *S. italica* is very important because of its several special properties: more resistant to drought conditions, higher-yielding, and richer in amino acids (Zhang and Meng 1990) and the B family vitamins (Wang et al. 1991) necessary for mankind than wheat, rice, or maize. With the recent advances in biotechnology of *Setaria italica* (Xu et al. 1984; Dong and Xia 1989; Lu and Sun 1990, 1992, 1994. Zhao et al. 1990, new methods for the conservation of callus and suspension cultures have to be developed. Thus cryopreservation as a means of germplasm preservation can be applied to *Setaria italica.*

[1]Department of Cell Biology, Institute of Botany, Academia Science, Beijing 100044, China

Biotechnology in Agriculture and Forestry, Vol. 32
Cryopreservation of Plant Germplasm I (ed. by Y.P.S. Bajaj)

Table 1. Collection of *Setaria* (Acheampong et al. 1985)

Origin/country	*Setaria italica*	S. *viridis*	*S.* spp.
China	3,206		
Czech and Slovak Republics	+	+	
Germany	+	+	
Hungary	2		
India (NBPGR)	+		
India (AICMIP) (Maharashtra)	3,588		
India (AICMIP) (Bangalore)	1,429		
India (ICRISAT)	1,160		
Japan (NIAS)	182		
Japan (PGI)	274		
Malawi			6
Nepal	3		
USSR	+		
USA (SRPIS)	16		
Upper Volta		1	
Total[a]	9880	1	6

[a]Totals include replicates.
+ Information on the number of samples not available.
AICMIP: All-India Coordinated Millets Improvement Project (India).
ICRISAT: International Crop Research Institute for the Semi-Arid Tropics (India).
NBPGR: National Bureau of Plant Genetic Resources (India).
NIAS: National Institute of Agriculture Science (Japan).
PGI: Plant Germplasm Institute (Japan).
SRPIS: Southern Regional Plant Introduction Station (Georgia, USA).

Table 2. Main characteristics of three types of storage. (Ma 1989)

Storage period	Seed water content (%)	Relative humidity of conservation (%)	Storage temperature (°C)	Survival (able to germinate)
Short-term storage	10 to 13	50 to 60	10 to 15	5 years
Medium-term storage	8	Below 60	0 to 10	15 years
Long-term storage	5 to 8	Below 50	−10 to −20	Several decades to 100 years

2 Cryopreservation

There is only one report on cryopreservation of *Setaria italica* (Lu and Sun 1992). Callus and cell suspension cultures survive freezing in liquid nitrogen (LN) and retain their morphogenetic potentials. Fertile plants were recovered from cryopreserved callus and cells, and seeds were harvested from these plants.

2.1 Materials and Methods

Initiation of Friable, Embryogenic Callus and Cell Suspension. Young inflorescences of *Setaria italica* L. about 0.5 to 1 cm long were cut into two or three pieces and cultured on initiation medium (N_6 solid medium containing 2 mg/l 2,4-D and 3% sucrose). Cultures were incubated at 25 to 27 °C in a 12-h light/12-h dark photoperiod (125 μmol/m^2/s, provided by cool white fluorescent light). Callus was formed from the segments of inflorescences within 14 days of culture initiation. The callus was subcultured on the initiation medium every 20 days, and friable, embryogenic callus was selected after 1 to 2 months of subculture. Some of the callus was used for testing regeneration capacity by transferring it to differentiation medium containing N_6 or MS salts and vitamins, 200 mg/l CH, 3% sucrose, 0.65% agar, 0.5 mg/l BA, and 0.2 mg/l NAA. The pH was adjusted to 5.6 to 5.8 before autoclaving.

About 1 g of friable, embryogenic callus with high regeneration capacity was transferred to flasks containing 40 ml of liquid N_6 medium supplemented with 2 mg/l 2,4-D, 3% sucrose, 200 mg/l CH, and 1 g/l MES. The pH was adjusted to 5.6 to 5.8 before autoclaving as above. The flasks were incubated on a gyratory shaker (120 rpm) in the dark at 26 °C and subcultured every 3 to 5 days using a selection subculture method. Small cell aggregates were selected every 10 to 15 days by permitting the large aggregates to sediment and then subculturing the supernatant. Finely dispersed cell suspensions were established in 2 to 3 months. Plant regeneration capacity of cell suspensions was evaluated by transferring 2 to 3 ml of packed fresh cells to the regeneration medium.

Cryopreservation of Callus and Suspension Cells. After 6 to 8 months of establishment, approximately 200 mg of friable, embryogenic callus or suspension cells at different days of subculture were collected and mixed with 4 ml of precooled (0 °C) cryoprotectants followed by an equilibration period of 30 min at 0 °C. The specimens in cryoprotective solution were frozen as follows: the temperature of specimens was lowered at the rate of 1 °C/min from 0 to −8 °C, where the temperature was held constant for 20 min. The cultures were then cooled at the same cooling rate from −8 to −36 °C, where the temperature was held constant for 2.5 h. The cultures were then transferred to liquid nitrogen.

Callus and suspension cells were stored in LN for 10 days and then thawed rapidly in a 40 °C water bath and washed three times with fresh liquid culture medium. The thawed callus and cells were divided into two parts: one part was transferred immediately to N_6 solid medium containing 2 mg/l 2,4-D and 3% sucrose for recovery in the dark at 26 °C. Another part was used to determine the viability of cryopreserved callus and suspension cells by the TTC method (Towill and Mazur 1975). Three ml of 0.6% TTC solution were added to 200 mg of callus or suspension cells and incubated in small assay tubes (10 × 75 mm) without shaking for 18 to 22 h in the dark at 22 °C. After incubation, the callus or cells were pelleted and washed once with distilled water. The red formazan was extracted from the pelleted callus or cells with 3 ml of 95% ethanol for 10 min in a 60 °C water bath. Absorbance of the extract was read at 485 nm with a 752-spectrophotometer (made in China).

Cryopreservation of Protoplasts. Protoplasts were isolated from 3-day-old suspension cultures. Approximately 1 g fresh weight of cells was added to 10 ml of filter sterilized enzymatic digestion medium (1% Cellulase Onozuko RS and 0.5% Pectolyase Y-23 dissolved in solution of 0.5 mM KH_2PO_4, 1.5 mM $CaCl_2 \cdot 2H_2O$, 10 mM KNO, 1 mM $MgSO_4 \cdot 7H_2O$, 5 mM glucose, KM-8p vitamins, 2 mg/ml bovine serum and 0.45 M Mannitol) and incubated at 26 °C in the dark for 4 h.

About $1–2 \times 10^6$ protoplasts were mixed with 3 ml precooled (0 °C) cryoprotectants (10% DMSO + 0.5 M sorbitol). The freezing and thawing methods were the same as those for callus and cell cultures. The viability of protoplasts was tested immediately after thawing by FDA staining (Widholm 1972). The thawed protoplasts were washed three times by centrifugation at low speed of 300 rpm with fresh culture medium and cultured in densities of 2×10^5/ml in a thin layer of KM-8p medium (Kao and Michayluk 1975) supplemented with 1 mg/l 2,4-D under atmospheric conditions in the dark at 26 °C.

Plant Regeneration. Plant regeneration from recovered callus and cells was induced by transfer to differentiation medium as described above. Rooted plants were potted in soil and grown under normal atmospheric conditions.

2.2 Results

Friable, embryogenic callus (Fig. 1A), and finely dispersed cell suspensions (Fig. 1B) used for cryopreservation were highly regenerable when transferred to differentiation medium (Fig. 1C and D, respectively). The viability test of thawed callus and suspension cells by the TTC method indicated that 10% DMSO plus 0.5 M sorbitol was most effective for cryopreservation of cultured callus and cells (Table 3).

Table 3. The effects of different kinds of cryoprotectants on survival of thawed callus and suspension cells of *Setaria italica* determined by TTC value[a]. (Lu and Sun 1992)

Cryoprotectants	TTC value (OD/g): 485 nm Callus	Suspension cells
10% DMSO + 0.5 M sorbitol	9.444 ± 0.402	10.158 ± 0.375
	18.900[b] ± 2.151	19.850[a] ± 1.763
10% DMSO + 0.5 M surcose	7.836 ± 0.308	8.950 ± 0.412
10% DMSO + 0.5 M glucose	8.063 ± 0.364	9.753 ± 0.389
10% DMSO + 0.5 M mannitol	7.433 ± 0.377	9.331 ± 0.399
5% DMSO + 0.5 M sucrose +0.5% glycerol	6.825 ± 0.443	8.110 ± 0.451
0	0.210 ± 0.098	0.190 ± 0.076

[a]Callus and suspension cells for cryopreservation were previously subcultured for 10 and 3 days, respectively.
[b]Represents the TTC value of fresh (unpreserved) callus or suspension cells previously subcultured for 10 or 3 days, respectively.

Table 4. The relationship among subculture time, recovery time and survival level of thawed callus and suspension cells of *Setaria italica* L.[a] (Lu and Sun 1992)

Days in subculture	Survival level			Recovery day after culture (initiation of regrowth)
	No. survived	Total no.	Survival frequency (%)	
Callus				
3	22	60	36.6	15
7	67	67	100	7
15	18	60	30	15
20	6	71	8.4	30
Cells				
3	57	57	100	4
5	54	63	85.7	5–6
7	19	65	29.2	10–15
9	8	70	11.4	30

[a]The cryoprotectant was 10% DMSO plus 0.5 M sorbitol. Each piece of callus or cell clump plated on medium for recovery was about 2 × 2 mm in size.

Therefore, 10% DMSO plus 0.5 M sorbitol was used to evaluate the effect of subculture duration of callus and suspension cells on survival following cryopreservation. The relationship among subculture time, recovery time, and survival level are shown in Table 4.

The recovery time of thawed suspension cells was shorter than that for thawed callus. Cryopreserved cells (3-day subculture treatment) regrew 4 days after thawing, and developed yellow embryogenic microcallus or cell clusters that grew over the thawed cells within 8 days.

The growth rate of the recovered cells is lower in the first culture period (0 to 8 drops) and slightly higher in the second culture period (8 to 15 days) than that of the fresh cells (Fig. 2). In addition, the morphogenetic potentials of recovered cells appeared to be increased slightly. Even in solid N_6 medium containing 2 mg/l 2,4-D, the somatic embryos were formed and germinated to plantlets occasionally.

The results of FDA stained protoplasts showed about 15.5% of the thawed protoplasts survived, but cell division was not observed. However, the fresh (unfrozen) protoplasts underwent division 5 days after culture.

Many plants were regenerated from retrieved cells or callus when transferred to differentiation medium (N_6 or MS basal components, 0.5 mg/l BAP, 0.2 mg/l NAA) (Fig. 1E). These plants were potted in soil (Fig. 1G). From ten potted

Fig. 1A–H. Regeneration of fertile plants from cryopreserved callus and cell suspension cultures of *Setaria italica* L. **A** Friable, embryogenic callus with high regeneration capacity. **B** Finely dispersed cell suspensions derived from friable, embryogenic callus. **C** High regeneration of shoots from friable, embryogenic callus. **D** The formation and germination of somatic embryos derived from suspension cells. **E** A large number of plants regenerated from cryopreserved suspension cells. **F** Plants regenerated from cryopreserved suspension cells grown to maturity in soil. **G** Plants regenerated from cryopreserved callus survived and grew in soil. **H** Fertile plants from cryopreserved callus

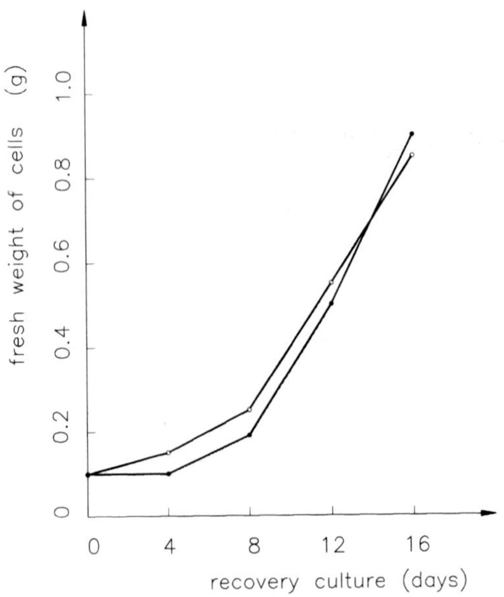

Fig. 2. Regrowth of thawed cells on N_6 solid medium supplemented with 2 mg/l 2,4-D. Each piece of cell clump plated on medium for recovery was about 2 × 2 mm in size. ○ growth of fresh cells; ● regrowth of thawed cells

plants, six survived to maturity. Though they were less vigorous than normal, each underwent self-pollination (Fig. 1F). Of the 173 harvested seeds, 48 germinated.

2.3 Discussion

One of the problems in protoplast culture is the loss of regeneration capacity in donor cells during subculture. This can now be partially overcome by means of cryopreservation. Suspension-cultured cells of maize were successfully stored in liquid nitrogen (Shillito et al. 1989). A large amount of viable protoplasts were isolated from recovered cells, and fertile plants of maize were regenerated. They reported that callus or cell clusters recovered from cryopreserved cells formed suspensions suitable for protoplast isolation more quickly than the original friable callus. The results obtained from maize, together with the present study on millet, demonstrated that cryopreservation is successful in overcoming the loss of regeneration capacity in donor cells in protoplast culture. Another important problem in tissue, cell, or protoplast culture is the fertility of regenerated plants. Although few reports are available, our results indicate that this difficulty can also be overcome by cryopreservation.

The cryoprotectant of 10% DMSO + 0.5 M sorbitol was confirmed to be most effective for cryopreservation of millet in this study. Similar results were obtained with sugarcane callus cryopreservation (Jian et al. 1987) and maize protoplast cryopreservation (Lu et al. 1989), while in cryopreservation of sugarcane excised meristem and callus cell suspensions, a cryoprotectant mixture of DMSO, sucrose, and glycerol gave good results (Bajaj et al. 1987). Changes in cell density and fresh mass are both useful in evaluating recovery potential. In many cases, growth is resumed rapidly, and the rate of growth closely resembles the normal growth rate of unfrozen cells; for example, in *Oryza sativa* (Sala et al. 1979). In the present study, cryopreserved callus and cells started regrowth 7 and 4 days after thawing, respectively. The growth rate of cryopreserved callus and cells resembled that of unpreserved callus and cells 8 and 15 days after culture, respectively. Cryopreservation might effectively select only strong, viable cells and recover quickly from cryopreservation and grow more rapidly than unselected cells. A cold-tolerant line of spring wheat has been obtained by cryopreservation (Kendall et al. 1990). Plants regenerated from cold-tolerant callus grew normally under cold conditions. In our studies fertile plants were obtained.

Protoplasts are different from cells in two important aspects; one is the absence of cell wall and the other is the lack of cytoplasmic connections with the adjacent cells. Thus, protoplasts, by not being interdependent, are, in fact, an ideal system to study the mechanism of cryopreservation at the cellular level. Protoplasts frozen in LN survived and retained their morphogenetic potentials as their control (unfrozen) protoplasts in several species, e.g., *Atropa belladonna*, *Datura innoxia*, *Nicotiona tabacum*, and *Zea mays.* (Bajaj 1988, 1989; Lu et al. 1989; Zhang et al. 1990). However, it has been realized that the thawed protoplasts are more fragile and more difficult to culture than the fresh ones. The culture failure of thawed protoplasts of *Setaria italica* is partially because of the low level of survival and the unsuitable culture system, while in maize the survival level of thawed protoplasts was about 80%, and successful nurse culture system was also employed (Lu et al. 1989).

3 Summary and Conclusions

Embryogenic callus, well-dispersed cell suspension cultures, and cell-suspension-derived protoplasts of Setaria italica were used for cryopreservation in liquid nitrogen. The results indicated that 3-day-old suspension cells and 7- to 10-day-old callus in subculture were most suitable for cryopreservation using a cryoprotectant of 10% DMSO plus 0.5 M sorbitol and a three-step freezing method. The cryopreserved cells recovered quickly at 4 days post-thawing on N_6 solid medium containing 3% sucrose and 2 mg/l 2,4-D. The survival of the thawed protoplasts was about 15.5%, but no cell division was observed in the culture. Plants were regenerated from recovered callus and suspension cells on a differentiation medium (N_6 or MS basal medium supplemented with 3% sucrose, 200 mg/l casein hydrolysate, 0.5 mg/l benzyladenine purine, and 0.2 mg/l naphthalene acetic acid). Ten plants were obtained from recovered callus and cells and were transferred to soil; six of them grew to maturity.

Acknowledgments. The authors are grateful to Professor Jian Ling-ching and Dr. Sun Long-hua, Institute of Botany, Academia Sinica, for help with this work.

References

Acheampong E, Anishetly NM, Williams JT (1985) A world survey of sorghum and millets germplasm. Int Board for Plant Genetic Resources, First Decade of Service (1974–1984), 39 pp

Bajaj YPS (1983) Cryopreservation of germplasm of cereals – progress and prospects. In: Sakamoto S (ed) Proc 6th Int Wheat Genet Symp. Kyoto University, Kyoto, pp 565–674

Bajaj YPS (1988) Regeneration of plants from frozen (−196 °C) protoplasts of *Atropa belladonna* L., *Datura innoxia* Mill and *Nicotiana tobacum* L. Indian J Exp Biol 26: 289–292

Bajaj YPS (1989) Cryopreservation of plant protoplasts. In: Bajaj YPS (ed) Biotechnology in agriculture and forestry, vol 8. Plant protoplasts and genetic engineering I. Springer, Berlin Heidelberg New York, pp 97–106

Bajaj YPS (1990) Cryopreservation of germplasm of wheat. In: Bajaj YPS (ed) Biotechnology in agriculture and forest, vol 13. Wheat. Springer, Berlin Heidelberg New York, pp 667–681

Bajaj YPS, Sala F (1991) Cryopreservation of germplasm of rice. In Bajaj YPS (ed) Biotechnology in agriculture and forest, vol 14. Rice. Springer, Berlin Heidelberg New York, pp 553–573

Bajaj YPS, Korneva SB, Gutierrez R, Maribona RH (1987) Freeze preservation of plantlets, excised meristems, and cell cultures of sugarcane (*Saccharum officinarum* L.) at −196 °C. In: Bose A, Sengupta P (eds) Proc Int Conf on Cryogenics. McGraw-Hill, New Delhi, pp 222–226

Blumer R (1964) Edible seeds and prehistoric stone mortars in the Highlands of New Guinea. Man 64: 147–150

De Wet JMJ, Oestry-Stidd IL, Cubero JI (1979) Origins and evolution of foxtail millets (*Setaria italica*). J Agric Tradit Bot Appl 26: 53–64.

Dong JJ, Xia ZA (1989) Plant regeneration from protoplasts of *Setaria italica.* Plant Physiol Commun 2: 56–57

Jian LC, Sun DL, Sun LH (1987) Sugarcane callus cryopreservation. Plant Biol 5: 323–337

Kao KN, Michayluk MR (1975) Nutritional requirements for growth of *Vicia hajastana* cells and protoplasts at a very low population density in liquid medium. Planta 126: 105–110

Kendall EJ, Qureshi JA, Kartha KK, Leung N, Chevrier E, Coswell K, Chen THH (1990) Regeneration of freezing-tolerant spring wheat (*Triticum aestivum*) plants from cryopreserved callus. Plant Physiol 94: 1756–1762

Lu TG, Sun CS (1990) Establishment of friable, embryogenic callus and cell suspension cultures with high regeneration capacity of millet (*Setaria italica*). In: Hu H (ed) Cell engineering and plant breeding. Beijing Industry Univ Press, Beijing, pp 357–361

Lu TG, Sun CS (1992) Cryopreservation of millet (*Setaria italica* L.). J Plant Physiol 139: 295–298

Lu TG, Sun CS (1994) Selection and characteristics of different types of callus during tissue culture of *Setaria italica.* Bot Res 7: 158–164

Lu TG, Sun CS, Jian LC, Sun LH (1989) Plant regeneration from cryopreserved protoplasts of maize (*Zea mays* L.). Chin J Bot 1: 165–166

Ma YS (1989) A survey of conservation for germplasm resources. In: Conservation techniques for crop resources. China Sci Technol Press, Beijing, pp 1–68

Sala F, Cella R, Rollo F (1979) Freeze-preservation of rice cells. Physiol Plant 45: 170–176

Shillito RD, Carswell GK, Johnson CM, Dimaio JJ, Harms CT (1989) Regeneration of fertile plants from protoplasts of elite inbred maize. Biotechnology 7: 581–587

Towill LE, Mazur P (1975) Studies on the reduction of 2,3,5-triphenyl tetrazolium chloride as a viability assay for plant tissue culture. Can J Bot 53: 1097–1102

Wang RQ, Gao JH, Sun JB (1991) Quality standards and measurement methods for good quality millet (*Setaria italica*) used as a food cereal crop. Millet Crops 3: 27–29

Widholm JM (1972) The use of fluorescein diacetate and phenosafranine for determining viability of cultured cells. Stain Technol 47: 189–194

Xu ZH, Wang DY, Yang LJ, Wei ZM (1984) Somatic embryogenesis and plant regeneration in cultured immature inflorescence of *Setaria italica.* Plant Cell Rep 3: 149–150

Zhang SB, Jian LC, Guo ZS, Qu GP, Qian YQ (1990) Differentiation of shoots and roots from cryopreserved protoplasts of maize (*Zea mays* L.). Acta Exp Biol Sin 23: 117–121

Zhang ZY, Meng SG (1990) Contents of eight kinds of amino acids necessary for mankind in *Setaria italica* L. Millet Crops 1: 46–47

Zhao LY, Ji Y, Duan SJ, Yan LL (1990) Establishment of effective culture and plant regeneration system from protoplasts of *Setaria italica* L. Millet Crops 4: 6–10.

III.3 Cryopreservation of Some Halophyte Grasses (*Puccinellia* Species)

Zs. Jekkel, G. Gyulai, and L.E. Heszky[1]

1 Introduction

The species of the genus *Puccinellia* Parl. (Atropis) are perennial halophyte grasses native to the Euro-Siberian region and adventive to North America. Besides the three widely popular species in Hungary, (1) *P. peisonis* (Beck.) Jáv., (2) *P. distans* (L.) Parl., and (3) *P. limosa* (Schur.) Holmbg., there have also been species like *P. pannonica* (Hack) which have disappeared from the Hungarian flora by this century (Simon 1992).

Puccinellia species, with their intensive tendency to interspecific hybridization, create a mobile genetic background, which explains the great adaptability of these species to extreme environmental conditions. *P. distans* (L.) Parl. (reflexed saltmarsh grass), and *P. limosa* (Schur.) Holmbg. (saltmarsh grass), are especially salt-tolerant grasses withstanding a very high level of salt and pH values of acidic and alkaline soil types. Because of this, *Puccinellia* sp. could play an important role in salt-tolerant breeding programs.

Although the seeds of these species, as with grasses in general, can be stored (after being desiccated to 6% seed water content) at 0 °C for a relatively long period (10 years) without decrease in viability, freeze preservation of their cells and organs has received increased attention in recent developments in the techniques of plant biotechnology. Genetic manipulation requires efficient protocols for the induction of suspension cultures, isolation of protoplasts, reestablishment of colonies of dividing cells, and the regeneration of plants.

The difficulty lies in maintaining embryogenic suspension cultures of the protoplast source, together with the problem of genetic instability. However, as soon as the cultures undergo the appropriate phase for protoplast isolation, they can be cryopreserved, and used again when the need arises.

In this chapter the results of cryopreservation of *P. distans* and *P. limosa* suspension cultures are reported.

[1]Department of Genetics and Plant Breeding, University of Agricultural Sciences, 2103 Gödöllö, Hungary

Biotechnology in Agriculture and Forestry, Vol. 32
Cryopreservation of Plant Germplasm I (ed. by Y.P.S. Bajaj)

2 Cryopreservation

2.1 Cell Suspension Culture

P. distans and *P. limosa* (wild-type) suspension cultures were established from the callus produced by seeds (received from the Research Center of Agrobotany, Tápiószele, Hungary) on agar MS medium containing 2 mg/l 2,4-D (2,4-dichlorophenoxy acetic acid) and 0.5 mg/l kinetin, and maintained in modified liquid B5 (Binh et al. 1989).

Cultures were incubated on a rotary shaker (120 rpm) at a photoperiod of 16 h light /8 h dark with a light intensity of 2000 lx at 26 °C and transferred weekly. For freezing, the exponentially growing cell aggregates were harvested on the 4th day after subculture, when the highly meristematic cell type with reduced vacuolar volume dominated in the growth phase. The size of the cell clumps varied from 1 to 2 mm in diameter.

Aggregates for plant regeneration were placed onto solid N_6 medium supplemented with 10 mg/l kinetin (Binh et al. 1989). The plant regeneration was analyzed by scanning electron microscopy, and intensive somatic embryogenesis was characterized (Fig. 1).

A large number of scutellar structures were developed from the 3rd week on the surface of actively dividing calli. Among the somatic embryos several twins were characterized (Fig. 1j). The regenerated plants were transferred to hormone-free medium for shoot multiplication.

2.2 Freezing and Thawing

In order to achieve the maximum number of surviving cell aggregates, a step-by-step experimental method was developed involving the determination of (1) freezing rate and (2) holding time, as a function of different transfer temperatures.

Freezing unprotected cells without holding them at transfer temperatures prior to immersion in LN

Polypropylene cryovials (2 ml) were filled with approximately 50 pieces of cell aggregates and 1 ml of liquid medium. Cooling rates ranging from 0.5 to 2 °C/min and 1000–2000 °C/min (direct immersion) were compared. Freezing was accomplished in a Sy-lab, Snowbal-1400 biological freezer. The freezing program began at +4 °C and continued down to −50 °C. Transfer temperatures −10, −20, −30, −40, and −50 °C were tested by the freezing program being interrupted at every subzero temperature −10, −20, −30, −40, −50 °C and samples being taken out of the freezing chamber and kept in liquid nitrogen for 10 min.

The effect of the cooling rates was studied without cryoprotectant treatments.

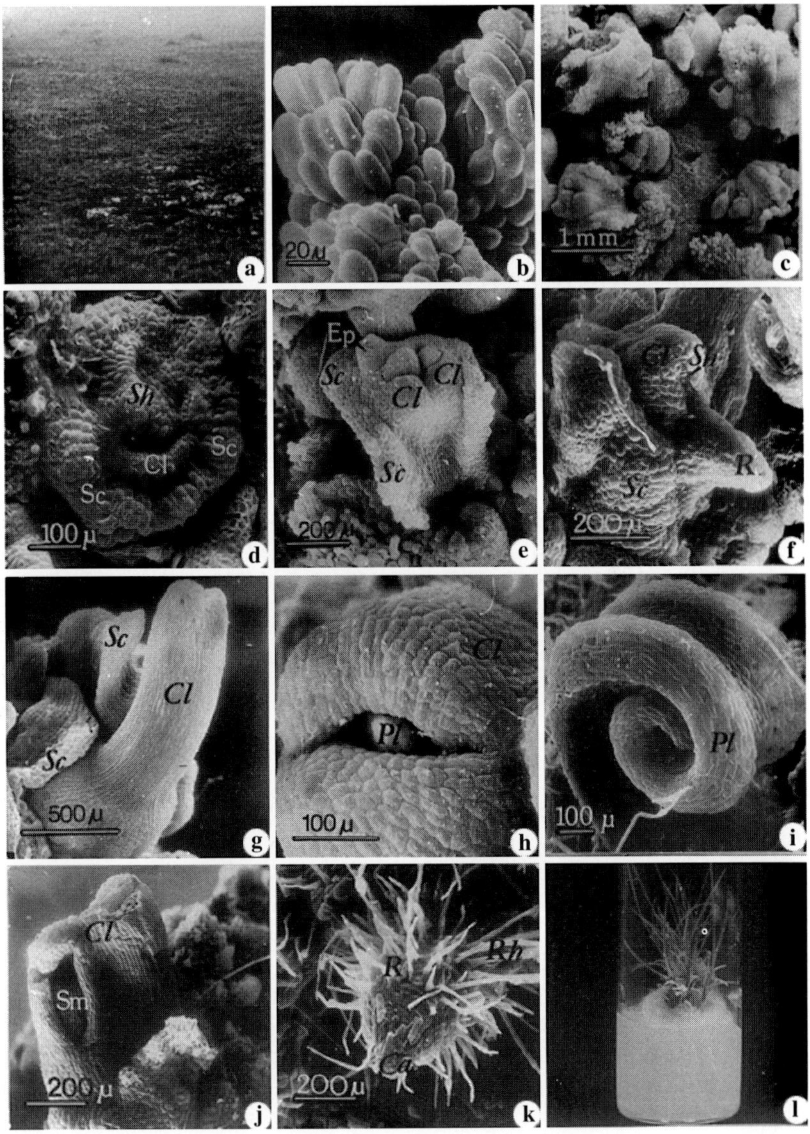

Fig. 1a–l. Scanning electron micrographs of the somatic embryogenesis of *P. limosa*. **a** Somatic embryogenesis in seed-callus culture. **b** Actively dividing callus cells. **c** Intensive somatic embryogenesis on the surface of calli. **d** Somatic embryo at the early scutellar stage (*Sc*) with differentiated shoot meristem (*Sh*) and coleoptile (*Cl*). **e** Epiblast (*Ep*) development of the later stage scutellar embryo with enfolded coleoptile (*Cl*). **f** Coleoptilar stage somatic embryo: *R* root tip; *Sh* shoot tip. **g** Coleoptile elongation. **h** Primary leaf (*Pl*) development emerging from coleoptile. **i** Fully developed primary leaf. **j** Twin embryos at the early coleoptilar stage. **k** Development of complete root system (*R*) with calyptra (*Ca*) and root hairs (*Rh*). **l** Multi-shoot culture of the regenerated plants. (unpubl. results)

Freezing cryoprotected cells without holding them at transfer temperatures prior to immersion in LN

Analytical grade of dimethylsulfoxide (DMSO) (v/v) and glycerol (v/v) sterilized by filtration were incorporated into the culture medium in concentrations of 12.5 and 15%. Double strength cryoprotectant solution was chilled on ice and the chemicals gradually added to an equal volume of cell suspension in four aliquots over a period of 1 h, then held in a refrigerator at 5 °C for 16 h. Freezing was carried out as described before but only the 1 °C/min cooling rate was used.

Freezing cryoprotected cells holding them at transfer temperatures prior to immersion in LN

Freezing protocol was improved by incorporating holding time periods into the freezing program. Prior to plunging into liquid nitrogen, vials were held at the current final temperature for up to 10, 20, 25, and 30 min.

Thawing was accomplished by using a +40 °C water bath. Vials were swirled in the bath until ice had just disappeared (50–60 s). Cell viability was determined either by the TTC test (triphenyl tetrazolium chloride) or by regrowth potential and plant regeneration.

2.3 Freezing Rates

The relationships between the freezing rates and viability of the freeze preserved *P. distans* cells are presented in Fig. 2. Viability observed under the current cooling rate is plotted against the five different transfer temperatures. Minimal survival was found in the case of all freezing rates at one or another of the subzero temperatures; there was, however, an optimal cooling rate for the cells at which maximum viability was observed.

Cell aggregates subjected to 1 °C/min freezing rate survived the LN in the highest number at every transfer temperature combination, with the exception of −10 °C. The best cooling rate and transfer temperature combination was the 1 °C/min with −40 °C, resulting in 45% survival. Aggregates showed considerable viability when frozen with the latter cooling rate and transferred at −20, −30, and −50 °C into LN, with survival of 22, 40, and 32%, respectively. Most of the published freezing programs are based on using 0.5–2 °C/min cooling rate intervals. Kartha et al. (1980) detailed the freezing rates and the related survivals, emphasizing the importance of exact determination throughout the freezing of strawberry and in pea (Kartha et al. 1979) meristems. A few cooling rate deviations brought dramatic changes to the survival. Viable cells were also observed by direct immersion of the vials into LN, although not at a satisfactory level. Bajaj achieved similar results in cassava (Bajaj 1977a), potato (Bajaj 1977b), and chickpea (Bajaj 1979), cryopreserved as meristems.

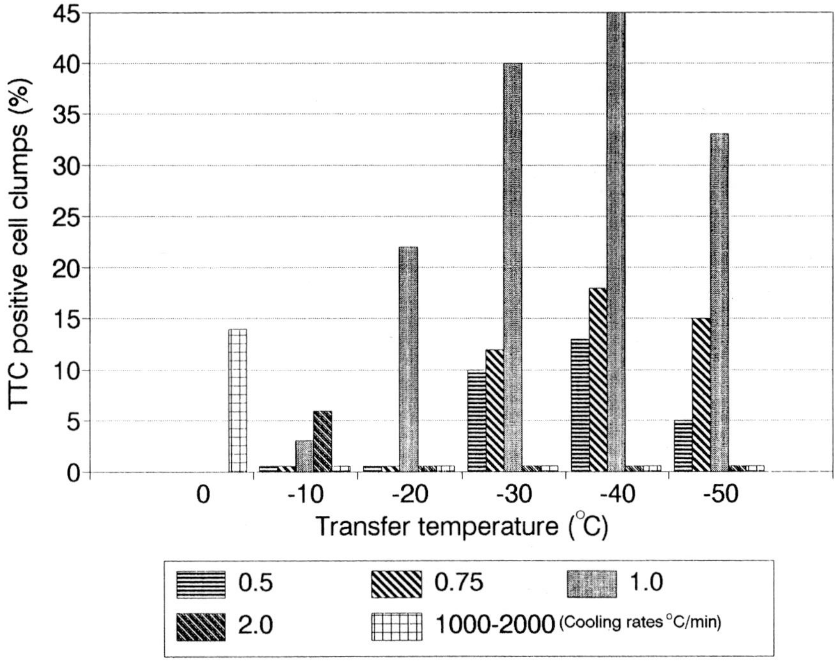

Fig. 2. Effect of cooling rate on survival of *P. distans* suspension cultures immersed in liquid nitrogen from different transfer temperatures. (Heszky et al. 1990)

2.4 Cryoprotectants

The data of survival of DMSO and glycerol cryoprotected *P. distans* cell aggregates frozen with 1 °C/min until the final subzero temperature, then plunged into LN, are presented in Fig. 3.

DMSO-treated cells showed normally higher viability than was observed using glycerol. Previous results (Heszky et al. 1990) indicated the importance of concentrations higher than 10% for survival. The maximum cryoprotective effect was obtained at 12.5% of DMSO, when the cells were prefrozen to −10 °C then plunged into LN (Fig. 3A). Similarly, 15% DMSO concentration resulted in a much higher survival rate than was found in the frozen untreated control group.

Dimethylsulfoxide is one of the most frequently applied penetrating cryoprotectants, which produces a useful amount of freezing point depression (Meryman et al. 1977) and stabilization of the cell membrane. The optimum DMSO solution, however, varies in the published results; some findings also suggest a direct connection between increased survival and higher DMSO content (Dereuddre et al. 1988). This anomaly seems to be demonstrated by the fact that the effect of DMSO is strongly related to its molar concentration, and the alterations in the optimum DMSO solution are based upon its toxic concentration, which obviously varies by species.

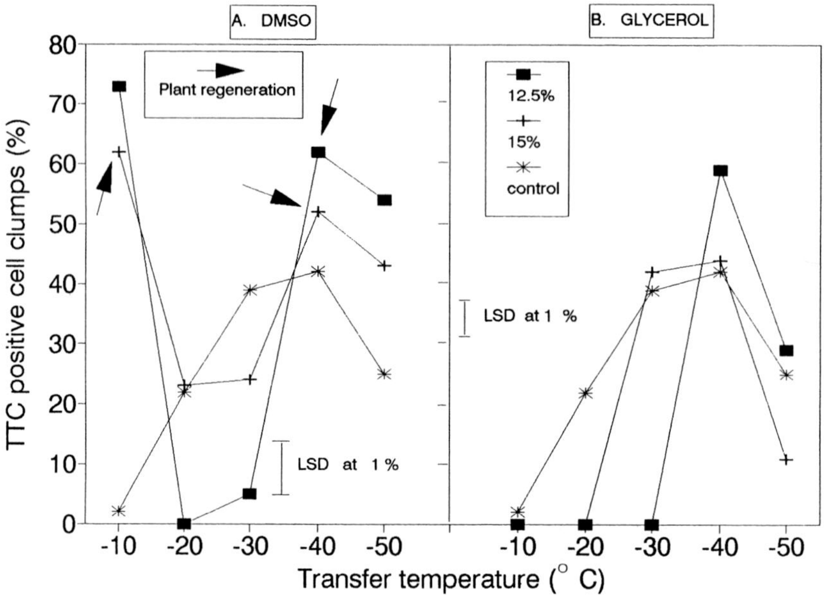

Fig. 3A, B. Effect of DMSO (**A**) and glycerol (**B**) on the viability and plant regeneration capacity of *P. distans* suspension cultures immersed in liquid nitrogen from different transfer temperatures. (Heszky et al. 1990)

Viability decreased below the control value or in the case of the 12.5% solution to about zero, when the cell aggregates were frozen until −20 or −30 °C prior to immersion; however, at the −40 and −50 °C transfer, the number of viable cells exceeded the value that was found in the control.

This gap in the viability curve disappeared when the glycerol cryoprotected cells were subjected to slow freezing, followed by immersion in LN (Fig. 3B). Until the −30 °C transfer temperature there were no viable cells independent of the glycerol concentration. Significantly, higher survival was found only among the cells treated with 12.5% glycerol and frozen to −40 °C (58%), but viability decreased again on using a final temperature of −50 °C. Glycerol has proved to be superior for animal cell systems, but is unsatisfactory for plant cells (Kartha 1985) when used alone. It is, however, referred to as an effective component of cryoprotectant mixtures (Finkle and Urlich 1978; Bajaj 1982).

2.5 Holding Time Periods

The viability of cryoprotected *P. distans* cells frozen in freezing programs involving holding time periods ranging from 10 to 30 min is illustrated in Figs. 4 and 5.

By holding the cell clumps cryoprotected with DMSO for 10–30 min prior to plunging in LN in subzero temperatures, a considerable improvement in their

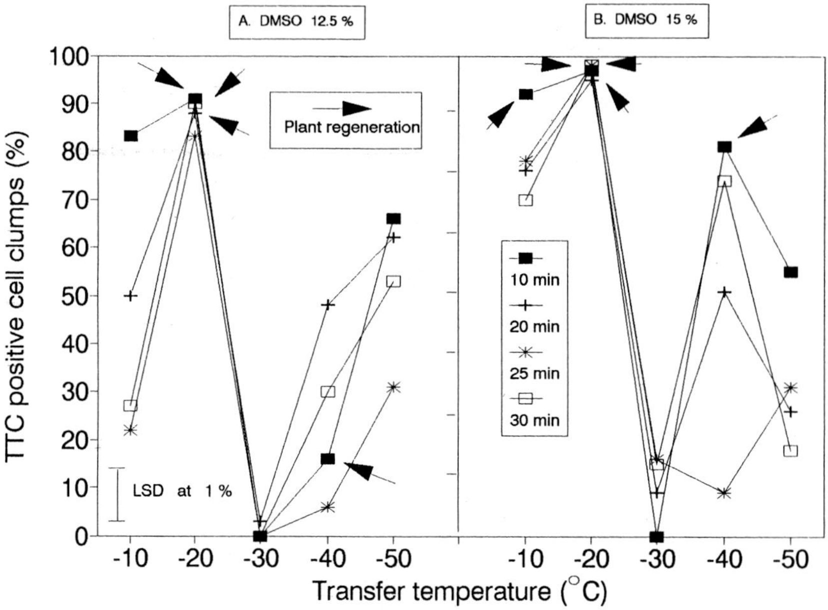

Fig. 4A, B. Effect of holding time periods on the viability and plant regeneration capacity of DMSO cryoprotected *P. distans* suspension cultures immersed in liquid nitrogen from different transfer temperatures. (Jekkel and Heszky 1991)

viability was observed (Fig. 4). This increase was especially striking among the cells frozen to −20 °C, using a 12.5% solution of DMSO, since a direct transfer from this temperature (Fig. 3A) resulted in viability toward zero. All of the holding periods tested, however, resulted in increased survival to ca. 90%. Similar to this observation, the cell reached maximum viability from this transfer temperature when a 15% DMSO solution was incorporated into the medium (more than 90% viability). As in Fig. 3, there was a well-definable subzero temperature, when the immersed cells completely lost their viability. However, while in the former case this occurred at both −20 and −30 °C, in this experiment, only the use of −30 °C resulted in zero viability and, in contrast, the −20 °C showed the highest survival.

This tendency appeared in the curve showing the results with improved holding time the freezing program of glycerol cryopreserved cell clumps (Fig. 5). The viability of directly immersed cells from the current transfer temperature, beginning with immersion at −40 °C in LN reached the maximum, and declined thereafter. The expanded freezing program, however, disrupted this clear running of the curve. A dramatic increase (from zero to about 80%) was observed when the cell clumps were protected with 15% glycerol and transferred from −10 and −30 °C into LN. While the duration of the holding time periods were scarcely different for the survival of the cell cryoprotected with DMSO, considerable dependency was found in the case of glycerol-treated cells. At a −10 °C

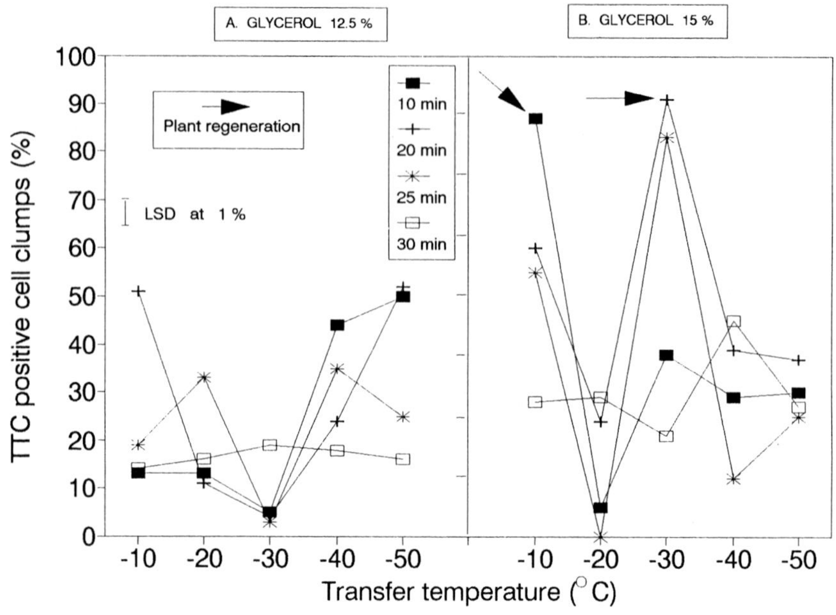

Fig. 5A, B. Effect of holding time periods on the viability and plant regeneration capacity of glycerol cryoprotected *P. distans* suspension cultures immersed in liquid nitrogen from different transfer temperatures. (Jekkel and Heszky 1991)

transfer, maximum survival was found after a 10 min holding time and decreased with the increasing of holding time durations. In contrast, −30 °C, the longer holding time periods, proved to be superior to shorter durations.

There is little evidence using a freezing program without holding the cells at subzero temperatures before immersion, which brought satisfactory survival only when the suspension is fine enough. Successful attempts at cryopreservation generally involved a method of slow cooling to −30 °C or to −40 °C temperatures, and holding specimens at those temperatures for up to 1 h before plunging into liquid nitrogen (Hauptmann and Widholm 1982; Withers and King 1980). Some of these reports involve the interaction of holding time periods and the nature and concentration of the cryoprotectant (Withers and King 1980), but at a constant subzero temperature, to date there is insufficient knowledge for a detailed discussion of their mutual effect.

2.6 Plant Regeneration

For plant regeneration, cell aggregates were transferred to the surface of solid culture media. After the freeze-thaw, most of the explants turned brown and showed no sign of growth for the first 3–4 weeks. A similar phenomenon was reported for thawed cotton callus (Bajaj 1982) and horse-chestnut somatic

Table 1. Green plant regeneration capacity of DMSO cryoprotected *P. limosa* suspension cultures frozen to −20 °C with 1 °C/min freezing rate and, after 10 min holding period, plunged into liquid nitrogen. (unpubl. results)

DMSO (%)	No. of frozen aggregates	No. of surviving aggregates	Survival (%)	No. of regenerants	No. of green plants
12.5	38	28	73	68	0
15.0	47	45	94	92	0
Nonfrozen	32	32	100	75	47

embryos (Jekkel et al. 1992) during the cryoshock phase. Therefore plant regeneration was assessed generally at the 6th week after thawing.

Plant regeneration was estimated by the proportion of surviving explants. The viability-determining method involved the TTC test and regrowth potential. There appears to be a close correlation between TTC estimates of viability and plant regeneration at 50% TTC. The number of shoots on a cell aggregate varied from 4 to 20; a similar result was found by Binh et al. (1989). Thus in the estimating procedure, one regenerable clump, regardless of the number of shoots, was considered as one plant regeneration value (Table 1).

The appropriate treatments resulting in plant regeneration are distinguished by arrows in Figs. 3–5. A high frequency (40–80% of the survival rate) of

Fig. 6. Albino *P. limosa* regenerants obtained from green (frozen) callus (unpubl. results)

regenerated plants was observed from the frozen DMSO cryoprotected cells (Fig. 3A) in −10 and −40 °C transfer temperature combination. Holding the DMSO-treated cells at −20 °C for 10 to 30 min regenerated plants appeared under almost every holding time regime (Fig. 4). In addition, even the glycerol cryoprotected cells produced plants when frozen in a freezing program containing these holding time periods. Without exception, the regenerants were albinos; in fact the albino plants appeared in high numbers among the regenerants in the nonfrozen control group. Heszky et al. (1989) overcame this problem by selecting greening calli, resulting in increased green plant regeneration in *P. limosa* suspension cultures.

Table 1 shows the results of *P. limosa* suspension cultures selected for greening clumps before freezing (Heszky et al. 1989). DMSO cryoprotected cells were frozen with 1 °C/min cooling rate to −20 °C and after a 10 min holding time period were immersed in LN. Although the viability ratio and the frequency of plant regeneration followed the tendency shown throughout the freezing of *P. distans*, the regenerants remained albinos (Fig. 6).

3 Summary and Conclusions

Experiments were performed sequentially with the purpose of achieving the highest survival rate of suspension cultures of some species of *Puccinellia*. Effects of freezing rates, cryoprotectants, and holding time periods were tested, resulting in ever higher survival rates at the different stages. *P. distans* survived the temperature of LN without cryoprotectant treatments when subjected to 1 °C/min freezing rate. This original viability was improved with a step-by-step freezing program modification. Though a number of combinations have been found to give similar survival rates, the protocol has been narrowed down. Cell aggregates reached their maximum viability (more than 90%) after 10-min storage in liquid nitrogen, following treatment in 15% DMSO solution frozen at 1 °C/min to −20 °C and being held at this temperature for 20–30 min before immersion. Plant regeneration was also observed, but when the freezing was repeated using selected greening calli suspension material, regenerants remained albinos.

A common feature of *Puccinellia* sp. is the increased tolerance to high level of salt and pH, acidic, and alkaline soils. The Hungarian name of *P. limosa* is "honey grass", referring to the strategy of surviving the high salt content of the soil by producing high osmotic intracellular conditions, resulting in a particular ability to withstand salt. Since the resistance mechanisms for abiotic stresses like drought, salt, and freeze tolerance are controlled in similar ways, the relatively high initial viability of the nontreated group is not a surprise. The observed survival rates varied remarkably on using the different combinations of the critical factors; on the other hand, some treatment combinations produced the same results. Therefore, it is assumed that all the studied parameters are based on their effect on cell dehydration. The same level of cell water reduction could be achieved in more ways, by modification of either of the components of the procedure. The experimental method allows the study of all the critical points as a function of transfer temperatures, and in this way, points out subzero temperatures resulting in low viability, presumably because the cell is under- or over-dehydrated at that value and lower or higher temperature must be chosen. We cannot explain, however, why between two transfer temperatures there is high viability and a temperature at which survival falls sometime to zero (Figs. 3, 4, 5). This is as if it were under the control of two different survival mechanisms. Furthermore, the relative independence of cryoprotectant concentrations and holding time periods are not clear.

References

Bajaj YPS (1977a) Clonal multiplication and cryopreservation of cassava through tissue culture. Crop Improv 4: 198–204

Bajaj YPS (1977b) Initiation of shoots and callus from potato-tuber sprouts and axillary buds frozen at −196 °C. Crop Improv 4: 48–53

Bajaj YPS (1979) Freeze preservation of meristems of *Arachis hypogaea* and *Cicer arientinum*. Indian J Exp Biol 17: 1405–1407

Bajaj YPS (1982) Survival of anther and ovule-derived cotton callus frozen in liquid nitrogen. Curr Sci 51: 139–140

Binh DQ, Heszky LE, Gyulai G, Kiss E, Csillag A (1989) Plant regeneration from callus of *Puccinellia ditans* (L.) Parl. Plant Cell Tissue Organ Cult 4: 101–109

Dereuddre J, Fabre J, Bassaglia C (1988) Resistance to freezing in liquid nitrogen of carnation (*Dianthus caryophyllus* 1: var. Eolo) apical and axillary shoot tips excised from different-aged in vitro plantlets. Plant Cell Rep 7: 170–173

Finkle BJ, Urlich JM (1978) Effect of combination of cryoprotectants on the freezing survival of sugarcane cultured cell. In: Li PH, Sakai A (eds) Plant cold hardiness and freezing stress mechanism and crop implications. Academic Press, New York, pp 373–388

Hauptmann RM, Widholm JM (1982) Cryostorage of cloned amino acid analog-resistant carrot and tobacco suspension cultures. Plant Physiol 70: 30–34

Heszky LE, Binh DQ, Kiss E, Gyulai G (1989) Increase of green plant regeneration efficiency by callus selection in *Puccinellia limosa* (Schur) Holmb. Plant Cell Rep 8: 174–177

Heszky LE, Jekkel ZS, Ali AH (1990) Effect of cooling rate, cryoprotectant and holding time at different transfer temperatures on the survival of cryopreserved cell suspension culture [*Puccinellia distans* (L.) Parl]. Plant Cell Tissue Organ Cult 21: 217–226

Jekkel ZS, Heszky LE (1991) Cryopreservation of *Puccinellia distans* suspension cultures: a model system for investigation the critical points of freezing-procedure. In: Proc EC-Hungary Joint Worksh Plant biotechnology, Gödöllö, pp 69–76

Jekkel ZS, Kiss J, Kiss E, Gyulai G, Heszky LE (1992) Freeze preservation of somatic embryos of recalcitrant seed species: horse-chestnut (*Aesculus hippocastanum* L.). In: Book for poster abstracts. XIIIth EUCARPIA Congr, Angers, pp 421–422

Kartha KK (ed) (1985) Cryopreservation of plant cells and organs. CRC Press, Boca Raton, pp 115–267

Kartha KK, Leung NL, Gamborg OL (1979) Freeze preservation of pea meristems in liquid nitrogen and subsequent plant regeneration. Plant Sci Lett 15: 7–13

Kartha KK, Leung NL, Pahl K (1980) Cryopreservation of strawberry meristems and mass propagation of plantlets. J Am Soc Hortic Sci 150: 481

Meryman HT, Williams RJ, Doughlas M St. Jr (1977) Freezing injury from "solution effects" and its prevention by natural or artificial cryoprotection. Cryobiology 14: 287

Simon T (1992) Taxonomy of higher plants of Hungarian flora. Tankönyvkiadó, Budapest, pp 1–892 (in Hungarian)

Withers LA, King PJ (1980) A simple freezing unit and cryopreservation method for plant cell suspension. Cryo Lett 1: 213

III.4 Cryopreservation of Germplasm of Sugarcane (*Saccharum* Species)

Y.P.S. BAJAJ[1] and L.C. JIAN[2]

1 Introduction

Sugarcane is the main sugar-producing tropical crop in the world. It belongs to the genus *Saccharum*, the tribe Andropogoneae, and the family Gramineae. It is grown in a number of countries (Table 1), on about 18 million ha (FAO 1992), and provides about 65% of the sugar produced (Heinz 1987). Brazil, India, China, and Cuba are the world's major sugar producers.

There are five species of *Saccharum*: *S. officinarum, S. sinense, S. barberi, S. spontaneum*, and *S. robustum*. Most commercial varieties are descendants of interspecific hybrids within the genus (Arcenaux 1967). Sugarcane plants are cold-sensitive and in the field do not tolerate +2 °C, or even brief exposure to −3 °C (Irvine 1978).

Saccharum officinarum (2n = 8x = 80) originated in New Guinea. Interspecific crosses between *S. officinarum* and *S. spontaneum*, followed by two or three backcrossings to the noble cane, yielded the first modern commercial varieties (*Saccharum* spp.) in the early 1900s in Java, and later in India (Daniels and Roach 1987). It is highly heterogeneous/polyploid, and new varieties are obtained by conventional breeding and selection. It is then propagated vegetatively by means of stalk cuttings ("setts", which bear one to three axillary buds). Planting in the field is done with new setts. During this process, the sugarcane breeder faces a number of problems:

1. To conserve the germplasm, the genetic stalks (setts) have to be grown in the nurseries periodically. Not only this is tedious, labor-intensive and time-consuming, but the germplasm is also vulnerable to hostile environments and attack by pests and pathogens.
2. Sugarcane pollen is short-lived and the breeder is not able to make sufficient crosses in a season (Tai 1993).
3. During exchange of germplasm, there is the risk of dissemination of diseases.

To overcome some of the above-mentioned problems, storage of in vitro cultured meristems/plantlets is being practiced. For instance, germplasm of more than 650 varieties is being conserved at CIRAD-CA in Montpellier, France

[1]Former Professor of Tissue Culture, Punjab Agricultural University, Ludhiana, India. *Present address*: A-137 New Friends Colony, New Delhi 110 065, India
[2]Laboratory of Cytology, Institute of Botany, Academia Sinica, Beijing, China

Biotechnology in Agriculture and Forestry, Vol. 32
Cryopreservation of Plant Germplasm I (ed. by Y.P.S. Bajaj)

Table 1. Area harvested, yield, and production of sugarcane in various countries. (FAO 1992)

Region/country	Area harvested (1000 ha)	Yield (kg/ha)	Production (1000 metric t)
World	17934	61 591	1 104 580
Africa	1243	57433	71 369
North-central America	2901	57 311	166 281
USA	356	74945	26 703
Cuba	1200	48 333	58 000
S. America	5251	66 210	347 649
Argentina	280	66 071	18 500
Brazil	4207	64 335	270 672
Asia	8110	59 911	485 854
China	1263	61 400	77 548
India	3882	62 219	249 300
Australia	348	84 195	29 300

(Paulet et al. 1993), and recently Taylor and Dukic (1993) have reported storage of cultures of over 200 *Saccharum* spp. clones at 18 °C at the Sugarcane Experimental Station, Indooroopilly, Australia. However, such cultures have to be looked after constantly, transferred to fresh media, and maintained under strictly controlled conditions.

Extensive studies have been conducted on various aspects of the in vitro culture of sugarcane, including regeneration (Maretzky 1987), somatic embryogenesis (Guiderdoni et al. 1995), androgenesis (Moore and Fitch 1990), somatic hybridization (Tabaeizadeh et al. 1986), genetic transformation (Birch and Maretzki 1993), and somaclonal variability of various types (Maretzki 1987). At present, cryopreservation appears to be the only method for the conservation of both the germplasm of vegetatively propagated crops and the in vitro induced variant cell lines (see Bajaj 1990, 1991).

2 Cryopreservation

Successful studies have been conducted on the freeze preservation of plantlets, meristems, and callus cell cultures of sugarcane, and meaningful literature has accumulated (Table 2). In the following, our published work (Bajaj et al. 1987; Jian et al. 1987) on freeze preservation of in vitro cultures of sugarcane is summarized, and discussed in the light of contemporary literature.

2.1 Cryopreservation of Plantlets

Although large plants are difficult to cryopreserve, attempts in the past with young plantlets of pea (Sun 1958), carrot (Withers 1979), and sugarcane (Bajaj et al. 1987) have shown encouraging results.

Table 2. Summary of work on cryopreservation of sugarcane

Explant used	Cryoprotectants[a]	Results	Reference
Cell suspension and callus	Glu + DMSO + PEG	Plants	Finkle and Ulrich (1979); Ulrich et al. (1979)
Cell suspension		Plants	Chen et al. (1979)
Cell suspension	Suc + Gly + DMSO	60% survival	Bajaj et al. (1987)
Meristems	Suc + Gly + DMSO	Plants	Bajaj et al. (1987)
Plantlets	not used	Plants	Bajaj et al. (1987)
Callus	DMSO (10%) + Sor 0.5 M	Plants	Jian et al. (1987)
Cell suspension	Combinations of Gly, DMSO, Suc, Sor	Plants	Gnanapragasam and Vasil (1990)
Callus	Suc + DMSO	Plants	Eksomtramage et al. (1992)
Shoot apices	Encapsulated in 3.4% suc + 3% alginate	Plants	Paul et al. (1993)

[a] *Abbreviations*: DMSO (dimethylsulfoxide), Glu (glucose), Gly (glycerol), PEG (polyethylene glycol), Sor (sorbitol), Suc (sucrose).

The plantlets of sugarcane (2–3 cm) grown in vitro were gently pulled out of the medium, and the agar around the roots was removed. Ten of them were pooled on dry filter paper in Petri dishes, and then placed in a desiccator to be partially dehydrated for 24, 48, and 72 h in a culture room. The desiccated plantlets were wrapped in aluminum foil and frozen by rapid immersion in liquid nitrogen (LN) for 10 min. They were thawed in warm water (40 °C), and recultured on a modified agar-solidified MS medium (Murashige and Skoog 1962) supplemented with GA (0.1 mg/l) + IBA (0.001 mg/l) + sucrose 2%.

Almost all the controlled plantlets, dehydrated for 48 h but not frozen, resumed growth after return to the culture. Likewise, some of the plantlets which were first dried for 48 h and then frozen also survived and started to grow within 2 days (Bajaj et al. 1987). However, plants not dehydrated and wet plants (treated with cryoprotectants) as well as the plantlets dehydrated for 72 h did not withstand freezing. Thus, desiccation of plantlets is very important, and its extent is critical. Below or above certain moisture limits the plants will not survive.

2.2 Cryopreservation of Meristems

The freeze preservation of sugarcane meristems offers a number of advantages over callus cultures: (1) being genetically more stable, they ensure the stability of clones; (2) they are easy to regenerate/develop into plants; (3) they yield pathogen-free stocks for propagation; and (4) they yield ideal material for the international exchange of germplasm.

About 3-cm-long segments of sugarcane shoots from the plants grown in vivo were sterilized with calcium chloride. The meristems (1–2 mm) were excised aseptically and 10–15 were pooled on moist filter paper. They were then transferred to 2-ml ampoules, treated with 1 ml solution of cryoprotectants (5% each of DMSO, sucrose, and glycerol) for 45 min, and subjected to sudden immersion in LN. After thawing at 38–40 °C and washing with the liquid medium, they were cultured on modified MS medium supplemented with growth regulators.

The meristems lay quiescent for about 4 days and then resumed growth, as evidenced by an overall swelling and general enlargement, callusing, greening, and elongation (Bajaj et al. 1987). Further observations were not recorded. Recently, Paulet et al. (1993) used meristems from the plants grown in vitro. The meristems were encapsulated in alginate beads, precultured in a high-sucrose concentration liquid medium, and partially desiccated before freezing in LN. Survival ranged from 38 to 91%.

2.3 Cryopreservation of Callus/Cell Suspensions

In view of the agriculturally useful properties of the somaclonal variant cell lines obtained in vitro, such as disease resistance (Heinz et al. 1977), high yield/sugar contents (Liu and Chen 1978), and salt tolerance (Korneva and Maribona 1984), it is highly desirable to develop methods for their long-term and stable storage. The work on the freeze preservation of sugarcane callus was initiated by Finkle and Ulrich in 1979; since then, extensive studies have been conducted (Table 2) on various factors affecting cryoability. It may be emphasized that for reproducibility of results, in addition to other factors, genotype and age/state of the cell cultures are of prime importance in cryopreservation studies. Some cultivars/clones are more sensitive than others. Some of the factors are discussed here:

1. *Effect of Age and Nature of the Cells.* Experience has shown that periodically transferred young suspensions of highly cytoplasmic, nonvacuolated, thin-walled cells in small aggregates are able to withstand freezing much better than relatively older cultures containing large and thick-walled single cells (Bajaj 1976).

Figure 1 shows the effect of age of culture on the survival of sugarcane callus stored in LN for 1 day (Jian et al. 1987). Recently, embryogenic calli of a number of commercial hybrids have produced good results (Eksomtramage et al. 1992).

2. *Effect of Cryoprotectants.* Cryoprotectants act as antifreeze agents by reducing the concentration of intracellular salts (Lovelock 1953) and the formation of ice crystals. In sugarcane a number of cryoprotectants have been applied (Table 2). In our work, mixtures of various cryoprotectants were used. Bajaj et al. (1987) treated callus cell suspensions with a mixture containing 5% each of DMSO, sucrose, and glycerol, and obtained a viability of 60%. The experimental results shown in Fig. 2 and Table 3 (Jian et al. 1987) indicate that DMSO was not very effective when used alone. However, the combination of 10% DMSO and 8% glucose markedly increased the TTC (triphenyl tetrazolium chloride) value. This is consistent with observations by Finkle and Ulrich (1979).

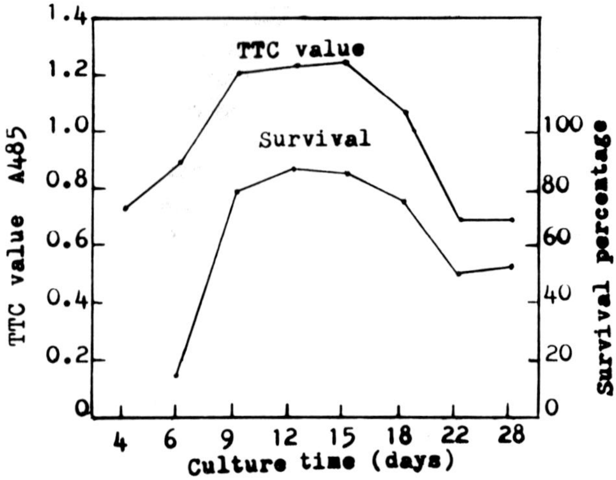

Fig. 1. Effect of the age of culture on the TTC value and survival of sugarcane callus after storage for 1 day in LN

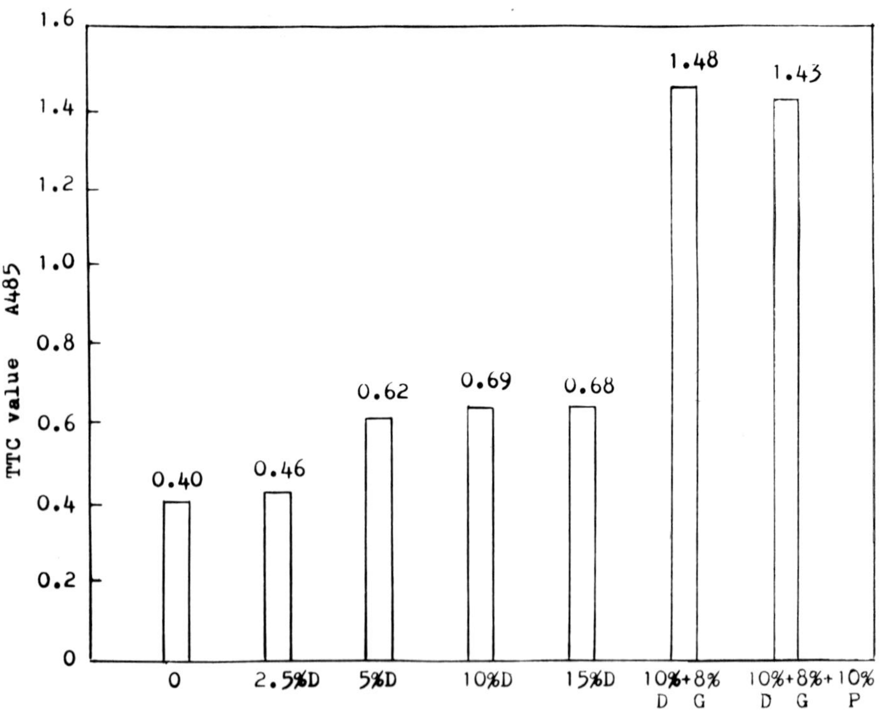

Fig. 2. Effects of various cryoprotectants on the TTC values of sugarcane callus stored in LN for 2 days. *D* DMSO; *G* glucose; *P* polyethylene glycol

Table 3. Effects of DMSO in combination with sugars and sugar alcohols on the TTC value, survival, and regrowth of sugarcane callus stored for 3 days at −196 °C. (Jian et al. 1987)

Cryoprotectant	TTC value	Survival (%)	Regrowth rate (%, fr. wt.)
10% DMSO + 0.5 M Glucose	1.33	79	153
10% DMSO + 0.5 M Sucrose	0.97	81	233
10% DMSO + 0.5 M Sorbitol	1.56	97	320
10% DMSO + 0.5 M Mannitol	1.26	85	104
10% DMSO + 0.5 M Maltose	0.86	72	77

The calli which had been subcultured for 10 days were frozen at a cooling rate of 1 °C/min to −40 °C, maintained for 2 h, stored in LN, and then thawed in a 40 °C water bath. Callus survival and regrowth rates were determined after reculture for 33 days.

Table 4. Effects of freezing methods on the TTC value and survival of sugarcane callus after freezing and storage in LN for 2 days. (Jian et al. 1987)

Freezing methods before storage in liquid nitrogen					
Cooling rate from 0 °C	Holding time at −40 or −10 °C	Further cooling rate	Holding time at −40 °C	TTC value	Survival (%)
1 °C/min to −40 °C	–	–	–	0.82	40.0
1 °C/min to −40 °C	2 h	–	–	1.06	85.5
1 °C/min to −10 °C	15 min	1 °C/min to −40 °C	–	0.88	56.0
1 °C/min to −10 °C	15 min	1 °C/min to −40 °C	1 h	1.38	77.5
1 °C/min to −10 °C	15 min	1 °C/min to −40 °C	2 h	1.17	83.5
1 °C/min to −10 °C	15 min	1 °C/min to −40 °C	5 h	1.26	80.0

Calli cultured for 12 days were frozen using 10% DMSO + 0.5 M sorbitol, and thawed in a 40 °C water bath.

3. *Effect of Freezing.* Table 4 shows the results (Jian et al. 1987) for specimens which were cooled slowly to −40 °C, kept for 1–5 h, and then put into LN. In most of the studies (Table 2), a slow freezing rate was used, however, survival was also obtained in cell suspensions (in a cryoprotectant solution of 5% each of DMSO, sucrose, and glycerol prepared in an MS medium containing 4% sucrose) subjected to sudden immersion in LN for 10 min (Bajaj et al. 1987). Survival was defined as the presence of growth. Almost all the cultures turned brown within a couple of days, and appeared to be dead, but after a long lag phase they started to grow. In view of the recent use of the vitrification method, i.e., with a high concentration of cryoprotectants followed by sudden immersion in LN (Paulet et al. 1993), the cumbersome method of regulated slow cooling seems redundant for future studies.

4. *Effect of Thawing.* Cryodamage can be avoided by adopting a method of thawing which discourages the growth of ice crystals. Thawing the frozen cultures in warm water at 40 °C has been used by most workers (Bajaj et al. 1987;

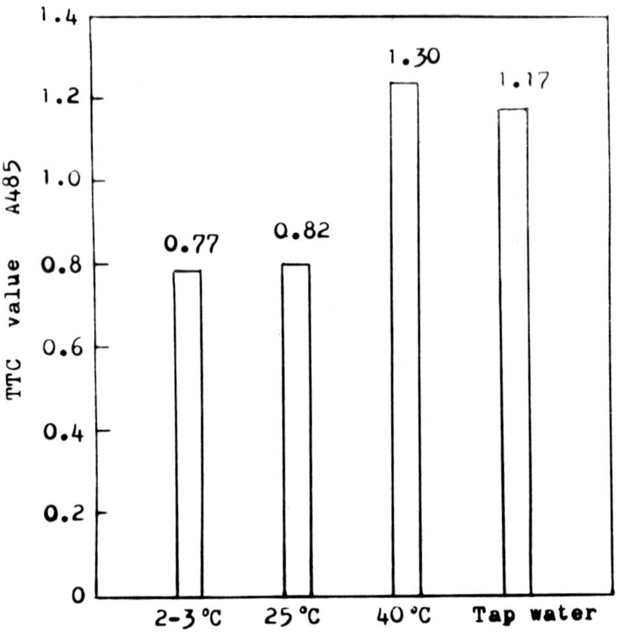

Fig. 3. Comparison of various methods of thawing. The calli cultured for 11 days were frozen using 10% DMSO + 8% glucose at a cooling rate of 1 °C/min to −40 °C (kept for 1.5 h), and stored in LN for 1 day

Gnanapragasam and Vasil 1990; Eksomtramage et al. 1992). A comparison of the four thawing methods, i.e., (1) 40 °C, (2) 2–3 °C in a refrigerator; (3) room temperature; and (4) running tap water showed that the effectiveness of thawing with running tap water was as good as at 40 °C; however, room temperature was not as effective (Fig. 3).

Table 5. Effect of light on the regrowth of sugarcane callus after freezing and storage for 2 days in liquid nitrogen. (Jian et al. 1987)

Treatment	Culture in light		Culture in dark	
Experiment	Survival (%)	Regrowth rate (%, fr. wt.)	Survival (%)	Regrowth rate (%, fr. wt.)
I	91.3		97	
II	67.0		100	
III	71.4	170	83	476
IV	69.6		95	
V	68.0		100	
VI	68.0		100	

Calli in culture for 14 days were frozen using the cryoprotectant mixture of 10% DMSO +0.5 M sorbitol at a cooling rate of 1 °C/min to −40 °C, kept for 2 h, stored in LN, and thawed in a 40 °C water bath.

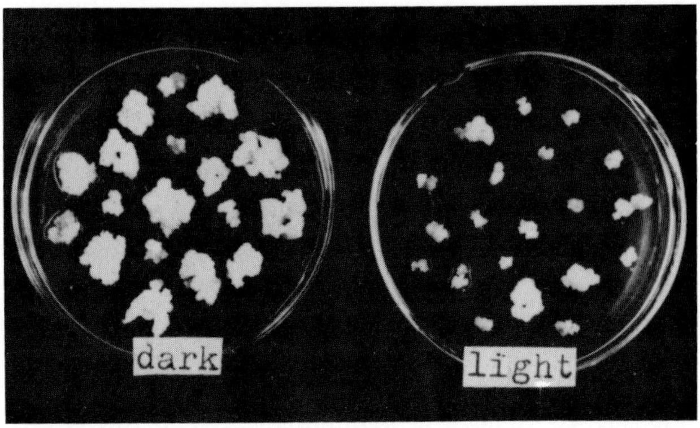

Fig. 4. Effect of light on regrowth of sugarcane callus after storage for 2 days in liquid nitrogen. Photographs were taken 34 days after culture

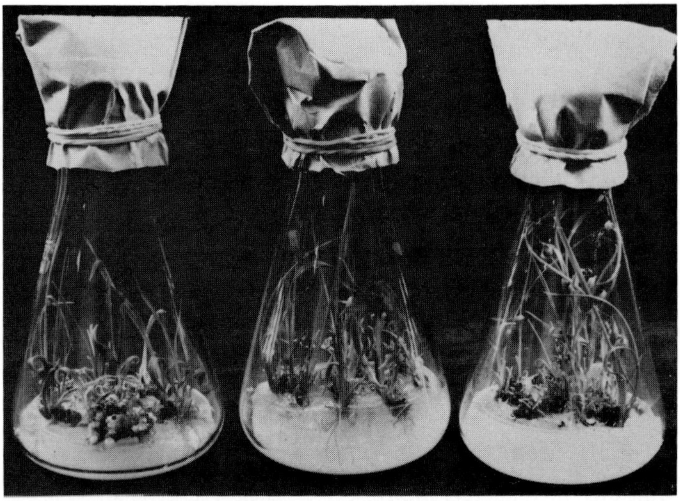

Fig. 5. Sugarcane plants regenerated from callus stored for 6 months in LN

5. *Effect of Light.* After the thawed calli were transferred to fresh medium, they were divided into two groups. One group was placed in the culture room in light at 2500 lx and the other in the dark. Significant differences were observed in their growth (Fig. 4, Table 5); illumination was not favorable. The percentage of survival was higher (Table 5) and the rate of regrowth much faster in the dark (Fig. 4). For cultures stored in the dark, the regrowth rate of callus fresh weight increased from 170 to 476% (Table 5 and Fig. 4; Jian et al. 1987).

6. *Regeneration of Plants.* The regeneration of sugarcane plants from cryopreserved callus was first observed by Ulrich et al. in 1979. Since then, a number

of reports have appeared (Table 2). When callus which had been cryopreserved in LN for 6 months was transferred to a 2,4-D-free medium, plantlets differentiated (Fig. 5). Some of these plants continued to grow when transferred to the field.

3 Summary

Sugarcane, a cold-sensitive tropical grass, is the source of about 65% of the sugar produced in the world. It is vegetatively propagated by stem cuttings ("setts"), and the germplasm is maintained in field nurseries. It not only has to be periodically repropagated, but is also constantly exposed to the hazards of pests, diseases, and hostile environments. To avoid this, in vitro cultured meristems/plantlets have been employed.

Recent developments in the in vitro culture of sugarcane, especially the induction of somaclonal cell lines resistant to disease and salt, as well as higher sugar contents, and yield, make it necessary to develop methods for their stable maintenance. At present, cryopreservation appears to be the only method. Successful studies have been conducted on the cryopreservation of in vitro cultures of sugarcane, and plants have been regenerated from callus, cell suspensions, excised meristems, and dehydrated in vitro plantlets subjected to freezing in liquid nitrogen. These studies would enable: (1) long-term storage and international exchange of sugarcane germplasm; and (2) stable maintenance of in vitro induced genetic variability.

References

Arcenaux G (1967) Cultivated sugarcanes of the world and their botanical derivation. Proc Congr Int Sugarcane Technol (Puerto Rico) 12: 844–854

Bajaj YPS (1976) Regeneration of plants from cell suspensions frozen at −20, −70, and −196 °C. Physiol Plant 37: 263–268

Bajaj YPS (1990) Cryopreservation of germplasm of vegetatively propagated crops. Bull Soc Bot Fr 137, Actual Bot 3: 99–114

Bajaj YPS (1991) Storage and cryopreservation of in vitro cultures. In: Bajaj YPS (ed) Biotechnology in agriculture and forestry, vol 17. High-tech and micropropagation I. Springer, Berlin Heidelberg New York, pp 361–381

Bajaj YPS, Korneva SB, Gutierrez R, Maribona RH (1987) Freeze preservation of plantlets, excised meristems, and cell cultures of sugarcane (*Saccharum officinarum* L.) at −196 °C. In: A Bose, Sengupta P (eds) Proc Int Conf on Cryogenics. McGraw-Hill, New Delhi, pp 222–226

Birch RG, Maretzki A (1993) Transformation of sugarcane. In: Bajaj YPS (ed) Biotechnology in agriculture and forestry, vol 23. Plant protoplasts and genetic engineering IV. Springer, Berlin Heidelberg New York, pp 348–360

Chen WH, Cockburn W, Street HE (1979) Preliminary experiments on freeze-preservation of sugarcane cells. Taiwania 24: 70–74

Daniels J, Roach BT (1987) Taxonomy and evolution. In: Heniz DJ (ed) Sugarcane improvement through breeding. Elsevier, Amsterdam, pp 7–84

Eksomtramage T, Paulet F, Guiderdoni E, Glaszmann JC, Engelmann F (1992) Development of a cryopreservation process for embryogenic calluses of a commercial hybrid of sugarcane (*Saccharum* sp.) and application to different varieties. Cryo-Letters 13: 239–252

FAO (1992) Production yearbook, vol 46. FAO, Rome

Finkle BJ, Ulrich JM (1979) Effects of cryoprotectants in combination on the survival of frozen sugarcane cells. Plant Physiol 63: 598–604

Gnanapragasam S, Vasil IK (1990) Plant regeneration from a cryopreserved embryogenic cell suspension of a commercial sugarcane hybrid (*Saccharum* sp.). Plant Cell Rep 9: 419–423

Guiderdoni E, Merot B, Eksomtramage T, Paulet F, Feldmann P, Glaszmann JC (1995) Somatic embryogenesis in sugarcane (*Saccharum* spp.). In: Bajaj YPS (ed) Biotechnology in agriculture and forestry, vol 31. Somatic embryogenesis and synthetic seed II. Springer, Berlin Heidelberg New York, pp 93–114

Heinz DJ (ed) (1987) Sugarcane improvement through breeding. Elsevier, New York

Heinz DJ, Krishnamurthi M, Nickell LG, Maretzki A (1977) Cell, tissue and organ culture in sugarcane improvement. In: Reinert J, Bajaj YPS (eds) Applied and fundamental aspects of plant cell, tissue, and organ culture. Springer, Berlin Heidelberg New York, pp 3–17

Irvine JE (1978) Identification of cold tolerance in *Saccharum* and related genera through refrigerated freeze-screening. Proc Int Soc Sugarcane Technol

Jian LC, Sun DL, Sun LH (1987) Sugarcane callus cryopreservation. In: Li PH (ed) Plant cold hardiness. Alan R Liss, New York, pp 323–337

Korneva SB, Maribona RH (1984) Salt resistant sugarcane somaclones. In: Novak FJ, Havel L, Dolezel J (eds) Proc Int Symp Plant tissue culture application to crop improvement. Czechoslovakia Acad Sci, Prague, pp 349–350

Liu MC, Chen WH (1978) Tissue and cell culture as aids to sugarcane breeding. II. Performance and yield potential of sugarcane callus derived lines. Euphytica 27: 273–282

Lovelock JE (1953) The haemolysis of human red blood cells by freezing and thawing. Biochim Biophys Acta 10: 414–426

Maretzky A (1987) Tissue culture; its prospects and problems. In: Heinz DJ (ed) Sugarcane improvement through breeding. Elsevier, New York, pp 343–384

Moore PH, Fitch MMM (1990) Sugarcane (*Saccharum* spp.): anther culture studies. In: Bajaj YPS (ed) Biotechnology in agriculture and forestry, vol 12. Haploids in crop improvement I. Springer, Berlin Heidelberg New York, pp 480–497

Murashige T, Skoog F (1962) A revised medium for rapid growth and bioassays with tobacco tissue cultures. Physiol Plant 15: 473–497

Paulet F, Engelmann F, Glaszmann JC (1993) Cryopreservation of apices of in vitro plantlets of sugarcane (*Saccharum* sp. hybrids) using encapsulation/dehydration. Plant Cell Rep 12: 525–529

Sun CN (1958) The survival of excised pea seedlings after drying and freezing in liquid nitrogen. Bot Gaz 119: 234–236

Tabaeizadeh Z, Ferl RJ, Vasil IK (1986) Somatic hybridization in the Gramineae: *Saccharum officinarum* L. (sugarcane) and *Pennisetum americanum* (L.) K. Schum. (pearl millet). Proc Natl Acad Sci USA 83: 5616–5619

Tai PYP (1993) Low temperature preservation of F_1 pollen in crosses between noble or commercial sugarcane and *Saccharum spontaneum* L. Sugar Cane No 5, 8–11

Taylor PWJ, Dukic S (1993) Development of an in vitro culture technique for conservation of *Saccharum* spp. hybrid germplasm. Plant Cell Tissue Organ Cult 34: 217–222

Ulrich JM, Finkle BJ, Moore PH, Ginoza H (1979) Effect of a mixture of cryoprotectants in attaining liquid nitrogen survival of callus cultures of a tropical plant. Cryobiology 16: 550–556

Withers LA (1979) Freeze-preservation of somatic embryos and clonal plantlets of carrot (*Daucus carota* L.). Plant Physiol 63: 460–467

Section IV
Legumes and Oilseed Crops
(for Coconut and Oil Palm see Section II)

IV.1 Cryopreservation of Peanut (*Arachis hypogaea* L.) and Chickpea (*Cicer arietinum* L.)

Y.P.S. Bajaj[1]

1 Introduction

The peanut, also known as groundnut (*Arachis hypogaea* L.), and the chickpea (*Cicer arietinum* L.) are the two oil-protein-rich legume seed crops. Peanut occupies third place in the world as a source of edible oil. It originated in South America and is now grown in many countries, India and China being the largest producers (Table 1). *Arachis* possesses both diploid (2n = 20) and tetraploid (2n = 40) species. The important diploid ones include: *A. benthamii*, *A. diogoi*, *A. duranensis*, *A. helodes*, *A. lutescens*, *A. repens*, *A. rigoni*, and *A. villosa*; whereas the tetraploid ones are: *A. glabrata*, *A. hagenbeckii*, *A. hypogaea*, *A. monticola*, *A. nambyquarae*, *A. rasteiro*, *A. salvagem*, *A. marginata*, and *A. prostrata.* In addition, *A. pusilla* is octaploid (Krapovickas and Rigoni 1952).

Chickpea (*Cicer arietinum* L.), also called gram or channa, is an annual plant with green pods which are eaten raw; the protein-rich seeds are dried. It is grown in a number of countries (Table 1), primarily in India, Pakistan, and Turkey (FAO 1992). The genus *Cicer* (2n = 16) comprises about 40 species of annuals and perennials, such as *C. arietinum*, *C. bijugum*, *C. chorassanicum*, *C. cuneatum*, *C. echinospermum*, *C. judaicum*, *C. pinnitifidum*, *C. reticulatum*, and *C. yamashitae* (Altaf and Ahmad 1990). There are two recognized races of cultivated chickpea: the microsperma and macrosperma groups (Moreno and Cubero 1978; Hawtin and Singh 1980). There is considerable genetic diversity within each group.

The seeds of both peanut and chickpea are "orthodox" and present no storage problems. Numerous accessions are conserved routinely in germplasm banks (Table 2) of various national and international institutes (Plucknett et al. 1983).

During the last decade, considerable progress has been made regarding various aspects of the in vitro culture, i.e., regeneration, somatic embryogenesis, androgenesis, protoplasts, transformation, genetic variability, and cryopreservation, both of the peanut (Bajaj 1983a,c, 1984; Bajaj et al. 1980, 1981a,b; Bajaj and Gosal 1988; Schnall and Weissinger 1993; Li and Jarret 1995; McKently 1995) and the chickpea (Bajaj 1979, 1982, 1983a; Gupta et al. 1984; Bajaj and Gosal 1987; Gosal and Bajaj 1988; Altaf and Ahmad 1990; Riazuddin and Husnain 1993; Barna and Wakhlu 1993); somaclonal variation has also been

[1]Former Professor of Tissue Culture, Punjab Agricultural University, Ludhiana (Punjab), India. *Present address*: A-137 New Friends Colony, New Delhi 110 065, India

Biotechnology in Agriculture and Forestry, Vol. 32
Cryopreservation of Plant Germplasm I (ed. by Y.P.S. Bajaj)

Table 1. Area under cultivation and production of peanut and chickpea in various countries. (FAO 1992)

Arachis hypogaea			*Cicer arietinum*		
Region	Area under cultivation (1000 ha)	Production (1000 metric t)	Region	Area under cultivation (1000 ha)	Production (1000 metric t)
World	20609	23506	World	9660	6887
Africa	6349	4670	Africa	443	265
Nigeria	1000	1214	N.C. America	122	195
North-central America	869	2166	Asia	8864	6218
USA	684	1943	India	6457	4500
S. America	327	574	Pakistan	1004	426
Asia	13019	16021	Turkey	860	810
China	2655	5580			
India	8600	8200			
Indonesia	655	1037			

Table 2. Accessions of peanut and chickpea in gene banks. (Plucknett et al. 1983)

Species	Accessions	Term of storage	Institution[a]	Location
Peanut (*Arachis hypogaea*)	8800	Medium	ICRISAT	Hyderabad, India
	4685	Short	USDA/SEA-AR	Experiment, Georgia, USA
	3925	Long	NSSL	Fort Collins, Colorado, USA
	2500	Medium	IITA	Ibadan, Nigeria
	1053	Medium, long	VIR	Leningrad, former USSR
Chickpea (*Cicer arietinum*)	13000	Medium	ICRISAT	Hyderabad, India
	4500	Short	ICARDA	Aleppo, Syria
	3100	Short	USDA/SEA-AR	Pullman, Washington, USA
	1685	Medium	VIR	St. Petersburg, former USSR
	1600	Medium	INIA	Mexico City, Mexico

[a]ICARDA, International Center for Agricultural Research in the Dry Areas; ICRISAT, International Crops Research Institute for the Semi-Arid Tropics; IITA, International Institute of Tropical Agriculture; INIA, Instituto Nacional de Investigaciónes Agricolas; NSSL, National Seed Storage Laboratory; USDA, United States Department of Agriculture; VIR, All-Union Institute of Plant Industry.

generated. At present, cryopreservation seems to be the only method for the conservation of in vitro induced variability (Bajaj 1991).

Earlier, the general progress made in the cryopreservation of germplasm of various legumes and oilseed crops was reviewed (Bajaj 1990); in this chapter, cryopreservation studies on peanut and chickpea are summarized.

2 Cryopreservation of Meristems

For the preservation of germplasm, meristem cultures have a number of advantages over callus cultures: (1) they are genetically more stable, (2) in some cases, it is difficult to regenerate an entire plant from callus cultures, whereas meristems are relatively easy and quick to grow, (3) cell cultures generally require a regulated rate of freezing, while meristems can withstand sudden immersion in liquid nitrogen, thus eliminating the need for elaborate and expensive cryostats.

2.1 Methodology

The meristems (approx. 1–2 mm) excised from the shoot tips of the 2-week-old in vitro grown seedlings of peanut (*Arachis hypogaea* L. cv. M 13), and chickpea (*Cicer arietinum* cv. G 543) were used. The details regarding the technique and culture media were outlined earlier (Bajaj, 1979). However, in later investigation (Bajaj 1983a) the meristems were precultured (before freezing) for 3–7 days on a medium supplemented with cryoprotectants (Table 3). The precultured meristems which showed signs of growth were selected and treated with various cryoprotectants (sucrose, glycerol, and dimethylsulfoxide). The cultures were frozen in liquid nitrogen and thawed at 25, 35 and 40 °C. The retrieved meristems were then recultured on a medium where kinetin was replaced by 0.2 mg/l benzylaminopurine (BAP).

The plantlets obtained from the frozen meristems were transferred to pots and reared to maturity. The percentage viability of the seeds was determined by germinating them in Petri dishes, and the chromosome number was studied in the root tips of 29 and 33 plants of *A. hypogaea* and *C. arietinum*, respectively.

2.2 Results

The extent of cryoability of the meristems, their revival, and eventual regeneration into complete plants depend upon a number of factors, the data on some of which, such as the preculture conditions, cryoprotectants, and the method of thawing are shown in Tables 3, 4, and 5 and are discussed below.

Table 3. Effect of various preculture treatments (1 week) on survival of excised meristems of *Cicer arietinum* freeze-preserved for 3 months at −196 °C. (Bajaj 1983a)

Preculture treatment (1 week)	Number of meristems cultured	Number of meristems showing growth	(%) Meristems showing growth
Basal medium (BM)	29	15	52
BM + DMSO (3%)	23	14	60
BM + DMSO (5%)	25	14	56
BM + DMSO (10%)	17	8	47

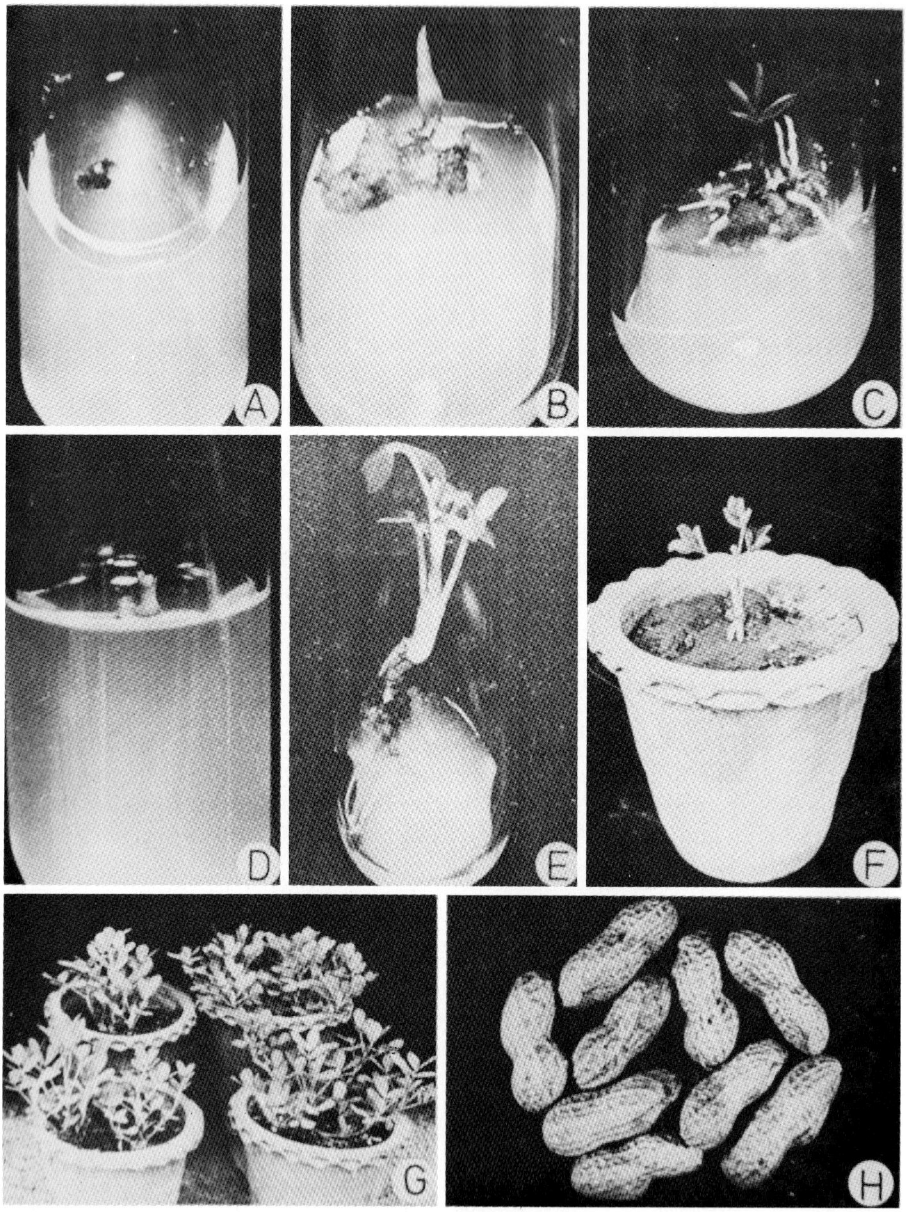

Fig. 1A–H. Various stages in the regeneration of plants of *Arachis hypogaea* from frozen (−196 °C) meristems. **A–C** Frozen-thawed meristems after 6, 9, and 11 weeks of culture, respectively. Note the initial all-round callusing (perhaps due to cryoinjury) of the meristem in **A**. **D**, **E** Normal growth of the retrieved meristem (preserved for 20 months), and regeneration of a complete plant, 4 and 12 weeks, respectively, after culture on MS + IAA + BAP. **F**, **G** Test-tube plants 3 and 7 weeks, respectively, after transferring to soil. **H** Normal seeds obtained from such plants (total time 24 weeks from the culture of excised meristems to seed setting). (Bajaj 1983a)

The excised meristems of both *A. hypogaea* and *C. arietinum*, when cultured on MS + IAA (2 mg/l) + BAP (0.2 mg/l) as the controls (unfrozen), started to enlarge within 1 week, and shoots developed in 70–80% of the cultures in 4–6 weeks. However, in the frozen meristems viability was considerably reduced. Earlier, 23–31% and 27–36% survival, respectively, was observed in *A. hypogaea* and *C. arietinum* meristems freeze-preserved for 27 days (Bajaj 1979). However, later, by manipulation of the above-mentioned factors, meristem survival was raised to 42 and 57% in *A. hypogaea* and *C. arietinum*, respectively (Bajaj 1983a). The plants regenerated from the retrieved meristems cryopreserved for up to 20 months set seeds (Figs. 1, 2). Although no detailed cytological studies were conducted, the chromosome number in the root tips of the germinated seeds obtained from the meristem-derived plants was normal, both in *A. hypogaea* (2n = 40) and *C. arietinum* (2n = 16).

Fig. 2A–D. Production of normal seeds in chickpea plants regenerated from freeze-preserved meristems. **A** Retrieved meristem 5 weeks after culture on MS + IAA (2 mg/l) + BAP (0.2 mg/l). **B** Same, showing multiple shoot formation. **C** Plant regenerated from freeze-preserved meristem, and transferred to soil. **D** Normal chickpea seeds obtained from such a plant. Total time from culture of retrieved meristem to seed production is 21 weeks. (Bajaj 1983b)

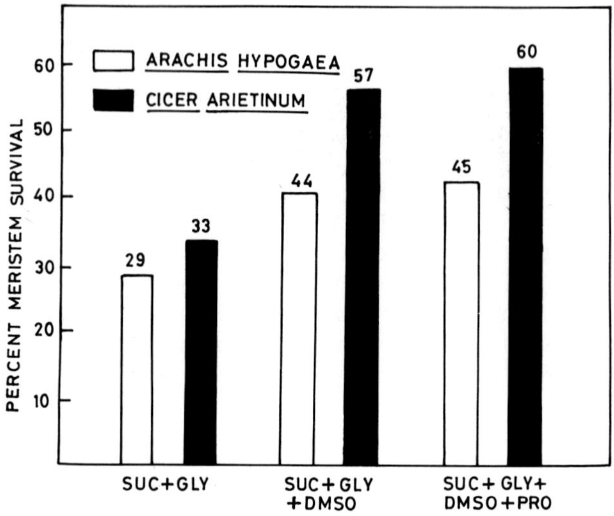

Fig. 3. Histogram showing the effect of mixture of various cryoprotectants on the survival of frozen meristems of *Arachis hypogaea* and *Cicer arietinum* (*SUC* sucrose; *DMSO* dimethylsulfoxide; *PRO* proline). (Bajaj 1990)

Effect of Preculture. Before freezing, the meristems of *C. arietinum* were precultured on a basal medium supplemented with various concentrations of DMSO for 3–7 days, and their cryoability was determined. Whereas the control (without DMSO) showed 52% survival, the addition of DMSO at 3 and 5% to the preculture medium increased it to 60 and 56%, respectively (Table 3). This effect was positive but not significant. DMSO at higher concentrations (10%) seemed inhibitory, as the survival was reduced. Thus, the preculture of the meristems on a medium containing an appropriate concentration of the cryoprotectants seemed to enhance the cryoability, and higher viability may be obtained.

Effect of Cryoprotectant. The results on the effect of various cryoprotectants on the precultured meristems are shown in Table 4. In both, *A. hypogaea* as well as *C. arietinum*, a combination of 5% each of sucrose, glycerol, and DMSO gave the highest survival, i.e., 42 and 57%, respectively. This was an improvement over sucrose (5%) + glycerol (5%) which gave 29 and 33% survival. The addition of proline further increased the survival to 60% (Fig. 3).

Effect of Thawing. The frozen cultures were thawed at 25, 35, and 40 °C (Table 5); the best results were obtained with cultures thawed at 35 °C (51%), followed by 40 °C (48%) and 25 °C (39%).

Table 4. Effect of various cryoprotectants on the survival[a] of meristems (not precultured) of *Arachis hypogaea* and *Cicer arietinum* freeze-preserved in liquid nitrogen for 7 months. (Bajaj 1983a)

Cryoprotectant	*Arachis hypogaea*			*Cicer arietinum*		
	Number of meristems frozen	Number of meristems showing growth	Survival (%)	Number of meristems frozen	Number of meristems showing growth	Survival (%)
Sucrose (5%) + glycerol (5%)	31	9	29	45	15	33
Suc (5%) + gly (5%) + DMSO (5%)	33	14	42	37	21	57

[a] Survival was judged by the capacity of the meristems to show greening, increase in size, and to form callus or shoot.

Table 5. Effect of various thawing temperatures on the survival of unfrozen and frozen (stored at −196 °C for 3 months) excised meristems of *Cicer arietinum.* (Bajaj 1983a)

Thawing temperature (°C)	Material	Number of meristems cultured	Number of meristems showing growth	(%) Meristems showing growth	(%) of Control
25	Unfrozen	29	23	80	–
	Frozen	32	10	31	39
35	Unfrozen	21	16	76	–
	Frozen	23	9	39	51
40	Unfrozen	24	17	70	–
	Frozen	26	9	34	48

For control purposes, before culture the unfrozen meristems were treated with sterile distilled water at the appropriate temperature (25, 35, 40 °C).

3 Cryopreservation of Pollen Embryos

Haploid cell cultures are known to be genetically unstable, and have a tendency to revert to their diploid form. In view of their importance in genetics and mutation studies, it is highly desirable to develop methods for their maintenance.

3.1 Methodology

Induction of Pollen Embryos. The anthers of cultivated tetraploid (2n = 40) peanut (*Arachis hypogaea* L. cv. M 13) and a wild diploid (2n = 20) species (*A. villosa*), containing uninucleate pollen and also undergoing mitosis, were cultured on agar-solidified modified MS medium (Murashige and Skoog 1962) + IAA (4 mg/l) + kinetin (2 mg/l) or BAP (2 mg/l) for 4–6 weeks. During this time, the anthers formed callus, and the pollen underwent segmentation and repeated nuclear and cell divisions to form pollen embryos (Bajaj et al. 1980,

1981b). Anthers were removed from the cultures, cut into two to four segments and used for the cryopreservation study (Bajaj 1983c).

Freezing and Thawing. The segments from 50 anthers were either pooled, treated with the cryoprotectant solution (5% each of sucrose, DMSO, and glycerol), and wrapped in aluminum foil, or were put in 1 ml of the cryoprotectant solution in an ampule. From some anthers a suspension of pollen embryos was prepared and frozen. The materials were either: (1) subjected to the vapors of liquid nitrogen

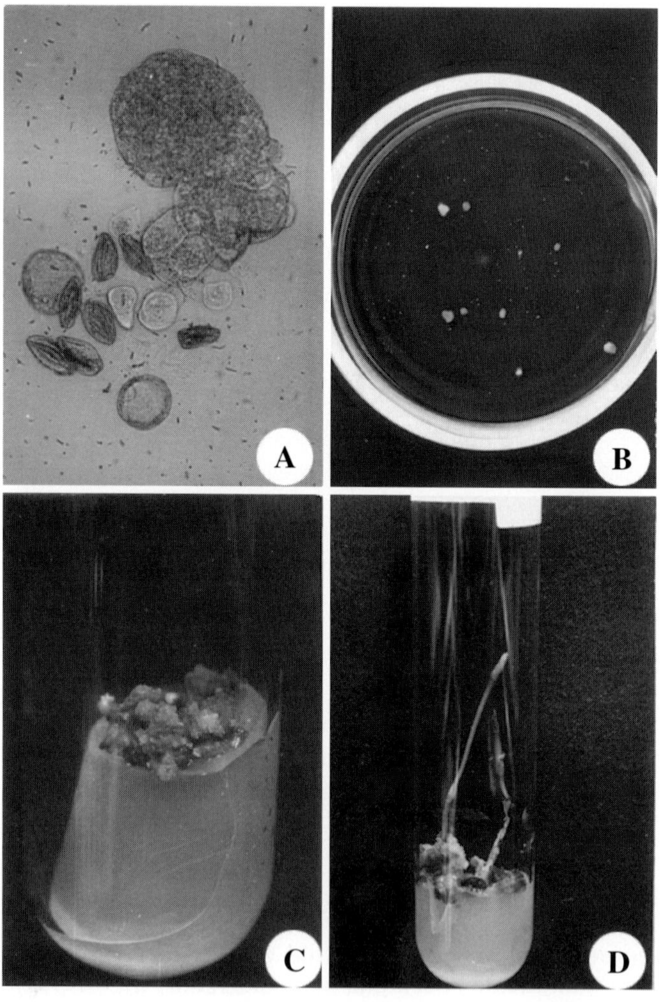

Fig. 4A–D. Survival of pollen embryos and anther segments of *Arachis hypogaea* and *A. villosa* stored in liquid nitrogen for 1 year. **A**, **B** A retrieved pollen embryo immediately after thawing, and **B** subsequent growth into callus colonies (*A. hypogaea*). **C**, **D** A mass of callus obtained from frozen anther segments of *A. villosa*, and its differentiation into shoots. (Bajaj 1983c)

Table 6. Survival of pollen embryos and segments of the androgenic anthers of *Arachis* frozen and stored at −196 °C for 2 months. (Bajaj 1983c)

Plant species	No. of pollen embryos frozen	No. of pollen embryos surviving	Survival (%)	No. of anther segments frozen	No. of anther segments resuming growth	Survival (%)
Arachis hypogaea	73	21	29	107	15	14
A. villosa	121	46	38	120	37	31

(LN) and immersed gradually, or (2) were quickly immersed in LN. After storage for 1 year, they were thawed at 35–40 °C, washed with liquid medium, and cultured.

3.2 Results

The survival was judged by: (1) increase in size of the embryos and the segments of the anthers, (2) callusing, (3) greening, and (4) the formation of roots and shoots. The pollen embryos underwent a lag period and showed no sign of growth for the first 4 weeks in culture; however, thereafter they started to proliferate and formed callus (Fig. 4 A–D). *Arachis villosa* showed higher survival as compared to *A. hypogaea* (Table 6). Viability in the two species varied. The callus occasionally differentiated malformed shoots and plants. The cultures freeze-preserved for 1 year revived.

The survival of anther segments was observed in both methods employed for freezing. However, anther segments wrapped in the aluminum foil (dry method) yielded better results than those frozen in ampules. In the latter case, the anthers had a tendency to become spongy.

4 Summary

Peanut (*Arachis hypogaea* L.) and chickpea (*Cicer arietinum* L.), two important legume crops, are rich sources of oil and protein. Their germplasm is traditionally conserved through “orthodox” seeds which are stored in a dry state and remain viable for years. Recently, considerable work has been done on various aspects of in vitro culture, i.e., regeneration, somatic embryogenesis, androgenesis, transformation, and genetic variability. However, cryopreservation methods need to be developed and refined for the long-term conservation of in vitro induced genetic variability.

The excised meristems were cryoprotected with a mixture of DMSO, sucrose, and glycerol, and subjected to freeze storage in liquid nitrogen. They survived, and the regenerated plants were capable of setting normal seed. Likewise, pollen embryos obtained from cultured anthers of peanut stored for 1 year in LN resumed growth and differentiation. These observations confirm the possibility of long-term conservation of in vitro induced variability in these crops.

References

Altaf N, Ahmad MS (1990) Chickpea (*Cicer arietinum* L.). In: Bajaj YPS (ed) Biotechnology in agriculture and forestry, vol 10. Legumes and oilseed crops I. Springer, Berlin Heidelberg New York, pp 100–113

Bajaj YPS (1979) Freeze preservation of meristems of *Arachis hypogaea* and *Cicer arietinum*. Indian J Exp Biol 17: 1405–1407

Bajaj YPS (1982) Cryopreservation of germplasm of potato, cassava, peanut and chickpea. In: Fujiwara A (ed) Plant tissue culture 1982. Maruzen, Tokyo, pp 799–800

Bajaj YPS (1983a) Production of normal seeds from plants regenerated from meristems of *Arachis hypogaea* and *Cicer arietinum* cryopreserved for 20 months. Euphytica 32: 425–430

Bajaj YPS (1983b) Cryopreservation and international exchange of germplasm. In: Sen SK, Giles KL (eds) Plant cell culture in crop improvement. Plenum Press, New York, pp 19–41

Bajaj YPS (1983c) Regeneration of plants from pollen-embryos of *Arachis*, *Brassica* and *Triticum* spp. cryopreserved for one year. Curr Sci 52: 484–485

Bajaj YPS (1984) Peanut. In: Ammirato PV, Evans DA, Sharp WR, Yamada Y (eds) Handbook of plant cell culture, vol 3. Crop species. Macmillan, New York, pp 193–225

Bajaj YPS (1990) Cryopreservation of germplasm of legumes and oilseed crops. In: Bajaj YPS (ed) Biotechnology in agriculture and forestry, vol 10. Legumes and oilseed crops I. Springer, Berlin Heidelberg New York, pp 49–62

Bajaj YPS (1991) Storage and cryopreservation of in vitro cultures. In: Bajaj YPS (ed) Biotechnology in agriculture and forestry, vol 17. High-tech and micropropagation I. Springer, Berlin Heidelberg New York, pp 361–381

Bajaj YPS, Gosal SS (1987) Pollen embryogenesis and chromosomal variation in cultured anthers of chickpea. Chickpea Newslett 17: 12–13

Bajaj YPS, Gosal SS (1988) Isolation and fusion of protoplasts of *Arachis hypogaea* and *Arachis villosa*. Int Arachis Newslett 3: 13–14

Bajaj YPS, Labana KS, Dhanju MS (1980) Induction of pollen-embryos and pollen-callus in anther cultures of *Arachis hypogaea* and *A. glabrata*. Protoplasma 103: 397–399

Bajaj YPS, Kumar P, Labana KS, Singh MM (1981a) Regeneration of plants from seedling-explants and callus cultures of *Arachis hypogaea* L. Indian J Exp Biol 19: 1026–1029

Bajaj YPS, Ram AK, Labana KS, Singh H (1981b) Regeneration of genetically variable plants from the anther-derived callus of *Arachis hypogaea* and *A. villosa*. Plant Sci Lett 23: 35–39

Barna KS, Wakhlu AK (1993) Somatic embryogenesis and plant regeneration from callus cultures of chickpea (*Cicer arietinum* L.). Plant Cell Rep 12: 521–524

FAO (1992) Production yearbook, vol 46. FAO, Rome

Gosal SS, Bajaj YPS (1988) Pollen embryogenesis and chromosomal variation in anther culture of three food legumes – *Cicer arietinum*, *Pisum sativum* and *Vigna mungo*. SABRAO J 20: 51–58

Gupta RP, Kalra MS, Bajaj YPS (1984) Nitrogen fixation in cell cultures of some legumes and non-legumes. Indian J Exp Biol 22: 560–563

Hawtin GC, Singh KB (1980) Kabuli-Desi introgression. Problems and prospects. Proc Int Worksh Chickpea Improv, ICRISAT, Hyderabad, pp 51–60

Krapovickas A, Rigoni VA (1952) Estidies citologicos en al Genero *Arachis*. Rev Invest Agric (Buenos Aires) 5: 289–294

Li Zhi-jian, Jarret RL (1995) Regeneration of plants from protoplasts of *Arachis* species (peanut). In: Bajaj YPS (ed) Biotechnology in agriculture and forestry, vol 34. Plant protoplasts and genetic engineering VI. Springer, Berlin Heidelberg New York (in press)

McKently AH (1995) Somatic embryogenesis in peanut (*Arachis hypogaea* L.). In: Bajaj YPS (ed) Biotechnology in agriculture and forestry, vol 31. Somatic embryogenesis and synthetic seed II. Springer, Berlin Heidelberg New York, pp 228–238

Moreno MT, Cubero JI (1978) Variation in *Cicer arietinum* L. Euphytica 27: 465–485

Plucknett DL, Smith NJH, Williams JT, Murthi Anishetty N (1983) Crop germplasm conservation and developing countries. Science 220: 163–169

Riazuddin S, Husnain T (1993) Transformation in chickpea (*Cicer arietinum* L.). In: Bajaj YPS (ed) Biotechnology in agriculture and forestry, vol 23. Plant protoplasts and genetic engineering IV. Springer, Berlin Heidelberg New York, pp 183–193

Schnall JA, Weissinger AK (1993) Culturing peanut (*Arachis hypogaea* L.) zygotic embryos for transformation via microprojectile bombardment. Plant Cell Rep 12: 316–319

IV.2 Cryopreservation of Alfalfa (*Medicago sativa* L.) and Clovers (*Trifolium* Species)

C.D. CACHIȚĂ[1] and C. CRĂCIUN[2]

1 Introduction

1.1 Plant Distribution and Important Species

Alfalfa (*Medicago sativa* L.) and clovers (white clover – *Trifolium repens* L. and red clover – *Trifolium pratense* L.) are the most important agronomic forage legumes and soil improvement crops in the temperate areas of the world. They are perennial, and are cultivated for their vegetable matter, fresh or preserved, which is used as feed for animals. The origin of alfalfa is in Iran (see Bolton et al. 1972), and of red and white clovers in Europe. Economically, white clover is more important than red clover. White clover is well adapted to cooler climates and to poor soils. *Medicago* and *Trifolium* species are the most useful culture as forage grasses, because their pasture value is much more important than that of red clover (after McCoy and Walker 1984; Phillips and Collins 1984).

The important species of *Medicago* and *Trifolium* are: *Medicago sativa* L., *M. falcata*, and *M. minima* (L.) Grufb; *Trifolium repens* L., *T. pratense* L., *T. campestre* Schreb., *T. strepens* Cr., *T. hybridum* L., *T. montanum* L., *T. arvense* L., *T. medium* L., *T. alpestre* L., and *T. pannonicum* Jacq. The species of *Medicago* and *Trifolium* which are endangered and threatened with extinction in Romania are: *Medicago prostrata* Jacq., *M. marina* L., and *M. arabica* (L.) All.; *Trifolium maritimum* Hdst, *T. angustifolium* L., *T. incarnatum* L., *T. scabrum* L., *T. alpinum* L., *T. vesiculosum* Savi, *T. ambiguum* M.B., and *T. michelianum* Savi (after Morariu 1957; Nyarady 1957).

1.2 Storage of Germplasm

Seed samples are maintained under controlled conditions of temperature and humidity. They can be stored at 2–5 °C, at −20 °C or in liquid nitrogen (LN) at −196 °C. Another method is the in vitro germplasm conservation, either in the growing state or by freeze preservation. Genetic variability may increase with time in the growing state, but in the frozen state the cultures are stable, especially

[1]Institute of Biological Research, 48 Republicii Street, 3400 Cluj-Napoca, Romania
[2]Department of Electron Microscopy, Faculty of Biology, "Babes-Bolyai" University, Clinicilor Street, 3400 Cluj-Napoca, Romania

Biotechnology in Agriculture and Forestry, Vol. 32
Cryopreservation of Plant Germplasm I (ed. by Y.P.S. Bajaj)

when stored in LN. Minimal growth conditions are suitable for meristems, but these are unsuitable for cell suspension cultures and callus culture. Therefore, in vitro culture techniques permit safe genetical storage of plant cells and tissue, and these can be used at any time for rapid propagation of the stored species (see Bajaj 1986; Withers 1987).

Rick (1984) indicated that in the USA the National Plant Germplasm System has a regional station, North-Eastern RPIS, Geneva, NY, – which specializes in perennial clover germplasm collection.

2 Cryopreservation

2.1 In Vitro Storage and Cryopreservation

Seed storage of forage legumes in a seedbank is mentioned by Stanwood and Bass (1978). Alfalfa and red clover were described as seeds whose germination was not adversely affected after 6 months' storage in LN. Stanwood and Bass (1978) mentioned the estimated cost of 1000 seeds stored in LN per year as: alfalfa – 2400 seeds, $0.03 and clover – 2600 seeds, $0.03. Stanwood (1985) pointed out that in the case of LN storage seed moisture content is probably the most critical factor influencing successful cryopreservation; for example, in clover with 25.6% seed moisture at −15 °C the germination capacity remained at 2%. Clover seeds showed a sharp drop in viability when the seed moisture content reached approximately 25%.

In alfalfa seeds stored in LN no kind of physical (breakage) damage is reported, and no loss of seed viability (Stanwood 1982, unpubl., after Stanwood 1985).

For seed storage in LN the procedure consists of ultrarapid freezing by seed immersion in LN. Table 1 illustrates studies carried out on alfalfa and clover seeds stored in LN.

Tomes (1979) cultivated shoots of birds-foot trefoil (*Lotus corniculatus* L.) in vitro and stored them at 2–5 °C for 1 month. In vitro storage of alfalfa and clover (red and white) shoots, for 15–18 months at 2–6 °C with 94–95% viability for alfalfa, and 70–92% viability for clovers, was reported by Cheyne and Dale (1980).

Bhojwani (1981) maintained *Trifolium repens* cultures shoot for 10 months at 5 °C, in the dark, and used this kind of storage as a slow-growth methods for shoot cultures.

Phillips and Collins (1984) observed that shoots of forage legume species, can be stored at 2–6 °C for 1–2 years (without appreciable loss of cell viability), and can then serve for rapid propagation of these species. Shoot tips may be the most desirable kinds of explant for long-term storage, as reported for *Arachis hypogaea* and *Cicer arietinum* (Bajaj 1983).

Considerable work has been done on the cryopreservation of in vitro cultures of legumes, and the subject has been reviewed (Bajaj 1990; see also Chaps. I.1, IV.1, IV.2, VI.3, this Vol.). Alfalfa plants were regenerated from

Table 1. Alfalfa and clover seeds frozen and stored in liquid nitrogen. (After Stanwood 1985)

Species/selection	Moisture (%)	Duration	Germination %		Comments
			Before	After	
Medicago sp.	Low	130 H	N.R.	No change	Liquid air immersion
M. sativa L.	6.5	3 years	86	90	−200 °C/min cooling, 30 °C/min rewarming, 4 cultivars
M. sativa L.	N.R.	9 months	95	92	Liquid air immersion
M. sativa L. cv. Grimm	N.R.	60 days	N.R.	No change	Rapid cooling, rewarming N.R., liquid air treatment on seedlings
M. sativa L. cv. Dupuits	5–17	ca. 1 h	N.R.	80	−50 °C/s cooling; 25 °C/s rewarming
M. sativa L. subsp.	7.0	180 days	92	89	−200 °C/min cooling, 30 °C/min rewarming
Trifolium alba	N.R.	N.R.	83	No change	−200 °C/min cooling, 37 °C H_2O rewarming
T. pratense L.	>12	24h	72	90	Rapid cooling, rewarming N.R., liquid air immersion
T. pratense L.	6.4	180 days	90	83	−200 °C/min cooling, 30 °C/min rewarming
Trifolium spp.	7.3	3 years	85	89	−200 °C/min cooling, 30 °C/min rewarming, 6 cultivars
T. subterraneum L.	N.R.	N.R.	97	No change	−200 °C/min cooling, 37 °C H_2O rewarming

N.R. = time not reported.

frozen callus by Finkle et al. (1979, 1985). Prior to immersion of callus in LN it was treated with a mixture of 10% PEG, 8% glucose, and 10% DMSO (w/v). The callus samples were then frozen at various temperatures down to −30 °C, at a rate of 1 °C/min to −196 °C by immersion of −30 °C samples in LN. After rapid thawing in a warm bath at 40 °C, the callus samples were recultured on nutrient agar culture. At all temperatures tested, the cultures grew rapidly, indicating little or no deterioration of cells by freezing treatment.

Yamada et al. (1991) reported the successful cryopreservation of apical meristem of *Trifolium repens*, using the vitrification method. The meristems from in vitro white clover were precultured on agar B5 medium supplemented by 1.2 M sorbitol; the meristems were maintained on this medium for 2 days at 4 °C under continuous light. They were then transferred into cryotubes. Prior to plunging these into LN, the meristems were covered by a highly concentrated cryoprotective mixture containing 30% (w/v) glycerol, 15% (w/v) ethylene glycol, and 15% (w/v) DMSO added in B5 medium containing 0.4 M sucrose. This treatment ensures sufficient dehydration of the meristematic cells. The meristems treated with cryoprotective mixture as above (designated PVS2) were maintained for 5 min at 25 °C or for 15 min at 0 °C. The cryotubes were then plunged directly into LN. The thawing consisted of a rapid warming of the cryotube which contained the meristems in water at 25 °C. In both experimental procedures, the average rate of plant regeneration was about 80%. The authors extended these procedures to other clover species, *T. pratense* (a hardy cultivar), and to *T. alexandrium* (a noncold-hardy cultivar). Both species showed moderate survival, about 50%.

As we investigated the ultrastructural aspects and cryoinjury produced in plant cells after cryopreservation under different conditions of cryoprotection, we must briefly analyze the results obtained by Sun Long-hua and Jian Ling-cheng (1990) in another legume, sainfoin (*Onobrychis viciaefolia* Scop), in tissue culture cryopreservation conditions and the ultrastructural observations made on normal or frozen tissue, cryoprotected by a mixture consisted of 10% DMSO, 10% glycerol, and 8% sucrose. The cooling rate in the freezing apparatus corresponded to from 0 °C to 1 °C/min to −35 to −40 °C. The temperature was kept at this level for 2 h. Then the biological material was plunged into LN. Thawing was carried out rapidly, at 40 °C in the water bath. The survival rate of the cultures was 60–70%, and the cells maintained their high differentiation ability. This percentage was also obtained when following the next procedure: from 0 °C lowering the temperature at a rate of 1 °C/min to −8 °C, staying for 20 min at this temperature, then continuing the decrease in temperature at the same rate to −35 °C. This temperature was maintained for about 30 min or 2 h, and then the material was plunged into LN. Freezing at the rate of 1 °C/min to −35 to −40 °C without keeping samples for a certain time produced lethal injury in the cells; this was reversible when the frozen cultures were kept for 30 min at the terminal temperature of −35 °C. When the frozen cultures were kept for 3 h at −35 to −40 °C the results were similar to those of the control material.

We examined the response of *Medicago* and *Trifolium* calli to various kinds of cryoprotectants, such as DMSO 0.5 M, glycerol 0.1 M and proline 10%, honey and liposomes, administered singly or in a mixture (in different proportions), and also the conservation of the regenerative capacity of the calli after freezing in LN, followed by thawing and subculturing the cryopreserved callus on optimal medium suitable for sustained cell growth and multiplication.

On the other hand, because much of the concern in biological membrane degradation by freezing injury is at the bilayered phospholipid level, we considered it necessary to study the influence of liposomal suspension (in an optimum composition), in biostructure protection, in the case of its freezing.

2.2 Experimental Protocol

In a first experiment, before submerging the callus in LN, it was cultured for 4 or 14 days on Murashige and Skoog (MS) (1962) mineral medium, plus vitamins (thiamine HCl, pyridoxine HCl, and nicotinic acid, 1 mg/l each), m-inositol 100 mg/l, sucrose 20 g/l with or without agar 0.7%; the medium was supplemented with different cryoprotectants, as follows: variant A_1 – DMSO 0.5 M; variant A_2 – glycerol 0.1 M; variant A_3 – a mixture of DMSO 0.5 M, glycerol 0.1 M, and sucrose 1 M; variant A_4 – the same mixture as for A_3, but supplemented with 10% proline. In one case, the ampules with inoculated calli were plunged directly into LN, and were maintained there for 2 days. In a second case, the ampules were put into an apparatus programmed to lower the temperature step by step as follows: at a rate of 10 °C/min to 4 °C; then maintained at 4 °C for 30 min; then gradually decreased at a rate of 0.5 °C/min, to −35 °C; the temperature was maintained at −35 °C for 40 min; the ampules with cryoprotected calli were then immersed directly in LN. In this case, the ampules were maintained submersed in LN for 40 days.

In another case, we studied the reaction of white clover callus cells cryoprotected by honey, and then frozen in LN. For this reason, untraumatized callus was submerged for 4 days at 1 °C in the dark in cryoprotectant agents as follows: variant V_1 – honey; variant V_2 – honey mixed with DMSO 0.5 M (v/v); variant V_3 – honey mixed with glycerol 0.1 M (v/v); variant V_4 – honey (1/3) mixed with DMSO 0.5 M (1/3), and glycerol 0.1 M (1/3); variant V_5 – glycerol 0.1 M; and variant V_6 – DMSO 0.5 M. After 4 days, the ampules containing calli were plunged directly into LN, then stored there for 24 days.

In experiments made with liposomes, we tried to prepare many kinds of liposomes of different composition, to obtain liposomes with the special capacity of producing slow physiological cell plasmolysis and deplasmolysis. In this case, we used only the vegetal (soya) lecitine, simple or in mixture with other additional cryoprotectant components such as honey, proline, sugar, DMSO, and glycerol.

In some experiments, different kind of calli were covered for 6 months, by liposomal suspension, or liposomes prepared from soya or egg lecitine, without freezing. In another experiment, white clover callus was covered by a simple liposomal suspension for 1, 4, or 7 days in the dark at 4 °C, after which the ampules with biological material were plunged into LN for 24 h.

A last experiment consisted of immersion of young alfalfa callus for 1 h in a new cryoprotectant composition such as: L_1 – liposome suspension, prepared from soya lecithin, mixed 1 : 1 with honey; L_2 — liposome mixed with honey and with L-proline solution 10% (1:1:1); L_3 – honey; L_4 – honey plus proline 10% (1:1); and L_5 – callus submersed in proline solution 10%. Then, the ampules containing calli were directly plunged into LN, and stored in it for 4 days; after which they were thawed and the cryoprotectant composition was diluted, step by step, gradually, in three steps (3:1, 2:2, 1:3 parts).

The thawing procedure, in all cases, consisted of ampules being plunged directly into warm water (40 °C), for 2 min. After thawing, the calli were immediately cultured on a suitable fresh media. The composition of the culture media was the same as that used generally for callus culture (in our case for its cryopreservation), without cryoprotectants, but with the addition of 2.4-D 2 mg/l and BAP 1 mg/l.

We also investigated ultrastructural aspects and cryoinjury produced in callus cells. The experimental protocol for electron microscopic examination was described previously (Cachiță and Crăciun 1990).

2.3 Results

Cachiță et al. (1991) obtained regenerative callus from alfalfa and white and red clover cryostored for 2 or 40 days in LN. Calli pregrown in media with only

DMSO or glycerol did not survive after thawing. The experiments proved that only calli cultured against freezing in LN from 4 or 14 days in media containing a mixture of cryoprotectants: DMSO 0.5 M, plus glycerol 0.1 M, and plus sucrose 1 M (v/v), with or without proline maintained their regenerative capacity after subsequent thawing and subculturing on suitable regenerative media. For alfalfa this was 81%, for clover 83%. When the calli were kept prior to submergence in LN for 4 or 14 days on media supplemented with a cryoprotectant mixture of DMSO 0.5 M, glycerol 0.1 M, and sucrose 1 M with or without proline, no modification of regenerative ability was observed. Preservation of calli in LN for 2 and 40 days also did not influence their regenerative capacity.

Electron microscopic examination revealed that cryoinjury was caused by LN in the uncryoprotected callus cells. Figure 1A shows normal white clover callus cells; Fig. 1B represents cells that were frozen in LN, without preliminary cryoprotection.

As can be seen in Fig. 1C and D and Fig. 2A in cells of callus cryoprotected for 4 days with prefreezing and different kinds of cryoprotectants (variants V_1–V_6), the ultrastructural injuries were less severe, especially in cells derived from callus cryoprotected exclusively by honey (Fig. 1C). As can be observed in Fig. 1C, the plasmalemma and the organelle structure was well preserved and upright only some scarcity of cytoplasm and some vacuolization can be seen. However, generally, the appearance of the cells is good. In the case of variants that were protected by a mixture of honey with different cryoprotectants, cellular damage after freezing in LN and thawing was more severe than in callus cryoprotected by honey only (cf. Fig. 1C and D and Fig. 2A).

In the same experiment, when the ampules were submitted to progressively decrease in temperature (at a rate of 1 °C/min) to −35 °C and then plunged into LN and maintained in it for 40 days, after thawing and subculturing the callus, the regenerative capacity was good only in callus cryoprotected by a mixture of DMSO plus glycerol, and especially in the presence of proline 10%.

The experiment, which consisted of studying the response of the callus cells on freezing in LN, when the callus was cryoprotected for 4 days at 1 °C by honey, alone or mixed with traditional cryoprotectants such as DMSO and glycerol (variants V_1–V_6) prior to immersion in LN, revealed that all the cryoprotectants tested caused plasmolytic damage, because the cryoprotectants have been able to influence the callus for a very long time. It is, however, important to emphasize that even in this situation, honey acted as a very good cell cryoprotectant. In experiments with liposomes, we observed that these can penetrate passively into the cells of the calli, crossing the cell wall if the calli were covered with a liposomal suspension for 6 months without freezing (Fig. 2B,D).

The liposomal composition affects the penetration of the liposomes into the cells, and then the liposome behavior. For example, egg lecitine-derived liposomes are not good for plant cells in long-term contact of the liposomal suspension with callus tissues. Liposomes in LN suffer a process of expansion (Cachiță and Crăciun, unpubl.). Liposome penetration into sinking liposomes – localized extra- or intracellularly – in LN is increased (accelerated). In our opinion, this method could be used to introduce genetic molecular information into the cell, without elimination of the cell wall.

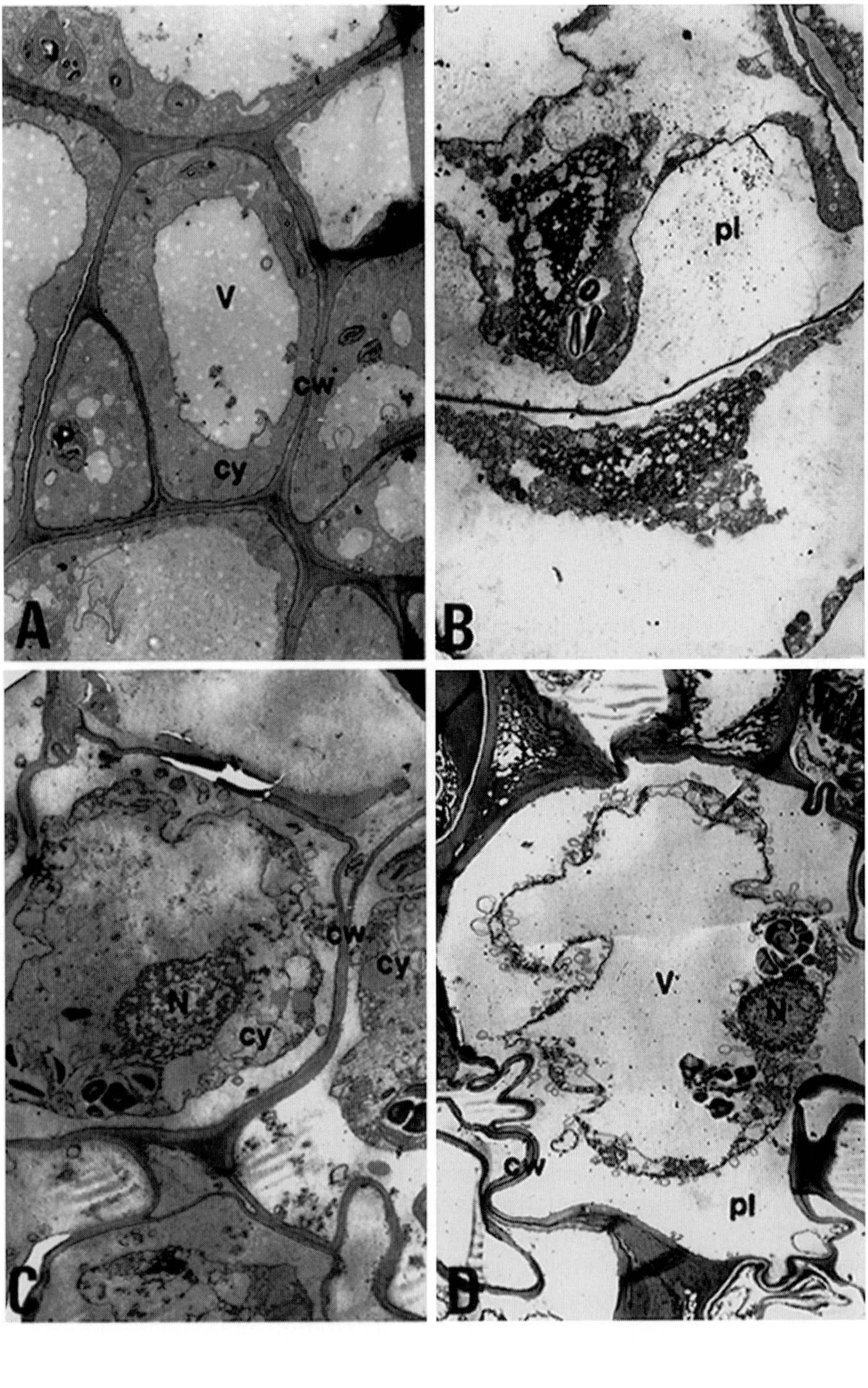
V
cw
cy
A
pl
B
cw
cy
N
cy
C
V
N
cw
pl
D

Our first aim in this respect was to study the ultrastructural aspects in white clover callus cells plunged for 1–4 or 7 days prior to freezing in LN in liposomes. In the case of this kind of liposome (after thawing and gradually deplasmolyzing the ampules), we observed liposome penetration into the cells through the cell wall (Fig. 3A–C). Even if the external callus cells were completely flooded by liposomes (as shown in Fig. 3A–C), the deeper-lying cells, especially those near the tracheary elements (Fig. 3D), showed perfect ultrastructure and were not influenced negatively by cryoprotection, LN, and the thawing process. This type of cell, however, can serve to obtain callus fragments that have preserved their regenerative capacity after being frozen in LN.

The cell biostructures were well protected by liposomes mixed with honey (L_1). An excellent ultrastructural situation and perfect protection was obtained in cryoprotection of alfalfa callus with mixture of liposomes, honey, and proline (variant L_2) (see Fig. 4A), for 1 h prior to freezing in LN. A similar situation, liposomes mixed with honey (variant L_1), was observed at variants L_3 and L_4 (Fig. 4B,C), but the ultrastructural aspects were very little altered. A very good situation was observed in cells which were derived from callus cryoprotected by proline 10% solution (Fig. 4D).

3 Summary and Conclusions

Alfalfa and clover (white or red) callus survived cryostorage in LN for 2 or 40 days. This callus prior to immersion in LN was cultured for 4 days on a medium with a mixture of DMSO 0.5 M, glycerol 0.1 M, sucrose 1 M, especially with addition of proline 10% (mixture in equal parts). After thawing and in subculturing, the callus rapidly proliferated.

Honey alone had a certain cryoprotective effect, but not as good as the previously described mixture. The ultrastructural image proved very good conservation of the biostructure in callus cryoprotected by a mixture of liposomes with honey and 10% proline, the mixture in equal parts. A good situation was observed in the case of callus cryoprotected in proline solution only. Keeping the callus in the cryoprotective agent for more than several hours before immersion in LN led to irreversible plamolysis and damage caused by plamolyzation and not by the freezing and defreezing.

The experiments demonstrated that liposomes can cross over the cell wall, penetrate into the cytoplasm, and consequently an accumulation in the cells

Fig. 1A–D. Ultrastructural aspects of cells derived from white clover callus treated or control for 4 days at 1 °C by different cryoprotectants, plunged into liquid nitrogen for 24 days, then thawed in warm water (40 °C); subsequently, callus fragments were fixed for electron microscopic examination. **A** Ultrastructural aspects of cells derived from control callus (2600x). **B** Cells derived from callus not treated with cryoprotectants, and stored in liquid nitrogen for 24 days; it can be observed that the cells were completely destroyed (2500x). **C** Cells derived from callus cryoprotected by honey only (V_1), prior to submerging callus in liquid nitrogen (2300x). **D** Cells derived by callus cryoprotected prior to submergence in LN in a mixture of honey and DMSO 0.5 M (v/v) (V_2) (3100x). *cy* Cytoplasm; *cw* cellular wall; *N* nucleus; *P* plastid; *pl* plasmolyze; *S* starch; *V* vacuole

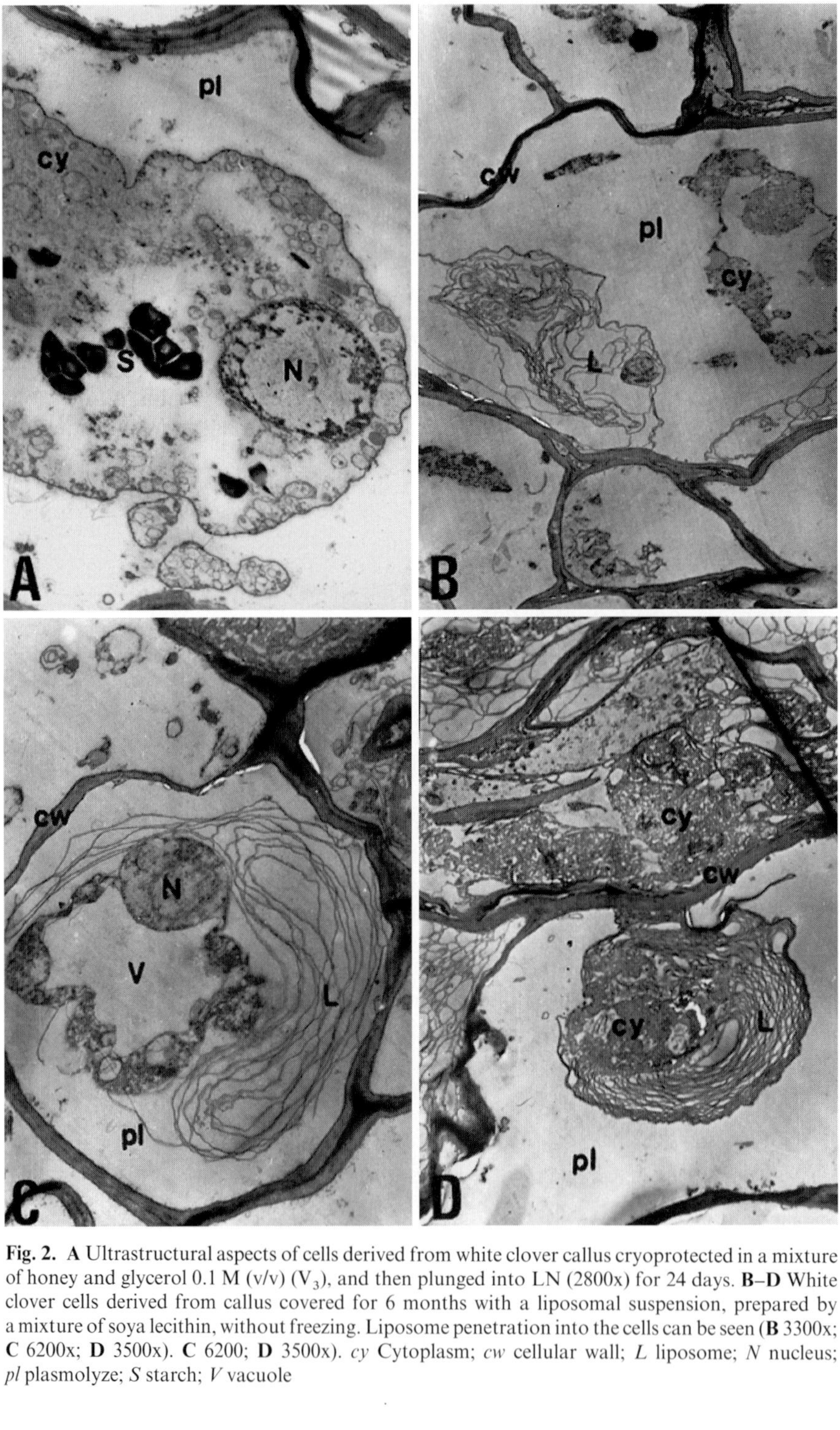

Fig. 2. **A** Ultrastructural aspects of cells derived from white clover callus cryoprotected in a mixture of honey and glycerol 0.1 M (v/v) (V_3), and then plunged into LN (2800x) for 24 days. **B–D** White clover cells derived from callus covered for 6 months with a liposomal suspension, prepared by a mixture of soya lecithin, without freezing. Liposome penetration into the cells can be seen (**B** 3300x; **C** 6200x; **D** 3500x). **C** 6200; **D** 3500x). *cy* Cytoplasm; *cw* cellular wall; *L* liposome; *N* nucleus; *pl* plasmolyze; *S* starch; *V* vacuole

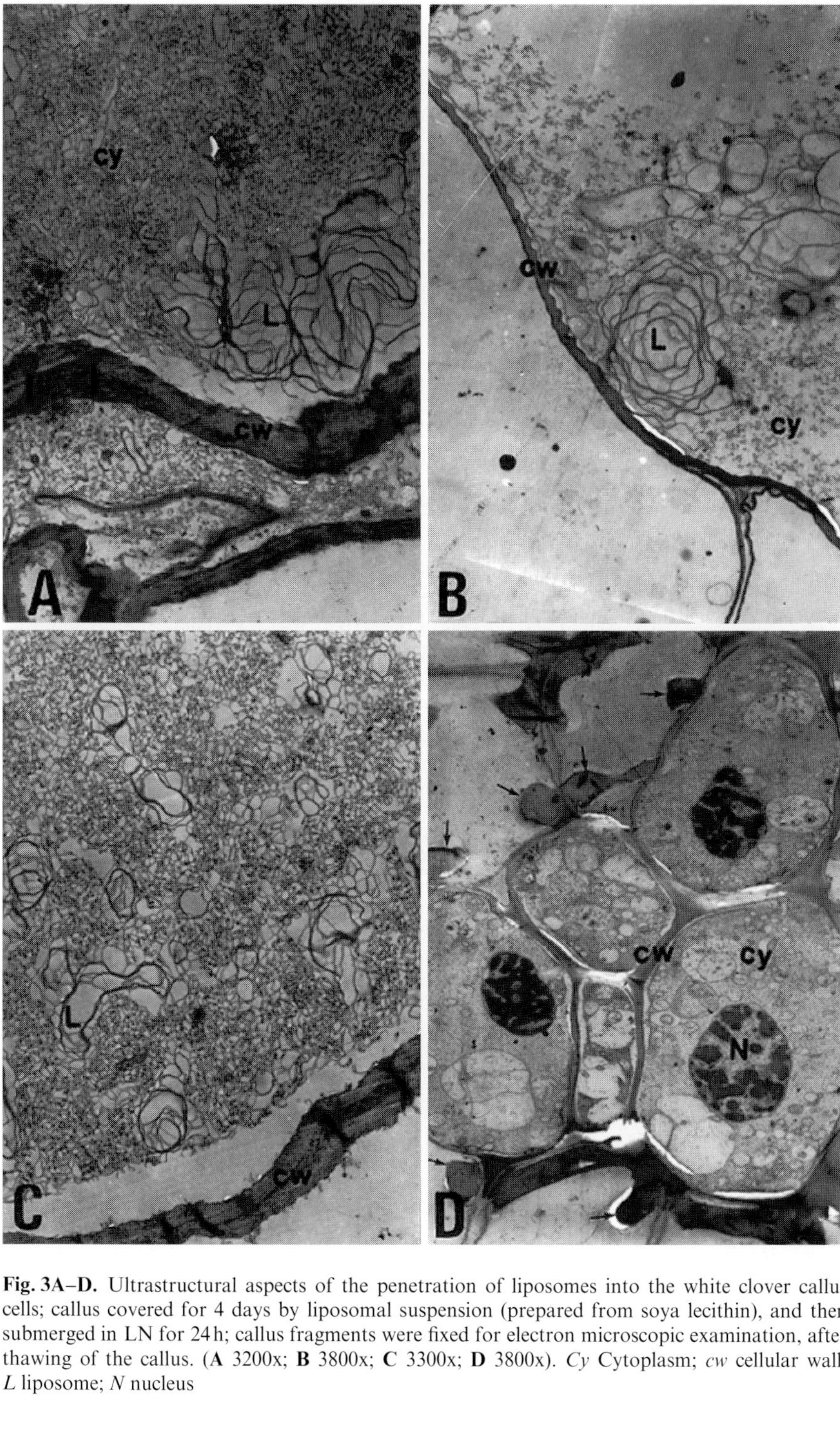

Fig. 3A–D. Ultrastructural aspects of the penetration of liposomes into the white clover callus cells; callus covered for 4 days by liposomal suspension (prepared from soya lecithin), and then submerged in LN for 24 h; callus fragments were fixed for electron microscopic examination, after thawing of the callus. (**A** 3200x; **B** 3800x; **C** 3300x; **D** 3800x). *Cy* Cytoplasm; *cw* cellular wall; *L* liposome; *N* nucleus

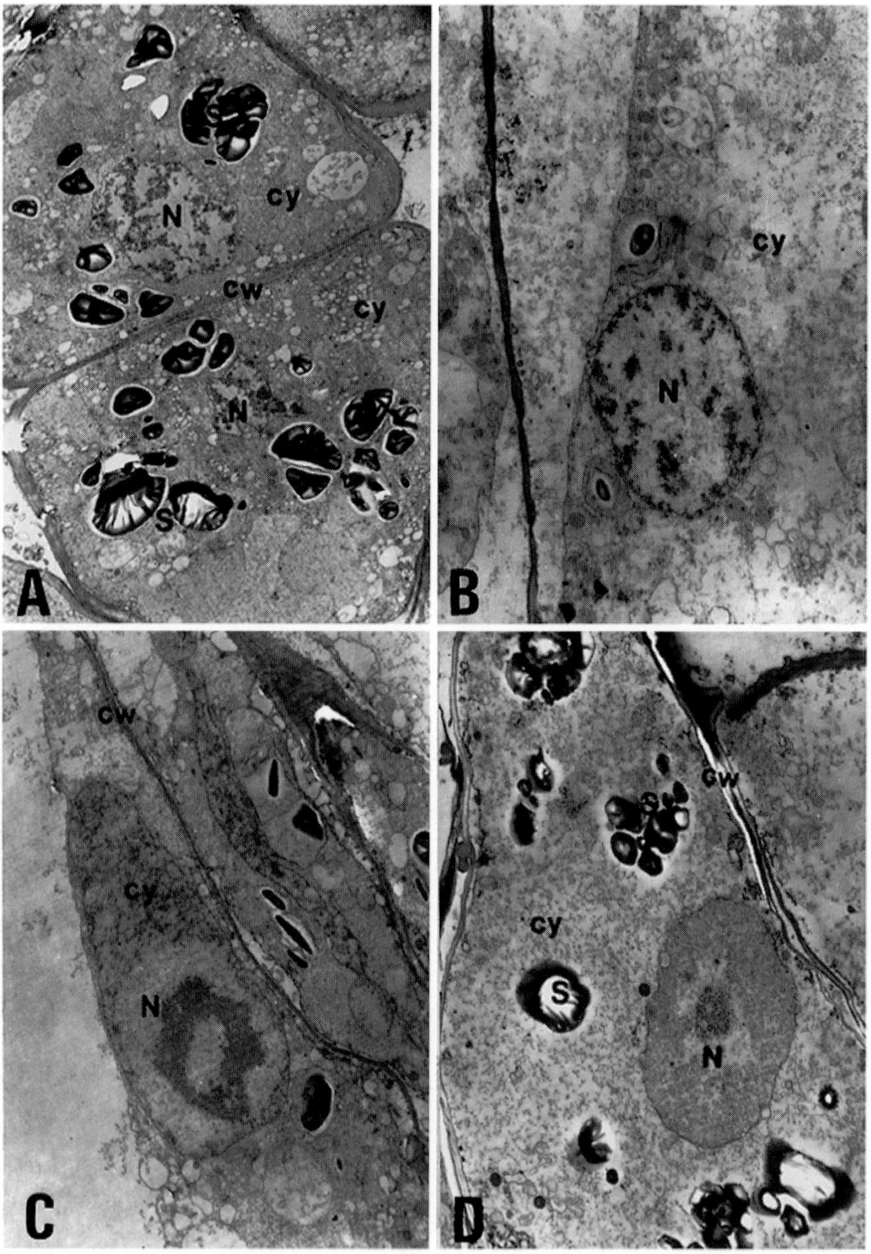

Fig. 4A–D. Ultrastructural aspects of the alfalfa callus cells cryostored in liquid nitrogen for 1h before being cryoprotected for 4 days in a liposomal mixture as follows: **A** Variant L_2, liposome prepared from soya lecithin mixed with honey and 10% proline (3400x). **B** Variant L_3, callus cryoprotected by honey (2500x). **C** Variant L_4, callus cryoprotected by a mixture 1:1 of honey with 10% proline (3500x). **D** Variant L_5, callus cryoprotected by proline, 10% solution (2500x). *Cy* Cytoplasm; *cw* cellular wall; *N* nucleus; *S* starch takes place (Figs. 2B–D, 3A–C) (Cachiţă and Crăciun, unpubl.). The ultrastructural image permitted liposomes localization in two to three external callus cell lines.

References

Bajaj YPS (1983) Production of normal seeds from plants regenerated from meristems of *Arachis hypogaea* and *Cicer arietinum* cryopreserved for 20 months. Euphytica 32: 425–430

Bajaj YPS (1986) In vitro preservation of genetic resources. Techniques and problems. Int Symp Nuclear techniques and in vitro culture for plant improvement, 1985 IAEA, Vienna, pp 43–57

Bajaj YPS (1990) Cryopreservation of germplasm of legumes and oilseed crops. In: Bajaj YPS (ed) Biotechnology in agriculture and forestry, vol 10. Legumes and oilseed crops. I. Springer, Berlin Heidelberg New York, pp. 49–62

Bhojwani SS (1981) A tissue culture method for propagation and low temperature storage of *Trifolium repens* genotypes. Physiol Plant 52: 187–190

Bolton JL, Goplen BP, Baenziger H (1972) World distribution and historical developments. In: Hanson CH (ed) Alfalfa science and technology. Am Soc Agron, Madison, pp 1–34

Cachiţă CD, Crăciun C (1990) Ultrastructural studies of some ornamentals. In: Ammirato PV, Evans DA, Sharp WR, Bajaj YPS (eds) Handbook of plant cell culture, vol 5, Ornamental species. McGraw-Hill, New York, pp 57–94

Cachiţă CD, Zăpîrtan M, Crăciun C, Rákosy TL, Vicol A, Cristea V, Varga P, Crăciun V (1991) The response of various types of cormophytoinocula to cryopreservation (−196 °C). In: The IVth Nat Symp on Plant Cell and Tissue Culture. Romania, Cluj-Napoca, 1989, pp 90–91

Cheyne VA, Dale PJ (1980) Shoot-tip culture in forage legumes. Plant Sci Lett 19: 303–309

Finkle BJ, Ulrich JM, Rains DM, Tisserat BB, Schaeffer GW (1979) Survival of alfalfa, rice and date palm callus after liquid nitrogen freezing. Cryobiology 16: 583

Finkle BJ, Ulrich JM, Rains WR, Stavarek SS (1985) Growth and regeneration of alfalfa callus lines after freezing in liquid nitrogen. Plant Sci 42: 133–140

McCoy T, Walker K (1984) Alfalfa. In: Ammirato PV, Evans DA, Sharp WR, Yamada Y (eds) Handbook of plant cell culture, vol 3, Crop species. MacMillan, London, pp 171–192

Morariu I (1957) *Genul Medicago*, In: Săvulescu T. (ed), Flora RP Române, Ed Acad Române, vol 5, pp 118–136

Murashige T, Skoog F (1962) A revised medium for rapid growth and bioassays with tobacco tissue cultures. Physiol Plant 15: 473–497

Nyarady A (1957) *Genul Trifolium*, In: Săvulescu T (ed), Flora RP Române, Ed Acad Române, vol 5, pp 145–220

Phillips GC, Collins GB (1984) Red clover and other forage legumes. In: Sharp WR, Evans DA, Ammirato PV, Yamada Y (eds) Handbook of plant cell culture, vol 2, Crop species. MacMillan, London, pp 169–210

Rick CM (1984) Plant germplasm resources. In: Ammirato PV, Evans DA, Sharp WR, Yamada Y (eds) Handbook of plant cell culture, vol 3. Crop species. MacMillan, London, pp 9–37

Stanwood PC (1985) Cryopreservation of seed germplasm for genetic conservation. In: Kartha KK (ed) Cryopreservation of plant cells and organs. CRC Press, Boca Raton, Florida, pp 199–226

Stanwood PC, Bass LN (1978) Ultracold preservation of germplasm. In: Li PH, Sakai A (eds) Plant cold hardiness and freezing stress. Mechanisms and crop implications. Academic Press, New York, pp 361–371

Sun Long-hua, Jian Ling-cheng (1990) Cryopreservation of sainfoin tissue cultures and their ultrastructural observation. Acta Bot Sin 32(4): 262–267

Tomes DT (1979) A tissue culture procedure for propagation and maintenance of *Lotus corniculatus* genotypes. Can J Bot 57: 137–140

Withers LA (1987) Long-term preservation of plant cells, tissues and organs. Oxford Surv Plant Mol Cell Biol 4: 221–272

Yamada T, Sakai A, Matsumura T, Higuchi S (1991) Cryopreservation of apical meristems of white clover (*Trifolium repens* L.) by vitrification. Plant Sci 78: 81–87

IV.3 Cryostorage of Pea (*Pisum sativum* L.)

P. BERJAK[1], D.J. MYCOCK[2], P. WATT[1], J. WESLEY-SMITH[1], and B. HOPE[1]

1 Introduction

The common pea (*Pisum sativum* L.) is probably one of the best known legumes used as a fresh vegetable (Langer and Hill 1982). It is not known in the wild state, but is very closely related to *P. arvense*, which occurs in Georgia, Russia (Purseglove 1968; Janick et al. 1981). Pea appears to have been first cultivated about 7000 B.C. in the south-western parts of Asia, the same general region from which wheat originated (Janick et al. 1981). It was spread west to the Greeks, who passed it on to the Romans and thence to the Germanic races of Europe. Prior to European global migration, pea cultivation had also spread east to India and China and south into central Africa (Purseglove 1968; Janick et al. 1981). During the 18th century, Gregor Mendel, in his classic work on the genetics of pea, recognized 32 different types. Today, about 10000 cultivars have been recorded, most from temperate areas of the world where the crop is grown best (Gantotti and Kartha 1986).

Pea protein has a high lysine content, making the crop superior nutritionally compared with most other legumes (Bajaj et al. 1971; Janick et al. 1981). Today it is grown as either field pea for harvest of the dry seed as fodder or as garden (green) pea mainly for vegetable use; the former type is grown in far greater quantities than garden pea (see for example, Gantotti and Kartha 1986).

1.1 Necessity for Genetic Conservation of *Pisum sativum*

For most of the Earth's history the genetic diversity of plants has been maintained in wild habitats and, latterly, also in the farmers' fields. However, it has been only in this century that the wild relatives of common crop plants have become threatened by widespread destruction of their natural habitats (Bajaj 1983a; Hawkes 1983). Indeed, today *P. sativum* as we know it is not found in the wild (Plucknett et al. 1987). Additionally, the trend towards genetically uniform crops has been particularly pronounced since the 1940s. *Pisum sativum* grown in the USA exemplifies this trend, where no more than four varieties account for

[1]Plant Cell Biology Research Unit, Department of Biology, University of Natal, Durban, South Africa
[2]Department of Botany, University of the Witwatersrand, Johannesburg, South Africa

Biotechnology in Agriculture and Forestry, Vol. 32
Cryopreservation of Plant Germplasm I (ed. by Y.P.S. Bajaj)

more than 50% of the acreage of the crop (Plucknett et al. 1987). Consequently, genetic diversity is being lost. Crop species, however, are only a small percentage of the plant species that are endangered; Mlot (1989) estimates that in excess of 9000 wild species are threatened in some way. Collectively, this emphasizes the urgent necessity for plant genetic conservation.

There are two broad ways in which plant genetic resources can be conserved: in situ and ex situ (see, for example, Plucknett et al. 1987). Conservation in situ involves the establishment of natural reserves where species are allowed to remain in optimal ecosystems with the minimum of management. Ex situ conservation involves the use of seed stores, gene banks, and field plantations.

1.2 Preservation Methods

Seed-bearing plants fall into two categories, those that produce what are called orthodox seeds, that are desiccation-tolerant, and those that have recalcitrant, desiccation-sensitive propagules (Roberts 1972; Berjak et al. 1992). Seeds of *P. sativum* are orthodox, but development of successful cryostorage techniques for hydrated pea germplasm is nevertheless important (see below). However, the primary reason for our undertaking studies on the hydrated zygotic axes of *P. sativum* was to establish a model system mimicking the properties of axes from newly shed, recalcitrant seeds. The finding that excised axes of recalcitrant seeds can withstand rapid drying to moisture contents akin to those of mature orthodox seeds, apart from being an enigma, has opened the way for cryopreservation of such germplasm (see Sect. 2.2) (Chin et al. 1981; Berjak et al. 1989, 1990).

As a part of their development, most orthodox seeds undergo a period of maturation drying, during which the seed tissues lose water, come to equilibrium with the ambient relative humidity, and enter a state of quiescence (Bewley and Black 1985). In this dry, quiescent state, the seed can survive unfavorable conditions and will germinate only when suitable conditions are established. The ability of orthodox seeds to dry and to withstand lowered temperatures has enabled storage of these seeds for protracted periods.

Justice and Bass (1978) maintain that only 50% of pea seeds can be expected to germinate after 3 to 5 years when stored under favorable ambient conditions. Although the germination rates reported by Haferkamp et al. (1953) are lower, certain pea cultivars stored under similar conditions have been reported to survive for several decades. Survival under germplasm storage conditions (for pea usually at 7% moisture content and at low temperatures) can therefore be expected to be far longer (Justice and Bass 1978). There are more than 20 000 accessions of pea in gene banks throughout the world (Lyman 1984). This is estimated to cover 70% of the landraces, but only 10% of the wild, closely related species of pea.

Under suboptimal storage conditions, pea, like all other (orthodox) seeds, can be destroyed by storage fungi (Christensen and Kaufmann 1974; Agarwal and Sinclair 1987). It has been the experience of the present authors that pea seeds produced in subtropical Africa, succumb to internal infection in a matter of only months, even in cold storage (4 °C). Field and King (1962) reported that *Aspergillus ruber*, a seed storage fungus, is able to reduce pea seed germinability to zero within 8 months if the seed is stored at 85% relative humidity (RH). Thus, successful cryostorage of pea germplasm, especially in the tropics and subtropics is a matter of some importance.

2 In Vitro Culture and Cryopreservation of Pea

Seeds of many plant species are stored at subzero temperatures in gene banks. In the frozen state, normal metabolic activity is halted while genetic stability should be optimized (see, for example Grout 1990). Moisture content and size play an important part in successful cryopreservation of whole seed: generally the smaller and [seemingly within limits; Vertucci and Roos 1990] the drier, the better (Stanwood 1986; Iriondo et al. 1992): consequently, all intact, orthodox seeds may not be amenable to storage in the frozen state. However, the size and equilibrium moisture content of whole pea seeds facilitate their storage in the frozen state, usually at −16 °C.

Plant tissue culture procedures have advanced to a level that most species can be regenerated in vitro. As a result, the cryopreservation of in vitro material (cells and/or small pieces of tissue) offers a practical alternative storage method, particularly for the embryonic axes of recalcitrant seed species (Grout 1979, 1986; Bajaj 1983a; Withers 1989; Mycock et al. 1991; Vertucci et al. 1991; Mycock and Berjak 1992; Wesley-Smith et al. 1992), and has the potential of conservation of genetic stability.

In vitro technology, in particular cryopreservation, has been applied to the conservation of *P. sativum* germplasm. Already in 1958, Sun had reported on the survival of hydrated, excised pea axes that had been partially dried, frozen in liquid nitrogen, and regenerated in vitro. That author made the significant link between moisture content and the extent of cellular vacuolation with subsequent successful cryostorage. Kartha et al. (1974, 1979) developed technique for the regeneration of virus-free pea plants from isolated shoot meristems subsequently used for the cryopreservation of meristems. A number of other tissues from pea have been used for in vitro studies, e.g., roots, stems, and leaves for callus culture and subsequent regeneration by embryogenesis or organogenesis (Gantotti and Kartha 1986). Protoplasts isolated from shoot apex tissue have also been used in the development of novel pea hybrids thus offering unique germplasm types (Gamborg et al. 1975).

Using mannitol and DMSO as cryoprotectants, Bajaj (1983b) successfully cryopreserved interspecific hybrid (*Pisum* × *Oryza* and *Pisum* × *Triticum*) protoplasts. McAdams et al. (1991) exposed pea shoot tips to cryoprotectants in the culture medium and related the subsequent survival rate to the terminal freezing temperature.

The present contribution outlines the methodology for the cryopreservation of zygotic embryonic axes of pea. Two approaches have been adopted, one which uses cryoprotectants and plunge-freezing (of axes in mesh baskets) and the second which uses fully hydrated or flash-dried (Berjak et al. 1990) axes, followed by extremely rapid freezing. These investigations on cryostorage of hydrated embryonic axes of pea were undertaken in the context of our work on recalcitrant seeds (Vertucci et al. 1991; Wesley-Smith et al. 1992), and therefore it is partly in this context that the methodology is discussed.

2.1 Cryostorage Using Cryoprotectants, Partial Drying, and Plunge-Freezing in Mesh Baskets (Mycock et al. 1991)

2.1.1 Pretreatment of Seeds and Isolation of Axes

Pisum sativum seeds were imbibed for 4 h in sterile distilled water. During this time, the seed coats are shed, making excision of the embryonic axes easier. The intact seeds were surface sterilized in 0.01% mercuric chloride for 20 min, followed by repeated washing in sterile distilled water to remove the fungicide. The embryonic axes were then aseptically removed and re-sterilized in 1% sodium hypochlorite for 20 min, the surface sterilant being removed by washing three times with sterile, distilled water. Axes imbibed for 4 h had not lost the desiccation tolerance characteristic of the air-dry condition. In keeping with our original objectives of developing a model system applicable to metabolically active, recalcitrant seeds, a parallel procedure was carried out on desiccation-sensitive axes from germinating pea seeds that had been imbibed for 72 h.

2.1.2 Cryoprotection, Cryopreservation, and Retrieval

The isolated axes were soaked in a 5% (v/v) glycerol solution containing 5% sucrose for 15 min, then transferred to a more concentrated cryoprotectant solution (10% glycerol containing 10% sucrose) for a further 15 min. After treatment with the cryoprotectant, the axes were dried to 50% or less of their moisture content after the 4 h imbibition treatment. The most rapid drying rate (within 1 h), that could be achieved in a laminar flow cabinet, was found to be the most successful. The axes were then placed in stainless steel mesh baskets (10 × 10 mm) and immersed directly in liquid nitrogen. The baskets containing the axes were retrieved from the liquid nitrogen and placed directly into a liquid medium consisting of Murashige and Skoog (1962) nutrients (MS) and 0.4% sucrose at 37 °C for 2–3 min. After this, axes were plated aseptically onto a germination medium consisting of MS nutrients, 4% sucrose, 0.1% casein hydrolysate at pH 5.6, solidified using 1% agar. Growth was generally initiated within a week.

Up to 50% survival has been repeatedly achieved for 4-h-imbibed embryonic axes cryopreserved in this manner. When applied to germinating seeds (72 h imbibition; axis length less than 20 mm) a 25% successful retrieval from liquid nitrogen resulted. The notable point of this protocol is that, whether desiccation-tolerant or not, plant material of relatively high water content can be successfully cryopreserved. Moisture contents at freezing were approximately 1.2 and 1.5 g

H_2O per g dry matter (g/g), respectively, for 4- and 72-h-imbibed axes (Table 1). According to Vertucci (1990), at this degree of hydration most of the intracellular water is freezable, and the cooling rate should have caused ice crystal formation within the tissues (see below). Consequently, successful cryopreservation of the tissue must be attributed to the glycerol/sucrose mixture. It is believed that both the cryoprotectants and the air drying exert a partial dehydration on the tissues, whilst the cryoprotectants alone suppress the freezing and supercooling points of the remaining intracellular bulk water during freezing. An additional factor contributing to survival of the pea axes in this work, was the freezing rate. Although this was not *very* rapid (see Sect. 2.2), the partially hydrated axes did not survive slower freezing rates achieved when axes were enclosed (in, e.g., small plastic tubes, aluminum foil envelopes, etc.), despite the cryoprotectant pretreatment (Mycock et al. 1991).

Application of similar technology (using mesh baskets) to the somatic embryos of coffee (*Coffea arabica*), a crop species which is said to exhibit seed behavior intermediate between truly orthodox and recalcitrant (Ellis et al. 1990), allowed for 75% successful retrieval from cryostorage (Mycock and Berjak 1992). It was interesting that, as the somatic embryos of *C. arabica* developed from globular, to heart, to the torpedo stage, the survival after freezing increased. A similar link between embryonic axis maturity and survival of freezing has been noted for the zygotic embryos of *Coffea* spp. (Abdelour-Esquivel et al. 1992).

2.2 Cryostorage Using Flash-Drying and Extremely Rapid Freezing

Using the protocol described above, no pea axes survived liquid nitrogen immersion if not pre-treated, but all cryoprotectant combinations used depressed subsequent axis development (Mycock et al. 1991). In this regard, cryoprotectants containing DMSO had more markedly adverse effects on the pea axes than glycerol-sucrose mixtures, which is in keeping with the documented toxic effects of DMSO (Kartha 1985). Our aim, therefore, was to achieve cryostorage in the absence of even the most benign of cryoprotectant pretreatments.

There are two other major variables that need to be optimized for any cryostorage protocol for hydrated plant tissues, viz. the degree of dehydration, and the cooling (freezing) rate. Previous work had established that desiccation-sensitive axes from recalcitrant seeds could withstand a remarkable degree of water loss, if that water was removed sufficiently rapidly (Berjak et al. 1990). This procedure, termed flash-drying, requires that the excised axes are placed upon a mesh within a small, vented container into which air is introduced at an appreciable rate. The air-stream is diffused below the mesh, and vented above the axes, through small holes in the lid of the container. Flash-drying provides a means by which the water content of hydrated axes may be more or less predictably diminished within a given time. In this regard, although desiccation-sensitive axes will survive the loss of much of their nonfreezable water (i.e., water that is not structure-associated), dehydration nevertheless imposes a certain stress that is proportional to the degree of desiccation imposed (Wesley-Smith et al. 1992). Thus, optimizing this variable means imposing the least dehydration

Table 1. Survival of pea axes in culture following liquid nitrogen immersion

Variety time	Imbibition (h)	Treatment	Freezing rate	Initial H_2O content (gH_2O/g dry mass)	H_2O content at freezing (i.e., after flash-drying; gH_2O/g dry mass)	Survival % (greening and growth) at 3 days	5 days
Greenfeast	27	Fully hydrated; control	Not frozen	2.5	Not dried; 2.5	75	100
		Fully hydrated; frozen	>550 °C/s		Not dried; 2.5	70	70
		Flash dried 80 min; control	Not frozen		0.38	83	100
		Flash dried 80 min; frozen	>550 °C/s		0.38	60	67
Knight	27	Fully hydrated; control	Not frozen	4.1	Not dried; 4.1	50	100
		Fully hydrated; frozen	>550 °C/s		Not dried; 4.1	18	18
		Flash dried 80 min; control	Not frozen		0.45	60	100
		Flash dried 80 min; frozen	>550 °C/s		0.45	22	63
[a]Dark Skin Perfection	4	Not cryoprotected; dried; frozen	≃200 °C/min	2.38	1.19	Survival % at 3 weeks: 0	
		Cryoprotected[b]; dried; frozen				50	
	72	Not cryoprotected; dried; frozen		2.90	1.50	0	
		Cryoprotected[b]; dried; frozen				25	

[a]Data from Mycock et al. (1991).
[b]Glycerol-sucrose.

stress that is commensurate with successful cryopreservation. In absolute terms, the optimal water content that allows an axis to survive liquid nitrogen exposure without obvious deleterious effects varies on a species basis (unpubl.) which, in turn, is probably a factor of the chemical composition of the tissues. A second factor which may be relevant, is the axis water content at the outset (cf. *P. sativum* var. Greenfeast with var. Knight, Table 1).

The remaining consideration is that of freezing rate. Slow cooling – i.e., temperature reduction at the rate of 0.1–10 °C/min (Nitzsche 1983) has generally been favored, as supercooling followed by extracellular ice crystal formation is promoted. What has hitherto been considered as fast freezing, i.e., temperature reduction at the rate of several hundred °C/min, is injurious in terms of the formation of usually lethal, intracellular ice crystals (Meryman and Williams 1985). However, the question must be asked: how fast *is* fast?

Prompted by this seemingly enigmatic question, and the observation that the more rapidly plant tissue can be frozen for electron microscopical observation, the better is the preservation of its ultrastructure, one of us (J.W-S) promoted the idea that we should be attempting to obtain similar cooling rates for successful cryopreservation for (partially) hydrated embryonic axes. Such cooling rates are in the order of hundreds (or even in the thousands) of °C/s and are made possible, in terms of the extremely small volumes of tissue that are cryoprepared for electron microscopy.

Although most axes (from recalcitrant seeds) are very small, they are nevertheless far larger than specimens prepared for electron microscopy. To counteract the negative effects of specimen size on cooling rate, a spring-loaded device was constructed, that would propel individual axes extremely rapidly into liquid nitrogen that was subcooled. This combination of immersion rate and subcooled liquid nitrogen achieved a freezing rate in excess of 550 °C/s for axes of the recalcitrant species, *Camellia sinensis*, with 100% survival of those flash-dried to the water content range of 1.6–1.1 gH_2O/g dry weight (Wesley-Smith et al. 1992). This approach succeeded on two counts: firstly, the cooling rate was sufficient to obviate formation of lethal intracellular ice (as could be demonstrated electron microscopically); and secondly, the degree of dehydration necessary was not injurious. Thus the answer to the enigmatic question posed above is: fast freezing can (and should) be several orders of magnitude more rapid than that hitherto considered as "fast"!

This same approach, detailed below, has been applied, with most encouraging results (Table 1) to hydrated axes of pea, at a germination stage when (according to Koster and Leopold 1988) desiccation tolerance had been lost. The axes used in this investigation were excised from seeds 27 h after the start of imbibition when their water content averaged 2.5 g/g (var. Greenfeast). At this stage, the radicle had grown out of the confines of the testa and the axes were, on average, 15 mm long.

2.2.1 Excision and Cryopreservation

Axes excised from pea seeds that had been imbibed for 27 h were maintained on moist filter paper in a closed Petri dish until all excisions had been done. Fully hydrated axes and those that had been flash dried for 80 min were individually propelled into the cryogen. The temperature of the liquid nitrogen was reduced below − 196 °C (subcooling) by its maintenance under slight vacuum which was released immediately before the introduction of an axis. Axes were retrieved individually from a plastic basket that was submerged in the liquid nitrogen and rapidly thawed by immersion in sterile water at 25 °C, in which they were retained for 20 min. Axes were then surface sterilized by immersion in 1% NaOCl for 10 min, followed by three rinses in sterile water, after which they were either placed into embryo culture, or prepared for electron microscopy.

2.2.2 *Embryo Culture*

Axes were plated individually into culture tubes containing solid germination medium as described above (Sect. 2.1.2) and their progress monitored daily. Germination was scored as having occurred when an axis showed either greening and/or root growth. It was found that axes not showing such development after 5 days in culture were either nonviable or appeared to have entered a static phase: thus assessment on days 3 and 5 is reported. For *P. sativum* var. Greenfeast, a survival rate in the region of 70% was obtained for both fully hydrated and flash-dried axes after freezing (Table 1). The flash-dried axes of a second variety, Knight, although developing more slowly, showed 63% survival after 5 days in culture. However, survival was low (18%) when these axes were fully hydrated. The explanation for the marked difference between survival figures for hydrated axes of the two varieties of pea may reside in the differing water contents: after 27 h of imbibition, axes of var. Greenfeast had an average water content of 2.5 g/g, while those of var. Knight were considerably more hydrated at 4.1 g/g (Table 1).

2.2.3 *Ultrastructural Studies*

The distal 1.5 mm was excised from the radicle of each of five axes per treatment, and processed as follows: immersed in 2.5% glutaraldehyde in Sørensen's phosphate buffer (0.1 M) for 2 h at 25 °C; rinsed and then postfixed in 0.5% OsO_4 for 1 h; dehydrated in an acetone series and embedded in low viscosity epoxy resin. Sections (1 µm) stained with toluidine blue were viewed with a light microscope and ultrathin sections for electron microscopy were post-stained with aqueous uranyl acetate followed by lead citrate. The apical meristematic, cortical and radicle cambial regions of both varieties (Knight and Greenfeast) were all examined microscopically. As the trends were similar in both varieties, the illustrations and discussion that follow are for var. Greenfeast and are confined to cells of the cambium or its immediate neighbors, unless otherwise indicated.

Cells of 27-h-imbibed, control material were characterized by mitochondria with small cristae and some electron-translucent matrical areas (Fig. 1). Although such mitochondrial morphology is not indicative of peak respiratory

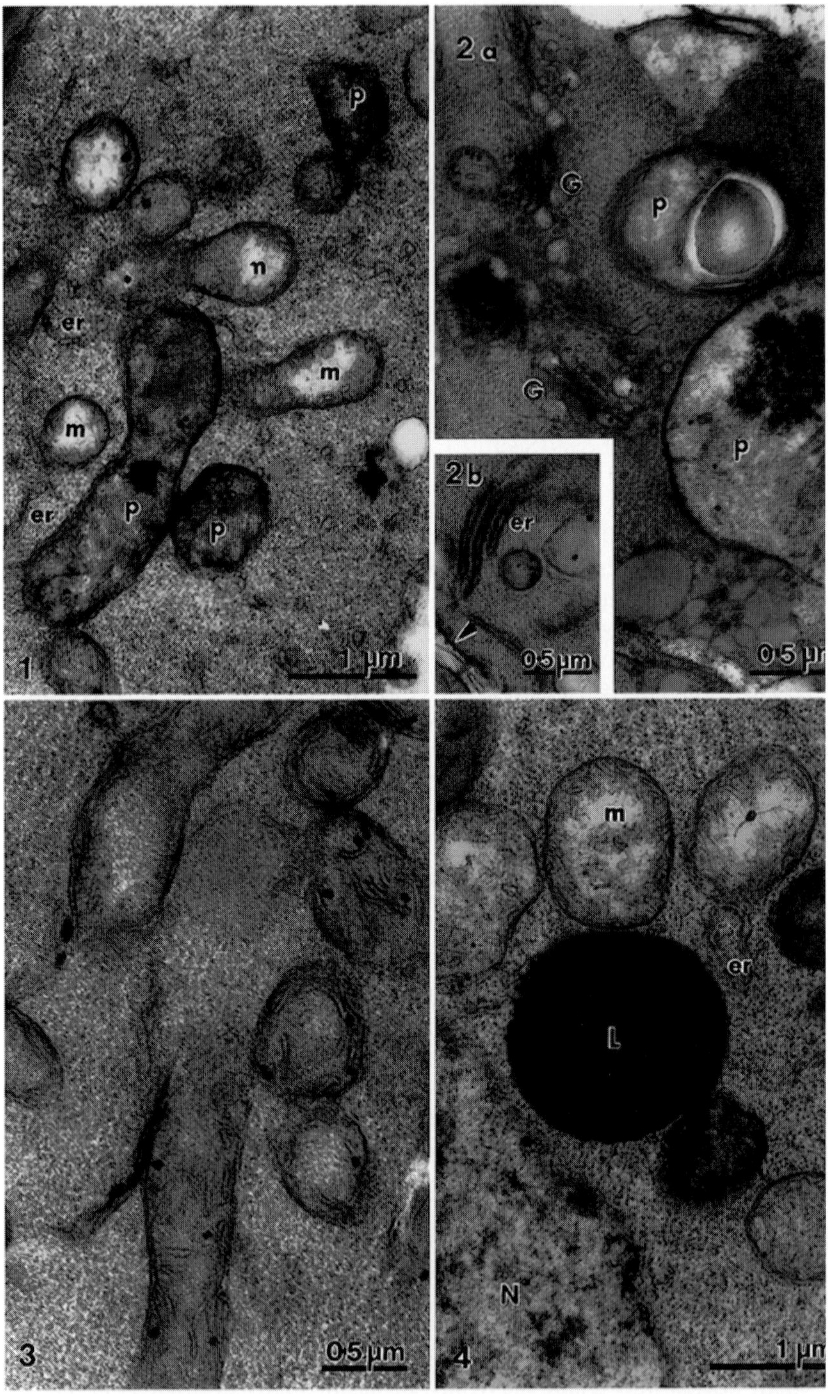
p
n
er
m
m
er
p
p
1
1 µm
2a
G
p
G
p
2b
er
0.5µm
0.5 µm
3
0.5µm
m
er
L
N
4
1 µm

activity, Vertucci (1990) has indicated that respiration can be reliably measured at far lower pea embryo (cotyledonary) water contents than the 2.5 g/g characterizing these axes. Additionally, as mitosis was occurring in the pea axes when they were sampled, there is little doubt of their active metabolic state. Plastids did not contain much starch, and generally showed a prolamellar body and many short thylakoid membranes (Fig. 1). This figure also shows the typically short rough endoplasmic reticulum (rER) profiles and polysome situation typifying the compactly vacuolate cells of the control material. Flash-drying generally compacted the cytomatrix, while retaining the morphology of organelles and endomembranes (Fig. 2, 2a). The success of this rapid dehydration in preserving functional integrity of intracellular components is demonstrated by the survival figures (Table 1). The compactly-vacuolated, cambial cells of flash-dried axes withstood freezing at a rate $>550\,^{\circ}C/s$ without any ultrastructurally visible signs of damage, judging from the appearance of organelles immediately after thawing (Fig. 3). Such cells from surviving axes sampled after 3 days in culture had an ultrastructure commensurate with the vigorous resumption of growth, and did not generally exhibit any subcellular abnormality (Fig. 4). A curious characteristic, however, was evident in most root tips irrespective of the experimental manipulation: this was visibly manifested by a sloughing of the outer tissues when growth resumed in culture. At the microscopic level it could be seen that cells immediately surrounding the regenerating cambial derivatives had undergone profound changes (Fig. 5). This developmental pattern was common also to control material of all the experiments (including those detailed in Sect. 2.1) and is presently ascribed to some stress imposed either by surface sterilization or by the culture conditions. Although this stress requires to be alleviated, it is relevant that in their work on cryopreservation of pea meristems, Haskins and Kartha (1980) found that surviving cells had the capacity to divide and regenerate new meristems. Those and the present observations imply that as long as a critical proportion of suitable cells is maintained viable, the structure (be it a meristem or entire root tip) can be regenerated.

Axes that were extremely rapidly frozen in the fully hydrated state (2.5 g H_2O/g dry mass) showed remarkable retention of ultrastructural integrity when sampled immediately after thawing. Membranes – including tonoplasts and the plasmalemma – were intact, most organelles appeared morphologically normal, and even Golgi bodies seemed unaffected by the axes having been

Figs. 1–4. All illustrations are of radicle cambial cells (unless otherwise indicated) from axes of *Pisum sativum* var. Greenfeast, excised from germinating seeds (water content 2.5 g/g) that had been imbibed for 27 h. **1.** The typical appearance of some organelles from a germinating pea axis immediately after excision (control material), is illustrated. *er* Endoplasmic reticulum; *m* mitochondrion; *p* plastid; **2.** Reduction of the water content from 2.5 g/g (**Fig. 1**) to around 0.4 g/g by flash-drying resulted in a general compaction of intracellular contents, with no visible membrane damage. **2a** *G* Golgi body; *p* plastid. **2b** *Insert*: *arrowhead* plasmalemma; *er* endoplasmic reticulum; **3.** Preservation of the fine structure of mitochondria after flash-drying, immersion in liquid nitrogen and 20 min equilibration in water at 25 °C is shown; **4.** After 3 days in culture, axes had resumed germinative development. Except for the relatively large lipid bodies (*L*), the intracellular appearance was similar in all respects to that of control material (not illustrated) at this stage. *er* Endoplasmic reticulum; *m* mitochondrion; *n* nucleus

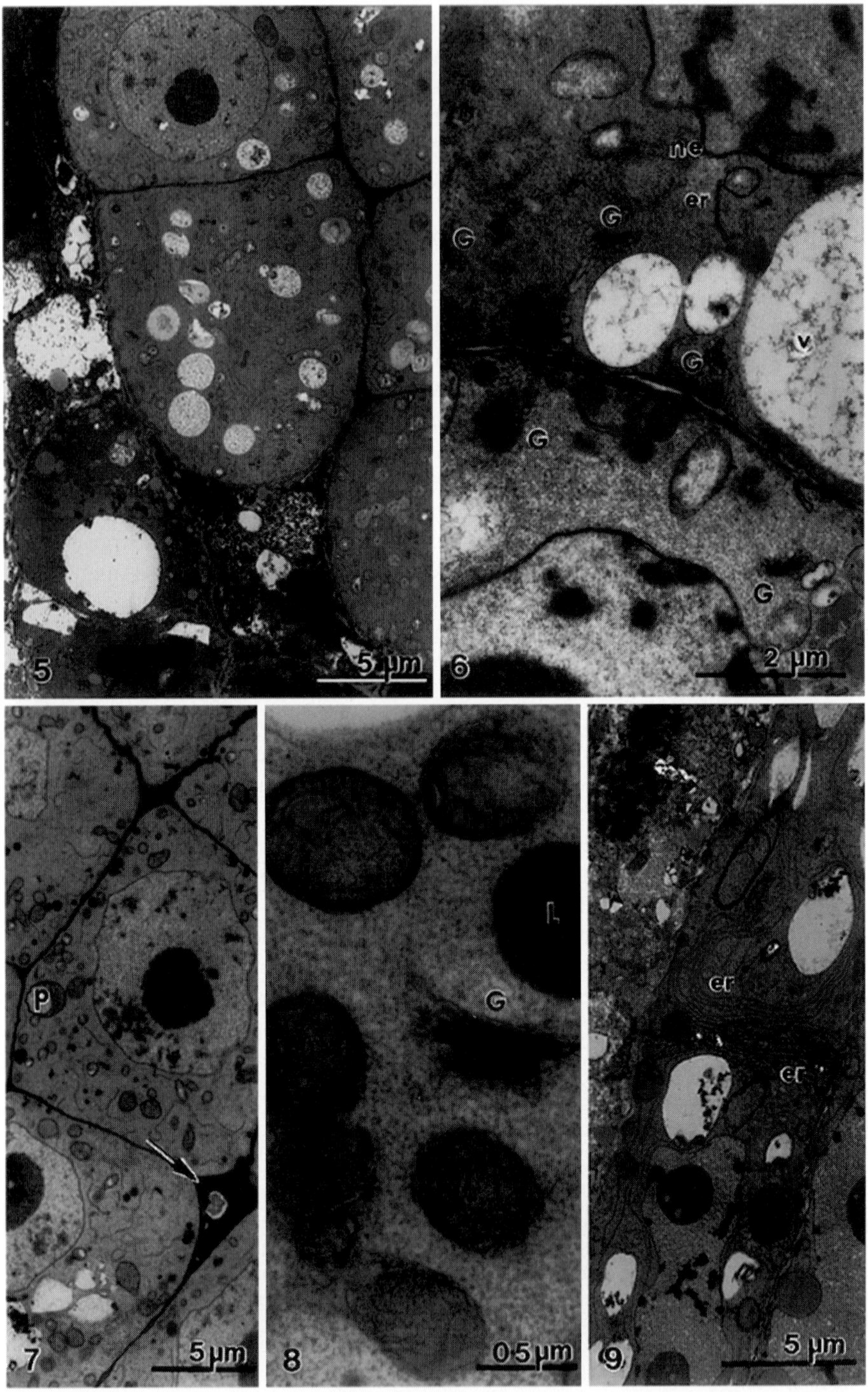
5
5 µm
6
ne
er
G
v
2 µm
7
p
5 µm
8
L
G
0.5 µm
9
er
5 µm

propelled into the subcooled liquid nitrogen (Fig. 6). Figure 7 illustrates the normal ultrastructural appearance of the regenerating tissue after axes had been in culture for 3 days. The only feature of these cells that was different from the control situation was the sporadic incidence of U-shaped plastids. However, when visualized at higher magnification, all organelles were seen to be extremely well developed and active (Fig. 8). One characteristic that exemplified cells from axes that had been frozen, and was still visible after 3 days in culture, was the incidence of rather large lipid bodies, which occurred whether or not the material had been flash-dried (Figs. 4, 8). This is indicative of some surface phenomenon (the nature of which can only be conjectural at this stage) brought about by freezing and/or thawing, that facilitated coalescence of several, originally discrete lipid droplets. Whatever its relevance, the formation of these larger deposits did not appear to have any adverse effects on the resumption of metabolism, as indicated by axis survival in culture. An aspect of the ultrastructure of the phenomenon of sloughing mentioned-above is illustrated after 3 days in culture for an axis that had been propelled into the cryogen when fully hydrated (Fig. 9). It can be seen that some cells at the interface between the regenerating tissue and that which lyses, show an extraordinary degree of rER development. It is not presently known whether these cells themselves are destined to repair or to lyse, or if the endomembrane development is somehow implicated in triggering the lysis of contiguous, centrifugally located cells or the sloughing of this entire ensheathing tissue.

3 Summary and Conclusions

The investigations presently reported show that hydrated axes may be frozen either following a treatment that lowers their water content, or at the original water content as long as this is not too high. Although the work with pea was initiated to provide a convenient model system applicable to axes from recalcitrant seeds, cryopreservation of pea germplasm is justified in its own right in view

Figs. 5–9. Ultrastructural aspects of cells from axes in the fully hydrated condition (2.5 g/g) that had been propelled into subcooled liquid nitrogen. **5.** The interface between the cambial cells and the ensheathing tissue which will be sloughed, is illustrated for an axis that was cultured for 3 days following flash drying and freezing. The high level of ultrastructural organization of the regenerating root contrasts sharply with the lysing contents of the ensheathing tissue; **6.** The fine structure of cambial cells is illustrated following retrieval from the cryogen and equilibration in water at 25 °C. Note particularly the preservation of membranous structures, including the tonoplast and plasmalemma. *er* Endoplasmic reticulum; *G* Golgi body; *ne* nuclear envelope; *v* vacuole; **7.** After 3 days in culture, the developing cells appear normal in all major respects. One of the U-shaped plastids occasionally seen is indicated (*p*) as is the initiation of lysigenous intercellular space formation (*arrow*) that comes to characterize the cortical and pith regions formed from the cambial derivatives illustrated; **8.** Details of the ultrastructure of the active mitochondria, a Golgi body (*G*) and the relatively large lipid bodies (*l*) apparent in germinating axes after 3 days in culture, are illustrated; **9.** A remarkable degree of development of the rough endoplasmic reticulum (*er*) characterizes some cells at the interface between the organized, developing root and the ensheathing, lysing tissue of the poor keeping qualities of the dry seeds in all but subzero storage. It is also desirable to be able to cryostore the germplasm with starting material other than zygotic axes, for example somatic embryos, meristems, or various tissues derived from in vitro systems.

It appears that there is a threshold water content above which pea axes cannot be frozen without lethal damage, irrespective of the freezing rate presently attainable. This is seen in the poor survival of the axes of *P. sativum* var. Knight (4.1 g/g) compared with those that had been flash-dried, and particularly with the 70% survival of axes of var. Greenfeast rapidly frozen at 2.5 g/g. A degree of dehydration, therefore, is apparently necessary if the water content of the fresh material exceeds some critical level (which remains to be accurately determined).

There are two ways in which dehydration may be achieved, having in common the desirability for rapid lowering of the water content. One method involves the use of cryoprotectants which seemingly can be avoided if flash-drying is combined with extremely rapid cooling (freezing). As discussed above, the use of cryoprotectants imposes undesirable effects as well as introducing an additional handling step, and therefore its avoidance may be considered an advance in the development of cryopreservation technology.

The success of flash-drying is considered to be a consequence of dehydration at a rate that imposes a stasis on both metabolism and deleterious changes that would occur during slow water removal (Berjak et al. 1989). However, flash-drying does not render originally desiccation-sensitive material desiccation-tolerant (pers. observ.), making immediate freezing a necessary adjunct for its conservation.

Although axes may be flash-dried to surprisingly low water contents without losing viability, there is a level below which dehydration stress is increasingly apparent. This appears to contribute to a decreased ability to withstand freezing (Wesley-Smith et al. 1992). These authors have suggested that the faster the freezing can be achieved, the higher is the water content range that facilitates successful cryopreservation – and consequently the less will be any deleterious effects of dehydration. Thus, for most hydrated tissues, a balance must be achieved between the duration of flash-drying giving the optimum water content (range) and the rate of freezing. Both parameters will be affected by the size of the specimens. In all probability, the water content at the outset, the level of subcellular differentiation, the proportion of the tissue composed of highly vacuolated cells, and the metabolic rate will also be factors in determining the success (or otherwise) of the flash-drying-rapid freezing protocol.

From the ultrastructure of the pea axes that had been frozen even at 2.5 g/g (var. Greenfeast, fully hydrated), the indications are that a success rate of 100% should be attainable. The development of the fast freezing methodology in our laboratory is still in its preliminary stages and certain technological problems are thought to underlie the fact that survival of entire samples was not attained. In particular, a problem exists with the attachment of axes to the vehicle for propulsion into the cryogen. Presently, an individual axis is affixed by means of two-sided adhesive to an aluminium pin, and the whole assembly is propelled into the liquid nitrogen. However, the problem (as sometimes happens with relatively large specimens) is that should an axis lose contact with the adhesive on propulsion into the cryogen, or when the assembly is introduced into the thawing water, then cooling or warming rates, respectively, are uncontrolled. We are of the opinion that this might be the major factor preventing a success rate of 100% and are presently investigating this further.

While the flash-drying-rapid freezing route is advocated as having the potential for the best yields of viable, cryopreserved material, it may have limitations depending on the species and/or tissues concerned. We have not had success in related investigations on jackfruit embryonic axes, and are of the opinion that use of (the most benign) cryoprotectant solutions may be unavoidable in this case. Thus, while we would suggest that cryoprotectants be avoided if similar or superior success rates are obtainable by using flash drying in conjunction with fast freezing, we are certainly not advocating that their use be summarily abandoned.

References

Abdelour-Esquivel A, Villalobos V, Engelmann F (1992) Cryopreservation of zygotic embryos of *Coffea* spp. Cryo Lett 13: 297–302

Agarwal VK, Sinclair JB (1987) Principles of seed pathology, vols 1 and 2. CRC Press, Boca Raton, FL

Bajaj S, Mickelson O, Baker LR, Markarian D (1971) The quality of protein in various lines of pea. Br J Nutr 25: 207–212

Bajaj YPS (1983a) Cryopreservation and international exchange of germplasm. In: Sen SK, Giles KL (eds) Plant cell culture in crop improvement. Plenum Press, New York, pp 19–41

Bajaj YPS (1983b) Survival of somatic hybrid protoplasts of wheat × pea and rice × pea subjected to −196 °C. Indian J Exp Biol 21: 120–122

Berjak P, Farrant JM, Pammenter NW (1989) The basis of recalcitrant seed behavior. Cell biology of the homoiohydrous seed condition. In: Taylorson RB (ed) Recent advances in development and germination of seeds. Plenum, New York, pp 89–108

Berjak P, Farrant JM, Mycock DJ, Pammenter NW (1990) Recalcitrant (homoiohydrous) seeds: the engima of their desiccation sensitivity. Seed Sci Technol 18: 297–310

Berjak P, Farrant JM, Pammenter NW, Vertucci CW, Wesley-Smith J (1992) Current understanding of desiccation-sensitive (recalcitrant) seeds: development, states of water and responses to dehydration and freezing. In: Côme D, Corbineau F (eds) Proc 4th Int Worksh on Seeds, pp 705–714

Bewley JD, Black M (1985) Seeds: physiology of development and germination. Plenum, New York

Chin HF, Aziz M, Ang BB, Samsidar H (1981) The effect of moisture and temperature on the ultrastructure and viability of seeds of *Hevea brasiliensis*. Seed Sci Technol 9: 411–422

Christensen CM, Kaufmann HH (1974) Microflora. In: Christensen CM (ed) Storage of cereal grains and their products. Am Assoc Cereal Chem, Minnesotta, USA, pp 158–191

Ellis RH, Hong TD, Roberts EH (1990) An intermediate category of seed storage behaviour? I. Coffee. J Exp Bot 41: 1167–1174

Field RW, King TH (1962) Influence of storage fungi on the deterioration of stored pea seed. Phytopathology 52: 336

Gamborg OL, Shyluk HP, Kartha KK (1975) Factors affecting the isolation and callus formation in protoplasts from shoot apices of *Pisum sativum* L. Plant Sci Lett 4: 285–292

Gantotti BV, Kartha KK (1986) Pea. In: Evans DA, Sharp WR, Ammirato PV (eds) Handbook of plant cell culture, vol 4. Techniques and applications. MacMillan, New York, pp 370–418

Grout BWW (1979) Low temperature storage of imbibed tomato seeds: a model for recalcitrant seed storage. Cryo Lett 1: 71–76

Grout BWW (1986) Embryo culture and cryopreservation for the conservation of genetic resources of species with recalcitrant seed. In: Withers LA, Alderson PG (eds) Plant tissue culture and its agricultural applications. Butterworths, London, pp 303–309
Grout BWW (1990) In vitro conservation of germplasm. In: Bhojwani SS (ed) Plant tissue culture: applications and limitations. Elsevier, Amsterdam, pp 394–411
Haferkamp ME, Smith L, Nilan RA (1953) Studies on aged seeds. 1. Relation of age of seed to germination and longevity. Agron J 45: 434–437
Haskins RH, Kartha KK (1980) Freeze preservation of pea meristems: cell survival. Can J Bot 58: 833–840
Hawkes JG (1983) The diversity of crop plants. Harvard Univ Press, Cambridge, Massachusetts
Iriondo JM, Pérez C, Pérez-García F (1992) Effect of seed storage in liquid nitrogen on germination of several crop and wild species. Seed Sci Technol 20: 165–171
Janick J, Schery RW, Woods FW, Ruttan VW (1981) Plant science. An introduction to world crops. Freeman, San Francisco
Justice OL, Bass LN (1978) Principles and practices of seed storage. Science and education federal research, agricultural handbook, 506, US Department of Agriculture, Washington
Kartha KK (ed) (1985) Cryopreservation of plant cells and organs. CRC Press, Boca Raton
Kartha KK, Gamborg OL, Constabel F (1974) Regeneration of pea (*Pisum sativum*) plants from shoot apical meristems. Z Pflanzenphysiol 72: 172–176
Kartha KK, Leung NL, Gamborg OL (1979) Freeze-preservation of pea meristems in liquid nitrogen and subsequent plant regeneration. Plant Sci Lett 5: 7–15
Koster KL, Leopold AC (1988) Sugars and desiccation tolerance in seeds. Plant Physiol 88: 829–832
Langer RHM, Hill GD (1982) Agricultural plants. Cambridge Univ Press, New York
Lyman JM (1984) Progress and planning for germplasm conservation of major food crops. Plant Genet Resour Newsl 60: 3–21
McAdams S, Ratnasabapathi D, Smith RA (1991) Influence of days of culture on cryoprotectant supplemented medium and of terminal freezing temperature on the survival of cryopreserved pea shoot tips. Cryobiology 28: 288–293
Meryman HT, Williams RJ (1985) Basic principles of freezing injury to plant cells: natural tolerance and approaches to cryopreservation. In: Kartha KK (ed) Cryopreservation of plant cells and organs. CRC Press, Boca Raton
Mlot C (1989) Blueprint for conserving plant diversity. Bioscience 39: 364–368
Murashige T, Skoog F (1962) A revised medium for rapid growth and bioassays with tobacco tissue cultures. Physiol Plant 15: 473–497
Mycock DJ, Berjak P (1992) Cryostorage of somatic embryos of coffee. In: Côme D, Corbineau F (eds) Proc 4th Int Worksh on Seeds, pp 879–884
Mycock DJ, Watt MP, Berjak P (1991) A simple procedure for the cryopreservation of hydrated embryonic axes of *Pisum sativum*. Plant Physiol 138: 728–733
Nitzsche W (1983) Germplasm preservation. In: Evans DA, Sharp WR, Ammirato PV, Yamada Y (eds) Handbook of plant cell culture, vol 1. Techniques for propagation and breeding. Macmillan, New York
Plucknett DL, Smith NJH, Williams JT, Anishetty NM (1987) Gene banks and the world's food. Princeton Univ Press, New Jersey
Purseglove JW (1968) Tropical crops: dicotyledons 1. Longmans, London
Roberts EH (1972) Viability of seeds. Chapman and Hall, London
Stanwood PC (1986) Dehydration problems associated with the preservation of seed and plant germplasm. In: Leopold AC (ed) Membranes, metabolism and dry organisms. Cornell Univ Press, Ithaca, New York
Sun CN (1958) The survival of excised pea seedlings after drying and freezing in liquid nitrogen. Bot Gaz 119: 234–236
Vertucci CW (1990) Calorimetric studies of the state of water in seed tissues. Biophys J 58: 1463–1471.
Vertucci CW, Roos EE (1990) Theoretical basis of protocols for seed storage. Plant Physiol 94: 1019–1023

Vertucci CW, Berjak P, Pammenter NW, Crane J (1991) Cryopreservation of embryonic axes of an homeohydrous (recalcitrant) seed in relation to calorimetric properties of tissue water. Cryo Lett 12: 339–350

Wesley-Smith J, Vertucci CW, Berjak P, Pammenter NW, Crane J (1992) Cryopreservation of desiccation-sensitive axes of *Camellia sinensis* in relation to dehydration, freezing rate and the thermal properties of tissue water. J Plant Physiol 140: 596–604

Withers LA (1989) In vitro conservation and germplasm utilisation. In: Brown AHD, Frankel OH, Marshall DR, Williams JT (eds) The uses of plant genetic resources. Cambridge Univ Press, Cambridge, pp 309–334

IV.4 Cryopreservation of *Brassica* Species

E.E. Benson[1]

1 General Account

1.1 Distribution and Importance of *Brassica* Species

Geographical locations of major, wild relative diversity in members of Brassicaceae such as *Brassica nigra*, *Sinapsis alba*, *Raphanus sativus*, and *Eruca sativa* are centered in the Mediterranean, particularly Spain. Other regions of diversity include France and southern Britain, a few *Brassica* species exhibit xerophytic adaptations to the Saharan region (Tsunoda 1980). Relationships between agronomically important *Brassica* species are complex as they are based on interspecific hybridizations and the phylogenetic relationships between *B. nigra*, *B. carinata*, *B. juncea*, *B. oleracea*, *B. napus*, and *B. campestris.* Within this system, a scheme termed the U-triangle has been proposed (see Sjodin 1992), thus, *B. napus* is the amphidiploid derived from the hybridization of *B. campestris* and *B. oleracea.* To add further complexity, different wild forms of *B. napus*, including crosses with cultivated *B. oleracea* and *B. campestris*, have been found at various geographical locations. In addition, Tsunoda (1980) reports a wide range of wild *Brassica* diversity within different geographical centers, for example: *B. fruticulosa* (Mediterranean coast), *B. oxyrrhina* (S Spain and Portugal and Morocco), *B. repanda* (Algeria), and *B. elongata* (SE Europe, W Asia). Reports of endangered, wild, *Brassica* species include *Coronopus navasii*, from Spain (Iriondo and Perez 1990a,b), *Lepidium hyssopifolium*, an endangered watercress located in Victoria, Australia (Cropper 1987), and *Braya humilis*, a native of the alpine tundra of central Colorado, (Neely and Carpenter 1986).

Brassica species are of considerable agronomic importance as vegetable, oil and fodder crops (Kalloo and Bergh 1993). Economically important species of *Brassica* can be grouped as: (1) vegetable and fodder crops: *B. oleracea* (broccoli, cabbage, cauliflower, Brussel sprouts, calabrese), *Raphanus sativus* (radish), *B. campestris* (turnip), *B. pekinensis* and *B. chinensis* (Chinese cabbage); or (2) oil crops: *B. campestris* (sarson, turnip rape), *B. juncea* (leaf mustard), *B. napus* (oilseed rape). The latter now ranks as a major world crop and is one of the main sources of northern European vegetable oil. In cooler geographical zones it is not possible to grow the traditional oil-producing species which are indigenous to the

[1]Department of Molecular and Life Sciences, Dundee Institute of Technology, Bell Street, Dundee, Scotland, DD1 1HG UK

Biotechnology in Agriculture and Forestry, Vol. 32
Cryopreservation of Plant Germplasm I (ed. by Y.P.S. Bajaj)

South Mediterranean and the Tropics. Major emphasis is presently placed on *B. napus* breeding programs which select for reduced erucic acid formation and improved nutritional oil quality.

1.2 *Brassica* Biotechnology

Biotechnological techniques have proved important as they offer considerable potential for *Brassica* crop improvement (see Sjodin 1992). As wide sexual hybridization can result in incompatibility, embryo rescue, ovule and ovary culture have been used to obtain wide *Brassica* crosses, including those involving wild relatives (Mohapatra and Bajaj 1987; Inomata 1990). Microspore and anther culture for haploid production have been exploited with major emphasis on regeneration via embryogenesis. Protoplast fusion and the production of somatic hybrids permit intergenic mixing which may be otherwise restricted by incompatibility barriers using conventional breeding. Transformation in conjunction with tissue culture manipulations offers the possibility of introducing beneficial genes, and a range of transformation techniques have been explored (Sjodin 1992). Introduction of cytoplasmic male sterility (CMS) and other means of self-incompatibility into *Brassica* species has received considerable attention in traditional breeding (Hinata and Nishio 1980; Shiga 1980). However, the manipulation of CMS using protoplast technology, the correction of undesirable incompatibility traits (chlorosis) and the regeneration of stable somatic hybrids provide useful adjuncts to conventional approaches (Pelletier 1990).

1.3 Importance of *Brassica* Genetic Resource Management

International seed collections of agronomically important *Brassica* species have been established (Innes 1985). Botanic gardens also play a major role in the conservation of *Brassica* wild relatives and endangered species (see Fay 1991). A considerable contribution has been made by the Universidad Politecnica, Madrid, in the collection, taxonomic characterization and storage of wild *Brassica* species of the Mediterranean and northern Spain (Tsunoda et al. 1980b). The conservation of endangered *Brassica* species and their habitats has received global attention (Gomez-Campo 1985; Neely and Carpenter 1986; Cropper 1987). Micropropagation techniques are also increasingly important for the multiplication of reduced populations of endangered *Brassica* species, as exemplified by the studies of Iriondo and Perez (1990a,b) which concern the conservation of *Coronopus navasii.* This *Brassica* is protected by the 1982, Berne Agreement for the Conservation of Wildlife and Natural Habitats in Europe. The entire population of the plant is located in two extremely vulnerable habitats located in the SE region of Spain. The previous workers concluded that micropropagation of endangered species is an important tool in both in vitro conservation and in facilitating the bulking up of plantlets which may, at a later stage, be reintroduced into their natural habitat.

The need to conserve a wide genetic base for *Brassica* is demonstrated by the use of germplasm from wild relatives, land races and agronomic collections in modern cultivar improvement. Thus, large numbers of accessions from collections have been screened and exploited in breeding programs to improve oil composition (Auld et al. 1989), produce resistance to club root (Crisp et al. 1989; Monteiro and Williams 1989), downy mildew, *Fusarium*, black leg, white rust and black rot (Monteiro and Williams 1989). Effective use of genetic resources, particularly those using biotechnological techniques, is dependent upon germplasm characterization. Members of the Brassicaceae present considerable classification problems and molecular approaches to solving taxonomic problems are important (Song et al. 1988; Kresovich et al. 1992).

1.4 Germplasm Storage Methods for *Brassica* Species

Brassica germplasm storage utilizes seeds, zygotic and somatic embryos, pollen, anthers and anther-derived embryos, shoot tips, callus, and cells (Table 1).

For standard conservation strategies, seed storage is still the usual method of choice for *Brassica* genetic resources. In general, seeds may be stored using orthodox methods of controlled dehydration to a minimum moisture content, followed by maintenance at subzero temperatures within the range of −5 to −20 °C (Takayanagi 1980; Tsunoda et al. 1980a; Innes 1985). Viability is greatly affected by the dry weight of the seeds as they reach maturity, thus, the storage longevity of *B. pekinensis* could be significantly increased if the seeds were harvested at the optimum stage of maturation (see Takayanagi 1980). The latter worker refers to long-term records of *Brassica* seed longevity in which a classical experiment demonstrated 8% germination of *B. nigra* seeds after they had been buried in soil for 50 years. By comparison, dehydrated seeds of *B. napus* and *B. campestris* maintained almost 100% viability at 2–4 °C when desiccated to ca. 3% moisture content and stored for up to 15 years (Takayangi 1980). *Brassica* germplasm is not recalcitrant to conventional storage methods and, at present, cryopreservation does not have a major role in *Brassica* seed conservation. However, there is an increasing interest in the use of ultralow temperature storage, even for long-term, orthodox seed conservation. At intermediate, subzero temperatures (−20 °C) there still exists the possibility of time-related deterioration, which may be reduced at ultralow storage temperatures (−196 °C) (Stanwood 1985). Seeds from a range of cultivars of *B. alba*, *B. campestris*, *B. carinata*, *B. napus*, and *B. oleracea* have survived exposure to liquid nitrogen (for periods of up to 3 years) and exhibited germination rates comparable to those of noncryopreserved control seeds. Similar results have been obtained with seeds derived from *Raphanus sativus* (Stanwood 1985). Cryogenic storage of *Brassica* seed generally requires a prefreeze desiccation treatment to achieve moisture contents in the range of 4.8 to 7.2%, followed by rapid freezing in liquid nitrogen, and thawing at 30–40 °C (see Stanwood 1985). Cryopreservation techniques have also been applied to immature zygotic embryos of *B. napus*, however, these proved difficult to preserve using conventional protocols involving chemical cryoprotection with dimethyl sulfoxide (Withers 1982).

Table 1. The application of cryopreservation to *Brassica* spp.

Species	Explant	Cryogenics[a]	Type of recovery	Survival (%)	Reference
B. napus	Shoot tips, zygotic embryos	C, CF, URF	Shoots/callus	Variable	Withers (1982)
B. campestris	Anthers/	C, CF, URF	Embryos/plants/	31	Bajaj (1983)
B. napus	pollen embryos		callus	44	
B. napus	Cell suspension	C, CF (−35 °C)	Cells	>20	Weber et al. (1983)
B. napus	Shoot tips	C, CF, URF (−40 °C)	Shoots/callus	50	Withers et al. (1988)
B. napus	Shoot tips	C, CF (0 °C)	Shoots/callus	24	Benson and Noronha-Dutra (1988)
B. napus	Microspores	C, CF (−15 °C)	Embryos/plants	5	Charne et al. (1988)
B. campestris	Cell suspension	V, URF	Cells	>40	Langis et al. (1989)
B. campestris	Cell suspension	C, CF (−50 °C)	Cells/protoplasts	75	Mei-zhong et al. (1990)
B. napus	Microspores	C, CF (−35 °C)	Embryos/plants	18–40	Chen and Beversdorf (1992a,b)
B. napus	Microspore embryos	E, D, URF	Plants	90	Uragami et al. (1993)

[a]C, Chemical cryoprotection; E, encapsulation; D, dehydration; V, vitrification; CF, controlled cooling at a specific rate to an intermediate transfer temperature (in parentheses) before immersion in liquid nitrogen; URF, ultrarapid freezing (direct plunge into liquid nitrogen).

Haploid production is of great importance to *Brassica* breeding and, as a result, the cryogenic storage of pollen and anthers has received considerable attention. Low temperature storage of pollen at −20 °C for 50 to 78 weeks has been reported for several *Brassica* species (*B. oleracea*, *B. napus*, *B. campestris*). Post-storage studies of the pollen demonstrated normal seed set in the majority of cases (see Towill 1985). Anther segments of *B. napus* and *B. campestris* have survived cryogenic storage at −196 °C, however, the method of freezing was a determinant of survival as dry-freezing procedures were more effective than those in which the anthers were frozen in a cryoprotectant solution (Bajaj 1983). Differential responses of pollen embryos to cryopreservation were also observed between genotypes of *B. campestris* and *B. napus* which showed survival frequencies of 31 and 44%, respectively (see Bajaj 1985). Whilst early studies of *Brassica* haploid production involved regeneration from anther cultures, embryo production in these systems was relatively low, and more efficient techniques have been developed using microspores (see Sjodin 1992). As a result, recent attention has focused on the conservation of pollen microspores in liquid nitrogen (see Table 1 and Sect. 2).

Regardless of the conventional role of seed storage in *Brassica* genetic resource management, the present and major emphasis of cryopreservation protocol development for these species concerns the possibility of storing tissues used in and derived from biotechnology programs; e.g., in the conservation of in vitro tissue containing novel gene combinations, achieved via transformation, somatic hybridization, microspores for haploid production, and in vitro CMS manipulations. An important aspect of these programs will be to use cryopreservation as a means of controlling the undesirable genetic changes that often accompany in vitro manipulations. This may be especially pertinent to genetically "labile" products such as haploids. With the increasing realization that tissue culture techniques have an important role in endangered species conservation (Iriondo and Perez 1990a,b), cryogenic storage may also be of potential importance. However, whether for biotechnological or endangered species applications, the future success of cryopreservation in conserving, in vitro, *Brassica* tissues will be greatly dependent upon optimizing both cryogenic and tissue culture factors.

2 Cryopreservation (Table 1)

2.1 Methodology

2.1.1 Prefreeze Factors

Post-thaw recovery of *B. napus* shoot tips after controlled freezing is enhanced by the application of 5% v/v dimethyl sulfoxide (DMSO) for 24–48 h before cryoprotection in DMSO at higher concentrations prior to freezing (Benson and Noronha-Dutra 1988; Withers et al. 1988). Similarly, a 4–6 week period of pregrowth was required before freezing tissue from excised anthers of *B. napus* and *B. campestris* (Bajaj 1983). Microspore cryopreservation requires the standardization of donor plant flowering, flower bud size, and microspore collection (Charne et al. 1988; Chen and Beversdorf 1992a). Storage density significantly affects post-thaw recovery in cryopreserved microspores. Chen and Beversdorf (1992a) were unable to regenerate embryos from *B. napus* microspores which had been cryopreserved at normal culture density. High cell populations were required to ensure post-thaw embryogenesis.

2.1.2 Chemical Cryoprotection and Controlled Freezing

These protocols are dependent upon the application of chemical cryoprotectants for up to 1 h, either alone (usually DMSO) or in combination (with sucrose and glycerol). Withers et al. (1988) demonstrated post-thaw survival of *B. napus* shoot tips using a cryoprotectant regime of 15% v/v DMSO for 1h at room temperature. Five percent mixtures of sucrose, DMSO and glycerol supported recovery of cryopreserved anther tissues (Bajaj 1983). At high concentrations (13% w/v) sucrose is also an effective cryoprotectant for microspore storage (Chen and Beversdorf 1992a). However, embryogenesis was inhibited when this cryoprotectant was combined with 0.5 M glycerol and 0.5 M DMSO (Charne et al. 1988), thus, supporting the need to perform adequate cryoprotectant toxicity trials. A range of controlled cooling/freezing regimes have been applied to *Brassica* tissues. The simplest approach is to gradually lower the tissues (contained in a suitable cryovial) into liquid nitrogen (Bajaj 1983). However, reproducibility is better ensured using programmable freezers or cooled alcohol baths, as a controlled freezing rate can be critical to survival and post-freeze regeneration. Thus, a rate of $-0.25\,^{\circ}$C/min to a terminal transfer temperature of $-35\,^{\circ}$C proved to be the most effective for maintaining the embryogenic capabilities of pollen microspores from two *B. napus* genotypes (Chen and Beversdorf 1992a). Post-freeze survival and regeneration of cryopreserved *B. napus* shoot tips were best achieved using either ultrarapid freezing or a prechill treatment (-1 to $0\,^{\circ}$C/min) before directly plunging them into liquid nitrogen (Benson and Noronha 1988; Withers et al. 1988). Neither of these regimes benefit from the affects of intracellular water withdrawal during external ice nucleation or colligative cryoprotection.

2.1.3 Vitrification, Encapsulation, and Dehydration

Vitrification involves the application of concentrated cryoprotectant mixtures to tissues. On exposure to liquid nitrogen a glass is formed as the high viscosity solution prohibits ice nucleation. Langis et al. (1989) achieved 40% survival rates in cell suspensions of *B. campestris* treated with vitrification cocktails based on high molarity ethylene glycol solutions. A major advantage of vitrification protocols is that the tissue may be directly plunged into liquid nitrogen. Attention must be given to the "unloading" of the cryoprotectants as the risks of osmotic injury are high; glass transitions on rewarming must also be avoided. This is usually achieved by gradually replacing, by dilution or washing, the potentially toxic osmotica of the vitrification solution with a high molarity solution of sucrose. Encapsulation is a novel technique involving the trapping of tissues in a calcium-alginate matrix (Fabre and Dereuddre 1990). Tissue-containing "beads" are transferred to high molarity sucrose solutions (0.75 M) for 24–72 h, followed by air desiccation in a sterile flow cabinet for 1–4 h. After these pretreatments the beads are directly plunged into liquid nitrogen. This technique has been successfully applied to microspore embryos of *B. campestris* (Uragami et al. 1993). An alternative approach for the same plant system involved the extreme dehydration of nonencapsulated embryos, first in high molarity sucrose (1.5 to 2.0 M) followed by desiccation over silica gel.

2.1.4 Post-Freeze Regeneration

A major requirement of cryopreservation protocols is the ability to retain morphogenetic competence. Regeneration of plants from cryopreserved microspore embryos has been successfully achieved (Table 1). By comparison, regeneration capabilities of cryopreserved *B. napus* shoot tips can be reduced and they are greatly dependent upon post-thaw tissue culture manipulations. For these tissues, genetic stability may be compromised if regeneration proceeds via an adventitious or callogenic route. Studies by Withers et al. (1988) demonstrated a significant interaction between mode of freezing (ultrarapid or controlled) and the hormone composition of the recovery medium. Thus, 6-benzylaminopurine (BAP) promoted organized recovery via normal shoot regeneration, but only after ultrarapid freezing. Auxins were required to support recovery after controlled freezing, for this regime, recovery proceeded via callus induction with greatly reduced shoot formation (Fig. 1). Subsequent studies have demonstrated a possible relationship between shoot regeneration, callogenic response and ethylene formation in shoot tips of *B. napus* recovering from cryogenic storage. Whilst high rates of ethylene production were correlated with good levels of survival, shoot proliferation was only associated with lower survival levels and diminished ethylene production (Fig. 2). Considerable differences in morphogenetic response and recovery are frequently observed in cryopreserved shoot tips (Fig. 1), which may be due in part to the effects of ethylene production. In the case of encapsulation studies, using *B. napus*, early removal of embryos 24 h after thawing was essential to ensure plantlet regeneration (Uragami et al. 1993).

2.2 Post-Freeze Stability

The most significant application of cryopreservation to *Brassica* species concerns the conservation of haploid systems for which the maintenance of post-freeze karyotype fidelity must be considered. Chen and Beversdorf (1992b) investigated spontaneous diploid production in cryopreserved microspores following a range of cryopreservation treatments. A high percentage (up to 50%) of diploid reversal

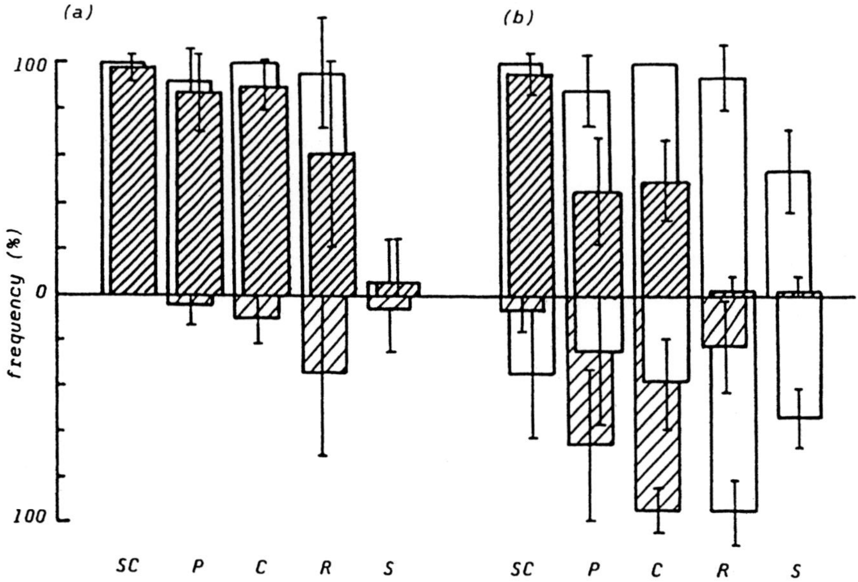

Fig. 1a,b. Survival profiles of *B. napus* shoot tips after cryoprotection (15% v/v DMSO) for 1 h and cryopreservation by ultrarapid (direct plunge into liquid nitrogen) or controlled freezing comprising a cooling profile (before transfer to liquid nitrogen) of $-1\,^\circ C.\ min^{-1}$ to $0\,^\circ C$ followed by a second cooling rate of $-0.5\,^\circ C.\ min^{-1}$ to $-40\,^\circ C$. Shoot tips were recovered on medium containing **a** 0.5 mg/l BAP or **b** 10 mg/l dichlorophenoxyacetic acid (2,4-D) and 10% v/v coconut milk. *SC* Undissected shoots; *P* dissected and pregrown shoots (5% DMSO for 24 h); *C* cryoprotected shoot tips; *R* cryopreserved shoot tips (rapidly frozen); *S* cryopreserved shoot tips (controlled freezing). *Above* the "X" axis: *open blocks* denote total survival; *cross-hatched blocks* denote shoot regeneration. *Below* the "X" axis: *cross-hatched blocks* denote callused leaves; *open blocks* denote fully disorganized callus. Errors are standard deviations, each treatment involved 50–80 shoot tips. (Withers et al. 1988)

was associated with the controlled cooling treatments and diploidy increased with a decrease in the terminal transfer temperature. Excursions to and from liquid nitrogen temperatures did not appear to influence doubling rates and it was proposed that the pre-chilling/freezing component of the two-step cryopreservation protocol exerted a selection pressure on haploid cells, diploid cells having a better chance of survival. Charne et al. (1988) also report post-freeze ploidy changes in *B. napus* after controlled cooling regimes. Thus, embryos regenerating from frozen microspores produced 88% diploids and 12% haploids, by comparison, unfrozen controls yielded 7% diploids and 93% haploids. Homozygosity is usually achieved in *Brassica* spp. by colchicine-induced doubling. It can take several months to retrieve suitable lines using this method; the previous workers suggested cryopreservation as a useful method for the rapid production of diploids. However, it must be cautioned that in the conservation of genetic resources, cryopreservation must not predispose tissue to genetic change or selection pressure. Ideally, those components which lead to ploidy changes, regardless of the mechanism, must be characterized. There is some evidence to

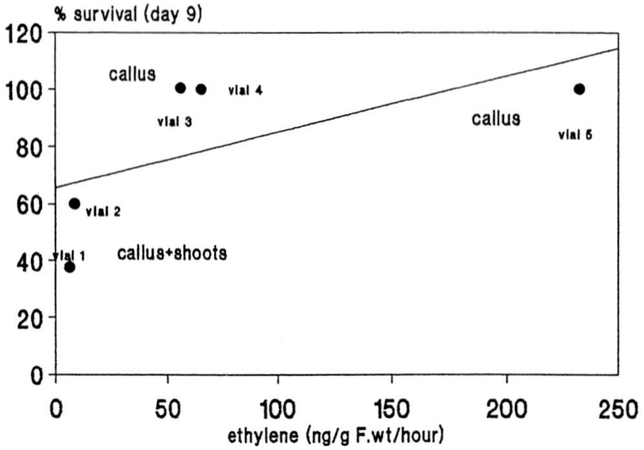

Fig. 2. Relationships between post-thaw survival, regeneration and ethylene evolution in cryopreserved shoot tips of *B. napus* recovered on medium containing 10 mg/l 2,4-D and 10% v/v coconut milk. Cryopreservation involved pregrowth in 5% v/v DMSO for 48 h, cryoprotection 15% v/v DMSO for 1 h and controlled chilling at $-1\,°C\,min^{-1}$ to 0 °C followed by direct immersion into liquid nitrogen. Data comprise ethylene production (9 days after thawing) by 4–5 shoot tips recovered in 5 replicate air-tight tissue culture vials. The regression line is statistically significant ($p = 0.1$, correlation coefficient 0.9623 for values >66 ng ethylene/g F. wt/h). (Benson, unpubl.)

suggest that ploidy change may be related to the cryopreservation protocol, as Uragami et al. (1993) did not observe any changes in the ploidy of plants regenerated from microspores of *B. napus* which had been cryopreserved using encapsulation/dehydration. Similarly, chromosomal numbers were stable in plants regenerated from shoot tips of unstable potato dihaploids which had been cryopreserved using ultrarapid freezing (Ward et al. 1993). To date, there is no evidence of genetic change per se occurring in *Brassica* species regenerated from cryopreserved tissues. Thus, regenerates of *B. napus* did not demonstrate any enhanced cold tolerance or variation in isoenzyme pattern (Chen and Beversdorf 1992a). However, stability studies within this area are limited and must be extended.

3 Summary and Conclusions

Using several different approaches, it is possible to cryopreserve and regenerate plants from a range of diverse tissues and cells derived from *Brassica* spp. Considerable emphasis has been placed on the cryogenic storage of pollen microspores, which are capable of retaining their embryogenic potential and thereby provide a strategic supply of haploid and diploid plants for biotechnology programs. The possibility that some components of the cryopreservation procedure can be manipulated to increase post-freeze recovery of diploid plants is an added advantage of using cryopreservation in *Brassica* species. However, this must be viewed with some caution. In the broader context, the application of a selection pressure to plant genetic resource conservation is not advantageous and other techniques (ultrarapid freezing, encapsulation), which do not appear to dispose cells to ploidy selection, are better suited to long-term germplasm conservation programs.

Acknowledgments. The author gratefully acknowledges Dr. L.A. Withers and M. Martin for their contributions to the production of the previously published and unpublished data presented in this chapter. Permission for the reproduction of Fig. 1 was kindly given by the editor of Cryo Letters.

References

Auld DL, Mahler KA, Letourneau DJ (1989) Evaluation of four *Brassica* germplasm collections for fatty acid composition, J Am Oil Chem Soc 66: 1475–1479

Bajaj YPS (1983) Regeneration of plants from pollen-embryos of *Arachis*, *Brassica* and *Triticum* spp. cryopreserved for one year. Curr Sci 52: 484–486

Bajaj YPS (1985) Cryopreservation of embryos. In: Kartha KK (ed) Cryopreservation of plant cells and organs. CRC Press, Boca Raton, pp 228–242

Benson EE, Noronha-Dutra AA (1988) Chemiluminescence in cryopreserved plant tissue cultures: the possible role of singlet oxygen in cryoinjury. Cryo Lett 9: 120–131

Charne DG, Pukacki P, Kott LS, Beversdorf WA (1988) Embryogenesis following cryopreservation in isolated microspores of rapeseed (*Brassica napus* L.). Plant Cell Rep 7: 407–409

Chen JL, Beversdorf WD (1992a) Cryopreservation of isolated microspores of spring rapeseed (*Brassica napus* L.) for in vitro embryo production. Plant Cell Tissue Organ Cult 31: 141–149

Chen JL, Beversdorf WD (1992b) Production of spontaneous diploid lines from isolated microspores following cryopreservation in spring rapeseed (*Brassica napus* L.). Plant Breed. 108: 324–327

Crisp P, Crute IR, Sutherland RA, Angell SM, Bloor K, Burgess H, Gordon PL (1989) The exploitation of genetic resources of *Brassica oleracea* in breeding for resistance to clubroot (*Plasmodiophora brassicae*). Euphytica 42: 215–226

Cropper S (1987) Ecological notes and suggestions for conservation of a recently discovered site of *Lepidium-hyssopifolium desv.* Brassicaceae at Bolwarrah, Victoria, Australia. Biol Conserv 41: 269–278

Fabre J, Dereuddre J (1990) Encapsulation-dehydration – a new approach to cryopreservation of *Solanum* shoot-tips. Cryo Lett 11: 413–426

Fay MF (1991) Labs working on tissue culture of rare species – a list. Bot Gard Microprop News 1: 49–50

Gomez-Campo C (1980) Morphology and morpho-taxonomy of the tribe Brassicaceae. In: Tsunoda S, Hinata K, Gomez-Campo C (eds) *Brassica* crops and wild allies. Biology and breeding. Japan Scientific Societies Press, Tokyo, pp 3–31

Gomez-Campo C (ed) (1985) Plant conservation in the Mediterranean area. Junk, Dordrecht

Hinata K, Nishio T (1980) Self-incompatibility in crucifers. In: Tsunoda S, Hinata K, Gomez-Campo C (eds) *Brassica* crops and wild allies. Biology and breeding. Japan Scientific Societies Press, Tokyo, pp 223–234

Innes NL (1985) The work of the NVRS in conservation and breeding of vegetables. Garden 110: 57–65

Inomata N (1990) Interspecific hybridization in *Brassica* through ovary culture. In: Bajaj YPS (ed) Biotechnology in agriculture and forestry, vol 10. Legumes and oilseed crops I. Springer, Berlin Heidelberg New York, pp 367–383

Iriondo JM, Perez C (1990a) Micropropagation of endangered plant species: *Coronopus navasii* (Brassicaceae). Plant Cell Rep 8: 745–748

Iriondo JM, Perez C (1990b) Application of in vitro culture techniques to the conservation of Iberian endemic plant species. Bot Gard Microprop News 1: 4–6

Kalloo G, Bergh BO (1993) Genetic improvement of vegetable crops. Pergamon Press, Oxford

Kresovich S, Williams JGK, McFerson JR, Routman EJ, Schall BA (1992) Characterization of genetic identities and relationships of *Brassica oleracea* L. via randomly amplified polymorphic DNA assay. Int Z Theor Angew Genet 85: 190–196

Langis R, Schnabel B, Earle ED, Steponkus PL (1989) Cryopreservation of *Brassica campestris* L. cell suspensions by vitrification. Cryo Lett 10: 421–428

Mei-zhong L, Shuang-Yin J, Hui-xian T (1990) Cryopreservation of suspension cell line derived from protoplast culture of *Brassica campestris* var. Pekinensis. Acta Bot Sin 32: 432–437

Mohapatra D, Bajaj YPS (1987) Interspecific hybridization in *Brassica juncea* x *Brassica hirta* using embryo rescue. Euphytica 36: 321–326

Monteiro AA, Williams PH (1989) The exploration of genetic resources of Portuguese cabbage and kale for resistance to several *Brassica* diseases. Euphytica 41: 215–226

Neely EE, Carpenter AT (1986) Size, structure and habitat characteristics of populations of *Braya humilis* var. *humilis* Brassicaceae an Alpine disjunct from Colorado USA. Great Basin Nat 46: 728–735

Pelletier G (1990) Cybrids in oilseed *Brassica* crops through protoplast fusion. In: Bajaj YPS (ed) Biotechnology in agriculture and forestry, vol 10. Legumes and oilseed crops I. Springer, Berlin Heidelberg New York, pp 419–433

Shiga T (1980) Male sterility and cytoplasmic differentiation. In: Tsunoda S, Hinata K, Gomez-Campo C (eds) *Brassica* crops and wild allies. Biology and breeding. Japan Scientific Societies Press, Tokyo, pp 205–221

Sjodin C (1992) Brassicaceae, a plant family well suited for modern biotechnology. Acta Agric Scand Sect B, Soil Plant Sci 42: 197–207

Song KM, Osborn TC, Williams PH (1988) *Brassica* taxonomy based on nuclear restriction fragment length polymorphisms (RFLPs) 2. Preliminary analysis of subspecies with *B. rapa* (syn. *campestris*) and *B. oleracea.* Theor Appl Genet 76: 593–600

Stanwood PC (1985) Cryopreservation of seed germplasm for genetic conservation. In: Kartha KK (ed) Cryopreservation of plant cells and organs. CRC Press, Boca Raton, pp 200–226

Takayanagi K (1980) Seed storage and viability tests. In: Tsunoda S, Hinata K, Gomez-Campo C (eds) *Brassica* crops and wild allies. Biology and breeding. Japan Scientific Societies Press, Tokyo, pp 303–321

Towill LE (1985) Low temperature and freeze-/vacuum-drying preservation of pollen. In: Kartha KK (ed) Cryopreservation of plant cells and organs. CRC Press, Boca Raton, pp 172–198

Tsunoda S (1980) Ecophysiology of wild and cultivated forms in *Brassica* and allied genera. In: Tsunoda S, Hinata K, Gomez-Campo C (eds) *Brassica* crops and wild allies. Biology and breeding. Japan Scientific Societies Press, Tokyo, pp 109–120

Tsunoda S, Hinata K, Gomez-Campo C (1980a) Preservation of genetic resources. In: Tsunoda S, Hinata K, Gomez-Campo C (eds) *Brassica* crops and wild allies. Biology and breeding. Japan Scientific Societies Press, Tokyo, pp 339–341

Tsunoda S, Hinata K, Gomez-Campo C (eds) (1980b) *Brassica* crops and wild allies. Biology and breeding. Japan Scientific Societies Press, Tokyo

Uragami A, Lucas MO, Ralambosoa J, Renard M, Dereuddre J (1993) Cryopreservation of microspore embryos of rapeseed *Brassica napus* L. by dehydration in air with or without alginate encapsulation. Cryo Lett 2: 83–90

Ward ACW, Benson EE, Blackhall NW, Cooper-Bland S, Powell JB, Davey MR (1993) Flow cytometric assessments of ploidy stability in cryopreserved dihaploid *Solanum tuberosum* and wild *Solanum* species. Cryo Lett 14: 145–152

Weber G, Roth EJ, Schweiger H (1983) Storage of cell suspensions and protoplasts of *Glycine max* (Merr.), *Brassica napus* L., *Datura innoxia* (Mill.) and *Daucus carota* (L.) by freezing. Z Pflanzenphysiol 109(1): 29–39

Withers LA (1982) The development of cryopreservation techniques for plant cell tissue and organ cultures. In: Fujiwara A (ed) Plant tissue culture. 1982. Maruzen, Tokyo, pp 793–794

Withers LA, Benson EE, Martin M (1988) Cooling rate/medium interactions in the survival and structural stability of cryopreserved shoot tips of *Brassica napus.* Cryo Lett 9: 114–119

Section V
Fruits, Vegetable Tuber and Root Crops
(for Fruit Trees see Section II)

V.1 Conservation of Germplasm of Kiwifruit (*Actinidia* Species)

P.L. Monette[1]

1 Introduction

1.1 Distribution and Important Species

The kiwifruit (syn. mihoutao, mao yangtao, Chinese gooseberry) is an important horticultural crop in New Zealand, Australia, Chile, South Africa, the United States, Italy, Japan, France, Greece, China, Canada, and elsewhere (Warrington 1990). It was previously known as *Actinidia chinensis* Planch., but has recently been renamed *A. deliciosa. A. chinensis* now refers to the smooth-skinned, rather than stiff-haired, fruit species (Liang and Ferguson 1984).

The genus *Actinidia*, of the family Actinidiaceae (van Tieghem 1899), consists of perennial, climbing or scrambling, deciduous vines, usually dioecious, from eastern and southern Asia. The genetic center of the genus *Actinidia* is in the hills and mountains of southern and central China, but species can be found from Siberia and Japan through China to Indochina, Thailand, the Indian subcontinent, Malaysia, and Indonesia (Ferguson 1984). *Actinidia* species often consist of a group of varieties that share distinct morphological traits and occupy discrete geographic areas. Recognition of the varieties within a species or the raising of these varieties to species status appears to be largely a matter of taxonomic interpretation (Li 1952; Liang 1982). The difficulties encountered in delimiting different species and the transitional forms between varieties (Liang 1982) suggest that considerable hybridization may occur naturally (Ferguson 1984). Liang (1983) recognizes over 50 *Actinidia* species and almost 100 distinct taxa. These have been listed by Ferguson (1990). Chromosomes in the genus *Actinidia* are small and very numerous; the range of numbers reported suggests a polyploid series with $x = 29$ (Ferguson 1984).

Kiwifruit has been used in traditional Chinese medicine for several thousand years (Huang and Tan 1988). A few species are grown for their edible fruit, which have a high content of vitamin C. The most important of these is undoubtedly *A. deliciosa*, with one cultivar, Hayward, accounting for 98% of commercial plantings (Sale 1983). Other valuable cultivars of *A. deliciosa* include Abbott/ Allison, Bruno, Monty, and Gracie, and the male pollinators Matua, Tomuri,

[1]Agriculture and Agri-Food Canada, Centre for Plant Health, 8801 East Saanich Road, Sidney, British Columbia, Canada V8L 1H3

Biotechnology in Agriculture and Forestry, Vol. 32
Cryopreservation of Plant Germplasm I (ed. by Y.P.S. Bajaj)

and Chico Male. The other species of economic importance is *A. chinensis.* Species currently being considered for utilization as fresh fruit or for processing into jam, candies, or wine include *A. arguta*, *A. chrysantha*, *A. eriantha*, *A. kolomikta*, *A. indochinensis*, *A. latifolia*, and *A. polygama*. The hardier species, *A. arguta*, *A. kolomikta*, and *A. polygama* are also grown extensively as ornamental plants in North America and Europe (Ferguson 1990).

1.2 Need for Conservation

The need to establish collections of kiwifruit germplasm is not quite so pressing as with certain other crops. Because kiwifruit occur largely in non-industrialized parts of the world, the opportunity still exists for developing and applying a rational conservation policy for kiwifruit germplasm.

The great genetic diversity of the wild populations of *Actinidia* species offers tremendous potential for the genetic improvement of cultivated species (Fig. 1). Germplasm collections can include selections with small (2–5 g) or large (over 200 g) fruit, with green, red, or brown fruit, with sweet or sour fruit, with fruit having low or high vitamin C content (30–1000 mg/100 g fruit), and with early- or late-ripening fruit (Huang and Tan 1988; Ferguson 1990). The preservation of an extensive gene pool for the *Actinidia* species can ensure that a wide variety of desirable attributes is available for breeding.

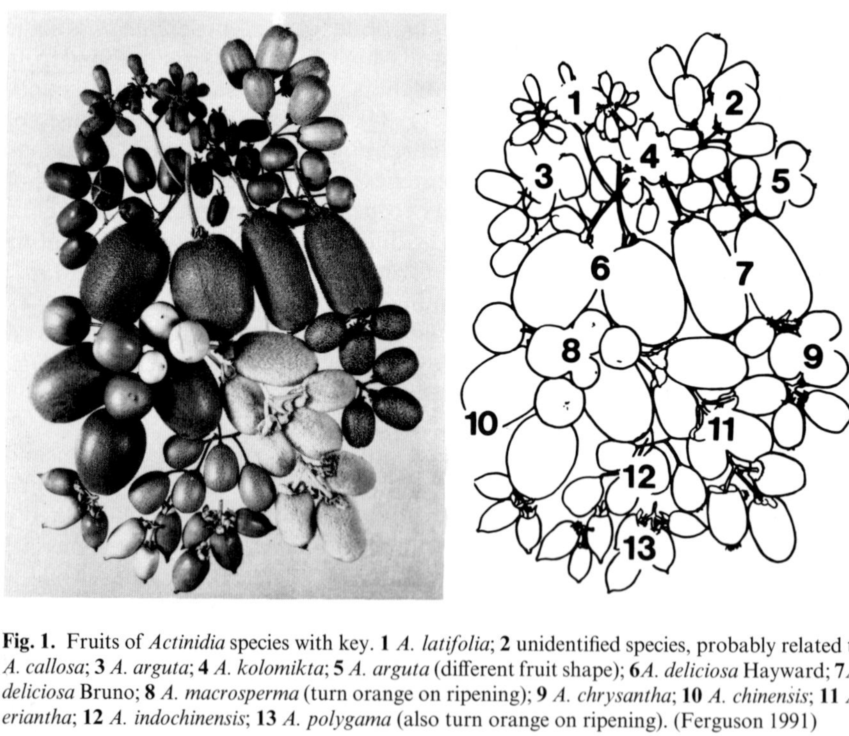

Fig. 1. Fruits of *Actinidia* species with key. **1** *A. latifolia*; **2** unidentified species, probably related to *A. callosa*; **3** *A. arguta*; **4** *A. kolomikta*; **5** *A. arguta* (different fruit shape); **6** *A. deliciosa* Hayward; **7** *A. deliciosa* Bruno; **8** *A. macrosperma* (turn orange on ripening); **9** *A. chrysantha*; **10** *A. chinensis*; **11** *A. eriantha*; **12** *A. indochinensis*; **13** *A. polygama* (also turn orange on ripening). (Ferguson 1991)

In addition to this requirement for crop improvement, there is also a need for the establishment and maintenance of repositories of commercially valuable cultivars and varieties. Repositories which can provide true-to-type kiwifruit plants can be used to satisfy the more immediate needs of the industry, and aid in its development.

1.3 Methods for Kiwifruit Germplasm Conservation

It is essential that the methods used for germplasm storage meet two basic criteria: they must preserve the genetic integrity of the stored material and they must require only minimal maintenance (Reed 1989). For most species, seed storage facilities have provided an acceptable solution. This approach cannot be used with kiwifruit, however, because their seedlings vary widely and are thus unsuitable for breeding purposes. Field plots are another method of maintaining germplasm. Field collections of kiwifruit have, in fact, been established, for example in Auckland, New Zealand (Ferguson 1990), and at Davis, California (Shands and Sisson 1989). Unfortunately, the maintenance requirements of such field collections are extensive. A further disadvantage of field collections is their vulnerability to insects and the climate. An alternative approach, which has considerable potential for the conservation of germplasm from kiwifruit and other vegetatively propagated crops, is in vitro storage (Bajaj 1990).

The methods used to date for in vitro germplasm conservation are of two types: slow-growth systems and cryopreservation. In slow-growth systems, culture conditions are modified so that the cultures can be stored for long periods before they need to be transferred to fresh medium. The storage of cultures in the dark at reduced temperatures has yielded promising results with a number of crops. Cultures growing in vitro at a slow rate inevitably accumulate a number of somaclonally variant cells. In order to avoid the production of these genetically variant cell types, cultures need to stop growing completely. This can be achieved by storing explants or in vitro cultures at −196 °C in liquid nitrogen (−150 °C in the vapor phase).

2 In Vitro Conservation of Kiwifruit

2.1 Brief Review

The possibility of storing cultures in vitro for the short- and medium-term under slow-growth conditions was examined by Monette (1986b), using shoot-tip cultures of *A. deliciosa* cv. Hayward. After developing media and procedures suitable for the micropropagation of kiwifruit, the suitability of two temperatures, 4 and 8 °C, for the storage of shoot-tip cultures in vitro for up to 1 year was compared (Monette 1986a, b). Then the organogenesis and plantlet regeneration from kiwifruit shoot-tip cultures stored for 1 year at 8 °C were examined

(Monette 1987). The possibility of storing kiwifruit stem segments in liquid nitrogen was examined by Jian and Sun (1989).

2.2 Slow-Growth System

2.2.1 Comparison of Two Storage Temperatures, 4 *and* 8 °C

2.2.1.1 Culture Initiation and Shoot Proliferation

Shoot tips about 3 cm in length were cut in late spring from 8-year-old field-grown female vines of *A. deliciosa* cv. Hayward grown from seed. Expanded leaves were removed and shoot tips were surface-sterilized by vigorous stirring for 20 min in a 0.6% solution of commercial sodium hypochlorite containing 0.1% Tween 20. Shoot tips were rinsed three times in sterile distilled water, the basal 1 cm was cut off, and the apical segments were placed on an initiation medium (Table 1) containing Murashige and Skoog (1962) salts. Eight weeks after initiation, the cultures were transferred to proliferation medium (Table 1). The liquid cultures were then placed on a device that tilted the flasks in opposite directions, 35 ° each way from the vertical, every 30 s (Harris and Mason 1983). Proliferating cultures were transferred routinely every 2 weeks into fresh medium. Cultures were maintained at 23 ± 2 °C with a 16-h photoperiod under 48 or 30 μE/m^2/s for agar and liquid media, respectively, provided by cool white fluorescent lights.

Table 1. Formulation of media used for culture initiation, shoot proliferation, and cold storage of kiwifruit shoot tips. (Monette 1987)

Components	Medium		
	Initiation[a]	Proliferation[b]	Storage[a]
Murashige and Skoog salts[c]	3/4-strength	Full-strength	1/4-strength
Sucrose[d]	22.5 g	30 g	27.5 g
Adenine sulfate[d]	60 mg	80 mg	–
$NaH_2PO_4 \cdot H_2O$[d]	128 mg	170 mg	–
Meso-inositol[d]	75 mg	100 mg	25 mg
Thiamine-HCl[d]	0.3 mg	0.4 mg	0.1 mg
IBA[de]	0.023 mg	0.03 mg	–
BAP[df]	2 mg	2 mg	4 mg
pH	5.7	5.0	5.7
Agar (Difco Bacto)[d]	7 g	–	7 g

[a] 15 ml per tube (25 × 150 mm); capped with Bellco Kaputs.
[b] 15 ml per 125-ml Erlenmeyer flask; capped with aluminum foil.
[c] Full-strength solution contains (mg/l): 1650 NH_4NO_3; 1900 KNO_3; 440 $CaCl_2 \cdot 2H_2O$; 370 $MgSO_4 \cdot 7H_2O$; 170 KH_2PO_4; 37.3 $Na_2 \cdot$EDTA; 27.8 $FeSO_4 \cdot 7H_2O$; 6.2 H_3BO_3; 16.9 $MnSO_4 \cdot H_2O$; 8.6 $ZnSO_4 \cdot 7H_2O$; 0.83 KI; 0.25 $Na_2MoO_4 \cdot 2H_2O$; 0.025 $CuSO_4 \cdot 5H_2O$; 0.025 $CoCl_2 \cdot 6H_2O$.
[d] per liter.
[e] Indole-3-butyric acid.
[f] N^6-benzylaminopurine.

2.2.1.2 In Vitro Cold Storage of Shoot Tips

Four hundred shoot tips about 15 mm long were cut from proliferating cultures 20 weeks after initiation, weighed aseptically, and transferred to storage medium (Table 1), one shoot tip per tube. The tubes were placed in darkness in two identical refrigerated incubators (Forma Scientific, Marietta, Ohio). Half were stored at 4 °C and half at 8 °C (±0.5 °C). Ten tubes were removed from each storage treatment every 2 or 4 weeks. All tubes were placed in the culture room (23 °C) for 3 days upon removal from storage. The shoot tips were then weighed aseptically and each was transferred to a 125-ml Erlenmeyer flask containing 7.5 ml of proliferation medium. The flasks were placed on a tilter apparatus and the cultures were kept for 2 weeks under the proliferation conditions described above. The number of shoots 5 mm or longer in each flask was then counted and the cultures were blotted dry with paper towels, and weighed. Mortality was assessed at this time. Cultures were considered dead if they were completely brown and if their fresh weight did not increase during the culture period at 23 °C.

2.2.2 Organogenesis and Plantlet Regeneration Following Storage at 8 °C

Ten randomly selected shoots (1 cm long) were cut from each of three types of cultures for this experiment: cultures stored at 8 °C for 1 year as above, cultures maintained at 23 °C under proliferation conditions throughout the storage interval, and proliferating cultures only 8 weeks old. The shoots were weighed aseptically and each was transferred to a 125-ml Erlenmeyer flask containing 7.5 ml of proliferation medium. Two weeks later they were transferred to fresh medium. After a further 2 weeks, the shoot cultures were blotted dry with paper towels and weighed. The percentage increase in fresh weight of the subcultured shoots was calculated. The number of shoots and roots 5 mm or longer in each culture were also counted. Finally, randomly selected shoots, 1 cm or longer, were cut from the cultures, dipped in 0.05% indole-3-butyric acid and placed in 5-cm peat pots containing a peat:vermiculite:perlite mix (2:1:1 by vol). Twenty-four, twenty-one, and nine shoots were cut from the cultures stored at 8 °C for 1 year, from the cultures maintained in the proliferation stage for 1 year, and from the 8-week-old proliferating cultures, respectively. The pots were placed in a mist chamber with bottom heat (26 °C) for 3 weeks and then moved to a greenhouse bench, where they received daily watering. Surviving plantlets were counted 5 weeks after removal from the mister.

2.2.3 Results

2.2.3.1 Storage at 8 °C vs. 4 °C

Almost all the shoot tips stored at 8 °C survived, including those kept in storage for 1 year (Fig. 2A). By contrast, the percentage survival of shoot tips stored at

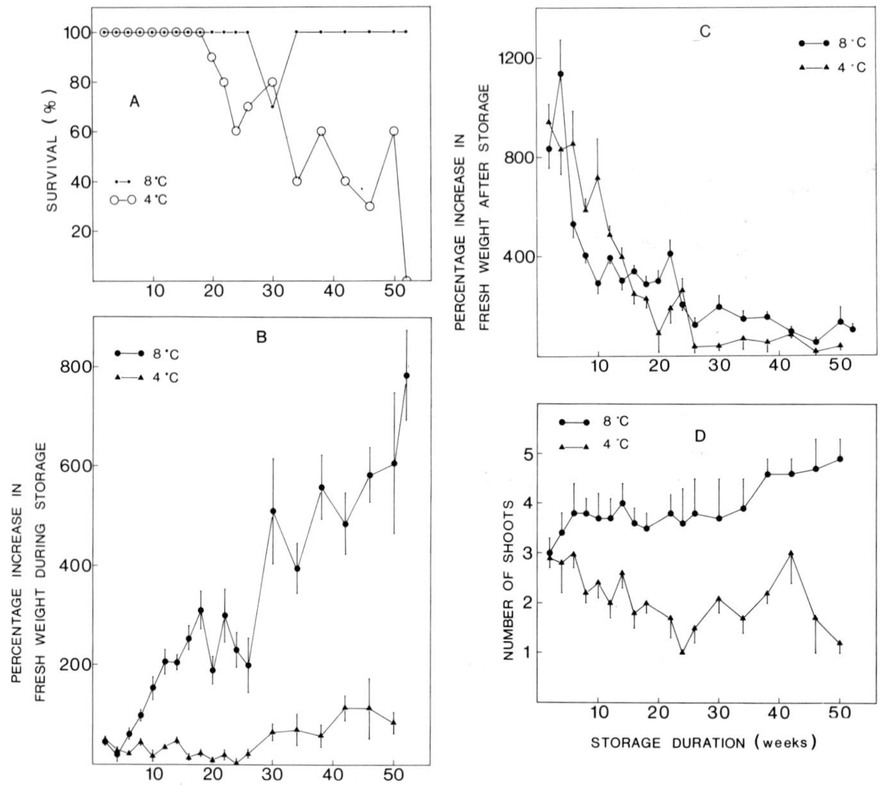

Fig. 2A–D. Survival (**A**), percentage increase in fresh weight during (**B**) and after (**C**) storage and shoot production (**D**) of kiwifruit shoot tips stored in vitro at 8 and 4 °C. Values in **B**, **C**, and **D** are the means of surviving shoot tips. *Vertical bars* denote SE. (Monette 1986b)

4 °C started declining after 20 weeks of storage and none of the shoot tips survived storage at 4 °C for 1 year. Storage temperature also had a marked influence on the fresh weight gain of the kiwifruit shoot tips during storage (Fig. 2B). The fresh weight of the shoot tips stored at 8 °C increased 800% after storage for 1 year, while that of the tips stored at 4 °C increased only 36% after 50 weeks. Storage duration had a marked effect on the growth of the shoot tips when they were transferred to proliferation conditions for 2 weeks after storage (Fig. 2C). The percentage increase in fresh weight of the tips decreased with storage duration at both temperatures. This observation was thought to reflect a requirement of the stored tips for a period of readaptation to growth room conditions (Monette 1986b). Alternatively, it may reflect the accumulation of an inhibitory compound within the tissues of the stored tips. Shoot tips stored at 8 °C produced more shoots 5 mm or longer than tips stored at 4 °C, when transferred to proliferation conditions (Fig. 2D). The number of shoots obtained from tips stored at 8 °C increased with storage duration.

Fig. 3. Kiwifruit shoot tips stored in vitro at 8 °C for 1 year. (Monette 1986b)

The quality of the shoot tips surviving storage at 8 or 4 °C was also different. Shoot tips surviving storage at 4 °C generally contained some necrotic tissue. Little or no necrosis was observed on the shoot tips stored at 8 °C, although physiological aberrations were sometimes apparent. Some of the tips stored at 8 °C were swollen, elongated, and etiolated (Fig. 3), and a few leaves on each tip were variegated.

2.2.3.2 Organogenesis and Plantlet Regeneration After Storage at 8 °C

The fresh weight gain of shoot cultures derived from shoot tips stored in vitro at 8 °C for 1 year was similar to that of cultures derived from shoot tips maintained at 23 °C under proliferation conditions for either 8 weeks or 1 year (Table 2). This percentage increase in fresh weight was much greater than that reported in Fig. 2C for shoot tips stored at 8 °C for 1 year and then maintained under proliferation conditions at 23 °C for 2 weeks. The data in Fig. 2C pertain to the entire stored culture, whereas the data in Table 2 pertain to shoot cultures consisting of a 1-cm shoot excised from the stored cultures. These observations suggest that a growth inhibitor was accumulating within the shoot tip cultures during storage and that the shoots cut from the stored cultures were removed from the influence of this inhibitor.

Storage of shoot tips at 8 °C for 1 year clearly did not impair their proliferation potential. Shoot cultures derived from the stored tips produced as many shoots as those derived from cultures maintained at the proliferation stage for either 8 weeks or one year (Table 2). While shoot production was similar for all three types of shoot cultures, an interesting observation was made concerning root production. Shoot cultures derived from the stored tips produced more

Table 2. Fresh weight increase and shoot and root production in vitro (±SE) of shoot cultures derived from cold-stored kiwifruit shoot-tip cultures and from cultures maintained at the proliferation stage.[a] (Monette 1987)

Origin of shoot culture	Percentage increase in fresh weight	No. of shoots ≥5 mm (per culture)	No. of roots ≥5 mm (per culture)
Shoot-tip cultures stored in vitro at 8 °C for 52 weeks	3630 ± 320 a	5.0 ± 0.4 a	2.2 ± 0.5 a
Shoot-tip cultures maintained at proliferation stage at 23 °C for 52 weeks	4030 ± 950 a	4.9 ± 0.4a	0.7 ± 0.2 b
Shoot-tip cultures maintained at proliferation stage at 23 °C for 8 weeks	3230 ± 890 a	3.7 ± 0.6 a	0.1 ± 0.1 b

[a]Numbers within columns followed by the same letter are not significantly different (ANOVA, $p < 0.05$; Student-Newman-Keuls multiple range rest, $p = 0.05$).

roots than those derived from the cultures maintained at the proliferation stage. These results suggest that root initiation was stimulated either by a dark treatment, a cold treatment, or both.

There was no significant difference in the survival of the plantlets regenerated from the shoot cultures, whether they were derived from the stored tips or from cultures maintained at the proliferation stage. All the regenerated plantlets survived, and plantlets from one treatment could not be distinguished from those from the other treatments.

2.2.4 Discussion

A temperature of 8 °C was clearly preferable to 4 °C for the storage of shoot tips over a 1-year period. The percentage survival was greater and the quality of the cultures was superior when the shoot tips were stored at 8 rather than at 4 °C. Some of the cultures stored at 8 °C exhibited physiological aberrations, but these were only transient.

Shoot cultures derived from shoot tips which had been stored at 8 °C for 1 year were compared to cultures derived from 8-week-old and 1-year-old proliferating cultures. There was no difference between the three types of cultures in fresh weight gain or in shoot production. The cultures derived from the cold-stored tips produced the greatest number of roots. All of the plantlets regenerated from all three treatments survived. Cold storage of shoot tips in vitro at 8 °C for 1 year clearly did not impair either the subsequent organogenesis of subcultured shoots or the survival of the regenerated plantlets. The method thus appears suitable for the establishment and maintenance of a repository of valuable kiwifruit plants. Based on the information available to date on kiwifruit tissue culture, there is no reason to believe that the culture of shoot tips (as opposed to callus) in vitro induces per se a higher frequency of "sports" than that observed in conventional propagation schemes. Because of the inevitable occurrence of somaclonally variant cells in growing cultures, however, slow-growth systems are acceptable only for short- to medium-term storage. To avoid the accumulation of somaclonal variants in a repository, it would be advisable to reinitiate cultures from representative selections and reestablish the repository every few years. While the experiments described above were conducted with only one cultivar, it is reasonable to assume that the media and procedures could be adapted for use with other cultivars and species.

2.3 Cryopreservation

Jian and Sun (1989) evaluated cryopreservation methods with kiwifruit tissues. The surface-sterilized shoots were cut into 1-cm-long segments, split longitudinally into four pieces, and then placed into test tubes. Cryoprotectants precooled to 0 °C were immediately added and the tubes were kept in an ice bath for 45 min. Two cryoprotectant solutions were tested: 10% DMSO (dimethylsulfoxide) + 0.5 M sorbitol, and 5% DMSO + 5% glycerol + 5% sucrose. After this pretreatment, the temperature was reduced to −10 °C, with a temperature reduction rate of 1 °C/min. The temperature was maintained at −10 °C for 20 min and then reduced to −40 °C. The explants were kept at −40 °C for 3 h and then placed in liquid nitrogen. After storage in liquid nitrogen for 120 days, the explants were thawed in a water bath at 40 °C and washed three times in a liquid medium consisting of Murashige and Skoog salts (MS; 1962). They were then transferred to MS medium solidified with agar and supplemented with 1 mg/l corn extract. Callus induction was allowed to take place for 40 days in this medium at 25 °C with an illumination of 1000 lx and a 12-h photoperiod. The cultures were then transferred to MS medium with 1 mg/l corn extract and 0.1% activated charcoal for plantlet regeneration.

The kiwifruit stem explants stored in liquid nitrogen for 120 days produced callus, derived from the cambium, after 25 days on MS medium supplemented with 1 mg/l corn extract. The frequency of callus induction (Table 3) after

Table 3. Frequency of callus induction on stem segments of Chinese gooseberry after cryopreservation. (Jian and Sun 1989)

Cryoprotectant	No cryopreservation			Cryopreservation for 120 days		
	No. of stem segments cultured	No. of calli produced	Induction frequency (%)	No. of stem segments cultured	No. of calli produced	Induction frequency (%)
10% DMSO[a] + 0.5 M sorbitol	31	23	77.4	57	44	77.2
5% DMSO + 5% glycerol + 5% sucrose	33	25	76.0	49	37	75.5

[a] DMSO = dimethylsulfoxide.

cryopreservation using either cryoprotectant solution, 76–77%, was similar to that obtained with control explants not subjected to cryopreservation. Similarly, the quality of the callus cultures was comparable for all treatments. All cultures were dense, firm, and greenish. Numerous adventitious shoots were produced on the callus cultures in MS medium with 0.1 mg/l corn extract and 0.1% activated charcoal. Regenerated plantlets were normal in morphology and color.

3 Summary and Conclusions

A slow-growth system using storage of kiwifruit shoot tips in vitro at 8 °C was evaluated by Monette (1986b, 1987) and found to be suitable for the year-long storage of *Actinidia deliciosa* cv. Hayward. With minor modifications, the media and procedures utilized should also be suitable for the storage of shoot tips from other cultivars and species. In vitro repositories based on this system could be valuable for the kiwifruit Industry. For example, promising selections could be stored in such a repository while they are being field-tested. If the results of the field trial indicated that the selections were commercially valuable, the cultures could be retrieved from storage and readily micropropagated for distribution.

A technique was developed by Jian and Sun (1989) for the cryopreservation of kiwifruit stem sections in liquid nitrogen. Callus was produced on the stem sections after they were removed from storage, and plantlets were regenerated from the callus. These promising results suggest that their cryopreservation technique may be applicable, with suitable modifications, for the long-term storage of kiwifruit meristems or shoot tips. These tissues are preferable to stem callus, as they are more likely to maintain the genetic integrity of the stored material.

The plantlets regenerated by both Monette (1987) and Jian and Sun (1989) from stored kiwifruit tissues were of normal appearance. No study has yet been reported concerning the yield or quality of fruit produced on kiwifruit plants regenerated from stored tissues. These data, which can be obtained from the appropriate field trials, can serve as a rough indication of whether the stored kiwifruit tissues have maintained their genetic integrity.

Acknowledgments. The author is grateful to Brent Warner, New Crop Specialist, Province of British Columbia, Ministry of Agriculture, Fisheries and Food, for helpful discussions. Thanks are also due to Wei Chao for translating articles from the Chinese and Sharon Godkin for photography.

References

Bajaj YPS (1990) Cryopreservation of germplasm of vegetatively propagated crops. Bull Soc Bot Fr Actual Bot 137 (3/4): 99–114

Ferguson AR (1984) Kiwifruit: a botanical review. Hortic Rev 6: 1–64

Ferguson AR (1990) The genus *Actinidia*. In: Warrington IJ, Weston GC (eds) Kiwifruit: science and management. Ray Richards, Auckland, pp 15–31

Ferguson AR (1991) Relatives of the kiwifruit. NZ Kiwifruit 80: 18–19
Harris RE, Mason EBB (1983) Two machines for in vitro propagation of plants in liquid media. Can J Plant Sci 63: 311–316
Huang ZG, Tan CY (1988) Chinese gooseberry, kiwifruit (*Actinidia* spp.). In: Bajaj YPS (ed) Biotechnology in agriculture and forestry, vol 6. Crops II. Springer, Berlin Heidelberg New York, pp 166–180
Jian LC, Sun LH (1989) Cryopreservation of the stem segments in Chinese gooseberry. Acta Bot Sin 31(1): 66–68
Li HL (1952) A taxonomic review of the genus *Actinidia*. J Arnold Arbor (Harvard Univ) 33: 1–61
Liang CF (1982) An addition to the infraspecific taxa of *Actinidia chinensis* Planch. Acta Phytotaxon Sin 20: 101–104
Liang CF (1983) On the distribution of actinidias. Guihaia 3: 229–248
Liang CF, Ferguson AR (1984) Emendation of the Latin name of *Actinidia chinensis* Pl. var. *hispida* C.F. Liang. Guihaia 4: 181–182
Monette PL (1986a) Micropropagation of kiwifruit using non-axenic shoot tips. Plant Cell Tissue Organ Cult 6: 73–82
Monette PL (1986b) Cold storage of kiwifruit shoot tips in vitro. HortScience 21(5): 1203–1205
Monette PL (1987) Organogenesis and plantlet regeneration following in vitro cold storage of kiwifruit shoot tip cultures. Sci Hortic 31: 101–106
Murashige T, Skoog F (1962) A revised medium for rapid growth and bioassays with tobacco tissue cultures. Physiol Plant 15: 473–497
Reed SM (1989) In vitro conservation of germplasm. In: Stalker HT, Chapman C (eds) IBPGR training courses: lecture series. 2. Scientific management of germplasm: characterization, evaluation and enhancement. Int Board Plant Genet Resour, Rome, pp 23–30
Sale PR (1983) Kiwifruit culture. Ashenden Williams D (ed) Gov Printing Office, Wellington, New Zealand
Shands HL, Sisson VA (1989) Plant germplasm maintenance: an example of a national program. In: Stalker HT, Chapman C (eds) IBPGR training courses: lecture series. 2. Scientific management of germplasm: characterization, evaluation and enhancement. Int Board Plant Genet Resour, Rome, pp 7–15
Van Tieghem P (1899) Sur les genres *Actinidia* et *Sauravia* considérés comme types d'une famille nouvelle, les actinidiacées. Ann Sci Natl Bot Biol Vég Sér 8(10): 137–140
Warrington IJ (1990) Areas and trends of kiwifruit production in New Zealand and around the world. In: Warrington IJ, Weston GC (eds) Kiwifruit: science and management. Ray Richards, Auckland, pp 511–525

V.2 Cryopreservation of Asparagus

A. URAGAMI[1]

1 Introduction

The genus *Asparagus* (Liliaceae) includes about 300 species found in Europe, Asia, Africa, and Australia (Everett 1980). Only *Asparagus officinalis* is cultivated as a vegetable, but some other species, such as *A. densiflorus* (syns. *A. sprengeri*; *A. myriocladus*; *A. sarmentosus*), *A. setaceus* (syn. *A. plumosus*), and *A. asparagoides* (syn. *A. medeoloides*), are grown and utilized as ornamental plants.

A. officinalis is a native of coastal regions in Europe and temperate regions in Asia. Approximately 140 000 ha of asparagus fields are distributed from subfrigid climates to the tropics (Nichols 1990).

A. densiflorus cv. Sprengeri has been classified as being immune to *Puccinia asparagi* (Thompson and Hepler 1956). *A. asparagoides*, *A. compactus*, *A. densiflorus* cv. Myers, *A. densiflorus* cv. Sprengeri, *A. larcinus*, *A. verticillatus*, and *A. virugatus* have been classified as being highly resistant to *Stemphyllium* sp. (Bansal et al. 1986). *A. densiflorus* cv. Sprengeri was found to be resistant to *Phoma asparagi* (Tu et al. 1988). *A. densiflorus* cv. Sprengeri and cv. Myersii received the lowest ratings for susceptibility to *Fusarium oxysporum* f. sp. *asparagi* and *F. moniliforme* among four tested species of *Asparagus* (Stephens et al. 1989). These species are possible donors of disease-resistance genes to *A. officinalis*.

Asparagus species are perennial, seed-propagated plants. There have been few reports on the long-term preservation of *Asparagus* seeds, but most species can be preserved as seeds.

At present, individual plants of *A. officinalis* are maintained in the field. Care is needed to maintain plants as genetic resources in the field, as follows. In order to avoid the mingling of crowns of adjoining plants as the crowns grow, an appropriately large area is required per plant. Seedlings from the seeds of female plants should be completely removed by hand every year. Control of diseases, insects, and weeds is required in all cases.

In the *Asparagus* collection of the USDA, ten tropical/subtropical species are currently maintained as clones in the greenhouse (K.R. Reitsma, pers. comm.). Other species have been grown using open pollination in isolated plantings. Maintenance of *Asparagus* germplasm in the field presents many problems there. Perennial species can grow into one another, and birds can contaminate plots by

[1]Hokkaido National Agricultural Experiment Station, 1 Hitsujigaoka, Sapporo, 062 Japan

Biotechnology in Agriculture and Forestry, Vol. 32
Cryopreservation of Plant Germplasm I (ed. by Y.P.S. Bajaj)

carrying seeds of another species. *Asparagus* seeds had been stored at 4 °C and 40% RH (relative humidity) until 1991, when they were moved into a new storage facility maintained at 4 °C and 25% RH. The germination rates of seeds in 1991 that have been stored since 1983 were 57–69%.

Thus, to overcome some of these problems, in vitro methods are now being developed for the storage of germplasm (see Bajaj 1986).

2 Need for Conservation and Cryopreservation

Asparagus plants are mostly propagated by seeds. However, most of the seeds of the newly released cultivars, such as UC157F1 (Benson and Takatori 1978), Jersey Giant (Ellison and Kinelski 1985), and Gynlim (Boonen 1987), were produced from parents propagated by tissue culture. Since asparagus is an almost completely allogamous plant, unique combinations of genes of individual plants cannot be preserved through their seeds.

Micropropagation has been very successful in asparagus (see Desjardins 1992). In the mid-1970s, Yang and Clore (1973) established a mass-propagation technique for asparagus using axillary buds of shoots that had been grown in vitro, and Chin (1982) subsequently improved their methodology. Regeneration of plants from somatic embryos, derived from cultured cells (Jullien 1974) or protoplasts (Kunitake and Mii 1990), has also been reported. Somatic embryogenesis is now utilized for mass propagation of asparagus, and tissue culture is also used to generate virus-free plants (see Desjardins 1992).

Genetically manipulated cell lines and cells with high regenerative ability should be stored before they lose their properties of interest as a result of genetic instability during continuous subculturing (Bajaj 1979). The genetic stability of biological materials over a long period of time can be achieved by preservation in liquid nitrogen (LN). Cultured plant cells and meristems can survive after slow freezing at 0.5 to 1 °C/min to about −40 °C, prior to direct immersion in LN, in the presence of suitable cryoprotectants, with subsequent rapid thawing (Withers 1985). However, such slow freezing is time-consuming and requires expensive rate-controlled freezing equipment. Thus, development of simple and reliable routine methods, which do not require special equipment for freezing, is necessary for the cryopreservation of a wide range of plant materials.

Cryopreservation of *Asparagus* is important for germplasm conservation, both for long-term preservation of materials in vitro and to complement preservation of seeds and maintenance of plants in vivo (in the field).

3 Cryopreservation

3.1 Review of Work (Table 1)

Kumu et al. (1983) reported the successful cryopreservation of shoot tips of asparagus (*A. officinalis*) that had been excised from greenhouse-grown young

Table 1. Summary of the work on cryopreservation of *Asparagus officinalis*

Explant	Pretreatment and Cryoprotectants	Freezing and storage/rewarming and dilution	Viability	Reference
Shoot tips (spear, 1.5–2 mm)	4% DMSO + 3% glucose 2 days preculture Frozen in 11% DMSO	Cooled 0.5 °C/min to −40 °C, LN storage Thawed in 40 °C water	100% survival	Kumu et al. (1983)
Cultured cells	12% EG + 0.5 M sorbitol 22 °C, 10 min	Direct immersion in LN	65% survival	Uragami et al.
Somatic embryos (globular)	Dropwise addition of PVS1, 0 °C, 10 min	Rewarmed in 22 °C water	48% survival (FDA staining)	(1989)
One node segments (in vitro, 5 mm)	0.7 M sucrose, 2 days preculture Air-dried	Direct immersion in LN Rewarmed in air, room temperature	63% shoot formation	Uragami et al. (1990)
Embryogenic callus	0.7 M sucrose and 10 μM ABA, 3 days preculture Air-dried	Direct immersion in LN Rewarmed in air, room temperature	80% survival	Uragami (1991)
	PVS1, 25 °C, 4 min	Direct immersion in LN 6 months storage Rewarmed in 25 °C water Diluted in 1.2 M sucrose	60% survival	
Cultured cells (embryogenic, (4-day-old)	2 M glycerol + 0.4 M sucrose, 25 °C, 10 min	Frozen at −30 °C, 1 h, then in LN Thawed in 40 °C water	80% survival (FDA staining)	Nishizawa et al. (1992)
Multiple bud clusters (3–5 mm)	PVS2 0 °C, 120 min	Direct immersion in LN Rewarmed in 38 °C water Diluted in 1.2 M sucrose	95% shoot formation	Kohmura et al. (1992)

Abbreviations: EG: ethylene glycol, FDA: fluorescein diacetate.

spears. Asparagus shoot tips were precultured with 4% DMSO (dimethylsulfoxide) and 3% glucose for 2 days, and then they were frozen slowly (0.5 °C/min) in 11% DMSO solution to −40 °C and finally immersed in LN. After rapid thawing in water at 40 °C, shoot tips were recultured. Survival was 100% and some of the shoot tips regenerated plantlets.

Cultured cells and embryos of asparagus have also been cryopreserved by vitrification (Uragami et al. 1989). Cultured cells or somatic embryos of about 0.5 ml in packed cell volume were pretreated with 12% ethylene glycol and 0.5 M sorbitol for 10 min at 22 °C. Plant vitrification solution 1 (PVS1), which consists of 19% glycerol, 13% ethylene glycol, 13% propylene glycol, and 6% DMSO in 0.5 M sorbitol, was added dropwise up to 5 ml at 0 °C over the course of 10 min. Fresh PVS1 was added to the pelleted material after centrifugation at 100 g and removal of the supernatant. Cultured cells and embryos suspended in PVS1 were held at 0 °C for another 10 min, placed in plastic straws and then immersed directly in LN. The straws were rewarmed rapidly in water at 22 °C and then expelled into 1.2 M sucrose medium that contained 8% ethylene glycol. The suspensions were then diluted stepwise. Sixty-five percent of cultured cells and 48% of the somatic embryos remained alive after this treatment. Plant regeneration occurred via normal embryogenesis.

Embryogenic callus of asparagus has been cryopreserved after air drying (Uragami 1991). The callus was dehydrated in air in the presence of dry silica gel, immersed directly in LN, and then rewarmed slowly in air. One-day preculture on medium that contained 0.7M sucrose increased the survival rates to 29 and 30% with abscisic acid (ABA), respectively. There was no significant difference between the survival rates of sucrose-pretreated cells and sucrose plus ABA-pretreated cells. The highest survival rate of 79.7% was obtained in the case of callus precultured with ABA and 0.7 M sucrose for 3 days. Embryogenic callus that had been cooled to the temperature of LN after air drying grew normally and plantlets were regenerated as a result of normal embryogenesis.

Embryogenic callus of asparagus has also been cryopreserved by the improved vitrification method (Sakai et al. 1990; Uragami 1991). Calli were treated with PVS1 for 4 min at 25 °C, loaded in straws, and then immersed in LN. The calli in the straws were stored in LN for 60 min, 1 day, 8 weeks, and 6 months. After storage, the straws were rewarmed in water at 25 °C and the calli were expelled in a medium containing 1.2 M sucrose. The survival rates after the storage were 69.2 ± 9.7%, 66.2 ± 10.5%, 57.9 ± 4.6%, and 61.5 ± 4.2%, respectively. There was no significant difference in survival among the treatments.

Clusters of multiple buds, obtained from a meristem culture of asparagus, were pretreated with PVS2 (30% glycerol, 15% ethylene glycol, 15% DMSO, and 0.4 M sucrose) at 0 °C for 120 min, and were then immersed directly in LN. Bud clusters were rewarmed rapidly in water at 38 °C and expelled into medium that contained 1.2 M sucrose. Approximately 90% of these bud clusters produced normal shoots (Khomura et al. 1992).

Four-day-old embryogenic cells of asparagus have been cryopreserved by a simple freezing method (Nishizawa et al. 1992). The cells were pretreated with 2 M glycerol and 0.4 M sucrose at 25 °C for 10 min. They were then frozen at −30 °C for 1 h and immersed in LN. The suspension of cells was thawed rapidly in water at 40 °C, and diluted in medium that contained 1.2 M sucrose. Eighty percent of the cells survived.

The axillary bud, which has not been through the callus phase, is considered to be a better material than undifferentiated cells for germplasm preservation and for the micropropagation of plants selected for their superior characteristics, because the tissue exhibits higher genetic stability. Single-node segments, which consist of a piece of stem with an axillary bud, have been commonly used for the micropropagation of asparagus. The next section describes the cryopreservation of the axillary buds from plantlets of asparagus that had been grown in vitro after air drying (Uragami et al. 1990).

3.2 Protocol for Cryopreservation of Axillary Buds

3.2.1 Plant Material

The asparagus female clone 873–32 was maintained in vitro as described by Yang and Clore (1973). The plantlets that developed in vitro from buds were transferred to fresh MS medium solidified with 0.25% Gelrite (Kelco Division, Merck and Co., USA) and supplemented with 5.4×10^{-7} M naphthalene acetic acid (NAA) and 4.6×10^{-7} M 6-benzyladenine (BA) at intervals of 4 to 6 weeks after harvesting of shoots. Plantlets were cultured at 25 °C under continuous cool white fluorescent light (37 $\mu E/m^2/s$). A microscope was used to aid in the preparation of segments with uniform buds.

3.2.2 Preculture

Stem segments were excised from the middle part of the shoots of 6- to 8-week-old plantlets that were approximately 10 cm tall. The diameter of the shoots was 0.5–0.8 mm. The excised shoots were cut into smaller segments of approximately 5 mm in length, with one bud in the center of each. The rank of the single node segments on the middle part of a stem had no effect on the growth rate or on the appearance of the new shoots that developed from the axillary buds. The segments were precultured on solidified MS medium supplemented with 0.7 M sucrose at 25 °C in the light for 2 days.

3.2.3 Desiccation

The pretreated segments were collected on a sheet of 500-μm nylon mesh in a small Petri dish (3 cm in diameter) which was then placed for dehydration in a larger Petri dish (6 cm in diameter) that contained 15 g dry silica gel. The larger Petri dish was sealed with Parafilm (American Can Company, USA) and kept at 25 °C. The segments inside the smaller Petri dish were gradually dried out.

3.2.4 *Cooling and Rewarming*

About 15 segments were placed in individual 1-ml plastic tubes which were then sealed and cooled rapidly by immersion in LN (cooling rate: 720 °C/min). Samples were stored in LN for 60 min, and were then rewarmed by placing the tubes in air at room temperature (rewarming rate: 120 °C/min). Samples were recultured on MS medium that had been solidified with 0.25% Gelrite and supplemented with 5.4×10^{-7} M NAA and 4.6×10^{-7} M BA.

3.2.5 *Analysis of Soluble Sugars*

Soluble sugars were extracted by homogenizing 25 of fresh stem segments (about 100 mg fresh weight) in distilled water. The homogenate was filtered through a nylon filter (45-μm pore size), and then four volumes of acetonitrile were added to the filtrate. The mixture was filtered again and analyzed by high-performance liquid chromatography (HPLC). Sugars were separated on aminopropyl silica column with a solvent of 75% acetonitrile and 25% water. The refractive of eluted fractions was monitored by index. A pumping system and detector from Nissei Sangyo Co. (Tokyo, Japan) were used. Quantitation was achieved by comparing areas of peaks of interest with the areas of peaks generated by standards. Standard curves, generated using sugars from Wako Pure Chemical Industries (Osaka, Japan), were used to calculate nanogram values from the ratios of measured peak areas. The results are expressed as the average, and standard deviations obtained from three experiments.

3.3 Results

3.3.1 *Time Course of Dehydration*

Figure 1 shows the time course of dehydration of precultured and nonprecultured single-node segments. The precultured segments were dehydrated more slowly than the nonprecultured.

3.3.2 *Effects of Preculture*

As shown in Table 2, the maximum survival rate of nonprecultured buds that had been immersed in LN after dehydration was only about 7%, while the precultured buds had a survival rate of 63%, which is equivalent to 71% of the survival rate of untreated controls since only 89% of the buds of untreated control developed shoots (Fig. 2). Thus, preculture for 2 days at 25 °C on a culture medium that contained 0.7 M sucrose was effective in increasing the survival rate of the buds after immersion in LN. During preculture with a high concentration of sucrose for 2 days, the water content decreased by about 16.4% (from 84.7% to 68.3%) and the weight of the dry matter increased approximately 2.7-fold.

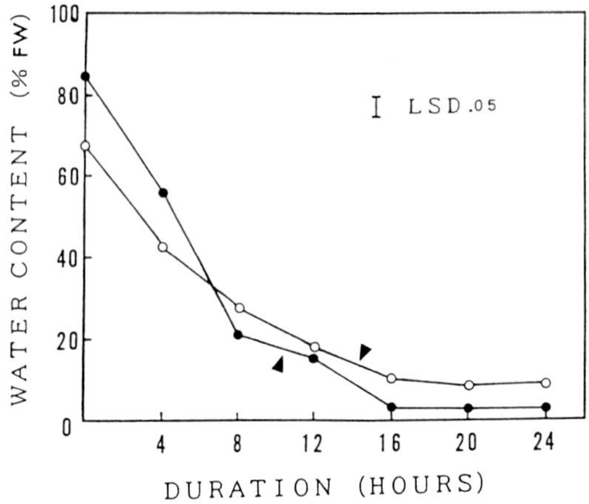

Fig. 1. The time course of dehydration of single-node segments. Segments were precultured on solidified MS medium that contained 0.7 M sucrose for 2 days in the light at 25 °C. ●, non-precultured; ○, precultured. *Arrows* indicate water contents at 50% lethality. (Uragami et al. 1990)

Table 2. Effects of pretreatment on the survival of axillary buds of *Asparagus* cooled to −196 °C, and changes in water content and dry matter during preculture (Uragami 1991)

	Nonprecultured	Precultured
Maximum survival (%)	6.7	63.2
Water content (%)	84.7	68.3
Dry matter (mg/segment)	1.0+0.34	2.7+0.49

Preculture: precultured on solidified MS medium that contained 0.7 M sucrose for 2 days at 25 °C in the light.

3.3.3 Effects of Water Content

Figure 2 shows that the threshold water content, below which the axillary buds remained alive after immersion in LN, ranged from approximately 20 to 17% fresh weight. However, the survival rate decreased rapidly at a water content below about 13% (fr. wt.), presumably as a result of desiccation injury, since the same trend was observed with the segments that had not been exposed to LN.

3.3.4 Increase in Amounts of Soluble Sugars in Stem Segments During Preculture

Levels of soluble sugars in stem segments increased significantly during preculture for 2 days (Table 3). Sucrose was the predominant sugar both before and after preculture. However, the levels of glucose and fructose also increased, suggesting that sucrose, which was absorbed during preculture by segments from the medium, had been converted into glucose and fructose.

3.3.5 Regeneration of Plants

Viable shoot primordia resumed growth on the regrowth medium within one week, and they developed shoots directly without formation of callus (Fig. 3).

Fig. 2. Survival of precultured axillary buds before and after exposure to LN at various water contents. ○, Before; ●, after exposure to LN. (Uragami et al. 1990)

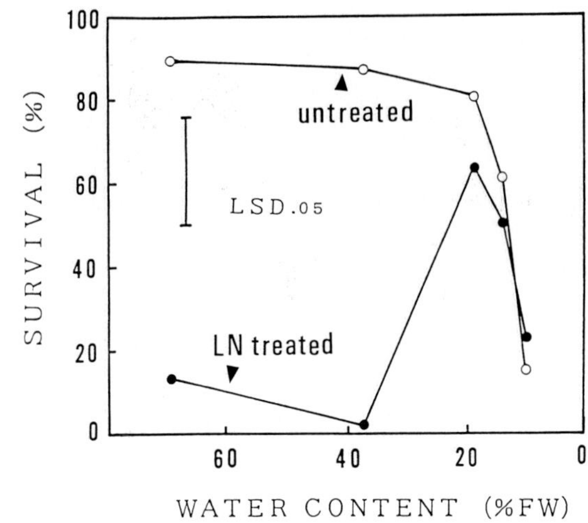

Table 3. Increases in concentrations of soluble sugars in stem segments during preculture. (Uragami 1991)

Sugars	Concentration (% fr.wt.)	
	Nonprecultured	Precultured
Glucose	0.36+0.07	1.26+0.22
Fructose	0.17+0.04	1.04+0.03
Sucrose	3.89+0.48	7.30+0.19
Maltose	0.46+0.99	0.66+0.03

Preculture: precultured on solidified MS medium that contained 0.7 M sucrose for 2 days at 25 °C in the light.

Approximately 70% of the cryopreserved buds regenerated plantlets with multiple shoots and roots (Fig. 4). Some axillary buds, which had sustained desiccation injury, resumed growth within 2 or 3 weeks after reculture. However, even these buds developed shoots without intermediate formation of callus.

Examination of chromosome revealed that all of the 20 regenerated plants examined were diploid and had 20 chromosomes.

4 Discussion

The induction or modification of dehydration tolerance may provide the key to successful cryopreservation by air drying. In the axillary buds of asparagus, 2-day preculture on solidified MS medium that contained 0.7 M sucrose was

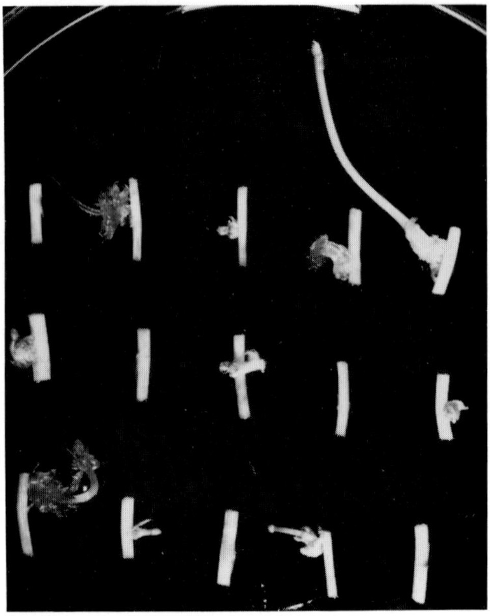

Fig. 3. Growth of shoots from lateral buds cooled to −196 °C after desiccation, 3 weeks after reculture. (Uragami 1991)

effective in enhancing dehydration tolerance. During the preculture, the water content decreased, the amount of dry matter increased considerably, and a remarkable increase in levels of sugars was observed. Koster and Leopold (1988) suggested that sucrose might serve as the principal mediator of desiccation tolerance in seeds of soybean, pea, and corn. In the case of embryogenic callus of asparagus, the water content of the callus decreased during preculture by about 20–30%, while the fresh weight decreased by less than 10% (Uragami 1991). This result suggests that the soluble sugar content of embryogenic callus may have increased during preculture, as seen in the stem segments of asparagus.

Pretreatment with ABA (Senaratna et al. 1989) or a high concentration of sucrose (Nitzsche 1978) has been reported as being effective in inducing desiccation tolerance in somatic embryos and callus. We adopted a 2-day preculture with sucrose that was much briefer than treatment with ABA, which is usually applied for about 2 weeks. It seems that pretreatment with sucrose has at least two effects, one of which is the induction of desiccation tolerance via accumulation of sucrose (Koster and Leopold 1988). Absorption of sucrose from the preculture medium was confirmed by analysis of single-node segments of asparagus by HPLC. The second effect is a slow reduction in water content of about 8% per day during preculture. This reduction may have an effect similar to that of slow drying, as reported in alfalfa somatic embryos (Senaratna et al. 1989). However, the nature of desiccation tolerance and its modification are still obscure. Further studies on the induction and the modification of dehydration tolerance in hydrated materials are required.

Fig. 4. Plantlets with multiple shoots and roots developed from lateral buds cooled to −196 °C after desiccation. Ten weeks after reculture. (Uragami et al. 1990)

It is particularly important to determine whether cryopreserved cells, meristems, and embryos are capable of producing plants with a genotype identical to that of the untreated plant. The patterns of growth of cryopreserved meristems or buds vary considerably among species and varieties, as well as with the cryogenic protocol employed. A callus phase prior to the initiation of shoots is undesirable, since callus tissue has the potential to increase the frequency of variants. However, in the present study, sufficiently dehydrated axillary buds of asparagus which had been cooled to −196 °C developed shoots and roots without any intermediate formation of callus, and they regenerated plantlets within about 2 months after reculture. The chromosome number of each of 20 regenerants was 20, which is that of the normal diploid.

In this study, survival rates before exposure to LN dropped significantly at water contents below 19%. This observation suggests that stem segments of asparagus are not fully tolerant to desiccation even after preculture with 0.7 M sucrose. After exposure to LN, the survival of segments was maximal at a water content of 19%. The survival rates of segments with water contents below 19% both before and after exposure to LN were similar. Thus, no freezable water, which causes freezing injury during cooling and rewarming, remained in the tissues of the asparagus segments.

There have been several reports of the successful cryopreservation of cultured shoot tips of *Solanum* (Fabre and Dereuddre 1990) and pear (Scottez et al. 1992), as well as of somatic embryos of carrot (Dereuddre et al. 1991), after significant dehydration with subsequent encapsulation in Ca-alginate beads. The materials were encapsulated, precultured with a high concentration of sucrose in the appropriate medium, dehydrated in air, immersed directly in LN, and then slowly rewarmed. Coating with alginate allows the size and the shape of the material to be unified so that it undergoes more uniform loss of water during the course of drying. An alginate coating can also work as a buffering agent by preventing the direct effects of environmental changes on the encapsulated material. The coating prevents rapid dehydration and the rapid absorption of water by reducing the difference between the water potentials of the outer and inner cells of embryos. This property of coating was confirmed by the observation that additional pretreatment was necessary for air drying of microspore-derived embryos of oilseed rape, in the absence of encapsulation, to obtain a survival rate similar to that of cryopreserved material after encapsulation and dehydration (Uragami et al. 1993). The standard deviations of the survival rates of asparagus axillary buds after cryopreservation by air drying were rather large as compared with those obtained after application of other techniques. The main cause of the difference seems to be the variation in size and shape among stem segments. The application of the encapsulation-dehydration technique may lead to higher and more consistent survival rates.

5 Summary and Conclusions

The cryopreservation protocols reported to date can be classified as being of four types according to the cell-dehydration procedure(s) applied prior to immersion in LN (Sakai 1991), as follows: (1) slow-freezing methods; (2) vitrification methods; (3) air-drying methods; and (4) combinations of the above methods (e.g., simple freezing methods). To date, successful cryopreservation of in vitro-cultured asparagus cells or organs by examples of each of these four types of method has been reported. For the practical application of these methods of cryopreservation, further studies on the uniformity of genotype and phenotype of regenerants obtained after long-term storage in LN are required.

References

Bajaj YPS (1979) Technology and prospects of cryopreservation of germplasm. Euphytica 28: 267–285

Bajaj YPS (1986) In vitro preservation of genetic resources – Techniques and problems. Int Symp Nuclear technique and in vitro-culture for plant improvement. IAEA, Vienna, pp 43–57

Bansal RK, Menzies SA, Broadhurst PG (1986) Screening of *Asparagus* species for resistance to *Stemphylium* leaf spot. NZ J Agric Res 29: 539–545

Benson B, Takatori F (1978) Meet UC 157. Am Vegetable Grower, pp 8–9

Boonen P (1987) The breeding and choice of asparagus in the Netherlands. Asparagus Res Newsl 5: 37–42

Chin C (1982) Promotion of shoot and root formation in asparagus in vitro by ancymidol. HortScience 17: 590–591

Dereuddre J, Blandin S, Hassen N (1991) Resistance of alginate-coated somatic embryos of carrot (*Daucus carota* L.) to freezing in liquid nitrogen. I. Effects of preculture. Cryo Lett 12: 125–134

Desjardins Y (1992) Micropropagation of Asparagus (*Asparagus officinalis* L.). In: Bajaj YPS (ed) Biotechnology in agriculture and forestry, vol 19. High-tech and micropropagation III. Springer, Berlin Heidelberg New York, pp 26–41

Ellison JH, Kinelski JJ (1985) 'Jersey Giant', an all-male asparagus hybrid. HortScience 20: 1141

Everett TH (1980) The New York botanical garden. Illustrated encyclopedia of horticulture. Garland, New York

Fabre J, Dereuddre J (1990) Encapsulation-dehydration: a new approach to cryopreservation of *Solanum* shoot-tips. Cryo Lett 11: 413–426

Jullien M (1974) La culture in vitro des cellules du tissu foliaire d'*Asparagus officinalis* L.: obtention de souches à embryogenèse permanente et régéneration de plantes entières. CR Acad Sci Paris Ser D 279: 747–750

Kohmura H, Sakai A, Chokyu S, Yakuwa T (1992) Cryopreservation of in vitro-cultured multiple bud clusters of asparagus (*Asparagus officinalis* L. cv Hiroshimagreen (2n = 20)) by the techniques of vitrification. Plant Cell Rep 11: 433–437

Koster KL, Leopold AC (1988) Sugars and desiccation tolerance in seeds. Plant Physiol 88: 829–832

Kumu Y, Harada T, Yakuwa T (1983) Development of a whole plant from a shoot tip of *Asparagus officinalis* L. frozen down to −196 °C. J Fac Agric Hokkaido Univ 61: 285–294

Kunitake H, Mii M (1990) Somatic embryogenesis and plant regeneration from protoplasts of asparagus (*Asparagus officinalis* L.). Plant Cell Rep 8: 706–710

Nichols MA (1990) Asparagus – the world scene. Acta Hortic 271: 25–31

Nishizawa S, Sakai A, Amano Y, Matsuzawa T (1992) Cryopreservation of asparagus (*Asparagus officinalis* L.) embryogenic cells and subsequent plant regeneration by a simple freezing method. Cryo Lett 13: 379–388

Nitzsche W (1978) Erhaltung der Lebensfähigkeit in getrocknetem Kallus. Z Pflanzenphysiol 87: 469–472

Sakai A (1991) The present and the future of the research on cryopreservation of cultured plant cells, organs, and embryos. Agric Hortic 66: 1223–1229 (in Japanese)

Sakai A, Kobayashi S, Oiyama I (1990) Cryopreservation of nucellar cells of navel orange (*Citrus sinensis* Osb. var. *brasiliensis* Tanaka) by vitrification. Plant Cell Rep 9: 30–33

Scottez C, Chevreau E, Godard N, Arnaud Y, Duron M, Dereuddre J (1992) Cryopreservation of cold-acclimated shoot tips of pear in vitro cultures after encapsulation-dehydration. Cryobiology 29: 691–700

Senaratna T, McKersie BD, Bowley SR (1989) Desiccation tolerance of alfalfa (*Medicago sativa* L.) somatic embryos. Influence of abscisic acid, stress pretreatments, and drying rates. Plant Sci 65: 253–259

Stephens CT, De Vries RM, Sink KC (1989) Evaluation of *Asparagus* species for resistance to *Fusarium oxysporum* f. sp. *asparagi* and *F. moniliforme*. Hortic Sci 24(2): 365–368

Thompson AE, Hepler PR (1956) A summary of resistance and suceptibility to *Puccinia asparagi* DC. within the genus *Asparagus*. Plant Dis Rep 40(2): 133–137

Tu CC, Cheng YH, Cheng AH, Chen CS (1988) Effect of environment on the outbreak of stem blight of asparagus and its control. In: Proc 4th Symp Asparagus Research in Taiwan. Taiwan District Agric Improvement Station, Republic of China, pp 285–298

Uragami A (1991) Cryopreservation of asparagus (*Asparagus officinalis* L.) cultured in vitro. Res Bull Hokkaido Natl Agric Exp Stn 156: 1–37

Uragami A, Sakai A, Nagai M, Takahashi T (1989) Survival of cultured cells and somatic embryos of *Asparagus officinalis* cryopreserved by vitrification. Plant Cell Rep 8: 418–421

Uragami A, Sakai A, Nagai M (1990) Cryopreservation of dried axillary buds from plantlets of *Asparagus officinalis* L. grown in vitro. Plant Cell Rep 9: 328–331

Uragami A, Lucas MO, Ralambosoa J, Renard M, Dereuddre J (1993) Cryopreservation of microspore embryos of oilseed rape (*Brassica napus* L.) by dehydration in air with or without alginate encapsulation. Cryo Lett 14: 83–90

Withers LA (1985) Cryopreservation of cultured cells and meristems. In: Vasil I (ed) Cell culture and somatic cell genetics of plants, vol 2. Academic Press, New York, pp 253–316

Yang HJ, Clore WJ (1973) Rapid vegetative propagation of asparagus through lateral bud culture. HortScience 8: 141–143

V.3 Cryopreservation of Germplasm of Sugarbeet (*Beta vulgaris* L.)

A.M. BRAUN[1]

1 Introduction

Sugarbeet [(*Beta vulgaris* L. var. Saccharifera (altissima)] is an important sugar crop that developed about 200 years ago. Although the genus *Beta* was known as a vegetable with high variability since the 5th century B.C., just when or where the sugarbeet developed is not known for certain (Oltmann et al. 1984). It is believed that the sugarbeet originated from fodder beet, an important forage crop in central Germany since the middle of the 18th century. Atanassov (1986) quotes Zosimovitsch when he says that sugarbeet is a result of natural interspecific hybridization between root beet of Anatolian origin and leaf beet from Western Europe.

Until World War II, only multigerm sugarbeet seed, an aggregate of two to six flowers fused together, was known. The multigerm had a major disadvantage in that rows of growing plants were difficult to thin. Cross-cultivation to planting orientation was often done to thin the beets. A source of monogerm seed (one true seed per seedball) was discovered in western United States shortly after World War II. From this plant, all present-day commercial sugarbeets have their origin.

The sugarbeet is a diploid biennial containing $2n = 18$ chromosomes. With colchicine treatment, the chromosome number can be doubled resulting in tetraploids. Tetraploids serve as pollinators in crosses with diploid male sterile monogerm sugarbeets to produce triploid hybrids with 27 chromosomes. Seed is harvested only from the male sterile (producing little or no pollen) monogerm diploid. Hardy beets of superior quality are produced from the seed obtained by this method.

Although triploids are usually seed-sterile and generally undesirable for plants where seeds are of commercial importance, sugarbeet is one of several economically important plants whose triploids are in commercial use.

1.1 Sugarbeet Distribution and its Important Species

The genus *Beta* comprises 15 species, which are grouped in four sections, namely: Vulgares, Patellares, Corollinae, and Nanae (Saunders et al. 1990). Besides sugarbeet, the Vulgares section is composed of the species *B. maritima* L.,

[1]American Crystal Sugar Company, Research Center, P.O. Box 1227, Moorhead, MN, 56561–1227 USA

Biotechnology in Agriculture and Forestry, Vol. 32
Cryopreservation of Plant Germplasm I (ed. by Y.P.S. Bajaj)

B. macrocarpa Guss., *B. patula* Ait., and *B. atriplicifolia* Rouy. Sugarbeet hybridizes readily with the four other Vulgare species, all of which are diploid with 2n = 18 chromosomes.

The sugarbeet is now grown over the entire world with most of the more than 9 million ha lying in the temperate zone of Europe. According to the FAO (1992), Europe in 1991 had more than one third of the world's area (3 215 000 ha) occupied by sugar beet. The former USSR had approximately the same area (3 160 000 ha) in sugarbeet production.

Sugarbeet, second only to sugar cane as a source of sugar, is an important industrial crop. Sugar is obtained from the large, tan-colored taproot which grows to a depth of 2 to 7 feet and weighs from $1\frac{1}{2}$ to 3 pounds. The upper part of the root is a whitish pulp which contains between 13 and 22% sucrose. Both the large water-storing cells and the small sugar-containing cells are arranged in what are called cambium rings, similar in appearance to the annual rings in a tree (Carlson 1988).

The crown of the beet, which is very short, sends out bunches of rich green, curly leaves about 14 inches long. Both the leaves and crown can be fed to cattle and other animals, as can the pulp and molasses that remain after processing the sugar. The molasses byproduct is also used as a substrate for commercial yeast production. Alcohol is another product of the sugarbeet.

1.2 Germplasm Storage Methods

Germplasm storage in the form of seed is economical and efficient but seed, unlike clones obtained from vegetative tissue, do not produce identical genotypes of the parent plants. Attempts at the in vitro storage of plant material have focused on fostering reduced growth by imposing some sort of stress on cultures in order to achieve reduced growth. Growth limitation can be achieved by lowering either temperatures or moisture, through use of osmoticums, and use of chemicals. Since the germplasm remains viable, it can be grown into plants in case of immediate demand.

Miedema (1982) incubated flower buds from sugarbeet on SF (shoot formation) medium for 1 day before placing them in cold storage at 5 °C. After 6 months, nearly all shoots had developed and were ready to be rooted. The rooted shoots obtained by this method were kept alive for 1 year at 5–10 °C.

Braun (1988) succeeded in growing sugarbeet plants from cultures stored at 4 °C under low-intensity incandescent light for up to 18 months. During the time in storage, the cultures lost all but a small fraction of photosynthetic green coloration. They appeared black and nearly dead. However, after trimming the cultures of dead material and placing them on fresh medium, growth resumed, and the plants grew normally into mature plants.

Disadvantages of stress-imposed growth reduction as a method for plant storage include (1) relatively short-term storage, (2) need for much storage space, and (3) the possibility of contamination.

Cultures can also be maintained for years as clones by transferring to new medium every 4–6 weeks. Besides being very time-consuming and requiring much storage space, regular transfer increases the risk of contamination. There is also the possibility of introducing unacceptable levels of genetic instability.

1.3 Sugarbeet Seed Storage

Most sugarbeet seed in the United States is grown in California and Oregon. The seed is shipped to eastern and midwestern sugar-producing areas for processing and planting. During transportation and storage, the moisture content of the seed rises, often to above 12% because of changes in temperature and humidity. The increased moisture makes decortication difficult, and is detrimental to seed viability.

Hall (1956) determined that sugarbeet seeds will go to a moisture content above 12% (wet basis) when the temperature goes below 25 °C and the relative humidity above 50%. To aid in solving the problem of excess moisture, Hall studied the effect on germination of sugarbeet seeds dried with heated air. He found that oven temperatures of 43 °C for 2 h did not impair 3- and 7-day germination tests, but that higher temperatures or longer drying time reduced germination.

To test the viability of sugarbeet in cold storage with time, Pack and Owen (1950) in 1928 placed seed samples of a German variety of sugarbeet in a commercial refrigeration cold-storage house at Salt Lake City, Utah. For the next 22 years, some of the seed was germinated each year and the results recorded. Temperatures during this time ranged from −12 to −23 °C. Percentage germination gradually fell from 83.5% in 1928 to 75.0% in 1950. Other varieties tested had an even higher rate of viability loss with time.

Seed from American Crystal Sugar Company's breeding stock in Moorhead, Minnesota, is stored in a temperature-controlled room. (20% relative humidity, temperature 50 °C).

2 Cryopreservation of Sugarbeet

Self-incompatible plants, if their individual genotypes are to be preserved without modifications, cannot be preserved as seed which are the result of outcrosses. Their germplasm needs to be stored as tissue. At ultralow temperatures, such as that of liquid nitrogen (−196 °C) metabolism is effectively stopped and the tissue preserved without deterioration. Cryopreservation is the only means of ensuring the availability of germplasm for years under conditions of genetic stability (see Bajaj 1991).

Cryopreservation methods for transformed root cultures of *Beta vulgaris* and *Nicotiana rustica* were developed by Benson and Hamill in 1991. Although both *Beta vulgaris* and *N. rustica* were viable after freezing, the two species exhibited different culture characteristics when stressed. The effects of pregrowth, age of culture, freezing rate, and postfreeze culture conditions were determined for both species.

Regeneration of sugarbeet plants from meristems using the slow freeze method are discussed in detail in this chapter.

2.1 Aseptic Preparation of Meristems for Culture

Inflorescences with closed flowers were taken from mature sugarbeet plants (Fig. 1). These inflorescences were cut into approximately 1-cm sections, and the small leaves protruding from between the closed flowers trimmed with a razor blade. The trimmed inflorescence pieces were surface sterilized for 1 min in 50-ml Erlenmeyer flasks (approximately 25–30 inflorescences/flask) containing 60% alcohol, and placed on a shaker for 1 min. This was followed by draining the alcohol from the flasks, replacing the alcohol with a 2.6% hypochlorite solution, and shaking for an additional 15 min. Then the hypochlorite solution was drained and the inflorescences rinsed four times with sterile water, shaking the flask manually for approximately 1 min for each rinse. The inflorescence pieces were then placed on mMS medium (Murashige and Skoog 1962) pH 5.8 containing 19.4 μM (4.4 mg/l) BA, 6 μM (3 mg/l) TIBA, 2% sucrose, and solidified with Difco Bacto agar. Four inflorescences were placed on each 60 × 15-mm plastic Petri dish. The cultures were kept at 22 °C under continuous light (120–130 lx) until shoots developed on the floral pieces, approximately 4 to 6 weeks.

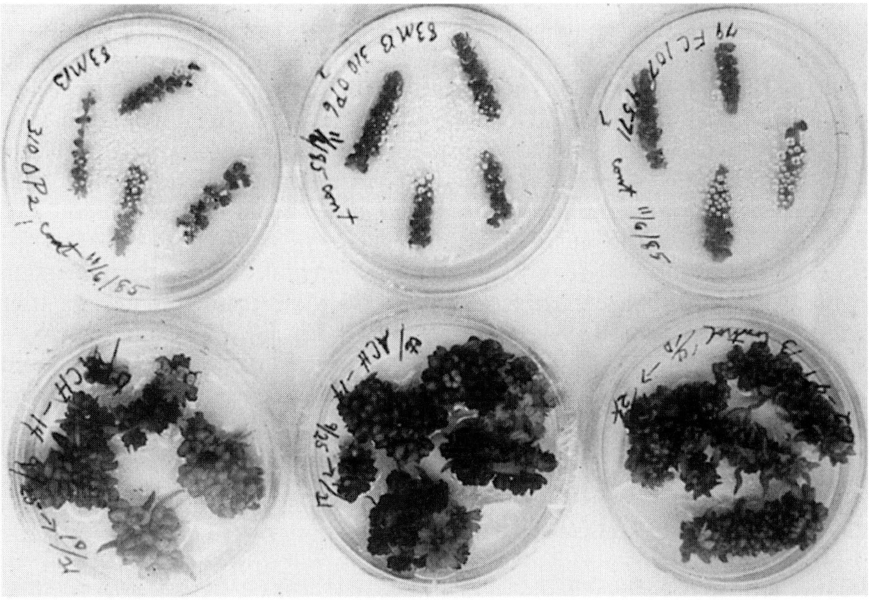

Fig. 1. Shoot development on inflorescence pieces of *Beta vulgaris* L. cultured on modified MS medium containing 19.4 μM BAP and 6 μM TIBA. Three cultures (*top row*) show newly plated inflorescence sections. Three cultures (*bottom row*) show inflorescence sections when ready for dissection, approximately 6 weeks after being placed in culture. Petri dishes are 60 × 15 mm

2.2 Meristem Technique

The meristem, either apical or axillary, is a dome of actively dividing cells protected by developing leaves and scales, and is about 0.1 mm in diameter. Larger pieces of meristematic pieces, between 0.1 and 1 mm in diameter, are generally referred to as shoot tips. While they have a better chance of developing than meristems, they are less likely to be free from virus.

Dissections of sugarbeet meristems were done in a laminar flow hood under aseptic conditions. A dissecting microscope and light were required. The meristems were dissected from shoots that developed from the base of each flower after the inflorescence was placed on culture. Under the microscope, the forceps and needle, the latter mounted in a suitable holder, were used to remove the primordia surrounding the meristem. A mounted and alcoholed safety razor blade piece (see diagram in Quak 1977 p. 602) was used to cut and remove the meristem. The meristems were placed in several concentric circles in a Petri dish by dipping the blade slightly into the medium. The meristems were precultured for 2 days on mMS medium containing DMSO.

2.3 Freeze Preservation

Methods for cryopreservation of meristems were adapted from Kartha (1985). After a 2-day preculture period, the meristems were transferred to liquid DMSO-free MS medium and sorbitol was added in ten equal increments over a period of 30 min to a final concentration of 0.75 M. After chilling this solution in an ice bath for 2 h, DMSO was added in ten equal increments over the period of 30 min to a final concentration of 10% DMSO (v/v). One ml of this freezing solution with approximately 20 meristems was put into each of several 2-ml polypropylene cryogenic vials and allowed to equilibrate in a salt-ice mixture bath. After approximately 20 min, the vials were placed into a liquid nitrogen freezer and frozen to $-40\,^{\circ}C$ at the rate of $0.8\,^{\circ}C/min$.

The frozen vials were removed from the freezer and stored in the immersion phase of a liquid nitrogen refrigerator. Samples were returned to culture after various lengths of time in LN to test viability. Upon removal from LN, the vials were quick-thawed in a $39\,^{\circ}C$ water bath for approximately 90 s. The thawed meristems were plated on mMS medium (see modified medium in Sect. 2.1) and placed in culture (22–24 °C 16 h day) for shoot development.

2.4 Propagation of Plantlets and Transfer to Soil

Shoots grown from meristems were transferred to vials containing medium developed at Maribo Breeding Station (Remick, unpubl.) to promote development and multiplication of shoots. After 3 to 4 weeks, when shoots had multiplied, the shoots were separated and rooted in half-strength mMS medium containing 16 μm (3 mg/l) NAA until ready for transplanting to soil.

Plantlets cultured in vitro require hardening and good root development in order to do well when transplanted to soil (Jensen 1977). Therefore cultures ready for transfer to soil should have a good shoot base on which roots can grow before being transferred to autoclaved jars containing a 50:50 mixture of sterile vermiculite and perlite. After transfer, the culture-containing jars were placed in a New Brunswick incubator (22 °C 18 h light, 6 h dark) until roots developed (about 2 weeks). The plants were watered, as necessary, with half-strength liquid MS medium containing 16 μm (3 mg/l) of the root-developing auxin, NAA.

At the end of this time, the plantlets were transplanted to soil-containing clay pots and placed in the greenhouse. The potted plants were covered with plastic for several days to prevent desiccation. The plantlets were checked twice daily and watered only as needed.

2.5 Results

Meristems of several varieties of *Beta vulgaris* L. survived the slow-freezing process when returned to culture after having been quick-thawed at 39 °C. Recovery growth resumed anywhere from 2 to 8 weeks after removal of the meristems from LN and placement on mMS medium at 22 °C under continuous light (120–130 lx). Rate of growth varied. Meristems beginning growth within 2 to 4 weeks after being thawed closely resembled the normal growth rate of unfrozen tissue. Phenotypically, these plantlets did not differ from controls. Chromosome counts at the time of transferring potted plants to the greenhouse showed the normal 2n number of chromosomes. Delayed initial recovery growth was accompanied by a significantly slowed rate of growth. Some of these plants did not survive.

Sugarbeet meristems began recovery growth in the presence of a mixture of DMSO and sorbitol in the following concentrations: sorbitol 0.25, 0.50, 0.70, and 0.75 M, DMSO 5, 7.5, and 10%. Cooling velocities ranged from 0.5 to 1 °C/min. Postthaw growth of meristems from individual vials varied from no growth to as high as 71% when compared to nonfrozen controls. The average highest recovery was 58.4% obtained at a slow freeze rate of 0.8 °C/min at a sorbitol concentration of 0.75 M (Table 1).

Green outgrowths on apparently dead tissue were followed by the development of small green shoots. Occasionally, growth appeared as a purplish red color, which gradually turned into small green shoots. The areas of growth, as noted visibly, appeared to arise at one or more points on the surface of the tissue mass (Fig. 2).

Sugarbeet meristems have shown regrowth along with differentiation upon return to culture medium after being in LN for up to 2 months. Because metabolic activities of cells are likely to be totally arrested at the temperature of liquid nitrogen, it is reasonable to postulate that these meristems could be stored indefinitely in liquid nitrogen. Rapid freezing of sugarbeet meristem resulted in no recovery growth.

Table 1. Rate of freezing and cryoprotectant concentrations that resulted in recovery growth of frozen *Beta vulgaris* meristems

Variety[a]	Sample	% Recovery[b]	Treatment		
			Rate of freezing (°/min)	DMSO concentration %	Sorbitol concentration (M)
A	1	58.4	0.8	10.0	0.75
	2	34.1	0.5	10.0	0.25
	3	11.4	1.0	10.0	0.25
B	1	11.0	0.8	5.0	0.70
	2	5.6	0.8	10.0	0.50
	3	3.0	1.0	7.5	0.50
	4	2.3	0.8	10.0	0.75
C	1	7.3	0.5	5.0	0.75
D	1	17.5	0.6	10.0	0.70
E	1	45.0	0.8	10.0	0.50

[a]Varieties are indicated by letter as a means of convenience.
[b]Each recovery value is the average of 4 replicates of 20 meristems, each expressed as a percentage relative to growth of control meristems which were treated similarly but subjected to no freezing. (Braun 1988)

2.6 Discussion

Fluctuating survival rates appear to be attributable to physiological variation among samples to a greater degree than to the inability of the meristems to withstand freezing (Kartha 1985). Some explants may be at an optimal stage of meristematic division at the time of dissection, while others may be at less than optimal stage. Actively growing cells with low freezable water are more likely to withstand freezing injury than older cultures with large thick-walled cells and little cytoplasm (Bajaj and Reinert 1977).

Varietal differences, which may be very pronounced between inbred and hybrid lines (Jarl and Bornman 1986), are a factor influencing viability rates. The more vigorous hybrids tend to show higher recovery rates than inbred lines. There also appears to be a seasonal effect, with higher recovery rates seen during late fall and early autumn.

A beneficial cryoprotectant effect is often attained by using a mixture of cryoprotective compounds (Finkle and Ulrich 1979; Bajaj 1981; Sakai and Sugawara 1984; Steponkus 1985; Withers 1985a). Chen et al. (1984) found that a combination of DMSO and sorbitol broadened the range of temperature from initiation to completion of ice crystallization. This group believes that during slow freezing, the extracellular medium freezes at a higher temperature and that the intracellular water begins freezing at lower temperatures so that there is a continuous flow of water out of the supercooled cytoplasm to the surrounding ice due to a vapor-pressure deficit. Thus the partially dehydrated cells are able to survive the LN temperature.

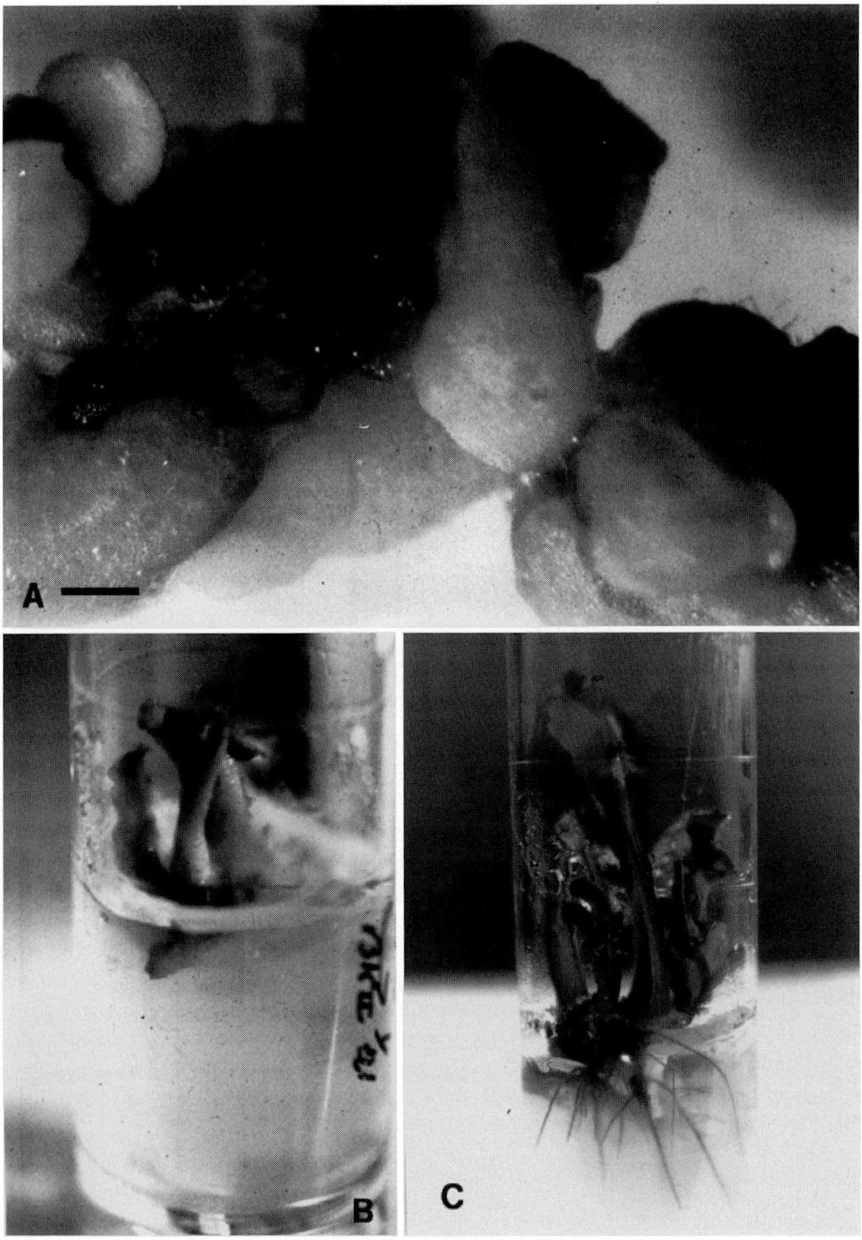

Fig. 2A–C. Stages in the recovery of plants from meristems frozen in liquid nitrogen. **A** Recovery growth of a meristem frozen to −40 °C at a rate of 0.8 °/min followed by immersion in LN for 1 h and subsequently thawed in a water bath at 39 °C. *Bar* 1 mm. **B** Shoot regeneration on medium containing 19.4 μM (w/v) BAP supplemented with 6 μM (w/v) NAA. **C** Plantlet showing rootlet development on medium containing 16 μM (w/v) NAA. (Braun 1988)

Shoot development from freshly dissected meristems was inhibited when placed on medium containing NAA. An acetamide and DMSO cryoprotectant added to ice-cooled sugarbeet meristems over a 30-min interval seemed to have the potential of reducing toxicity to the meristems. However, more experimentation with this combination is needed. Washing of cells upon thawing and before plating was not necessary for the sugarbeet meristem.

Preculture of meristems on 5% DMSO-containing MS medium for 2 days before placing them in a freezing solution of DMSO and sorbitol decreased the freezing sensitivity of the meristem (Braun 1988). Benson and Hamill (1991), on the other hand, obtained only poor recovery of sugarbeet root tips after pregrowth.

Beginning growth of thawed sugarbeet meristem on what appeared to be dead tissue may have resulted from nonuniform cooling of the cells in the interior of the meristem relative to cells on the outside of the meristem at the time of freezing. Withers (1985b) also reported massive damage to the cryopreserved shoot apex as seen by electron-microscopical examination. Seemingly, cells had survived only at the points of growth, and it was these cells that were actively dividing. Renewed growth led to differentiation and whole plantlet regeneration.

3 Summary and Conclusions

The successful regeneration of viable sugarbeet plants upon return to culture of meristems frozen in liquid nitrogen suggests the feasibility for long-term preservation of sugarbeet germplasm without deterioration. Cell cultures of sugarbeet, though not studied, may also be capable of being cryopreserved. Cryopreservation would allow superior varieties of sugarbeet to be maintained in a stable state, available at any time as a source for regeneration of new plant material.

References

Atanassov AI (1986) Sugarbeet (*Beta vulgaris* L.). In: Bajaj YPS (ed) Biotechnology in agriculture and forestry, vol 2. Crops I. Springer, Berlin Heidelberg New York, pp 462–470

Bajaj YPS (1981) Regeneration of plants from potato meristems freeze-preserved for 24 months. Euphytica 30: 141–145

Bajaj YPS (1991) Storage and cryopreservation of in vitro cultures. In: Bajaj YPS (ed) Biotechnology in agriculture and forestry, vol 17. High-tech and micropropagation I. Springer, Berlin Heidelberg New York, pp 361–382

Bajaj YPS, Reinert J (1977) Cryobiology of plant cell cultures and establishment of gene-banks. In: Reinert J, Bajaj YPS (eds) Applied and fundamental aspects of plant cell, tissue, and organ culture. Springer, Berlin Heidelberg New York, pp 757–777

Benson EE, Hamill JD (1991) Cryopreservation and post freeze molecular and biosynthetic stability in transformed roots of *Beta vulgaris* and *Nicotiana rustica*. Plant Cell Tissue Organ Cult 24: 163–172

Braun A (1988) Cryopreservation of sugarbeet germplasm. Plant Cell Tissue Organ Cult 14: 161–168

Carlson D (1988) So this is farming – a journal. Adventure, Cambridge, MN

Chen THH, Kartha KK, Constabel F, Gusta LV (1984) Freezing characteristics of cultured *Catharanthus roseus* (L). G. Don cells treated with dimethylsulfoxide and sorbitol in relation to cryopreservation. Plant Physiol 75: 720–725

FAO (1992) Production yearbook. FAO, Rome

Finkle BJ, Ulrich JM (1979) Effects of cryoprotectants in combination on the survival of frozen sugarcane cells. Plant Physiol 63: 598–604

Hall CW (1956) Drying temperatures and storage problems of sugar beet seeds. J Am Soc Sugar Beet Technol IX (2): 161–165

Jarl J, Bornman CH (1986) Observation on genotypic variation in *Beta vulgaris* (sugarbeet) tissues cultured in vitro. Hereditas 105: 55–59

Jensen CJ (1977) Monoploid production by chromosome elimination. In: Reinert J, Bajaj YPS (eds) Applied and fundamental aspects of plant cell, tissue, and organ culture. Springer, Berlin Heidelberg New York, pp 299–330

Kartha KK (1985) Meristem culture and germplasm preservation. In: Kartha KK (ed) Cryopreservation of plant cells and organs. CRC Press, Boca Raton, pp 115–134

Miedema P (1982) A tissue culture technique for vegetative propagation and low temperature preservation of *Beta vulgaris*. Euphytica 635–643

Murashige T, Skoog F (1962) A revised medium for rapid growth and bioassays with tobacco tissue cultures. Physiol Plant 15: 473–497

Oltmann W, Burba M, Bolz G (1984) Sugarbeet quality: importance, criteria for evaluation and breeding for its importance. Adv Plant Breed (Suppl) 12: 135–147, Parey, Berlin

Pack DA, Owen FV (1950) Viability of sugar beet seed held in cold storage for 22 years. In: Proc Am Soc Sugar Beet Technol, 6th General Meet. Beet Sugar Dev Foundation, Fort Collins, CO, pp 127–129

Quak F (1977) Meristem culture and virus-free plants. In: Reinert J, Bajaj YPS (eds) Applied and fundamental aspects of plant cell, tissue, and organ culture. Springer, Berlin Heidelberg New York, pp 598–615

Sakai A, Sugawara Y (1984) Survival of plant germplasm in liquid nitrogen. In: Li PH, Sakai A (eds) Plant cold hardiness and freezing stress, vol 1. Academic Press, New York, pp 345–359

Saunders JW, Doley WP, Theurer JC, Yu MH (1990) Somaclonal variation in sugarbeet. In: Bajaj YPS (ed) Biotechnology in agriculture and forestry, vol 11. Somaclonal variation in crop improvement I. Springer, Berlin Heidelberg New York, pp 465–490

Steponkus PL (1985) Fundamental aspects of cryoinjury as related to cryopreservation of plant cells and organs. In: Zaitlin M, Day P, Hollaender A (eds) Biotechnology in plant science. Academic Press, New York, pp 145–159

Withers LA (1985a) Cryopreservation of cultured cells and protoplasts. In: Kartha KK (ed) Cryopreservation of plant cells and organs. CRC Press, Boca Raton, pp 243–267

Withers LA (1985b) Long-term storage of in vitro cultures. In: Schafer-Menuhr A (ed) In Vitro Techniques. Nijhoff Junk, Dordrecht Brown, Lancaster, pp 137–148

V.4 Conservation of Germplasm of Strawberry (*Fragaria* Species)

B.M. REED and K.E. HUMMER[1]

1 Introduction

The genus *Fragaria* L. is a member of the Rosaceae and subfamily Rosoidea. The base chromosome number is x = 7, and the genus includes 17 generally accepted diploid, tetraploid, hexaploid, and octoploid species (Table 1). This genus is native to temperate locations in the northern hemisphere and South America. There are no endangered species at this time. *Fragaria* can be found at elevations from sea level through 4000 m. *Fragaria vesca* L., the wood strawberry, is the most extensively distributed species of this genus. Hybridization between *Fragaria* and a related genus *Potentilla* is possible, although most hybrids are sterile.

The common cultivated strawberry, *F.* x *ananassa* Duch., is grown in most of the arable regions of the world (Hancock et al. 1990). This crop is economically significant, with world production close to 2 000 000 metric tons (Scott and Hancock 1989). Consumption of this fruit has increased by 10 to 15% during the last 10 years (Hancock and Luby 1993). One of the parents of the cultivated strawberry was *F. chiloensis.* The parental *F. chiloensis* stock was brought to France from Concepcion, Chile, by Lt. Col. Amédée François Frézier, sent by King Louis XIV to study the Spanish colony. Frézier was careful to secure vigorous plants with selected individuals bearing the largest fruit. By choosing these he inadvertently selected primarily pistillate flowering forms. Only five plants survived the 6-month trip and were planted in the King's gardens. The runners of these plants were distributed throughout Europe by avid botanists, horticulturists, and nurserymen. The other parent *F. virginiana*, which was brought to England in the late 1500s from the Virginia Colonies, probably by Sir Francis Drake (Wilhelm and Sagen 1974) and had already spread to the botanical gardens of Europe so it was available to supply pollen to the pistillate Chilean strawberry. *F.* x *ananassa* was first recognized and described by Antoine Duchesne in France in 1766. He suggested that large fruited clones of *F. chiloensis* L. from Chile, crossed with pollen from *F. virginiana*, known as the Carolina strawberry, produced the large fruited strawberry which was called the "Pineapple" strawberry or "Pine" for short (Wilhelm and Sagen 1974). Duchesne's recognition of *F.* x *ananassa* was the beginning of strawberry crop

[1]National Clonal Germplasm Repository, USDA/ARS, 33447 Peoria Road, Corvallis, OR 97333-2521, USA

Biotechnology in Agriculture and Forestry, Vol. 32
Cryopreservation of Plant Germplasm I (ed. by Y.P.S. Bajaj)

Table 1. *Fragaria* species, ploidy, and distribution. (Hummer, unpubl.; Luby et al. 1991)

Name and polidy level	Distribution
Diploid	
F. daltoniana Gay	Himalayas
F. hayatai Makino	Japan
F. iinumae Makino	Japan
F. mandschurica Staudt	China
F. nilgerrensis Schldl. ex Gay	S.E. Asia
F. nipponica Makino	Japan
F. nubicola Lindley ex Lacaita	Pakistan, Nepal
F. vesca L.	N. Hemisphere
F.v. subsp. *americana* (Porter) Staudt	N. America
F.v. subsp. *bracteata* (A.A. Heller) Staudt	Western N. America
F.v. subsp. *californica* (Cham. & Schldl.) Staudt	California
F.v. var. *crinita* (Rydb.) C. Hitchc.	N. America
F.v. subsp. *vesca*	Europe, Asia
F. viridis Duchesne	Europe, Asia
Tetraploid	
F. moupinensis (Fr.) Card.	China
F. orientalis Los.-Losinsk.	W. Asia
Hexaploid	
F. moschata Duchesne	N. Europe
Octoploid	
F. chiloensis (L.) Duchesne	N. and S. America
F.c. subsp. *pacifica* Staudt	Pacific Coast, N. America
F. iterupensis Staudt	Juril Island
F. virginiana subsp. *virginiana* Duchesne	N. America
F. v. subsp. *glauca* (S. Watson) Staudt	N. America
F.v. subsp. *platypetala* (Rydb.) Staudt	N. America
F. x *ananassa* Duchesne	
F. x. *a.* var. *cuneifolia* (Nutt. ex Howell) Staudt	Western N. America
F. x *bringhurstii* Staudt	California
F. x. *hagenbachiana* Lange ex Koch	

improvement through breeding. Since that time strawberry breeding programs have developed in many countries throughout the world as the crop has developed to significant economic production.

The United States ranks first in world strawberry production (Hancock et al. 1990). The value of the total production of fresh market and processing strawberries for the USA in 1989 was $520 million (Anonymous 1990). California produces 80% of the US crop. Japan, Poland, Italy, Russia, France, Mexico, and Spain are also significant world producers, while Israel, Chile, Brazil, Peru, and Argentina are looking to increase their strawberry production. Conservation of selected cultivated genotypes is significant for future strawberry improvement. Cultivated and wild *Fragaria* germplasm are preserved ex situ as seeds, and clonally as plants in fields, under screen, in greenhouses, as in vitro cultures, and as cryopreserved tissues (Fig. 1). The choice of preservation

Fig. 1. Storage of *Fragaria* collections at the National Clonal Germplasm Repository, Corvallis, OR. **A** Screenhouse collection. **B** In vitro cold storage. (Reed and Hummer, unpubl.)

technique depends on the purpose of the collection, available technology and expertise, and funding. In situ preservation, while providing future possibilities of conservation of *Fragaria* species in designated native locations, may not save the highly vulnerable genetic diversity on the fringe of the species ranges.

2 Germplasm Storage Methods

2.1 Seed and Pollen Storage

Seed of most *Fragaria* species are resistant to desiccation and cold and can survive for many years. Although Scott and Draper (1970) observed that the germination of several *Fragaria* seed lots stored at 1 to 4 °C for 23 years was 89% or greater, strawberry seed germination can be irregular. It may be improved by stratifying the seed for 3 months (Bringhurst and Voth 1957) or exposing the seed to light (Scott and Drapper 1967). Seed storage is excellent for preserving collections of species populations and lines from specific crosses in breeding programs. However, because outcrossing is prevalent in some species, seed does not exactly reproduce the specific parental genotypes, clonal germplasm preservation techniques must be implemented to conserve important cultivars and selections.

Fragaria germplasm can also be stored as pollen when preservation of a specific genotype is not important. A high percentage of pollen stored in closed vials at −20 °C will remain viable for at least several years (Hancock et al. 1990). Strawberry pollen or anther fragments have been cryopreserved in liquid nitrogen for 2 years with high viability (65–75%) following air drying for 24 h (Craddock 1987).

2.2 Field Gene Banks

Fragaria germplasm throughout the world is commonly preserved in field gene banks as growing plants. Field plantings are of greatest use for evaluation of descriptive traits and qualities, morphological verification of taxonomic and horticultural identity, and as a backup for other germplasm storage techniques. This simple preservation method has some disadvantages, such as periodic replanting and naturally spread viruses or viruslike diseases. Virus diseases are present wherever cultivated strawberries are grown and cause major losses in the quantity and quality of the crop (Converse 1987).

Aphids, leafhoppers, and nematodes can transmit various virus and viruslike diseases in strawberries (Converse 1987). Thus, once a pathogen-negative clone has been produced, it could be maintained either with the regular application of a systemic insecticide or without the presence of insect vectors, or the clone will become reinfected. *Fragaria* germplasm collections can be stored as potted plants under insect-proof screens to reduce the risk of virus contamination (Hummer 1991). Aphids must not be allowed to survive near the pathogen-negative strawberries. This ideal can be accomplished on a practical level in a screenhouse setting with an active integrated pest management program to observe and react to insect vectors or a chemical control program. Periodic retesting for virus is performed on stock plants stored in screenhouses. Screenhouse storage must allow for natural seasonal acclimation of the plants and the acquisition of the chilling hours needed for normal bud break. Pots may need to be protected from extreme cold as well since roots can be killed at warmer temperatures than the upper portions of the plant. Containerized strawberries must be regularly and carefully monitored to insure that runners from one clone do not become established in the pot of another clone. Identity of plants in screenhouses can be checked by chemical analyses or morphological examination in field trials (Hummer 1991). Flowers and fruit are generally removed from primary collections of strawberries under screen, to improve runnering for propagation, and to reduce the possibility of contamination from seedling genotypes.

2.3 In Vitro Storage and Cryopreservation

Micropropagation is an excellent germplasm storage technique for strawberry clones. Plantlets of most genotypes can be readily cultured in vitro and stored under refrigerated conditions for several years. Plantlets are retrieved from cold storage, recultured, and acclimatized to greenhouse conditions. Studies examining field performance of micropropagated strawberry subclones (Swartz et al. 1981: Moore et al. 1991) have documented epigenetic differences, such as increased runnering in field-grown plants following in vitro culture. Moore et al. (1991) concluded that differences observed in Olympus were probably "transient responses to the micropropagation environment, not genetic changes" and reputed clonal differences in yield were not substantiated in field trials. Swartz et al. (1981) examined tissue cultured transplants of Earliglow, Redchief, and Guardian for off-types. They suggested that changes in performance were caused by increased vigor and that off-types could be traced to an original off-type meristem from which they were produced (Swartz et al. 1981). No changes in cultivar-specific properties were found with in vitro depot plants grown on phytohormone-free medium (Jungnickel 1988). Storage techniques for in vitro *Fragaria* at the USDA-ARS National Clonal Germplasm Repository in Corvallis (NCGR-Corvallis) use minimal levels of cytokinins and auxins for culturing and no phytohormones in storage media. Off-types have not been observed in plants following in vitro storage of *Fragaria* germplasm (Reed, unpubl.).

Plant materials can be cultured and preserved as callus. The higher frequency of mutation in callus cultures growing with strong plant growth regulator concentrations makes this technique useful for inducing somaclonal variation and developing new genotypes, but callus cultures are not appropriate for preservation of specific cultivars.

Meristems can be excised from runners or from in vitro plantlets and preserved in liquid nitrogen (Sakai et al. 1978; Kartha et al. 1980). Research on cryogenic preservation of *Fragaria* meristems is ongoing at several locations; at present this technique is not used for routine long-term conservation in any world gene bank.

2.4 Status of *Fragaria* Germplasm

There are no currently endangered *Fragaria* species. Rather, locations representing distinct ecotypes or perhaps varieties are experiencing habitat changes due to human encroachment. As development of beach-front property occurs in California, many sites no longer support wild strawberries. In Chile, changes caused by forestry and farming have affected wild strawberry populations at some forest edge and coastal locations (Scott Cameron, pers. comm.).

The germplasm base for the cultivated strawberry is quite narrow. Cultivars-introduced from North American breeding programs since 1960 are derived from 53 founding clones (Sjulin and Dale 1987). A narrow germplasm base can predispose a crop to genetic vulnerability to diseases, pests, and environmental stresses. *Fragaria* is host to many pathogens and pests, including fungi, bacteria, viruses, nematodes, insects, and arthropods (Maas 1984). Some of these diseases are becoming major obstacles to maintaining or developing production levels of the cultivated strawberry. Development of resistant cultivars is a commonly used strategy to overcome production difficulties in regions where diseases, such as red stele, verticillium, or anthracnose have become established. The continued development of genetically resistant cultivars requires the availability of a broad spectrum of germplasm, both within and among species.

One of the strategies proposed by Sjulin and Dale (1987) to increase the genetic diversity in strawberry includes using germplasm from wild *Fragaria* species. Luby et al. (1991) list desired traits that could be derived from wild *Fragaria* germplasm collected during recent USDA-sponsored collection expeditions to obtain wild material from many sites throughout Chile and the Northern Rocky Mountains of the USA. *Fragaria* species have also recently been col lected from China, Japan, and Pakistan. Collections in Canada, Minnesota, Michigan, Kentucky, North Carolina, Washington, Oregon, and California have been made by public and private breeders in the USA. (Luby et al. 1991).

In 1956, George Darrow traveled through Chile to collect native Chilean strawberries for the USDA and some clones that he collected showed promise for his breeding program. He made crosses and eliminated the majority of the plant material that he collected because of lack of funds to preserve the germplasm. Because breeders were again emphasizing the need for wild material, two more USDA plant-collecting expeditions were sent to Chile in 1990 and 1992. This *Fragaria* germplasm is housed at the NCGR-Corvallis. Although only a portion of the recently collected material may be initially incorporated into the present breeding efforts throughout the world, the remaining genotypes will be preserved for future use.

3 Conservation/Cryopreservation

3.1 Review of Germplasm Conservation of Strawberry

Active strawberry research programs exist in Australia, Belgium, Canada, China, Czechoslovakia, Denmark, England, Egypt, France, Finland, Germany, Ireland, Israel, India, Italy, Japan, Lebanon, Poland, Russia, Scotland, The Netherlands, South Africa, Spain, Sweden, and the USA (Moore 1991; Maas and Galletta 1993). Most research collections are stored as plants in the field while some are preserved under screen or in vitro. Specific germplasm collections are located at gene banks in the USA, Canada, Denmark, India, and Germany (Bettencourt and Konopka 1989).

Research on in vitro storage of strawberry cultures was first reported by Mullin and Schlegel (1976) using liquid medium with filter paper bridges. Plantlets were maintained at 4 °C for up to 6 years with medium added every 3 months. Damiano (1979) reported the storage of 19 genotypes grown on solid medium and stored at 2 °C. These cultures survived for up to 27 months without transfer, depending on the genotype. Storage of mother plants used in virus-free strawberry production was successful on basal medium under weak illumination for 12–18 months (Boxus 1976). A strawberry germplasm collection at the Freidrich-Schiller Universitat has now been maintained for more than 17 years in vitro with yearly transfers on phytohormone-free medium (Jungnickel 1988). More recent studies by Reed (1991, 1992) use gas-permeable heat-sealable polyethylene bags for germplasm storage. Bettencourt and Konopka (1989) report that strawberries are stored in vitro at facilities in Belgium, Germany, Ireland, Sweden, Italy, and the United States. Many additional countries with strawberry research programs have developed in vitro collections since 1989 (Hummer, unpubl.).

Sakai et al. (1978) cryopreserved excised runner apices of Hokowase, using a cryoprotective solution of DMSO and sucrose with slow freezing to −20 or −30 °C before plunging in liquid nitrogen. Later, Kartha et al. (1980) developed a two-step method for Redcoat strawberry involving a 2-day preculture on 5% DMSO medium, addition of a cryoprotectant, and slow freezing at a controlled rate with rapid thawing. The importance of cryopreserved storage has been emphasized (Bajaj 1991); however, some authors (Withers 1986; Hancock et al. 1990; Sullivan 1991) consider cryogenic preservation as experimental because few cryopreserved plants have been analyzed for trueness to type.

3.2 Cold Storage of In Vitro *Fragaria* at the USDA National Clonal Germplasm Repository (NCGR-), Corvallis, OR

At NCGR-Corvallis, some *Fragaria* germplasm has been stored at refrigerated temperatures since 1983. Plantlets were stored in glass tubes and plastic boxes until 1989 (Gunning and Lagerstedt 1985) when gas-permeable tissue-culture bags became available. These bags provided an opportunity for contamination-free medium-term storage (Reed 1991, 1992).

3.2.1 Methodology

Plantlets were multiplied on NCGR-FRA, a modified MS (Murashige and Skoog 1962) medium pH 5.7 (Reed 1991). Cold storage was at 4 °C in the dark in a walk-in cold room. Duration of storage differed but the same storage and growth room conditions were used.

Tubes. Single plantlets were transferred to 16 × 100 mm glass tubes with 5 ml agar-solidified basal medium without growth regulators and capped with plastic caps, grown for 2 weeks, sealed with Parafilm and stored. Each accession was stored as two cultures in individual tubes.

Boxes. Sixteen plantlets of each accession were transferred to Magenta GA7 plastic boxes (6 × 6 × 9 cm) with 40 ml basal agar-based medium without growth regulators and grown for 2 weeks. Boxes were sealed with Parafilm to decrease dehydration and contamination and stored.

Bags. Plantlets were transferred to bags (CultuSAK, Becton Dickinson, Lincoln Park, NJ) which were then heat-sealed with an impulse sealer. Cultures were grown for 1 week in the growth room and 1 week under cold-hardening conditions of 8 h, 22 °C day and 16 h, −1 °C night before storage (Angelo 1939; Reed 1991). For each accession, ten plantlets were stored, each in an individual section (15 × 150 mm) of a five-section bag with 10 ml per section of basal medium without growth regulators. A firmer medium (3 g agar and 1.5 g Gelrite per liter) compared to that in boxes and tubes was used in the bags to compensate for the low level of water loss through the bag walls (Reed 1991).

Cultures were ranked for condition at 3-month intervals on a scale of 0–5: 0, dead, brown; 1, etiolated, pale tan, no green color; 2, etiolated, pale green color; 3, etiolated, retaining medium green color; 4, not etiolated, medium green color; 5, not etiolated, dark green color. Cultures with a rating of 2 or 1 are considered to be at the end of their storage life and are removed for repropagation.

3.2.2 Results

Longevity and contamination rates were compared for *Fragaria* in vitro cultures cold stored using three different storage methods. These conditions were chosen to reflect storage systems in use for plant germplasm throughout the world.

Storage in polyethylene bags with cold hardening and Gelrite as part of the medium was examined at three month intervals over a 2-year period. At each three month inventory, bags with all plants rated 2 or lower on a scale of 1–5 were removed for repropagation. Of the original 96 accessions stored in bags, 99% remained in storage after 6 months and 47% at 15 months (Fig. 2). These levels are 12 to 15% higher than those of plants stored in boxes for the same time periods. At each 3-month inventory a larger percentage of accessions stored in bags remained in good condition than did those in boxes. Five percent of the individual accessions stored in bags maintained high condition ratings for 24

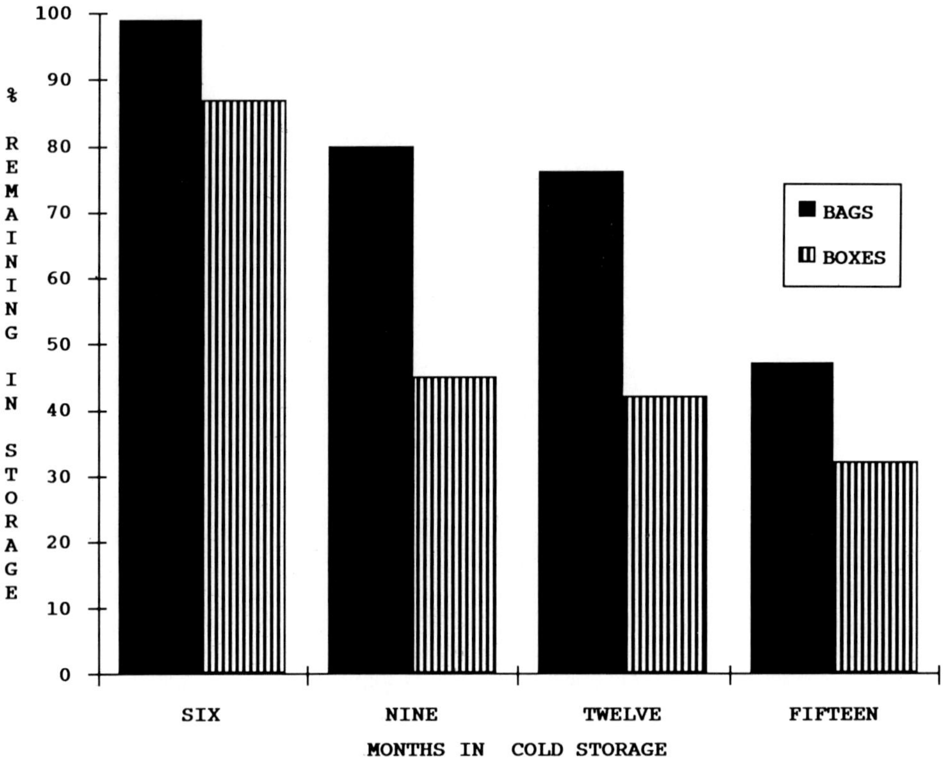

Fig. 2. Percentage of uncontaminated accessions in bags and boxes remaining in storage during a 15-month period. (Reed 1992)

months or more. No accessions were lost due to contamination. Contamination of a single chamber in a five-chamber bag occurred in 10 to 15% of the bags due to cracks or improper sealing, but did not spread to other chambers. Newer versions presently on the market do not crack in storage (Reed, unpubl.).

Cultures in boxes were stored as long as 15 months. Of the original 127 accessions stored 87% had high condition ratings at 6 months but only 32% at 15 months (Fig. 2). Contamination of cultures in boxes (30%) was primarily fungal and occurred at twice the level of that in bags, usually resulting in the loss of the accession. Contamination was not always evident until repropagation was attempted. A greater number of boxes for each accession would decrease the number lost but would greatly increase the storage space, labor, and material costs required for the collection. Contamination was a major factor in the decline of cultures stored in boxes. Losses of large numbers of plant tissue cultures due to contamination have also been reported in storage jars (Marino et al. 1985) and tubes (Nord and Hanniford 1989).

From a collection of 130 accessions stored in tubes for 24 months, 40% were in fair or good condition (rating of 2 or 3), 31% were in very poor condition (rating of 1), and 29% were dead. Contamination (47% of the tubes stored) was not always evident until the plants were repropagated, and reduced the survival rate to 20%. This study found that standard storage in tubes was not reliable due to high contamination rates thoughout the storage period.

In general, uncontaminated *Fragaria* accessions stored well for 12 to 15 months in any of these systems (Table 2). Genotype survival varied and the length of time an individual accession remained viable in storage ranged from 6 to 24 months. Individual cultivars, Aliso, Francesco, and Pocahontas, had good or very good ratings using the bag system (rating 3 or 4) after 12 months and fair ratings (rating 2) at 15 months with 100% survival. Damiano (1979) reported 90% survival for Pocahontas at 12 months, 80% for Aliso at 16 months, and 70% for Francesco at 17 months, for cultures stored in jars or large tubes. The advantages of the bag system over that of Mullin and Schlegel's liquid medium system are that the bags minimize contamination, and do not require the time and effort needed to replenish liquid medium quarterly.

The percentage of accessions remaining in storage and contamination rates varied greatly among the three storage systems. Judging by the large percentage of accessions with very low contamination rates and high average health ratings remaining in storage after 1 year, as well as the reduced need for storage space, we

Table 2. Ratings for individual *Fragaria* accessions stored in both plastic boxes (Magenta GA7) and in gas-permeable polyethylene bags (CultuSak) at 4 °C in the dark. The rating scale is 0 = dead, 5 = excellent. Mean of ten plants per treatment. (Reed 1992)

Accession	Rating	
	Form stored	
	Box	Bag
12 months' storage		
Fou Chu	4	3
F. chiloensis Darrow 72	3	3
F. virginiana	4	4
F. virginiana subsp. *glauca*	4	4
Klondike	3	3
Komsomalka	4	4
Kurume 103	3	4
MDUS 3022	3	4
Podnyaya Zagorya	4	3
Robinson	4	3
Tufts	4	4
Vantage	4	4
15 months' storage		
Fou Chu	3	2
Komsomalka	3	3
MDUS 3022	2	3
Surecrop	2	2

concluded that bags provided more secure and healthy storage for strawberry germplasm than either tubes or boxes. Bags have the additional advantages of easy handling, reduced storage space, resistance to breakage, ease of inventory, and safer shipping of germplasm.

The present storage system at NCGR-Corvallis employs the plastic bags described here, cold hardening for 1 week and evaluation at 3-month intervals to identify declining cultures. Over 350 *Fragaria* accessions are stored in bags. Additional studies of photoperiod and cytokinin effects are in progress.

3.2.3 Discussion

Researchers, breeders and certification programs have an increased interest in medium-term germplasm storage for many crop plants. The number of accessions presently in cold storage is increasing as additional specimens are collected from around the world, and cultivated genotypes are added to the germplasm core. In vitro collections will play an important part in the distribution of pathogen-negative plants throughout the world. The desire for correctly identified, pathogen-negative germplasm is increasing. Many certification programs have adopted in vitro culture as standard technique in the production of pathogen-negative plants from stock collections, and virus and viruslike pathogens do not spread between in vitro cultures. Improvements in storage containers, conditions, and media are needed to make in vitro culture less labor-intensive and more secure as a storage system for important germplasm.

3.3 Cryopreservation of *Fragaria* Germplasm at the USDA National Clonal Germplasm Repository, Corvallis, OR

The cryopreservation protocols used at the Repository are modifications of the methods of Sakai et al. (1978), Kartha et al. (1980) and Reed (1990). Although cryogenic storage of clonal germplasm is still experimental, the US National Plant Germplasm System is planning a cryogenic base collection at a site remote from the active collection.

3.3.1. Cryopreservation Methodology

Tissue-cultured plants were multiplied on modified MS medium. In vitro plants were cold-hardened for 1 week prior to excision of meristems. Dissected meristems (0.8 mm) were grown for 48 h on MS medium with 0.8% agar and 5% DMSO under cold-hardening conditions (Reed 1988). After 48 h, meristems were transferred to 0.25 ml of liquid MS medium in 1.2-ml plastic cryotubes on ice and the cryoprotectant PGD [a mixture of 10% each of polyethylene glycol (MW 8000), glucose, and DMSO in medium (Finkle and Ulrich 1979)] was added dropwise to 1.2 ml over 30 min. A 30-min equilibration period at −1 °C was followed by removal of excess cryoprotectant.

Samples were frozen in 1 ml of cryoprotectant at 0.8 °C/min to −40 °C in a programmable controlled-temperature cooling chamber and then immersed in LN for 1 hour. Vials were thawed in 40 °C water for 1 min then transferred to 23 °C water. The cryoprotectant was drained from the vials and replaced with liquid medium. Meristems were drained on sterile filter paper and plated on growth medium.

3.3.2 Results

Fifty-six *Fragaria* accessions were screened using cold hardening for 1 week, slow freezing, and PGD as the cryoprotectant (Table 3). The wide range of germplasm

Table 3. Cryopreservation screening of 56 *Fragaria* accessions using cold hardening for 1 week, slow freezing, and PGD as the cryoprotectant. Freezing rate was 0.8 °C/min to −40 °C followed by storage in liquid nitrogen for 1 h. (Reed, unpubl.)

Accession	Percent regrowth	General survival rating[a]
Albritton	30	+
Allstar	78	+++
Atlas	14	+
Beaver	60	++
Berkeley	30	+
Blakemore	0	−
Bounty	88	+++
Capron (117) *F. moschata*	0	−
Capron (210) *F. moschata*	22	+
Cardinal	30	+
Catskill	70	++
Cavalier	100	+++
Comet	0	−
Del Norte	0	−
Donner	45	++
Dunlap	90	+++
Earliglow	30	+
Empire	17	+
F. moschata	20	+
F. chiloensis Darrow 5	38	+
F. chiloensis Yaquina B	88	+++
F. chiloensis RCP-19	38	+
F. chiloensis D6	33	+
Fletcher	25	+
Florida Belle	60	++
Florida Ninety	22	+
Fresno	80	+++
Garnet	0	−
Gorella	13	+
Holiday	40	+

(*Contd.*)

Table 3. (*Contd.*)

Accession	Percent regrowth	General survival rating[a]
Jerseybelle	30	+
Kaiser's Samling	82	+ + +
Kurume	70	+ +
Kurume 103	80	+ + +
Marlate	0	–
Massey	91	+ + +
Orland	40	+
Pai Yu	0	–
Pocahontas	67	+ +
Podnyaya Zagorya	92	+ + +
Primella	25	+
Profumata di Tortona	10	+
Red Gauntlet	33	+
Redchief	70	+ +
Redcoat	70	+ +
Redstar	20	+
Robinson	80	+ + +
Scott	56	+ +
Sierra	30	+
Tennessee Beauty	11	+
Titan	60	+ +
Tribute	0	–
Tufts (70)	20	+
Tufts (231)	20	+
Veestar	60	+ +
White Pine	56	+ +

[a] – No survival.
+ = 1–40%.
+ + = 41–70%.
+ + + = 71–100%.

tested showed an equally wide response to cryopreservation. A frequency distribution of the genotypes by percent survival shows that 25 genotypes had greater than 50% survival from slow freezing (Fig. 3). Survival rates may also be dependent on genotypic variation and can be improved by varying the freezing speed (Table 4). To be useful for germplasm storage, any new method must produce high survival rates in diverse genotypes. Newer techniques may also prove effective for cryopreservation of strawberry germplasm (Table 5). Initial tests of vitrification methods (Yamada et al. 1991) with several *Fragaria* accessions produced varied results. Additional modifications to the currently published techniques may increase the survival of vitrified tissues. The addition of a gradual dehydration with PGD to the vitrification technique of Yamada et al. (1991) improved the survival of two *Fragaria* accessions. The alginate bead encapsulation-dehydration technique (Dereuddre et al. 1990) has also shown some survival in preliminary trials (Table 5).

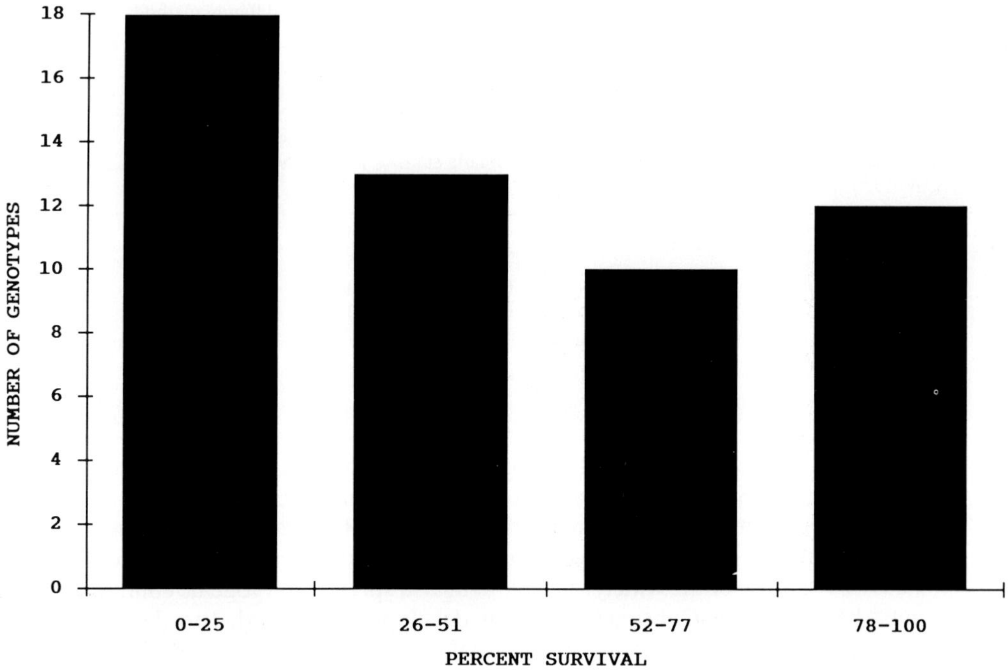

Fig. 3. Frequency distribution of survival rates of 53 strawberry genotypes following cryopreservation using cold hardening, PGD as the cryoprotectant, and slow freezing (Reed, unpubl.)

Table 4. Effect of freezing rate on the percent survival of diverse genotypes of cryopreserved *Fragaria*. (Reed, unpubl.)

Genotype	Percent survival		
	Freezing rate		
	0.3 °C/min	0.5 °C/min	0.8 °C/min
F. chiloensis Darrow 72	25	10	0
F. chiloensis CA 1466	4	7	2
F. virginiana UC 10	64	58	45
LaPush Beach	60	26	–
Vermilion	50	5	–

–, Not tested

3.3.3 Discussion

Little published information is available about actual long-term storage of cryopreserved *Fragaria* germplasm (Kartha 1985). Several *Fragaria* genotypes are stored for experimental purposes at NCGR-Corvallis, but a cryogenic base collection has not yet been established. Once research procedures become

Table 5. Survival of *Fragaria* accessions from cryopreservation using four different methods. (Reed, unpubl.)

Genotype	Percent survival			
	PGD→PVS2 Reed (unpubl.)	PVS2 Yamada et al. (1991)	Slow Freeze Reed (1990)	Alginate Dereuddre et al. (1990)
F. virginiana UC 11	60	9	35	–
Guardian	–	17	25	0
La Sans Rivale	23	0	0	–
Profusion	–	0	0	8

–, Not tested

routine, this base will reside as a remote back-up collection in cryogenic storage at the National Seed Storage Laboratory in Fort Collins, Colorado.

Two problems remain before routine cryogenic storage can be implemented and accepted as a standard technique for strawberries. First, the differential survival response of the widely variable *Fragaria* species to cryogenic conditions must be addressed. Continued methods development is needed to determine techniques applicable to both tender and cold-hardy genotypes. Second, a protocol for identity confirmation should be established because an insufficient amount of regenerants obtained from cryogenically treated buds or shoot tips have been analyzed (Hancock et al. 1990; Sullivan 1991). Field plantings of 20 plants each of 15 strawberry cultivars which had undergone cryopreservation and micropropagation procedures at the NCGR-Corvallis displayed no off-types (Reed, unpubl.), and plants of other species regenerated following cryopreservation have been true to type (Bajaj 1985; Towill 1988). The authors have not observed exposure to cryogenic conditions to cause an increase in mutation rate over that normally experienced in a field collection.

4 Summary and Conclusions

Field gene banks are important for the evaluation of horticultural traits; screenhouses provide working and certified collections available for immediate use and direct distribution; in vitro culture in refrigerated conditions provides medium-term (several years') storage for back-up collections and increasingly provides material for plant distribution. Long-term (more than 10 years) storage of *Fragaria* species can be accomplished through seed storage; however, cryopreservation of meristems will be the key for efficient long-term preservation of clonal base collections. Protocols for clonal preservation of *Fragaria* germplasm as plants in field, screenhouse, and in vitro culture are routine in gene banks throughout the world with identity verification as a critical component in each of these systems, though only a few locations are presently researching cryogenic techniques for the preservation of *Fragaria* germplasm.

References

Angelo E (1939) The development of cold resistance in strawberry varieties. Minn Agric Exp Stn Tech Bul 135: 7–19

Anonymous (1990) Agricultural Statistics. US Dep Agric, Gov Printing Office, Washington, DC

Bajaj YPS (1985) Cryopreservation of germplasm of potato (*Solanum tuberosum* L.) and cassava (*Manihot esculenta* Crantz): Viability of excised meristems cryopreserved up to 4 years. Indian J Exp Biol 23: 285–287

Bajaj YPS (1991) Storage and cryopreservation of in vitro cultures. In: Bajaj YPS (ed) Biotechnology in agriculture and forestry, vol 17. High-tech and micropropagation I. Springer, Berlin Heidelberg New York, pp 361–381

Bettencourt EJ, Konopka J (eds) (1989) Directory of germplasm collections 6.II Temperate fruits and tree nuts. Int Board Plant Genet Resour, Rome

Boxus P (1976) The production of virus-free strawberry by in vitro culture. Acta Hortic 66: 35–38

Bringhurst R, Voth V (1957) Effect of stratification on strawberry seed germination. Proc Am Soc Hortic Sci 92: 369–375

Converse R (1987) Detection and elimination of virus and viruslike diseases in strawberry. In: Converse R (ed) Virus diseases of small fruits. USDA Agric Handbook No 631: 1–100

Craddock WJH (1987) Cryopreservation of pollen. MS Thesis, Oregon State University, 77 pp

Damiano C (1979) Cold storage of in vitro strawberry cultures and the resumption of multiplication. Ann Ist Sper Fruttic 10: 53–58

Dereuddre J, Scottez C, Arnaud Y, Duron M (1990) Resistance d'apex de vitroplants de poirier (*Pyrus communis* L. cv. Beurré Hardy) enrobés dans l'alginate, à une deshydration puis à une congélation dans l'azote liquide: effet d'un endurcissement préalable au froid. CR Acad Sci Paris 310: 317–323

Finkle BJ, Ulrich JM (1979) Effects of cryoprotectants in combination on the survival of frozen sugarcane cells. Plant Physiol 63: 598–604

Gunning J, Lagerstedt HB (1985) Long-term storage techniques for in vitro plant germplasm. Proc Int Plant Prop Soc 35: 199–205

Hancock JF, Maas JL, Shanks CH, Breen PJ, Luby JJ (1990) Strawberries (*Fragaria*) In: Moore JN, Ballington JR Jr (eds) Genetic resources of temperate fruit and nut crops II. Acta Hortic 290: 489–546

Hancock JF, Luby JJ (1993) Genetic resources at our doorstep: the wild strawberries. Science 43(3): 141–147

Hummer K (1991) *Fragaria* at the National Clonal Germplasm Repository at Corvallis, Oregon. In: Dale A, Luby JJ (eds) The strawberry into the 21st century. Timber Press, Portland, OR, pp 106–107

Jungnickel F (1988) Strawberries (*Fragaria* spp. and hybrids). In: Bajaj YPS (ed) Biotechnology in agriculture and forestry, vol 6. Crops II. Springer, Berlin Heidelberg New York, pp 38–103

Kartha KK (1985) Cryopreservation of plant cells and organs. Newsl Int Assoc Plant Tissue Cult 45: 2–15

Kartha KK, Leung NL, Pahl K (1980) Cryopreservation of strawberry meristems and mass propagation of plantlets. J Am Soc Hortic Sci 105(4): 481–484

Luby J, Hancock J, Cameron JS (1991) Expansion of the strawberry germplasm base in North America. In: Dale A, Luby JJ (eds) The strawberry into the 21st century. Timber Press, Portland, Oregon, pp 66–75

Maas J (1984) Compendium of strawberry diseases. Am Phytopathol Soc, St. Paul, MN

Mass J, Galletta G (1993) eds 2nd Int Strawberry Symp. Acta Hortic 348: 1–520

Marino G, Rosati P, Sagrati F (1985) Storage of in vitro cultures of *Prunus* rootstocks. Plant Cell Tissue Organ Cult 5: 73–78

Moore J (1991) The strawberry into the 21st century: a symposium overview. In: Dale A, Luby JJ (eds) The strawberry into the 21st century. Timber Press, Portland, Oregon

Moore P, Robbins J, Sjulin T (1991) Field performance of olympus strawberry subclones. Hortsci 26(2): 192–194

Mullin RH, Schlegel DE (1976) Cold storage maintenance of strawberry meristem plantlets. HortScience 11(2): 100–101

Murashige T, Skoog F (1962) A revised medium for rapid growth and bioassays with tobacco tissue cultures. Physiol Plant 15: 473–497

Nord SM, Hanniford GG (1989) In vitro storage of *Pelargonium* x *domesticum.* Am Soc Hortic Sci Annu Meet Program and Abstr 158 (Abstr): 78

Reed BM (1988) Cold acclimation as a method to improve survival of cryopreserved *Rubus* meristems. Cryo Lett 9: 166–171

Reed BM (1990) Survival of in vitro-grown apical meristems of *Pyrus* following cryopreservation. HortScience 25: 111–113

Reed BM (1991) Application of gas-permeable bags for in vitro cold storage. Plant Cell Rep 10: 431–434

Reed BM (1992) Cold storage of strawberries in vitro: a comparison of three storage systems. Fruit Var J 46(2): 98–102

Sakai A, Yamakawa M, Sakata D, Harada T, Yakuwa T (1978) Development of a whole plant from an excised strawberry runner apex frozen to −196 °C. Low Temp Sci Ser B 36: 31–38

Scott D, Draper A (1967) Light in relation to seed germination of blueberries, strawberries and *Rubus.* HortScience 2(3): 107–108

Scott D, Draper A (1970) A further note on longevity of strawberry seed in cold storage. HortScience 5(5): 439

Scott D, Hancock JF (1989) Strawberry cultivars and worldwide patterns of strawberry production. Fruit Var J 42: 102–108

Sjulin T, Dale A (1987) Genetic diversity of North American strawberry cultivars. J Am Soc Hortic Sci 112: 375–386

Sullivan JA (1991) Improvement of strawberry using tissue culture techniques. In: Dale A, Luby JJ (ed) The strawberry into the 21st century, Timber Press, Portland, OR pp 84–90

Swartz H, Galletta G, Zimmerman R (1981) Field performance and phenotypic stability of tissue culture-propagated strawberries. J Am Soc Hortic Sci 106(5): 667–673

Towill L (1988) Survival of shoot tips from mint species after short-term exposure to cryogenic conditions. HortScience 23: 839–841

Wilhelm S, Sagen J (1974) A history of the strawberry. Univ California, Berkeley

Withers LA (1986) In vitro approaches to the conservation of plant genetic resources. In: Withers LA, Alderson FG (eds) Plant tissue culture and its agricultural applications. Butterworths, London, pp 261–276

Yamada T, Sakai A, Matsumura T, Higuchi S (1991) Cryopreservation of apical meristems of white clover (*Trifolium repens* L.) by vitrification. Plant Sci 78: 81–87

V.5 Cryopreservation of Germplasm of Tomato

B.W.W. Grout and P.C. Crisp[1]

1 General Account

1.1 Production and Origins

Produced as an annual vegetable the tomato can be considered as much a field as a horticultural crop under current growing practices. The edible fruit is being used worldwide as both a fresh vegetable and, in the processing industries, for juices, sauces and chutneys, soups, canned tomatoes, concentrates and powders. The fruit provides both colour and flavour to the diet and is a valuable source of vitamin C, which may exceed 25 mg per 100 g (Langer and Hill 1982). The domestic tomato has forms adapted to a wide range of soils and climates and its cultural range extends from tropical to cold temperate regions. The importance of the tomato crop can be seen from agricultural statistics for 1990 (FAO 1991) which indicate that a global production area of some 3 million ha produced more than 70 million tonnes of fruit (Table 1).

Currently, the edible tomato is known systematically as *Lycopersicon esculentum* Miller., having previously been described as both *Solanum lycopersicon* L. and *Lycopersicon lycopersicum* (L.) Karsten (Broome et al. 1983). The genus *Lycopersicon* is relatively small, comprising eight species and one further variety, and is divided into two complexes. The *esculentum* complex contains:

Lycopersicon esculentum	*L. pimpinellifolium*
L. cheesmanii Riley	*L. parviflorum**
*L. chmielewskii**	*L. hirsutum*
L. esculentum var. *cerasiforme*	

*two closely related species previously known as a single species, *L. minutum.*

Separated from these species by natural hybridisation barriers are the two species in the *peruvianum* complex: *L. peruvianum* and *L. chilense.*

The region of origin of the genus *Lycopersicon* is probably the western coastal areas of Chile, Perud and Ecuador. The weed species *L. esculentum* var. cerasiforme is believed to be the ancestral species of the domestic tomato and, having spread into Central America, was taken into domestication in Mexico (Simmonds 1976; Langer and Hill 1982; Taylor 1986). The crop tomato (var. cerasiforme) was taken to Europe by the Spanish after their Mexican conquest in

[1]Crisp Innovar Ltd, Glebe House, Station Road, Reepham, Norfolk NR10 4NB, UK

Biotechnology in Agriculture and Forestry, Vol. 32
Cryopreservation of Plant Germplasm I (ed. by Y.P.S. Bajaj)

Table 1. Worldwide production of tomato, *L. esculentum* Miller, (FAO 1991)

	Area harvested (1000 ha)		Yield (t/ha)		Production (1000 MT)	
	1979–81	1990	1979–81	1990	1979–81	1990
World	2494	3027	23.39	23.74	52 666	71 867
Africa	339	462	14.54	18.15	4923	8389
N. America	321	365	29.64	38.33	9517	13 995
S. America	133	163	22.60	27.66	3003	4512
Asia	811	1133	16.15	17.42	13 099	19 749
Europe	484	492	29.95	36.85	14 483	18 150
Oceania	10	11	26.23	33.96	270	372
USSR	397	400	18.25	16.75	7371	6700

1523, and then spread to the rest of the world. Consequently, the commercially available varieties of tomato were, until very recently, derived from a very limited genetic base that arose from the limited number of accessions of var. cerasiforme that made their way into Europe in this initial introduction.

1.2 Germplasm Storage

The germplasm of tomato is distributed throughout the conventional categories of:

modern cultivars,
obsolete/older cultivars,
commercial breeding stocks,
landraces and primitive cultivars,
wild forms of cultivated species,

and there is also the valuable genetic material that can be integrated via breeding techniques contained within:

a) closely related species,
b) species with potential for transgenic hybridisations.

This latter grouping includes the other species in the genus and various closely related species in the genus *Solanum*. Much of this material is stored primarily as seed in working and informal collections used for research and commercial purposes. There are also 28 formal, national base and active germplasm collections of tomato, based primarily on seed, that conserve, characterise and distribute material, e.g. the USA Tomato Genetics Stock Centre, Davis, California, the laboratories of the National Germplasm System, also in the USA and laboratories in the International Board for Plant Genetic Resources (IBPGR) network. These international collections contain, in total, some 32000 acquisitions of tomato distributed between the above categories (Garanko 1991). The tomato seed shows storage behaviour categorised as *orthodox* (Ellis and Roberts 1981) and can be maintained at conventional deep-freeze temperatures (−20 °C) for years without any significant threat to viability or genetic integrity. With time, however, some loss of viability and/or deterioration in performance of seed populations stored under such conditions is inevitable due to biochemical activity that, although slowed, is not completely eliminated (Stanwood and Bass 1978). This decline typically occurs over a time span of decades for orthodox seeds (Roberts and Ellis 1984). The changes are seen at one level as an increasing number of non-viable seeds within a recovered population, but of perhaps greater concern is non-lethal injury, that may include heritable genetic alterations, present in damaged but still viable seeds. This includes injury to the genetic material of stored seed that can result from free radical activity (Benson 1990). It is possible that without strict monitoring and controls, altered genes could be introduced from previously stored seed into plant populations if this material is brought into use. A second, significant loss of seeds from stored accessions is also due to the seeds that are taken for intermittent viability testing to monitor the progress of viability loss within the stored population. Each of these areas of loss can be brought to a minimum by the storage of seed samples at ultra-low temperatures, e.g. liquid nitrogen at $-196\,^{\circ}C$, where the basic chemical reactions that underlie damage to the stored material are, for thermodynamic reasons, reduced to their absolute minimum (Stanwood and Bass 1978; Roberts and Ellis 1984; Grout 1991). Change within the genetic composition of a seed population may also occur, with time, as a result of the periodic regeneration that is necessary to replenish seed stocks (Rick 1984; Withers 1987). As cryogenic storage dramatically reduces the need frequency of this regeneration, this aspect of genetic risk is similarly reduced.

Conservation of pollen can be used to maintain a significant portion of the crop gene pool in a compact and convenient way, and provides significant benefits in breeding programmes as the timing and location, of fertilisations can be manipulated (see Bajaj 1987). Refinement of techniques in pollen culture may also make this a convenient way of storing material from which to generate haploid and dihaploid plants of tomato (Sharp et al. 1972; Kut et al. 1984).

1.3 The Need for Cryopreservation of Clonal Material

Much of the work in germplasm conservation and crop improvement with tomato is adequately sustained using genetic material maintained as seed populations. However, with the increasing successes of techniques in plant cell, tissue and organ culture such as somatic hybridisation and gene transfer, some valuable, unique material is being generated that needs to be maintained as single genotype lines to be used to its best advantage. For example, specific, characterised genotypes are being used in breeding programmes aimed at genome transfer into interspecific hybrids (Wijbrandi and Koornneef 1990; Ratushnyak et al. 1994). Techniques of somatic hybridisation are also producing intergeneric and interspecific hybrids of potential value that are sterile or very poor performers in terms of fruit or seed set. The generation of clonal populations of this material, using in vitro techniques such as meristem culture (Kartha, et al. 1977) makes further investigation and application more practicable. Somaclonal variation is increasingly used to generate novel genetic variation (Buiatti and Morpurgo 1990). Once generated, much of this material will need to be stabilised with respect to further genetic variation, and in the most cost- and labour-effective fashion, for which cryopreservation is ideally suited (Withers 1991; Bajaj 1991).

2 Cryopreservation

2.1 Storage of Seeds at −20 °C

The orthodox behaviour of tomato seeds means that, at relatively low moisture contents, few storage problems over practical periods are to be expected. An acceptable storage moisture content will be above 5.5%, as a reduction in viability can be expected if the seeds are dried below this (Nutile 1964). Above 18.5% moisture, the High Moisture Freezing Limit (Stanwood 1985), a major freezing event can be recorded at −20 °C which is lethal to the seeds and eliminates germination completely. However, if a moisture content is selected between these limits, and for preference below 10%, then storage periods in excess of 15 years with germination above 90% of control levels can be expected (James et al. 1964). It is important that the seeds are sealed into suitable packets or vials, and that these are then held in plastic boxes with lids to prevent them from picking up moisture from the atmosphere, as the storage freezer is likely to be opened many times during the storage period.

2.2 Liquid Nitrogen Preservation of Tomato Seeds

The orthodox properties of tomato seeds, and their relatively small size, means that they also pose few problems for cryopreservation using conventional liquid nitrogen (LN) storage technology (Stanwood and Bass 1978; Grout 1980), and they are in a category formally described as both desiccation and liquid nitrogen tolerant (Stanwood 1985). A range of cultivars have been reported as surviving cryopreservation in LN following cooling at 200 °C/min into the cryogen. Warming rates, where quoted, were at 30 °C/min to 0 °C. Storage periods ranging from 180 to 1095 days resulted in germination rates of 99% of control where moisture contents were below 7.2% of fresh weight. Where the moisture content was somewhat higher, at 8.7%, the germination rate of recovered seeds fell to 84% of control levels.

In a simple protocol routinely used by the author, samples of 100 seeds at a water content of 6–7% wet weight are sealed into aluminium foil packets and frozen by direct immersion in liquid nitrogen. Subsequently, seeds are thawed by immersion of the packets in water at 40 °C for 30 s and germinate without loss of viability. This protocol provides a cooling rate, measured in the cotyledonary tissues, of 940 °C/min from 0 °C to the base temperature and a warming rate in excess of 700 °C/min (Grout 1980). Similar results have been obtained using the

Table 2. Germination of seeds of *Lycopersicon esculentum* Miller. cv. Ailsa Craig following direct immersion in liquid nitrogen. (After Grout 1980)

Water content (% wet wt.)	Germination (% unfrozen control) DMSO conc (% v/v)				
	0	2.5	5	10	15
6.6	100	–	–	–	–
17.6	99	100	100	100	99
22.6	91	98	100	99	100
27.8	62	71	83	98	100
33.4	0	0	0	39	81

slower, but still relatively rapid, cooling rate of 200 °C/min (Stanwood 1985). With the inclusion of a 1-h incubation in an aqueous solution of dimethylsulphoxide (DMSO), germination, and subsequent growth performance at 100% of unfrozen control values, can be maintained even when seeds have a water content in excess of 25% of wet weight (Table 2).

2.3 Cryopreservation of Seedling Meristems

The successful cryopreservation techniques for tomato meristems and/or shoot tips have not, as yet, been refined to a point where 100% survival and regrowth after thawing is ensured. Consequently, any attempts at cryopreservation of this type of material must be preceded by a period of clonal propagation to produce a sufficiently large population of meristems for the procedure. If the material to be preserved is already maintained as plantlets in vitro, then multiplication by nodal subculture is the preferred option. If seed is the starting material, then in vitro germination on hormone-free MS medium (Murashige and Skoog 1962), followed by nodal multiplication, provides the sterile starter material. For cryopreservation, shoot-tip explants of *L. esculentum* Miller. cv. Ailsa Craig, comprising a small piece of hypocotyl and the meristem itself, were dissected from germinated seedlings with a 2-cm radicle (Grout et al. 1978; Grout and Tay, unpubl.) or from older, 7-day seedlings germinated and grown in vitro (Grout, unpubl.). This material is comparable to the axillary meristems of plantlets raised in vitro from sources other than seed, e.g., adventitiously from callus, from somatic embryos or from nodal multiplication of plantlets. The tissue pieces were incubated in MS medium, without growth regulators, cooled to 0 °C and containing 3% w/v sucrose for 20 min, and then transferred to similar medium containing increasing concentrations of DMSO as a cryoprotectant. A further 20-min incubation was allowed in solutions containing 2.5, 5 and 10% v/v DMSO and the explants were finally transferred to medium containing 15% v/v DMSO for a final 40 min.

Explants were transferred, in units of ten contained in 1 ml of the final incubation medium, to small, cylindrical vials (4 mm width) simply fashioned from a single thickness of domestic aluminium foil. To cool, the vials were suspended in vapour above liquid nitrogen held in a wide-necked Dewar. This provided a cooling rate, recorded using a copper/constantan micro-thermocouple embedded in an explant, that varied continuously between 0 and −120 °C/min. After storage in LN for a minimum of 24 h, the samples were thawed by immersion of the vials in water at 40 °C for 1.5 min. The thawed explants were rinsed three times in the original MS medium without cryoprotectant additive.

Subsequent culture in vitro was made using MS supplemented with a range of plant growth regulators. The standard formulation used in the laboratory for tomato seedling shoot-tip culture employed a supplement of 0.01 mg/l naphthalene acetic acid and 1 mg/l of dimethylallyl-amino purine (MSNP). Additional formulations of the MS medium were investigated using supplements of 3 mg/l GA (MSG), 3 mg/l GA plus 1 mg/l casein hydrolysate (MSGC) and 3 mg/l GA plus 370 mg/l $MgSO_4 . 7H_2O$ (MSGM). The cultures were maintained at 25 °C with a 16 h photoperiod from warm white fluorescent tubes (Atlas, UK) providing 3500 lx at shelf level.

The standard medium (MSNP) produced shoot growth directly from the apical meristem in 100% of unfrozen explants, including those that had been treated with DMSO as if they were to be frozen. Following freezing and thawing, the 4MSNP medium sustained growth of callus tissue from the recovered explants (Table 3), from which spontaneous, adventitious shoot regeneration was frequently observed, but no direct outgrowth of the organised shoot meristem was recorded for either age of explant. On the range of media containing GA that were investigated, direct growth of the shoot meristem was observed, in addition to the callusing observed with the (MSNP) medium (Table 3). It should be noted that trials with a rapid cooling rate (800 °C/min) produced surviving cultures that gave rise to callus only, indicating a non-lethal injury that affects the regulation of growth and development, and that a slow (2 °C/min) cooling rate failed to produce any surviving explants whatever.

It is also evident that even where direct meristem growth occurred, the control of ordered growth in the explant had been altered in some way by the freezing and/or thawing procedures and that the recovery medium needed to have a GA supplement. The modification of the post-thaw environment for in

Table 3. The percentage[a] of explants of *Lycopersicon esculentum* Miller cv. Ailsa Craig producing callus, and showing apical meristem outgrowth to produce shoots, following thawing from liquid nitrogen. (Data taken, in part, from Grout et al. 1978)

Growth medium	Initial explant from seedling with 2-cm radicle		Initial explant from 7-day seedling
	Shoots	Callus	Shoots
NSNP	0	49	0
NSG	52	24	18
MSGC	51	12	38
MSGM	37	12	27

[a]Based on an unfrozen control. n = > 100 in all treatments.

vitro material may be of considerable significance when considering the complete repair of sub-lethal freeze/thaw injury and the optimisation of cryopreservation protocols (Benson 1990; Grout 1990).

This technique has potentially broader applications than simply the meristems culture systems for very young seedlings illustrated above, as meristem culture techniques for tomato with high frequency shoot regeneration have been reported for a range of growth media (Kartha et al. 1977; Kut et al. 1984). Interestingly, none of the media used for this unfrozen material has a requirement for gibberellic acid as an additive.

2.4 Cryopreservation of Pollen

Freeze-dried tomato pollen stored at room temperature for 2 years was still capable of seed set, albeit at a reduced level, on recovery (King 1965). A comparable freeze-drying method successfully employed by the author involved prefreezing of fresh pollen by immersion of the vials containing the material up to their necks in liquid nitrogen. After 1 min, the material was transferred to a conventional bench-top freeze-dryer and dried with 30 min of vacuum. The freeze-dried material was stored at −20 °C for up to 9 months, and recovered by warming the vials in water at 35 °C for 1 min prior to using the material for artificial pollinations. No precise data were recorded at the time of the investigations, but seed set was perceived as acceptable.

Practical, cryogenic storage of the apparently robust tomato pollen can be achieved at the −20 °C temperature conveniently supplied by the laboratory/domestic deep-freeze. Germination of pollen stored in this way for 3 years has been recorded as increasing from 47 to 63%, followed by normal seed set (Towill 1985). Increased pollen germination following storage is not an uncommon phenomenon (Towill 1985) and has been attributed, without confirmation, to an after-ripening process or the release of a needed, and previously compartmentalised, nutrient. Freshly collected pollen of *L. esculentum* Miller has also previously been stored at −196 °C for 1 year with a reduction of 25% in post-thaw germination, presumably due to damage incurred either during freezing or thawing (Towill 1985). In the authors' laboratory freshly collected tomato pollen has also been successfully cryopreserved. Samples, without any pretreatment, were wrapped in small packets made of a single layer of domestic aluminium foil and cryopreserved by plunging directly into liquid nitrogen (Withers 1987). In a single trial the viability of thawed tomato pollen, determined by fluorescence in fluorescein diacetate (Heslop-Harrison and Heslop-Harrison 1977), was 85% of control values, but no attempt was made to set seed with this material. Caution should be adopted with regard to indirect viability tests for frozen/thawed pollen (Towill 1985) and the only completely reliable indicator of pollen undamaged by the cryopreservation procedure is the satisfactory performance of the pollen with regard to seed production and subsequent seed performance. An example of the problems that can arise from the use of a sub-optimal cryopreservation procedure is the poor storage characteristics observed in otherwise normal seeds of *Brassica oleracea* var. italica produced using the frozen/thawed pollen (Crisp and Grout 1984). This material had high viability (assessed by in vitro germination) when thawed, comparable to control material, and gave an acceptable level of seed set. The inference must be that there was, however, non-lethal and non-repairable damage sustained by the pollen during freezing and/or thawing that was expressed in the storage properties of the seed in some way.

3 Summary and Conclusions

Conservation of the greater portion of the gene pool of tomato and its relatives is probably best achieved by the storage of orthodox seed. For many purposes, storage at −20 °C in a conventional deep-freeze will be adequate, but it is essential that base collections are cryopreserved in liquid nitrogen to ensure the maximum genetic stability (Stanwood 1984; Grout 1991). Tomato seeds at a moisture content within the range 6–18% can be cryopreserved, without pretreatment, by direct immersion in the cryogen and without reduction in viability. Storage of pollen is also valuable for conventional breeding (Bajaj 1987) and, with advances in biotechnology, will also have a role in the production of homozygous, diploid breeding material. Again, conventional freeze-drying or direct immersion in cryogen preserves pollen at a high level of germinability, and with a high level of seed set (Towill 1985). Caution must be exercised, however, and the success of preservation should be judged only on the performance of seeds produced using the preserved pollen. Meristem culture allows for the propagation of clones of specific genotypes, and the successful application of cryopreservative techniques enables a collection of valuable, characterised genotypes to be maintained with complete genetic integrity in a cost- and labour-effective way (Withers 1991). A successful protocol for cryopreservation of tomato meristems is based on incubation in 15% DMSO and cooling in the vapour above liquid nitrogen, to −175 °C, before eventual immersion in liquid nitrogen. There is also a role in crop improvement studies for suspension cultures of tomato that have specific properties or genes (Kut et al. 1984). Given the orthodox seed behaviour of the species and the apparently conventional behaviour of tomato pollen with regard to cryopreservation, it is likely that the suspension cultures will be amenable to the cryopreservative procedures.

References

Bajaj YPS (1987) Cryopreservation of pollen and pollen embryos, and the establishment of pollen banks. In: Giles KL, Prakash J (eds) Pollen development and cytology. Academic Press, London, pp 361–381

Bajaj YPS (1991) Storage and cryopreservation of in vitro cultures. In: Biotechnology in agriculture and forestry, vol 17. High-tech and micropropagation I. Springer, Berlin Heidelberg New York, pp 361–381

Benson EE (1990) Free radical damage in stored germplasm. Int Board Plant Genet Resour (IBPGR), Rome

Broome CR, Terrell EE, Reveal JL (1983) Proposal to ensure *L. esculentum* Miller as the scientific name of the tomato. TGC Rep 33: 55–56

Buiatti M, Morpurgo R (1990) Somaclonal variation in tomato. In: Bajaj YPS (ed) Biotechnology in agriculture and forestry, vol 11. Somaclonal variation in crop improvement I. Springer, Berlin Heidelberg New York, pp 400–415

Crisp PC, Grout BWW (1984) Storage of broccoli pollen in liquid nitrogen. Euphytica 33: 819–823

Ellis RH, Roberts EH (1981) The quantification of ageing and survival in orthodox seeds. Seed Sci Technol 9: 373–409

FAO (1991) Bull Stat, 4(3), FAO, Rome, pp 226

Garanko IB (1991) Germplasm resources in tomato. In: Kallo G (ed) Genetic improvement of tomato. Springer, Berlin Heidelberg New York, pp 51–57

Grout, BWW (1980) Low temperature storage of imbibed tomato seeds: a model for recalcitrant seed storage. Cryo Lett 1(2): 71–76

Grout BWW (1990) Genetic preservation in vitro. In: Nijkamp HJJ, Van Der Plas LHW, Von Aartrijk J (eds) Progress in plant cellular and molecular biology. Kluwer Academic, Dordrecht, pp 13–22

Grout BWW (1991) Cryopreservation of plant cells and organs. In: Zakri AH, Normah MN, Abdul Karim AG, Senawi MT (eds) Conservation of plant genetic resources through in vitro methods. FRIM/MNCPGR, Malaysia, pp 43–56

Grout BWW, Westcott RJ, Henshaw GG (1978) Survival of shoot meristems of tomato seedlings frozen in liquid nitrogen. Cryobiology 15: 478–483

Heslop-Harrison J, Heslop-Harrison Y (1977) Evaluation of pollen viability by enzymatically induced fluorescence; intracellular hydrolysis of fluorescein diacetate. Stain Technol 45: 115–120

James E, Bass LN, Clark DC (1964) Longevity of seeds stored 15–30 years at Cheyenne, Wyoming. Am Soc Hortic Sci 84: 527–534

Kartha KK, Champoux S, Gamborg OL, Pahl K (1977) In vitro propagation of tomato by shoot apical meristem culture. J Am Soc Hortic Sci 102: 346–349

King JR (1965) The storage of pollen – particularly by the freeze-drying method. Bull Torrey Bot Club 92: 270–287

Kut SA, Bravo EJ, Evans DA (1984) Tomato. In: Ammirato PV, Evans DA, Sharp WR, Yamada Y (eds) Handbook of plant cell culture, vol 3. Crop Species. Macmillan, New York, pp 247–289

Langer RHM, Hill GD (eds) (1982) Agricultural plants. Cambridge Univ Press, Cambridge

Murashige T, Skoog F (1962) A revised medium for rapid growth and bioassays with tobacco tissue cultures. Physiol Plant 15: 473–497

Nutile GE (1964) Effects of dessication on the viability of seeds. Crop Sci 4: 325–328

Rick CM (1984) Plant germplasm resources. In: Ammirato PV, Evans DA, Sharp WR, Yamada Y (eds) Handbook of plant cell culture, vol 3. Crop Species. Macmillan, New York, pp 9–37

Roberts EH, Ellis RH (1984) The implication of the deterioration of orthodox seeds during storage for genetic resources conservation. In: Holden JHW, Williams JT (eds) Crop genetic resources, conservation and evaluation. Allen and Unwin, London, pp 18–36

Ratushnyak YI, Cherep NN, Latypov SA (1994) Somatic hybridization between *Lycopersicon esculentum* Mill. and *Lycopersicon peruvianum* var. *dentatum* Dun. In: Bajaj YPS (ed) Biotechnology in agriculture and forestry, vol 27. Somatic hybridization in crop improvement I. Springer, Berlin Heidelberg New York, pp 224–243

Simmonds NW (ed) (1976) Evolution of crop plants. Longman, London

Sharp WR, Raskin RS, Sommer HE (1972) The use of nurse culture in the development of haploid clones in tomato. Planta 104: 357–361

Stanwood PC (1984) Cryopreservation of seeds. In: IBPGR Advisory committee on seed storage. Rep 2nd Meet, Int Board Plant Genet Resour, Rome, pp 8–27

Stanwood PC (1985) Cryopreservation of seed germplasm for genetic conservation. In: Kartha KK (ed) Cryopreservation of plant cells and organs. CRC Press, Boca Raton, pp199–226

Stanwood PC, Bass LN (1978) Seed germplasm cryopreservation. In: Li PH, Sakai A (eds) Plant cold hardiness and freezing stress – mechanisms and crop implications. Academic Press, New York, pp 361–372

Taylor IB (1986) Biosystematics of the tomato. In: Atherton JG, Rudich J (eds) The tomato crop. Chapman and Hall, London, pp 1–34

Towill L (1985) Low temperature and freeze/vacuum-drying preservation of pollen. In: Kartha KK (ed) Cryopreservation of plant cells and organs. CRC Press, Boca Raton, pp 172–198

Wijbrandi J, Koornneef M (1990) Partial genome transfer in interspecific tomato hybrids. In: Nijkamp HJJ, Van Der Plas LHW, Van Aartrijk J (eds) Progress in plant cellular and molecular biology. Kluwer Academic Dordrecht, pp 280–285

Withers LA (1987) The low temperature preservation of plant cell, tissue and organ cultures and seed for genetic conservation and improved agricultural practice. In: Grout BWW, Morris GJ (eds) The effects of low temperatures on biological systems. Edward Arnold, London, pp 389–409

Withers LA (1991) Tissue culture in the conservation of plant genetic resources. In: Zakri AH, Normah MN, Abdul Karim AG, Senawi MT (eds) Conservation of plant genetic resources through in vitro methods. FRIM/MNCPGR, Malaysia, pp 1–18

V.6 Cryopreservation of Germplasm of Banana and Plantain (*Musa* Species)

B. PANIS[1] and R. SWENNEN

1 Introduction

1.1 Distribution and Importance of Banana and Plantain

Banana and plantain (*Musa* spp.) are the fourth largest food commodity after rice, wheat, and milk in terms of the gross value of production (CGIAR 1992). They are produced in the humid tropics of many developing economies; about 74 million tons are produced annually (FAO 1991).

The export trade deals almost entirely with dessert bananas and accounts for only 10% of the production. Most of the remaining 90%, consisting of cooking and beer bananas, plantains, and local dessert bananas, is grown on small subsistence farms and in backyard gardens. The cost of producing bananas is lower than that of any other staple crop grown in the humid tropics, such as yam, maize, sweet potato, or cassava. This is of particular significance to the marginal growers and the markets they serve (INIBAP 1987, 1988). In North America and western Europe, average consumption figures around one banana a week (30 g/person/day). In parts of Africa and Latin America, average consumption is five to ten times as high and may even reach 250 kg/person/year in the countries around the great lakes of Eastern Africa (INIBAP 1988).

Locally consumed bananas are eaten fried, boiled, steamed, or uncooked. They also can be brewed and made into an alcoholic beverage. Nearly 1000 accessions are known to exist. The majority of those of immediate importance for mankind produce their fruits parthenocarpically and are sterile. They evolved from two wild species within the genus *Musa*, i.e., *acuminata* (AA) *balbisiana* (BB). Both are diploid (2n = 22) and native to Southeast Asia, where the center of origin is located.

From the wild ssp. *acuminata*, man selected over time the first edible diploid bananas which produce pulp. Later, varieties which combine pulp formation and high sterility were selected. Such varieties were seedless. Through a process of outcrossing in combination with restitution at meiosis, man selected triploids which produce longer and heavier fruits than diploids. Because of the different interference of the two subspecies, edible cultivars of the following genomic configuration are currently cultivated: diploids AA and AB; triploids AAA,

[1]Laboratory of Tropical Crop Husbandry, Catholic University of Leuven (K.U. Leuven), Kardinaal Mercierlaan 92, 3001 Heverlee, Belgium

Biotechnology in Agriculture and Forestry, Vol. 32
Cryopreservation of Plant Germplasm I (ed. by Y.P.S. Bajaj)

AAB, and ABB; and the tetraploids AAAA, AAAB, AABB, and ABBB. Tetraploids developed apparently from an unreduced triploid gamete and a reduced diploid gamete or from diploids with each a nonreduced gamete. Plantain cultivars form a subgroup in the AAB group (Simmonds 1987).

Other more or less important species are *Musa textilis*, which yields Manila hemp and some species grown as ornamentals (*Musa velutina* and *Musa coccinea*). The related genus *Ensete* is grown also as a staple food crop and is locally very important in Ethiopia (Purseglove 1972).

Many pests and diseases are threatening banana and plantain production, among which are black sigatoka, a foliar fungal disease caused by *Mycosphaerella fijiensis*; Panama disease caused by *Fusarium oxysporum* f. sp. *cubense*, which blocks the vessels; bunchy top, a virus disease causing plant distortion; nematodes, and banana weevils.

1.2 Need for Conservation of *Musa*

Over 50 priority crops have been defined by the International Board for Plant Genetic Resources/International Plant Genetic Resources Institute (IBPGR/IPGRI) as needing urgent action, among which are starchy bananas and plantains (Ford-Lloyd and Jackson 1986).

Most cultivated bananas and plantains have not yet been systematically evaluated for their yield potential or resistance to pests and diseases. There is also a need to preserve their wild ancestors, *M. acuminata* and *M. balbisiana*, with their primary centers of diversity respectively in the Malaysia-Indonesia region and the Indian subcontinent, as they are important in breeding programs.

1.3 Methods for *Musa* Germplasm Storage (Table 1)

1.3.1 Seed Storage

Until recently, it was thought that *Musa* seeds could not be stored; however, research at the Universiti Pertanian Malaysia has shown that *Musa* seeds could be stored for a long period if properly treated (Williams 1987). Consequently, the storage of the wild *Musa* genepool by seed can be considered as a cheap alternative to field collections. However, the slow and erratic seed germination is a great hindrance to its application.

1.3.2 Field Gene Banks

There exist at least 42 field collections situated in 33 countries (The Philippines, Honduras, Jamaica, Nigeria, Cameroon, etc.) (INIBAP 1990). Most of these are to be viewed as working collections and as complementary to each other (Williams 1987). Such collections demand a large labor input. In addition, they are continuously under risk of genetic erosion, since they are subject to pests,

Table 1. Necessary conditions for the storage of 1000 banana accessions

	Conditions of storage			
	Field collection	In vitro collection Normal growth	Limited growth	Cryopreservation
Amount per accession	5 plants	20 multiple buds	20 multiple buds	Millions of cells
Area	30 000 m^2	15 m^2	15 m^2	2 m^2
Labor per annum (technicians only)	6 men	9 men	1.5–2 men	< 0.5 men
Conditions				
Ambient temperature	+ 25 to + 30 °C	+ 28 °C	+ 15 °C	− 196 °C
Risk of losing germplasm	High	Medium	Low	None
Pests and diseases	High risk	No risk	No risk	No risk
Human error	Moderate	Moderate	Low	Extremely low
Maintenance	Replanting every 3–5 years, application of fertilizers, pesticides, and weed control	6 Subcultures/year	1 Subculture/year	No subculture, refill with liquid nitrogen

diseases, and adverse weather conditions. Since each accession is represented by a few plants only, supply of plant material upon request is very slow because multiplication rates are small. Accessions are distributed as suckers, which pose serious risks with respect to pathogen transfer, especially of viruses, bacteria, and fungi. Hence, long quarantine periods are necessary.

1.3.3 In Vitro Gene Banks

The majority of culture work which has been done on *Musa* has focused on the production of aseptic shoot cultures (Cronauer and Krikorian 1986). In vitro proliferating shoot tips can also be applied for the maintenance of *Musa* germplasm (Berg and Bustamente 1974; Dore Swamy et al. 1983; Cronauer and Krikorian 1984; Banerjee and De Langhe 1985; Jarret et al. 1985; Vuylsteke and De Langhe 1985).

Banerjee and De Langhe (1985) described shoot-tip cultures (Fig. 1) grown on a medium based on MS (Murashige and Skoog 1962) mineral salt mixture supplemented with 10 μM (2.25 mg/l) N6-benzylaminopurine (BA) and 1 μM (0.175 mg/l) indole-3-acetic acid (IAA), which is called a proliferation medium. In vitro cultures are placed under continuous light of about 5000 lx at 30 ± 2 °C.

Fig. 1. In vitro proliferating shoot tips of *Musa*, 6 weeks after subculture on MS medium supplemented with 10 μM BA and 1 μM IAA (proliferation medium)

Under these conditions, cultures are transferred to fresh medium every 4–6 weeks. It is evident that this procedure is too labor-intensive for the conservation of large collections. In addition, it poses the threat of losing accessions due to contamination at every subculture.

The maintenance of such in vitro cultures under "limited" growth conditions reduces the need for frequent subculturing. This can be achieved by modification of the nutrient medium (use of osmotics or growth retardants) or of the environment in which cultures are incubated. Banerjee and De Langhe (1985) significantly reduced the growth of proliferating shoot tips through storage at $15 \pm 2\,^{\circ}C$ and light intensity of about 2000 lx. Under these conditions, cultures can be kept for 13–17 months on average (depending on the cultivar). The cultivars tested also seemed to vary in their ability to withstand minimal growth temperature.

The largest *Musa* germplasm collection is currently stored this way (De Smet and Van den houwe 1991). It is situated at the INIBAP (International Network for the Improvement of Banana and Plantain) *Musa* germplasm Transit Center (ITC) at K.U. Leuven in Belgium. This collection consists of 1032 accessions (INIBAP 1991) containing wild varieties, cultivars, and hybrids from all over the world. For safety reasons, all accessions are preserved for maximally 12 months. The germplasm exchange at the Transit Center follows the *Technical Guidelines for the Safe Movement of Musa Germplasm* (Frison and Putter 1989), a document prepared by FAO, IBPGR, and INIBAP.

1.4 Need for Cryopreservation

Although in vitro collections under limited growth conditions have proven their value, the occurrence of somaclonal variation is a serious impediment (Vuylsteke et al. 1991). In addition, there is always the risk of losing germplasm due to contamination or human error during subculture. Lastly, the preservation of a large collection in vitro is still a very labor-intensive endeavor.

Since all *Musa* cultivars are vegetatively propagated, their preservation under cryopreserved conditions would be an attractive proposition for the long-term conservation (Bajaj 1990). The material of choice for the establishment of germplasm collections under long-term conditions is cell suspensions and proliferating meristems in vitro.

Cell suspensions offer the advantage of storing material with a very high multiplication potential (Dhed'a et al. 1991; Panis et al. 1993a) and which can readily be used for protoplast culture (Panis et al. 1993b). These embryogenic cells and protoplasts are used for genetic manipulation (Panis et al. 1992a, 1993a; Sagi et al. 1992). The fact that it takes 4–6 months before a suitable embryogenic cell suspension can be established from a meristem is a major drawback (Dhed'a et al. 1991).

Since genetic transformation requires true-to-type material (suspension cells), the storage of these competent cultures by cryopreservation could overcome the problem of somaclonal variation.

2 Cryopreservation Studies on *Musa*

In *Musa*, research on cryopreservation has been carried out with zygotic embryos (Mora et al. 1991; Abdelnour-Esquivel et al. 1992a; Villalobos and Abdelnour 1992), embryogenic suspension cultures (Panis et al. 1990, 1992a, b), and in vitro meristems (Panis 1986)

2.1 Cryopreservation of Zygotic Embryos

Mature zygotic embryos of *M. acuminata* ssp. *burmannicoides* type Calcutta 4 (AA) and *M. balbisiana* type Honduras (BB) were excised and cultured for 3 to 5 h on a semisolid medium based on MS mineral nutrients (Abdelnour- Esquivel et al. 1992a). Subsequently, embryos were desiccated in the sterile air flow of a laminar flow cabinet for periods ranging from 0 to 3 h, and then transferred to sterile 2-ml cryotubes and plunged into liquid nitrogen. Thawing took place rapidly by warming sample vials in a 40 °C water bath whereafter they were cultured on the above-mentioned medium, supplemented with 0.5 mg/l benzyladenine (BA). At the end of each desiccation period, water contents of the embryos were determined through oven-drying at 101 °C.

The results of these experiments are indicated in Table 2. Embryos of *M. acuminata* and *M. balbisiana* had a slightly different water content, but their reaction to dehydration and freezing was similar. Survival of cryopreserved embryos was highest when dessicated to 11–15% water content.

Similar experiments were carried out with excised zygotic embryos of *Veitchia* and *Howea* palm (Chin et al. 1988) and *Coffea* spp. (Abdelnour-Esquivel et al. 1992b). Embryos could be successfully cryopreserved after partial desiccation to about 10% (for *Veitchia* and *Howea*) and 20% water content (for *Coffea* spp.).

Table 2. Water content of *Musa* zygotic embryos and their survival (%) after 0–3 h dehydration [Surv (– LN)] and subsequent freezing [Surv (+ LN)]. Average of three experiments with three replicates of 12 embryos each. (Abdelnour-Esquivel et al. 1992a)

	M. acuminata			*M. balbisiana*		
Time (h)	Water content	Surv (– LN)	Surv (+ LN)	Water content	Surv (– LN)	Surv (+ LN)
0	64	100	0	79	100	0
0.5	26	98	31	37	100	22
1.0	20	96	53	28	98	44
1.5	15	95	83	19	94	74
2.0	11	87	80	14	94	92
2.5	8	70	59	11	88	80
3.0	6	48	35	9	70	66

2.2 Cryopreservation of Embryogenic Cell Suspension Cultures

2.2.1 Protocol

Embryogenic cell suspensions were obtained from explants of the upper part (scalp) of proliferating meristematic shoot tips according to the method developed by Dhed'a et al. (1991). This protocol, previously developed for the cv. Bluggoe (ABB), was also successfully applied to four other cultivars, i.e., the ABB cooking bananas Saba and Cardaba, the plantain cultivar Three Hand Planty, and the wild banana *M. balbisiana* (BB) (Dhed'a et al. 1992). The suspensions were maintained in half-strength MS medium supplemented with 5 μM 2,4-D, 1 μM zeatin, standard MS vitamins, 10 mg/l ascorbic acid, and 30 g/l sucrose. The pH was adjusted to 5.8 before autoclaving. In the beginning, suspension cultures are very heterogeneous and contain large translucent cells as well as small dense cells. When subcultured every 2 weeks, cell suspensions become more uniform and consist of clusters of small tightly packed cells with dense cytoplasm. This indicates embryogenic competence (McWilliams et al. 1974).

These embryogenic suspensions cultures were examined for their reaction towards cryopreservation (Panis et al. 1990, 1992a, b, 1993a).

A pregrowth phase was carried out by subjecting *Musa* cells to osmotic stresses by culturing in 6% mannitol for 2 or 7 days.

Cryoprotectants used were 5, 7.5, 10, or 15% (v/v) DMSO, 5, 10, or 15% glycerol, 10% (v/v) proline, and a mixture of 0.5 M DMSO (dimethylsulfoxide), 0.5 M glycerol, and 1 M sucrose. In each case, an equal volume of chilled sterile medium containing the cryoprotectant at twice the concentration was added gradually over 1 h to a suspension chilled in an ice bath. Part of each suspension was kept as an unfrozen control.

Samples of 1.5 ml were transferred to 2-ml cryotubes. These tubes were introduced in a stirred methanol bath (Cryocool CC-60, Exatrol and agitainer from NESLAB, Portsmouth, New Hampshire, USA) and cooled at a rate of 1 °C/min to − 40 °C. This temperature was maintained for 30 min, after which the tubes were plunged into liquid nitrogen. The samples were thawed rapidly in a stirred water bath of 40 °C for about 1 min until most of the ice had melted in order to prevent damaging recrystallization.

The viability of the frozen suspension was microscopically determined with the fluorescein diacetate (FDA) test (Widholm 1972) and/or the 2,3,5-triphenyl tetrazolium chloride (TTC) reduction test (Dixon 1985). The success of the cryopreservation was also assessed by regrowing plantlets on semisolid medium containing half-strength MS salts supplemented with MS vitamins, 10 mg/l ascorbic acid, 30 g/l sucrose, 100 mg/l myoinositol, and 10 μM BA.

2.2.2 Results and Discussion

A chemically induced pregrowth in the presence of 6% mannitol did not influence viability of banana cells after freezing (results not shown). The treatment

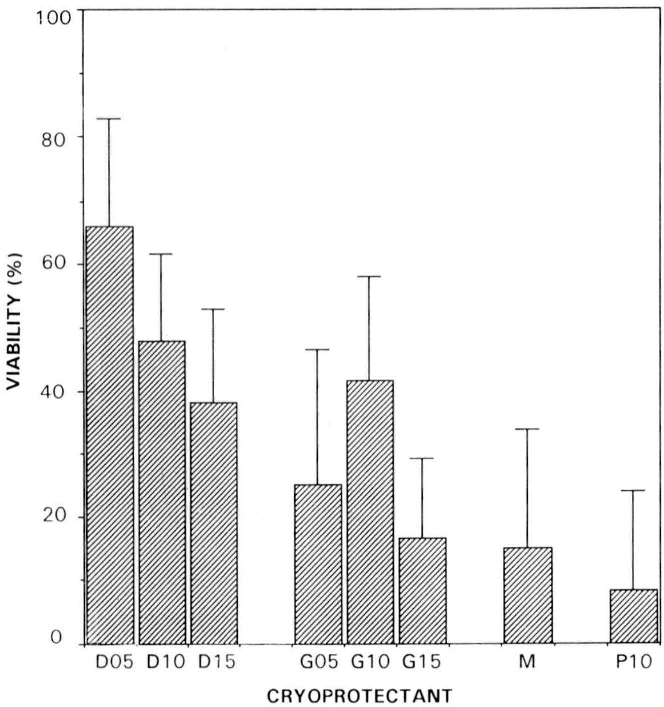

Fig. 2. Cryoprotective effects as assessed by the fluorescein diacetate (FDA) test of DMSO, glycerol, proline, and a cryoprotectant mixture. The procedure for freezing and thawing is described in the text. *Vertical bars* represent standard deviations. *D05, D10*, and *D15* respectively 5, 10, and 15% (v/v) dimethylsulfoxide; *G05, G10*, and *G15* respectively 5, 10, and 15% (v/v) glycerol; *M* cryoprotectant mixture containing 0.5 M DMSO plus 0.5 M glycerol plus 1 M sucrose; *P10* 10% (v/v) proline. (Panis et al. 1990)

normally decreases the amount of cellular water by osmotic dehydration. This indicates that embryogenic *Musa* cells already have a low water content and thus do not need any extra dehydration. Consequently, no special pregrowth phase was included in further experiments.

According to the FDA test (Fig. 2), DMSO as sole cryoprotectant provided optimal cryoprotection. However, survival decreased with increasing concentration of DMSO. Glycerol showed an optimal cryoprotective activity at a concentration of 10% with a viability comparable to 10–15% DMSO.

Although the FDA test is very rapid and convenient, it has its shortcomings, as FDA viability estimates are in some cases considerably higher than in regrowth tests. For example, suspensions cryoprotected with 5% glycerol resulted in 24% viability according to the FDA test but they failed to regrow on semisolid medium.

Frozen *Musa* cell suspension after transfer onto semisolid medium manifested:

1. No regrowth: the suspension became white and showed no sign of life.
2. Partial regrowth: some of the cells grew, divided, and formed colonies, while others became black or white and died.
3. Full regrowth: cells regrew, individual colonies were difficult to distinguish because a high number of cells resumed growth.

The removal of the cryoprotectant solution directly after thawing and its replacement by cryoprotectant-free liquid medium before cell transfer to a semisolid medium, resulted in a complete loss of regrowth capacity. Direct transfer of cells to a liquid medium, which subjects the cells to similar post-thaw wash stresses, likewise resulted in regrowth failure. Benson and Withers (1987) reported that DMSO can act as a free radical scavenger in the stabilization of cryopreserved cultures. This could help to explain the beneficial effect of leaving the cells in contact with DMSO during their recovery phase.

Regrowth rates of frozen suspensions were increased from 64 to 92%, when ice crystallization by seeding took place during the slow freezing process (Table 3). This was realized by plunging the cryotubes in liquid nitrogen for 3 s as a temperature of −7.5 °C was reached. Thereafter, they were further cooled to −40 °C. It is speculated that viability was increased due to the avoidance of excessive supercooling, thus allowing protective dehydration of the cells.

Histological observations and FDA staining of cryopreserved *Musa* cell suspension cultures (Fig. 3) revealed that only the most embryogenic cells survived freezing. Since the cell characteristics are related to age, suspensions were taken at different intervals after their last subculture and subjected to the optimal freezing method. After thawing, they were transferred to semisolid medium. Figure 4 illustrates the importance of suspension age for surviving success during the freezing process. Indeed, survival rate of a frozen-thawed cell suspension was highest (87.5%) with a 2-week-old culture. In addition, all Petri dishes exhibited regrowth on the entire medium surface. Younger and older suspension, however, showed a mixed response, since there were large differences among the repetitions. This was clearly correlated with the characteristics of the cells in suspensions. Microscopic examinations revealed that the majority of cells of a 2-week-old culture were embryogenic and had a dense cytoplasm. Older cells were more vacuolated, larger, and contained distinct starch grains.

Table 3. Influence of crystallization at −7.5 °C during the slow freezing process on the regrowth of *Musa* suspensions on semisolid medium 4 weeks after thawing. The procedure of freezing and thawing is described in Sect. 2.2.1. (Panis et al. 1990)

	Regrowth of frozen-thawed suspensions after inoculation on semisolid medium (25 repetitions)		
Ice initiation	No regrowth (%)	Partial regrowth (%)	Full regrowth (%)
−	36	36	28
+	8	48	44

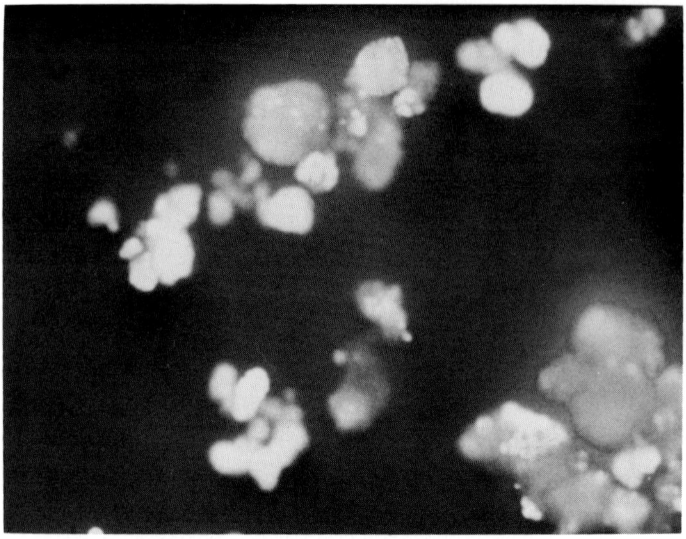

Fig. 3. Suspension culture, cryoprotected with 10% DMSO, and frozen in liquid nitrogen, stained with FDA and observed under ultraviolet light. The small embryogenic clumps fluoresce very brightly (alive), while larger structures show more diffuse fluorescence (dead). (Panis et al. 1990)

The growth of frozen-thawed cells on semisolid medium was studied through fresh weight measurements. Growth of a 2-week-old frozen cell suspension was compared with a cryoprotected suspension which was not frozen and a noncryoprotected, nonfrozen suspension. Growth fitted in all three cases an exponential curve. Since their linear models run parallel after log transformation (Fig. 5), it is concluded that cryoprotection and freezing kill a limited number of cells but that surviving cells regrow normally. Indeed, 4 to 6 weeks after the frozen suspension had been spread onto semisolid regeneration medium, a layer of loose cells and small globules formed (Fig. 6). These globules further developed into somatic embryos upon transfer to a fresh semisolid medium (Fig. 7). After separation, these somatic embryos were inoculated onto a hormone-free medium, where they developed into normal plantlets (Fig. 8).

Through histological examinations it was demonstrated that surviving cells regenerated through direct somatic embryogenesis into normal plantlets. Since no intervening callus phase occurs, it is anticipated that the rate of somaclonal variation is minimized.

Recently, we applied the freezing method which was initially optimized for the cryopreservation of embryogenic cell suspensions of the cv. Bluggoe on other cultivars. As such, it was possible to regenerate frozen suspensions of the wild diploid *Musa balbisiana* (BB group), Monthan (ABB group), and Three Hand Planty (AAB group) (Panis et al. 1994). This indicates that the capacity of a suspension for preservation in liquid nitrogen depends rather on the quality of the suspension than the genotype.

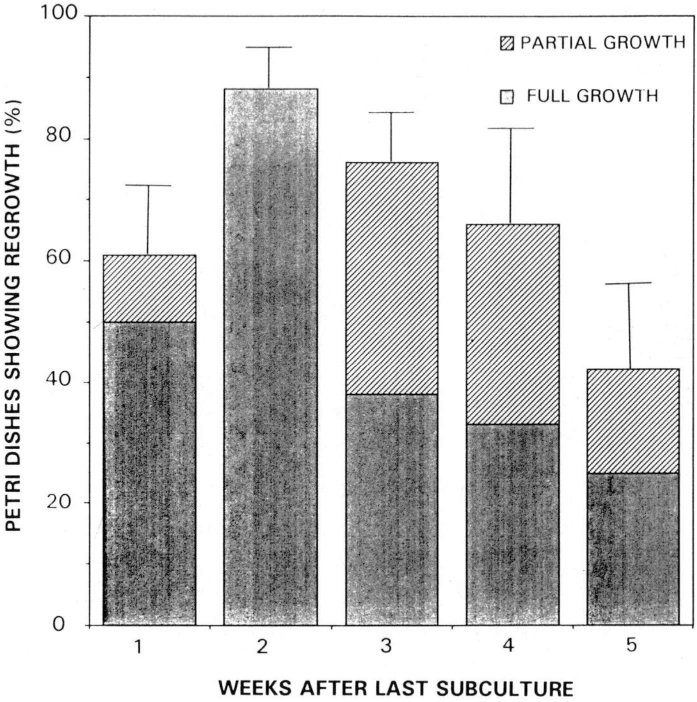

Fig. 4. Influence of the period (in weeks) between the last medium refreshment and freezing of an embryogenic cell suspension on regrowth rates on semisolid medium. *Vertical bars* represent standard deviations based on total regrowth

2.3 Cryopreservation of In Vitro Meristems

The cryopreservation method developed for cell suspensions is not applicable to 1-mm sized in vitro meristems obtained from proliferating meristematic clumps, irrespective of the concentrations of cryoprotectants (Panis 1986). Meristems were killed as soon as ice crystallization of the surrounding cryoprotective solution took place, based on differential scanning calorimetry (DSC) studies. The crystallization temperature was recorded to be as low as $-30\,^{\circ}\mathrm{C}$ when 15% DMSO was used. Below this freezing point, meristems were killed.

Two methods are now being tested for the cryopreservation of *Musa* meristems, i.e., encapsulation-dehydration and a simplified freezing method.

The encapsulation-dehydration method was developed by Dereuddre et al. (1990) for pear meristems. Shoot tips and somatic embryos were embedded in calcium alginate beads and precultured for 1 to 3 days in media containing 0.5 to 1 M sucrose. Subsequently, they were dehydrated in the laminar air-flow bench, then plunged into liquid nitrogen. This modified vitrification procedure does not rely upon a slow freezing apparatus or cytotoxic cryoprotectants. Recently, it was observed that banana meristems could be successfully encapsulated in

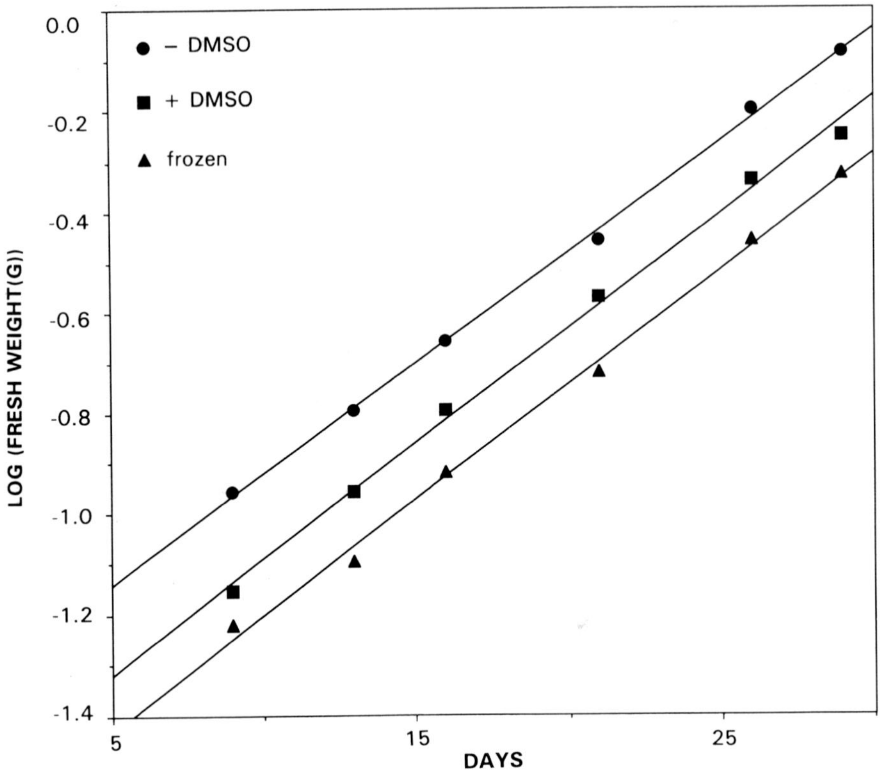

Fig. 5. Fitted linear models (after logarithmic transformation) of growth curves of (●) a noncryoprotected, nonfrozen suspension, (■) a cryoprotected, nonfrozen suspension, and (▲) a cryoprotected, frozen suspension. Regression equations are:
(●): $Y = -1.3646 + 0.044147 \times (R^2 = 0.999)$
(■): $Y = -1.5496 + 0.045880 \times (R^2 = 0.996)$
(▲): $Y = -1.6610 + 0.045803 \times (R^2 = 0.996)$

alginate beads and used in clonal propagation Ganapathi et al. (1992). The cryopreservation of such beads is currently under investigation at the Laboratory of Tropical Crop husbandry (K.U. Leuven) and preliminary experiments have been very encouraging. Encapsulated meristems of the cv. Bluggoe were incubated in liquid media containing sucrose levels which were increased daily in 0.25 M steps from 0.25 M to 0.75 M. Beads were then subjected for 3–5 h to the sterile air of a laminar airflow cabinet to obtain moisture contents of 20 to 30%. After rapid freezing about 5% of the meristems regrew and regenerated into plantlets (Fig. 9; Panis and Swennen 1994; Panis et al. 1994). However, taking the low regrowth rate, into consideration this method needs considerable improvement before it can be applied to germplasm preservation of banana.

An alternative and simple method was developed using a 2-week preculture on a proliferation medium containing high (0.4–0.5 M) sucrose concentrations. Precultured meristems were plunged directly into liquid nitrogen for storage.

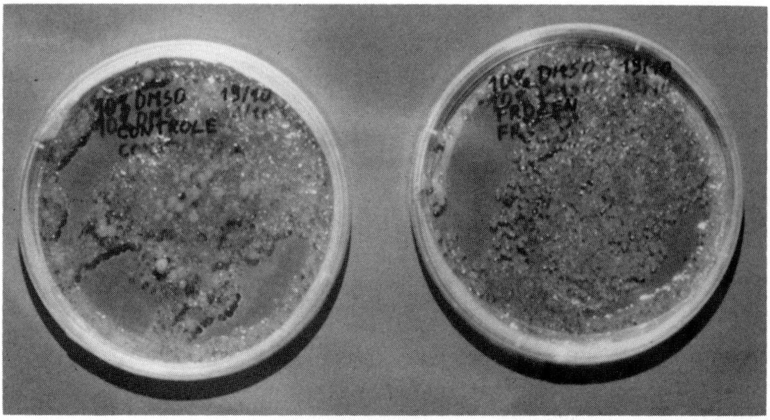

Fig. 6. Regrowth after 4 weeks on semisolid MS medium supplemented with 1 μM BA of unfrozen (*left*) and frozen (*right*) cell cultures, cryoprotected with 10% DMSO. (Panis et al. 1990)

After thawing, 50% of the meristematic buds developed further (Panis and Swennen 1994). An additional reduction of the water content by drying, as successfully applied to oil palm somatic embryos (Dumet et al. 1993), did not improve our results.

3 Summary and Conclusions

A cryopreservation protocol was developed for embryogenic cell suspension cultures of *Musa.* Slow freezing (1 °C/min) in the presence of 7.5% DMSO followed by storage in liquid nitrogen and rapid thawing results in high regrowth rates. The cells surviving the cryopreservation protocol are able to regenerate into normal plants through somatic embryogenesis. Our results also indicate that the applicability of this method depends on the quality of the embryogenic cell suspension (Panis et al. 1992b).

Although zygotic embryos of wild diploid varieties can be cryopreserved, this method is discouraged, since the exact genetic nature of such embryos is always unknown.

For germplasm preservation of *Musa*, attention needs to be paid to the cryopreservation of in vitro meristems. The slow freezing method in the presence of DMSO proved to be unsuccessful for in vitro banana meristems.

The encapsulation-dehydration method is a novel approach in which the material is frozen at ultra-rapid cooling rates and stored in the vitrified (glassy) state. Damaging ice crystallization, which is the most important injury factor in organized structures like meristems and embryos, is thus avoided. In addition, the encapsulation-dehydration method does not rely on the use of cryoprotectants.

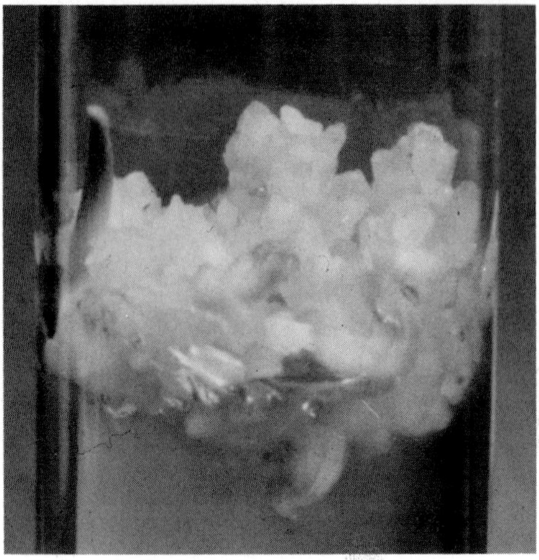

Fig. 7. Mass of somatic embryos originating from a frozen cell culture, 10 weeks after cryopreservation. Medium: hormone-free MS

Fig. 8. Outgrowth of a plantlet through somatic embryogenesis 15 weeks after cryopreservation. Medium: hormone-free MS. (Panis et al. 1990)

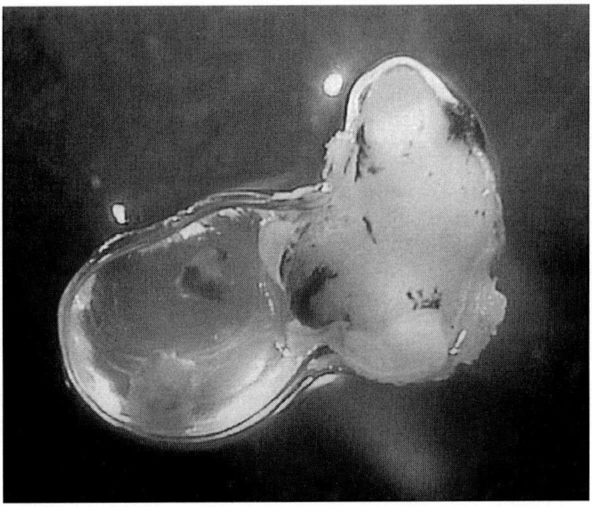

Fig. 9. Regeneration after 6 weeks of an encapsulated banana meristem which was treated with 0.75 M sucrose, dehydrated for 4 h, and frozen in liquid nitrogen

To date, results using the encapsulation-dehydration method are very poor since only 5% of the frozen banana meristems survived. A more promising technique, resulting in 50% survival, involves a 2-week preculture on high sucrose-containing media. The advantages of the latter system are numerous: potential for all *Musa* cultivars, simplicity of the technique, few manipulations, regrowth without an intervening callus phase, avoidance of controlled slow freezing, and relatively high frequency of regeneration. Further research on the cryopreservation of banana meristems deserves high priority.

Although medium-term conservation techniques are now routinely employed for *Musa* meristem cultures in many laboratories and several international centers, for example, INIBAP (Belgium), CIRAD/FHLOR (Guadeloupe), CATIE (Costa Rica), IITA (Nigeria), QDPI (Australia), cryopreservation remains indispensable in view of long-term storage.

Acknowledgments. The authors gratefully acknowledge Dr. D. Dhed'a for providing the banana cell suspensions, Wim Dillemans for his technical assistance, and Liesbeth Vloeberghs for her excellent experimental work. This study was partially financed by INIBAP (International Network for the Improvement of Banana and Plantain) through a grant of ABOS/AGCD (Belgian Administration for Development Cooperation).

References

Abdelnour-Esquivel A, Mora A, Villalobos V (1992a) Cryopreservation of zygotic embryos of *Musa acuminata* (AA) and *Musa balbisiana* (BB). Cryo Lett 13: 159–164

Abdelnour-Esquivel A, Villalobos V, Engelmann F (1992b) Cryopreservation of zygotic embryos of *Coffea* spp. Cryo Lett 13: 297–302

Bajaj YPS (1990) Cryopreservation of germplasm of vegetatively propagated crops. Bull Soc Bot Fr Actual Bot 137 (3/4) 99–114

Banerjee N, De Langhe E (1985) A tissue culture technique for rapid clonal propagation and storage under minimal growth conditions of *Musa* (Banana and Plantain). Plant Cell Rep 4: 351–354

Benson EE, Withers LA (1987) Gas chromatographic analysis of volatile hydrocarbon production by cryopreserved plant tissue cultures: a non-destructive method for assessing stability. Cryo Lett 8: 35–46

Berg LA, Bustamente M (1974) Heat treatment and meristem culture for the production of virus-free bananas. Phytopathology 64: 320–322

CGIAR (1992) Review of CGIAR priorities and strategies, Part 1, Section 5.3.5. Consultative Group on International Agricultural Research (CGIAR), Washington DC

Chin HF, Krishnapillay B, Alang ZC (1988) Cryopreservation of *Veitchia* and *Howea* palm embryos: non-development of the haustorium. Cryo Lett 9: 372–379

Cronauer SS, Krikorian AD (1984) Rapid multiplication of bananas and plantains by in vitro shoot-tip culture. HortScience 19: 234–235

Cronauer SS, Krikorian AD (1986) Banana (*Musa* spp.). In: Bajaj YPS (ed) Biotechnology in agriculture and forestry, vol 1. Trees I. Springer, Berlin Heidelberg New York, pp 233–252

Dereuddre J, Scottez C, Arnaud Y, Duron M (1990) Resistance of alginate-coated axillary shoot tips of pear tree (*Pyrus communis* L. cv Beurré Hardy) in vitro plantlets to dehydration and subsequent freezing in liquid nitrogen: effects of previous cold hardening. CR Acad Sci Paris 310 Sér III: 317–323

De Smet K, Van den houwe I (1991) The banana germplasm collection at the INIBAP Transit Center. In: Annual Report of INIBAP 1991. Int network for the improvement of banana and plantain. Montpellier, France, pp 35–37

Dhed'a D, Dumortier F, Panis B, Vuylsteke D, De Langhe E (1991) Plant regeneration in cell suspension cultures of the cooking banana cv. Bluggoe (*Musa* spp., ABB group). Fruits 46 (2): 125–135

Dhed'a D, Panis B, Swennen R, Vuylsteke D (1992) The applicability of embryogenic cell suspension cultures from vegetative tissue to different banana varieties. Banana Newsl 15: 43–44

Dixon RA (1985) Isolation and maintenance of callus and cell suspension cultures. In: Dixon RA (ed) Plant cell culture: a practical approach. IRL Press, Oxford, pp 1–20

Dore Swamy R, Srinivasa Rao NK, Chacko EK (1983) Tissue-culture propagation of banana. Sci Hortic 18: 247–252

Dumet D, Engelmann F, Chabrillange N, Duval Y (1993) Cryopreservation of oil palm (*Elaeis guineensis* Jacq.) somatic embryos involving a desiccation step. Plant cell Rep 12: 352–355

FAO (1991) Production year book 1991. Food and Agriculture Organisation of The United Nations, Rome

Ford-Lloyd B, Jackson M (1986) Plant genetic resources: an introduction to their conservation and use. Edward Arnold, London

Frison EA, Putter CAJ (eds) (1989) FAO/IBPGR Technical guidelines for the safe movement of *Musa* germplasm. FAO/IBPGR/INIBAP, Rome

Ganapathi TR, Suprasanna P, Bapat VA, Rao PS (1992) Propagation of banana through encapsulated shoot tips. Plant Cell Rep 11: 571–575

INIBAP (1987) Annual Report 1987. Int network for the improvement of banana and plantain. Montpellier, France

INIBAP (1988) Annual Report 1988. Int network for the improvement of banana and plantain. Montpellier, France

INIBAP (1990) *Musa* conservation and documentation: Proc workshop held in Leuven, Belgium, 11–14 Dec 1989. Int network for the improvement of banana and plantain. Montpellier, France

INIBAP (1991) Annual Report 1991. Int network for the improvement of banana and plantain. Montpellier, France

Jarret RL, Rodriguez W, Fernandez R (1985) Evaluation, tissue culture and propagation and dissemination of Saba and Pelipita plantains in Costa Rica. Sci Hortic 25: 137–147

Lu TG, Sun CS (1992) Cryopreservation of Millet (*Setaria italica* L.). J Plant Physiol 139: 295–298

McWilliams AA, Smith SM, Street HE (1974) The origin and development of embryoids in suspension cultures of carrot (*Daucus carota*). Ann Bot 38: 243–250

Mora A, Abdelnour A, Villalobos (1991) Cryopreservation of zygotic embryos of *Musa*. In: Nabors M (ed) Proc 4th Conf Int Plant Biotechnol Network (IPBNet), San José, Costa Rica, 14–18 Jan 1991, p 81

Murashige T, Skoog F (1962) A revised medium for rapid growth and bioassays with tobacco tissue cultures. Physiol Plant 15: 473–497

Panis B (1986) Cryopreservatie van banaanmeristemen. Thesis, Catholic Univ Leuven, Belgium

Panis B, Swennen R (1994) Cryopreservation of banana meristems using encapsulation dehydration and a simplified freezing method. Proc Int Meeting Soc Low Temp Biol. Leuven, Belgium, 19–23 July 1994 (in press)

Panis B, Withers LA, De Langhe E (1990) Cryopreservation of *Musa* suspension cultures and subsequent regeneration of plants. Cryo Lett 11: 337–350

Panis B, De Smet K, Dhed'a D, Swennen R (1992a) The use of embryogenic *Musa* suspension cultures in biotechnology. Banana Newsl 15: 45–46

Panis B, Dhed'a D, Swennen R (1992b) Freeze-preservation of embryogenic *Musa* suspension cultures. In: Adams RP Adams JE (eds) Conservation of plant genes: DNA banking and in vitro biotechnology. Academic Press, New York, pp 183–195

Panis B, Dhed'a D, De Smet K, Sagi L, Cammue B, Swennen R (1993a) Cell suspensions from somatic tissue in *Musa*: applications and prospects. In: Ganry J (ed) Breeding banana and plantain for resistance to diseases and pests. Proc Int Symp Genetic improvement of bananas for resistance to diseases and pests. CIRAD, Montpellier, France, 7–9 Sept 1992, pp 317–325

Panis B, Van Wauwe A, Swennen R (1993b) Plant regeneration through direct somatic embryogenesis from protoplasts of banana. Plant Cell Rep 12: 403–407

Panis B, De Smet K, Van den houwe I, Swennen R (1994) In vitro conservation of *Musa* germplasm: prospects of cryopreservation. Proc XIth ACORBAT Meet, San José, Costa Rica, 13–18 Jan 1994 (in press)

Purseglove JW (1972) Tropical crops. Monocotyledons. Longman, London

Sagi L, Remy S, Volckaert G, Panis B, Swennen R (1992) Transient expression in banana protoplasts. Banana Newsl 15: 42

Sakai A, Kobayashi S, Oiyama I (1990) Cryopreservation of nucellar cells of navel orange (*Citrus sinensis* Osb. var. *brasiliensis* Tanaka) by vitrification. Plant Cell Rep 9: 30–33

Simmonds NW (1987) Classification and breeding of bananas. In: Persley GJ, De Langhe EA (eds) Banana and plantain breeding strategies: Proc Int Workshop held at Cairns, Australia, 13–17 Oct 1986. ACIAR Proc 21: 69–73

Villalobos VM, Abdelnour A (1992) Cryoconservation of *Musa* spp. and its potential for long-term storage of other tropical crops. In: Adams RP, Adams JE (eds) Conservation of plant genes: DNA banking and in vitro biotechnology. Academic Press, New York, pp 197–210

Vuylsteke D, De Langhe E (1985) Feasibility of *in vitro* propagation of bananas and plantains. Trop Agric (Trinidad) 62: 323–328

Vuylsteke D, Swennen R, De Langhe E (1991) Somaclonal variation in plantains (*Musa* spp, AAB group) derived from shoot-tip culture. Fruits 46 (4): 429–439

Widholm JM (1972) The use of fluorescein diacetate and phenosafranine for determining viability of cultured plant cells. Stain Technol 47: 189–194

Williams JT (1987) Banana and plantain germplasm conservation and movement and need for research. In: Persley GJ, De Langhe EA (eds) Banana and plantain breeding strategies: Proc Int Workshop held at Cairns, Australia, 13–17 Oct 1986. ACIAR Proc 21: 177–181

Withers LA, King PJ (1980) A simple freezing unit and routine cryopreservation method for plant cell cultures. Cryo Lett 1: 213–220

V.7 Cryopreservation of Germplasm of Potato (*Solanum tuberosum* L.) and Cassava (*Manihot esculenta* Crantz)

Y.P.S. Bajaj[1]

1 Introduction

The potato (*Solanum tuberosum* L., family Solanaceae) and the cassava (*Manihot esculenta* Crantz, family Euphorbiaceae) are two important vegetatively propagated crops which provide a rich source of carbohydrate in the human diet.

The center of origin of the potato is the Andean region of Peru, Bolivia, Ecuador, and Colombia at altitudes of 2000 m. It was around 1570 that this crop was introduced to Europe, in Ireland (see Bronk 1975), and then spread to many countries (Table 1). It is an annual plant, essentially an autotetraploid ($2n = 4x = 48$), but over the years dihaploid and monohaploid clones have also been produced. It is propagated through tubers which are a source of food in numerous countries. Recently, experiments have been carried out on the potato for the large-scale production of ethanol, and a Danish distiller in Aalberg has developed a semi-continuous process for the production of ethanol from potatoes (see Dellweg 1983). The germplasm of potato is traditionally conserved through storage of tubers. Numerous clones are being maintained by this method at the International Potato Centre, in Lima, Peru, and at other institutes (Table 2). However, novel methods for the induction of genetic variability and its preservation are also being worked out.

Cassava, also known as tapioca, is an important root crop. It is a perennial autotetraploid (36 chromosomes) shrub which originated in South America (Rogers 1963) and then spread to Africa and Asia (Greenway 1944). Most of the world's production (65%; Table 3) is used for human consumption; the rest is used for cattle feed, industrial starch, alcohol, etc. (Cock 1985). It is one of the cheapest sources of calories for human consumption, and has lately assumed considerable importance as a source of bioenergy for the production of fuel. There are 98 wild species (Jennings 1976). Cassava can be crossed with wild *Manihot* species (Nasser 1980); and *M. dichotoma*, *M. saxicola*, and *M. melanobasis* have been used as a source of disease resistance. Efforts are being made not only to produce new and high-yielding cultivars, but also to preserve the available stocks. Numerous accessions are preserved at various international institutes, especially at the International Centre of Tropical Agriculture (CIAT) in Cali, Colombia (Table 2).

[1]Former Professor of Tissue Culture, Punjab Agricultural University, Ludhiana (Punjab), India
Present Address: A-137 New Friends Colony, New Delhi 110 065, India

Biotechnology in Agriculture and Forestry, Vol. 32
Cryopreservation of Plant Germplasm I (ed. by Y.P.S. Bajaj)

Table 1. Area harvested, yield, and production of potato in various countries. (FAO 1992)

Region/country	Area harvested (1000 ha)	Yield (kg/ha)	Production (1000 metric t)
World	18031	14890	268492
Africa	769	9791	7531
North-central America	761	31377	23885
Canada	124	28482	3529
USA	527	35429	18671
S. America	836	12727	10645
Asia	5108	13477	68848
China	3002	11305	33937
India	957	16196	15500
Japan	122	29918	3650
Turkey	198	22727	4500
Europe	4239	19609	83119
Poland	1757	13309	23388
Former USSR	6262	11661	73024

To conserve the germplasm of vegetatively propagated crops such as the potato, the tubers ("seeds") have to be grown and multiplied in nurseries year after year. Likewise, maintenance of cassava germplasm is through vegetative field propagation of freshly cut stem pieces (stakes). The stakes can only be stored for a short time, as there is premature sprouting. Stakes treated with chemicals have been reported to be preserved for about 6 months, but the vigor of the material deteriorated (Sales-Andrade and Leihner 1980). Thus the preservation of tubers and stakes and their planting/maintenance in the fields are not only time-consuming, labor-intensive, expensive, and cumbersome, but above all, the material is exposed to the hazards of pests and diseases. This may result in a total loss of germplasm. The problem of the conservation of genetic resources of vegetatively propagated plants has encouraged attempts to develop unconventional methods. As an alternative, callus cultures, meristem cultures, and test tube plants have been stored (Schilde-Rentschler and Roca 1987; Dodds et al. 1992). However, the main drawback of callus culture is that it undergoes genetic erosion, and therefore the clones cannot be maintained. Although this difficulty can be surmounted by culturing the excised meristems, they still need to be transferred to the fresh medium. They require extra space and manpower, and there is still the risk of contamination during frequent transfers. Thus the problem of conserving germplasm of vegetatively propagated crops has encouraged attempts to look for other unconventional methods. In this regard cryopreservation of excised meristems is a promising approach (Bajaj 1990). Moreover, during the last decade, significant work has been done on various aspects of in vitro culture of potato (see Table 4), as well as that of the cassava (Bajaj 1985; Schilde-Rentschler and Roca 1987; Raemakers et al. 1993; Schöpke et al. 1993). Increased variability has been generated, some of which needs to be conserved.

Table 2. Potato and cassava accessions in gene banks. (Plucknett et al. 1983)

Species	Accessions	Storage	Institutions[a]	Location
Potato (*Solanum* species)	13 000	Medium, long	CIP	Lima, Peru
	9 435	Medium, long	VIR	Leningrad, USSR
	6 000	Medium	SPBS	Roslin, UK
	5 000	Short, medium	EMBRAPA	Brasilia, Brazil
	3 400	Short	INIPA	Peru
	2 800	Medium	Inter-Regional Potato Introduction Station	Sturgeon Bay, USA
	2 370	Short, medium	FAL	Braunschweig, FRG
Cassava (*Manihot esculenta*)	1 000	Short	AVRDC	Shanhua, Taiwan
	3 000	Medium	CIAT	Cali, Colombia
	2 922	Medium	IITA	Ibadan, Nigeria
	1 800	Short	Central Tuber Crop Res Inst	Kerala, India
	1 500	Short	CENERGEN	Brazil
	1 060	Short	National Cassava Centre	Umuahia, Nigeria

[a]CIP, International Potato Centre; VIR, All Union Institute of Plant Industry; SPBS, Scottish Plant Breeding Station; EMBRAPA, Empresa Brasiliera de Pesquira Agropecuaria; INIPA, Instituto Nacional de Investigacion y Promocion Agropecuaria; FAL, Institut für Planzenbau und Pflanzenzüchtung; AVRDC, Asian Vegetable Research and Development Centre; CIAT, Centre Internacional de Agricultura Tropical; IITA, International Institute of Tropical Agriculture; CENARGEN, Central National Plant Genetics Resources Agency.

Table 3. Area harvested, yield and production of cassava in various countries. (FAO 1992)

Region/country	Area harvested (1000 ha)	Yield (kg/ha)	Production (1000 metric t)
World	15 757	9660	1 52 218
Africa	9303	7572	70 444
Ghana	540	7407	4000
Mozambique	973	3329	3239
Nigeria	1800	11 111	20 000
Uganda	430	8791	3780
Zaire	2400	7631	18 300
North-central America	196	5102	999
S. America	2453	11 963	29 343
Brazil	1884	12 021	22 652
Asia	3788	13 525	51 237
China	231	14 563	3358
India	247	21 053	5200
Indonesia	1333	12 242	16 318

Table 4. Some examples of in vitro induced genetic variability in potato

Resistance incorporated		Reference
	Somaclonal variation	
Alternaria solani (early blight)		Matern et al. (1978)
Phytophthora infestans (late blight)		Behnke (1980), Shepard et al. (1980), Cassells et al. (1991)
Fusarium (dry rot)		Wenzel et al. (1987)
Streptomyces scabies (scab)		Thomson et al. (1986)
Nematodes and viruses		Wenzel and Uhrig (1981)
Insect		Lentini et al. (1990)
	Genetic transformation	
Late blight	Exogenous DNA	Yamamoto (1989)
Insect	*Bacillus thuringiensis*	Cheng et al. (1992)
	Somatic hybrids/cybrids	
Potato virus X	*S. chacoense* + *Solanum tuberosum*	Butenko and Kuchko (1980)
Atrazine	*S. nigrum* + *S. tuberosum*	Gressel et al. (1984)
Leaf roll virus	*S. brevidens* + *S. tuberosum*	Austin et al. (1985)
Late blight	*S. brevidens* + *S. tuberosum*	Helgeson et al. (1986)
Oligomycin	*Nicotiana sylvestris* + *S. tuberosum*	Perl et al. (1991)
Verticillium	*S. torvum* + *S. tuberosum*	Jadari et al. (1992)

2 Cryopreservation of Potato

Work on the cryopreservation of potato germplasm in liquid nitrogen was initiated by the author in 1977 (Bajaj 1977a); since then, considerable improvements in methodology have been made and literature has accumulated (Table 5). This work has been carried out with excised meristems, pollen, root tips, callus cultures, and protoplast-derived cell cultures (Bajaj 1987a), and is summarized below.

2.1 Cryopreservation of Meristems

The degree of success with frozen meristems depends on a number of factors which, if carefully manipulated, result in higher revival. Some of the factors influencing cryoability are discussed below.

2.1.1 Genotype and Source of the Meristems

Various species of *Solanum*, *S. tuberosum* (Bajaj 1977a), *S. gonicalyx* (Grout and Henshaw 1978), *S. etuberosum* (Towill 1981), and *S. phureja* (Fabre and Dereuddre 1990) have responded differently. Differences were observed among cultivars; the cv. Kufri Alankar had a slightly higher viability than Kufri

Table 5. Cryopreservation of in vitro cultures of Potato and Cassava

Crop species	Explant	Response	Reference
		Potato	
Solanum tuberosum	Meristems from sprouts and axillary buds	Callus and shoots	Bajaj (1977a)
S. tuberosum	Meristems from sprouts and axillary buds	Plants with normal tubers	Bajaj (1978)
S. tuberosum	Root tips	Resumed growth	Bajaj (1978)
S. gonicalyx	Shoot meristems	Plants	Grout and Henshaw (1978)
S. tuberosum	Meristems from three cultivars	2 years of storage, normal plants	Bajaj (1981)
S. etuberosum	Meristems from shoot tips	Multiple shoots	Towill (1981)
S. tuberosum	Meristems	Plants	Towill (1984)
S. tuberosum	Meristems from axillary buds	4 years of storage, normal tubers	Bajaj (1985)
S. tuberosum	Somaclones (protoplast-derived callus colony)	Regeneration of Plants	Bajaj (1986)
S. tuberosum	Shoot tips	Plants, tubers	Benson et al. (1989)
S. phureja	Encapsulated shoot tips	Plants	Fabre and Dereuddre (1990)
S. tuberosum	Shoot tips	Shoot regeneration	Harding et al. (1991)
		Cassava	
Manihot utilissima Pohl	Meristems	Callus, plants	Bajaj (1977b, 1982)
M. esculenta	Meristems	Plants	Kartha et al. (1982)
M. esculenta Crantz	Meristems	Survival improved, plants from 3 years storage	Bajaj (1983)
M. esculenta	Meristems	Plants from 4 years of storage, no loss in viability	Bajaj (1985)
M. esculanta	Zygotic embryos	Plants	Marin et al. (1990)
M. esculenta	Somatic embryos	Survival	Sudarmonowati and Henshaw (1990)

Sindhuri and Kufri Chandramukhi (Bajaj 1981). Likewise, cv. Golden Wonder and Desiree reacted differently to freezing (Benson et al. 1989). The axillary buds provided better material in comparison to the sprouts. The shoot tips, though difficult to isolate, were as good as the axillary buds (Table 6). Potato meristems taken from long-term in vitro cultures showed less cryoability than those taken from recently grown in vitro plantlets (Harding et al. 1991).

2.1.2 Preculture

The culture of meristems on a medium containing low concentrations of cryoprotectants for a couple of days before freezing has been observed to be

Table 6. Survival of meristems from tuber sprouts, axillary buds, and shoot tips of three cultivars of potato subjected to sudden freezing and stored in liquid nitrogen for 31 days. (Bajaj 1981)

Cryoprotectant	Survival % of the control				
	cv. Kufri Chandramukhi		cv. Kufri Sindhuri		cv. Kufri Alankar
	Tuber sprout (3–5 mm)	Axillary bud (1–2 mm)	Tuber sprout (3–5 mm)	Axillary bud (1–2 mm)	Shoot tip (0.8–1.5 mm)
DMSO 10%	11	19	12	21	19
DMSO 5% + glycerol 5% + sucrose 5%	13	23	14	26	27

beneficial (Table 7). The potato meristems were precultured on a basal medium containing 2–7% DMSO. After 2 days, the healthy meristems were selected and frozen. Of the various concentrations tried, 2% DMSO yielded the best response (Table 7). Towill (1981) observed that meristems of *S. etuberosum* precultured on basal medium (without cryoprotectant) gave better results than the freshly harvested ones. Fabre and Dereuddre (1990) precultured meristems of *S. phureja* in high sucrose solution for 24–72 h.

2.1.3 Cryoprotectants

The function of the cryoprotectant is to act as an "antifreeze"; it avoids intracellular freezing of cells, which, when it occurs, is invariably lethal. A number of compounds have been employed for the cryopreservation of plant cell cultures. In the potato, initially glycerol (15%) or a mixture of glycerol (10%) and sucrose (5%) was used for the meristems (Bajaj 1977a). This was later improved by using a mixture containing 5% each of DMSO, sucrose, and glycerol; the results were better than on DMSO (10%) alone (Bajaj 1978).

Grout and Henshaw (1978) used 10% DMSO for *S. gonicalyx*. Of the various concentrations of DMSO (10, 15, and 20%) employed by Towill (1981) for *S. etuberosum*, maximum viability was obtained with 10% DMSO; 15 and 20% were

Table 7. Effect of preculture (2 days) on survival of potato (cv. Kufri Alankar) meristems freeze-preserved for 3 months in liquid nitrogen

Preculture medium	No. of meristems cultured	No. of meristems surviving	Survival (%)
1. Basal medium (BM)	31	12	38.6
2. BM + DMSO 2%	39	17	43.5
3. BM + DMSO 5%	44	15	34.0
4. BM + DMSO 7%	40	13	32.5

inhibitory, however, he did not use any other cryoprotectant. Benson et al. (1989) also used 10% DMSO for *S. tuberosum*.

2.1.4 Freezing

Although the method and the rate of freezing are crucial for cell cultures, the meristems of a number of plant species have survived sudden immersion in liquid nitrogen (LN). In 1977, Bajaj reported for the first time the revival of frozen potato meristems using an ultra-fast method of freezing (Bajaj 1977a). Later, similar observations were made by Grout and Henshaw (1978) with *S. gonicalyx*. These findings were further confirmed by Benson et al. (1989) and Harding et al. (1991), who reported that ultra-rapid freezing improved total survival and conserved the ability of potato shoot tips to develop into plants.

In another species, *S. etuberosum*, the meristems were first subjected to −30 and −40 °C, and then immersed in LN (Towill 1981). The viability of meristems was higher than that of *S. tuberosum*. It cannot be ascertained whether high viability is due to the different freezing method, or to the differences in cryosensitivity of the species, as no comparison of the two species was made.

2.1.5 Storage Temperature

The injury caused to the system may be due to freezing or storage, the latter occurring when the cells are not stored at sufficiently low temperatures. No work has been done to compare the effect of storage temperature on survival of potato meristems, however, studies conducted on the in vitro system of other plants show that there is a deterioration of material stored at −20 and −70 °C (Bajaj 1976). Thus, for long-term storage a temperature below −130 °C, preferably −196 °C, is recommended.

2.1.6 Thawing

Based on previous experience with other plants in the studies conducted by Bajaj, as well as those by Grout and Henshaw (1978), thawing was routinely done at 35–40 °C. Towill (1981), however, warmed the frozen cultures either by briskly shaking in a 37 °C water bath, or by suspension in air at room temperature; no difference between the two treatments was observed. Benson et al. (1989) also thawed the shoot tips at 40 °C.

2.1.7 Culture

Although the culture medium and the incubation conditions are important, there does not seem to be any difference in the requirements of the retrieved and the control cultures, or at least no critical work has been done on this aspect. Plants are regenerated from the retrieved meristems of potato cultured on an MS medium (Murashige and Skoog 1962) supplemented with BAP or kin (1 mg/l) and IAA (0.1 mg/l) in an agar-solidified or on liquid medium in wick cultures, and incubated at 24–26 °C in a 12-h fluorescent light regime (Bajaj 1981)

2.2 Tuberization in Plants Regenerated from Frozen Meristems

Most of the frozen meristems did not show any sign of growth for the first 1–2 weeks. Some of them lost chlorophyll and turned brown. On prolonged culturing, however, they had a tendency to show localized proliferation at the basal end, and an elongation of leaf primordia. The callus varied from white to light-green and friable to nodular. Like the untreated callus (Bajaj and Dionne 1967), the callus obtained from frozen meristems underwent morphogenesis when transferred to the differentiation medium of Roest and Bokelmann (1976), and produced shoots.

In other cultures the meristems, without undergoing callusing, developed directly into plants. Initially, the elongation and unfolding of the leaves were slow; however, in about 10–12 weeks, shoots and plantlets fit to be transferred to pots were obtained. Some of the shoots which did not develop roots were transferred to a medium containing IAA. Occasionally, such plants in the test tubes produced microtubers (Bajaj 1978). The plantlets, when transferred to pots or to the field, underwent tuberization (Fig. 1F).

2.3 Cryopreservation of Root Tips

Actively growing root tips (5 mm–1 cm) of two cultivars (Kufri Sindhuri and Kufri Chandramukhi) survived subjection to −196 °C in a mixture of 5% each of DMSO, sucrose, and glycerol (Bajaj 1978). Unlike the untreated controls (Bajaj and Dionne 1968) in which root tips started to elongate after 1 day of culture, in the frozen cultures the tip, though it remained white, showed a lag phase of 2–5 weeks. Thereafter, on the basal medium, it elongated and formed a mass of thin roots, whereas on 2,4-D (2 mg/l) medium the roots underwent sparse proliferation (Montezuma-de-Carvalho and Guimaraes 1976). In some cases, as in the untreated roots (Bajaj and McAllan 1969), growth was very slow; the roots became thick, developed small protuberances, and turned green when transferred to light.

The root tips, like the shoot meristems, are free of viruses and therefore could be employed for the storage of disease-free materials. Likewise, the opposite would also be applicable, i.e., virus-infected roots would serve as a host for virus multiplication and conservation (White 1934).

2.4 Cryopreservation of Cell Suspensions and Somaclones

The potato possesses very little or no frost tolerance, and normally cannot survive at −3 °C. Since potato plants can be regenerated with relative ease from

Fig. 1A–F. Regeneration of plants from potato meristems frozen in liquid nitrogen. **A–C** Various stages in the development of retrieved meristems freeze-preserved for 14 months and cultured on agar medium. **D–F** Transfer of plants from test tube to pot (**D**) and the field (**E**), and the production of tubers (**F**). (After Bajaj 1981)

Fig. 2. Differentiating callus obtained from protoplast-derived cell colonies frozen in LN. Note the differentiation of shoot buds; the plants regenerated from such cultures underwent normal tuberization. (Bajaj 1986)

callus cultures, the development of frost-tolerant cell lines/plants should be helpful. Chen et al. (1979) observed that leaf callus from three cultivars of *S. tuberosum* showed no increase in frost hardiness in response to low temperature, and the callus, like the plants, was killed at −3 °C. However, leaf callus of *S. acaule* could be hardened to −9 °C after 15 days at 3 °C, a 3 °C increase in frost hardiness. Potato cell suspension subjected to −14 °C in the presence of proline revived (Hellergren and Li 1981); however, no attempts were made to regenerate plants from the retrieved cells. Such studies would be rewarding for obtaining frost-tolerant plants.

The selection of novel somaclones and their preservation is another area of immense importance. Bajaj (1986) successfully cryopreserved somaclones of potato. Twenty-five to 50 protoplast-derived cell colonies (about 1 mm) pooled in a cryoprotectant mixture containing 5% each of DMSO, sucrose, and glycerol were frozen in LN. When returned to the culture, they remained quiescent for about 1 week, then resumed growth and underwent morphogenesis (Fig. 2) to form shoot buds and plantlets. The plants thus obtained produced normal tubers.

2.5 Viability and Stability During Storage

Potato and cassava are the two plant species in which long-term storage, i.e., up to 4 years, has been attempted (Bajaj 1985). The periodic viability studies on meristems cryopreserved for periods ranging from 3 months to 4 years showed no significant change in survival (Table 8). The plants produced from such cultures appeared morphologically normal. Although no detailed cytological studies were undertaken, the chromosome number was normal (2n = 48). Upon transfer to pots and field plots, the plants underwent normal tuberization. The tubers were fairly uniform in size, shape, and color, and virtually no differences were observed as compared to controls. The tubers in turn produced normal plants.

Table 8. Studies of viability of excised meristems of cassava and potato freeze-preserved for various lengths of time. (Bajaj 1985)

Preservation period	Cassava cv. GS-1				Potato cv. Alankar			
	No. of meristems cultured	No. of cultures resuming growth	Survival (%)	Survival (%) of control	No. of meristems cultured	No. of cultures resuming growth	Survival (%)	Survival (%) of control
Control (unfrozen)	38	31	81.5	—	43	37	86.0	39.7
3 months	23	8	34.7	42.5	35	12	34.2	39.7
6 months	26	9	34.6	42.4	27	10	37.0	43.0
1 year	30	10	33.3	40.8	24	9	37.5	43.6
2 years	31	11	35.4	43.4	32	11	34.3	39.8
3 years	25	9	36.0	44.1	25	9	36.0	41.8
4 years	29	10	34.4	42.2	27	9	33.3	38.7

Likewise, cassava produced starchy roots. These studies preclude the possibility of loss of viability, and avoid the selection pressure during storage.

2.6 Cryopreservation of Pollen

The preservation of pollen is important for a number of purposes, especially for breeding and crop improvement programs, and additionally to help to reduce the incidence and spread of pollen-transmitted diseases. Pollen is stored by a number of methods such as refrigeration, freeze- and vacuum-drying, the use of organic solvents, and others (see Bajaj 1987b). However, with these methods pollen viability is reduced or lost with the passage of time. Thus, for meaningful long-term conservation of germplasm, pollen storage in LN is recommended.

The cryopreservation of pollen is rather simple as compared to that of cells and tissue cultures. This is due to low water content, highly packed storage materials such as sugars, starches, and oils, the nonvacuolated nature, and resistant exine. Moreover, they do not require cryoprotectants or a specific rate of cooling; they can be directly immersed in LN, and thus sophisticated and expensive cryostats are not required.

Potato pollen is shed at the binucleate stage, and retains its viability for a few days. At low temperatures, for instance at −20 and −34 °C, the potato pollen retains high viability and sets seeds after storage for 1 year (King 1955; Howard 1958). Weatherhead et al. (1978) successfully stored pollen from three tuber-bearing species of *Solanum* for 1 year in LN, and found this superior to −20 °C. Likewise, Towill (1987) worked on pollen from several *Solanum* species and his data showed that a moisture content of less than 35–40% allowed for high survival of pollen in LN; rehydration was necessary for some species. The cryopreserved pollen produced normal fruit and seed set.

3 Cryopreservation of Cassava

Cryopreservation of meristems has yielded some promising results, as complete plants have been regenerated from meristems which had been freeze-preserved for 4 years (Bajaj 1985). Studies have also been conducted on seeds and embryos (Table 5).

3.1 Cryopreservation of Meristems

3.1.1 Methodology

Ten to 15 cm long pieces of stakes (stem) are planted in the soil; the lateral buds begin to sprout within 1 week. They are excised, surface sterilized, and washed. The meristems are excised from the sterilized buds and precultured on modified MS medium + BA (0.1 mg/l) + NAA (0.2 mg/l) + GA_3 (0.05 mg/l) containing 3%

DMSO. After 5 days the cultures showing signs of growth are selected; 25 of them are placed on filter paper soaked in a mixture of cryoprotectant (5% each of DMSO, sucrose, and glycerol), and are kept for 2–4 h. The meristems are either transferred to the vials, or they are frozen along with the filter paper wrapped in aluminum foil (Bajaj 1983). The freezing is then done by either direct immersion in LN or by initial exposure to vapors of liquid nitrogen and then gradual lowering and immersion in LN. The meristems subjected to slow freezing and cultured in microdrops also revived (Kartha et al. 1982). Recently, cassava meristems were encapsulated in alginate beads, precultured in a high sucrose concentration, and partially dehydrated before freezing (Benson et al. 1992).

The meristems are generally thawed at 35–40 °C and cultured on MS + BA + NAA + GA medium. The cultures are maintained at 23–27 °C in diffuse light. The plantlets obtained from the retrieved cultures are transferred to pots and then to the field (Bajaj 1983).

3.1.2 Results

The survival of the retrieved meristems was judged by their ability: (1) to increase in size, (2) to turn green, (3) to proliferate to form callus, and (4) to develop into shoots and plantlets.

The response of the retrieved cultures depended on the physiological state of the material, preculture, cryoprotectant used, and the method of freezing and thawing. These factors are critical in obtaining optimal viability. On reculturing, the frozen-thawed meristems (Fig. 3A) lost their green pigmentation, turned whitish to light brown, and did not show any visible sign of growth during the first 1–2 weeks. After that period, when observed under the dissecting microscope, they showed overall swelling. The time taken to resume growth depended on the extent of cryoinjury. The meristems which were damaged during excision or freezing invariably proliferated to form callus, whereas others developed shoots or normal-looking plants. Sometimes the cultures did not show any response, were quiescent for as much as 4 months, and were regarded as dead. However, thereafter there was a spurt in growth activity and a mass of callus was formed. Stages of growth, development, and regeneration are shown in Fig. 3. The plantlets obtained from the retrieved cultures, on transfer to the soil, produced starchy roots (Fig. 3I).

The effect of various cryoprotectants was studied on the frozen cultures and was compared with the controls (Table 9). The combination of various cryoprotectants was more effective than were the chemicals used singly. The highest response was obtained with a mixture of various cryoprotectants, containing 5% each of DMSO, sucrose, and glycerol. This was an improvement on earlier work (Bajaj 1977b) in which only sucrose and glycerol were used, and only 21% of the retrieved cultures resumed growth. In later studies, viability was improved (Bajaj 1983, 1985). The meristems cryopreserved for various lengths of time ranging from 3 months to 4 years (Table 8) showed no significant difference in viability; the survival varied from 40.8 to 44.1%. However, this viability can be further increased by manipulating various factors. It is perhaps of interest to point out

Table 9. Effect of various cryoprotectants on the growth and development of unfrozen and frozen meristems of cassava cultured on MS + BA + NAA + GA medium. (Bajaj, 1983)

Treatment	Unfrozen culture			Frozen culture			
	No. of cultures raised	No. of cultures growing	Percentage growing (control)	No. of cultures raised	No. of cultures growing	Percentage survival	Survival percent of control
Control (distilled water)	34	31	91.1	23	0	0	0
(1) Sucrose (10%)	27	23	85.2	34	6	17.6	20.7
(2) Glycerol (10%)	31	27	87.1	39	5	12.8	14.6
(3) DMSO (10%)	24	19	79.1	27	5	18.5	23.3
(3) Sucrose (5%) + glycerol (10%)	43	38	88.3	51	9	17.6	19.9
(5) Sucrose (5%) + glycerol (5%) + DMSO (5%)	61	54	88.5	69	18	26.0	29.3

that the meristems excised from the plants in winter (November–December) generally withstood freezing better than those obtained in summer (June–July). Moreover, a difference in the cultivars was also noticed: the cv. M-4 was more sensitive than GS-1. Cassava is a tropical plant and sensitive to low temperature.

3.2 Cryopreservation of Embryos

Since somatic embryos can now be produced in large numbers in a number of plant species, cryopreservation would be ideal for their storage (Bajaj 1995a). Preliminary studies have been conducted on the cryopreservation of somatic embryos of cassava (Sudarmonowati and Henshaw 1990). The somatic embryos were precultured in 7% sucrose, treated with 10% DMSO for 12 h, then subjected to slow freezing. Up to 92% of the cultures callused, and 25% underwent secondary embryogenesis. Experiments have also been conducted on zygotic embryos of cassava (Marin et al. 1990).

4 Gene Banks and the International Exchange of Germplasm

In addition to many other advantages, the in vitro storage and exchange of germplasm ensure the supply of virus-free materials, and also reduce the chances of dissemination of other diseases which normally occur through the conventional method of exchange of tubers. The international exchange of disease-free germplasm can be affected by the following methods:

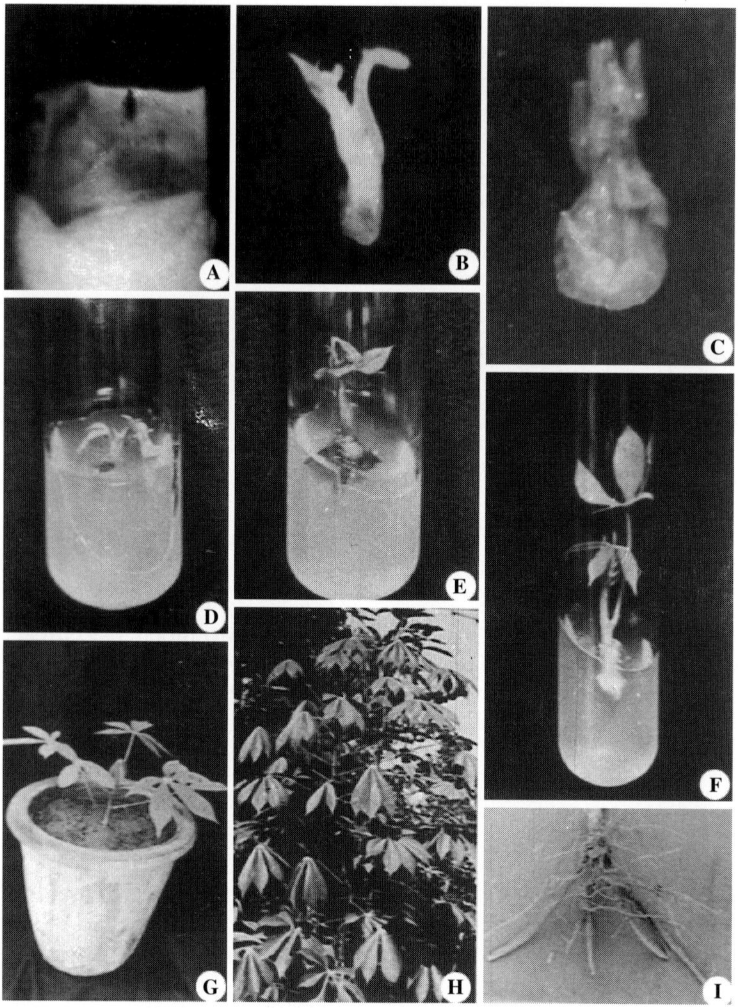

Fig. 3A–I. Regeneration of cassava plants from frozen meristems. **A** Frozen-thawed shoot meristem at the time of culture on filter-paper wick in a liquid medium. **B, C** 7-week-old retrieved culture showing growth. **D–F** Development of shoot and plantlet from meristem cryopreserved for 3 years. **G** Transfer of a plant from test tube to a pot, and then to the field (**H**). **I** Starch roots. (Bajaj 1983)

1. Test-Tube Plants. In vitro, meristem-derived, disease-free potato and cassava plantlets in test tubes packed in polystyrene boxes have been exported to a number of countries with success (Schilde-Rentscheler and Roca 1987). The exchanged plants can be temporarily stored, or directly transferred to the field.

2. Microtubers. Microtubers have certain advantages over test tube plants for storage and exchange: they can withstand longer periods in the dark, and they also tolerate the changes in outside temperature better. They can be easily transported in polythene bags or plastic dishes to various destinations, stored,

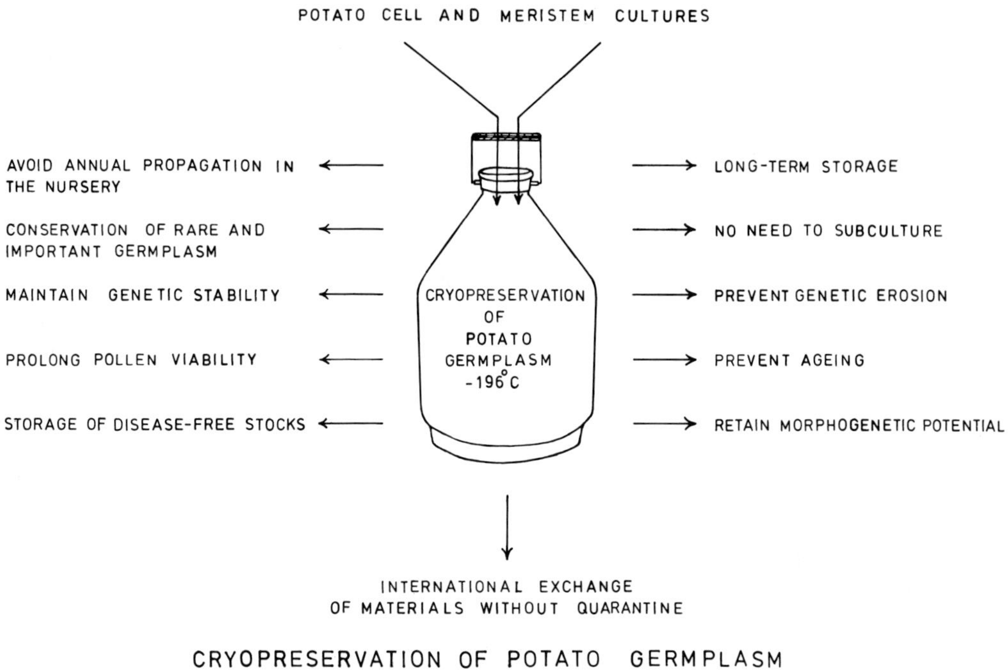

Fig. 4. Prospects for the cryopreservation of cell, tissue, and organ culture of potato and cassava and the establishment of "germplasm banks". (Bajaj 1982)

planted, and multiplied. In fact, microtubers are already being commercially exported by some private biotechnology firms, however, exact data and details are not available.

3. Synthetic Seed. The encapsulation of the meristems or the somatic embryos would perhaps be the easiest method to store and transport clonal material (see Bajaj 1995b). The meristems of *Solanum phureja*, coated in calcium alginate beads and subjected to freezing, formed shoots when thawed and recultured (Fabre and Dereuddre 1990).

4. Cryopreservation. Since normal potato and cassava plants have been obtained from cultures freeze-preserved for 4 years (Bajaj 1985), cryopreservation can be used as an effective tool for the establishment of gene banks and for the international exchange of germplasm.

5 Summary and Conclusions

Excised meristems of two vegetatively propagated crops, potato and cassava, were freeze-preserved in liquid nitrogen for various lengths of time. They revived and regenerated plants capable of undergoing tuberization and starchy root formation, respectively. Potato somaclones obtained from protoplast cultures subjected to LN also resumed growth and underwent differentiation, showing that cryopreservation can be applied for the conservation of in vitro induced genetic variability.

To conserve the germplasm of potato the plants have to be grown in nurseries year after year. This is time-consuming, labor-intensive and also involves hazards of pathogens, pests, and adverse environments. In this regard cryopreservation assumes great importance for the conservation of germplasm of potato and cassava and other vegetatively propagated crops, and for the establishment of gene banks (Fig. 4).

References

Austin S, Baer MA, Helgeson JP (1985) Transfer of resistance to potato leaf role virus from *Solanum brevidens* into *Solanum tuberosum* by somatic fusion. Plant Sci 39: 75–82

Bajaj YPS (1976) Regeneration of plants from cell suspensions frozen at −20, −70 and −196 °C. Physiol Plant 37: 263–268

Bajaj YPS (1977a) Initiation of shoots and callus from potato tuber-sprouts and axillary buds frozen at −196 °C. Crop Improv 4: 48–53

Bajaj YPS (1977b) Clonal multiplication and cryopreservation of cassava through tissue culture. Crop Improv 4: 198–204

Bajaj YPS (1978) Tuberization in potato plants regenerated from freeze-preserved meristems. Crop Improv 5: 137–141

Bajaj YPS (1981) Regeneration of plants from potato meristems freeze-preserved for 24 months. Euphytica 30: 141–145

Bajaj YPS (1982) Cryopreservation of germplasm of potato, cassava, peanut and chickpea. In: Fujiwara A (ed) Plant tissue culture 1982. Maruzen, Tokyo, pp 799–800

Bajaj YPS (1983) Cassava plants from meristem cultures freeze-preserved for three years. Field Crop Res 7: 161–167

Bajaj YPS (1985) Cryopreservation of germplasm of potato (*Solanum tuberosum* L.) and cassava (*Manihot esculenta* Crantz): viability of excised meristems cryopreserved up to four years. Indian J Exp Biol 23: 285–287

Bajaj YPS (1986) Cryopreservation of potato somaclones. In: Semal J (ed) Somaclonal variations and crop improvement. Nijhoff, Dordrecht, pp 244–250

Bajaj YPS (1987a) Cryopreservation of potato germplasm. In: Bajaj YPS (ed) Biotechnology in agriculture and forestry, vol 3. Potato. Springer, Berlin Heidelberg New York, pp 472–486

Bajaj YPS (1987b) Cryopreservation of pollen and pollen embryos, and the establishment of pollen banks. In: Giles KL, Prakash J (eds) Pollen development and cytology. Academic Press, New York, pp 397–420

Bajaj YPS (1990) Cryopreservation of germplasm of vegetatively propagated crops. Bull Soc Bot Fr 137, Actual Bot (3/4): 99–114

Bajaj YPS (1995a) Cryopreservation of somatic embryos. In: Bajaj YPS (ed) Biotechnology in agriculture and forestry, vol 30. Somatic embryogenesis and synthetic seed I. Springer, Berlin Heidelberg New York, pp 221–229

Bajaj YPS (ed) (1995b) Biotechnology in agriculture and forestry, vols 30, 31. Somatic embryos and synthetic seed I, II. Springer, Berlin Heidelberg New York

Bajaj YPS, Dionne LA (1967) Organ formation in potato callus by plating cell suspensions. Am J Bot 54: 640

Bajaj YPS, Dionne LA (1968) The continuous culture of excised potato roots. N Z J Bot 6: 386–394

Bajaj YPS, McAllan W (1969) Effect of various light treatments on chlorophyll formation in excised potato roots. Physiol Plant 22: 25–28

Behnke M (1980) General resistance to late blight of *Solanum tuberosum* plants regenerated from callus resistant to culture filtrates of *Phytophthora infestans*. Theor Appl Genet 56: 151–152

Benson EE, Harding K, Smith H (1989) Variation in recovery of cryopreserved shoot tips of *Solanum tuberosum* exposed to different pre- and post-freeze light regimes. Cryo Lett 10: 323–344

Benson EE, Chabrillange N, Engelmann F (1992) A comparison of cryopreservation methods for the long-term in vitro conservation of cassava (*Manihot esculenta*). In: Proc Soc Low Temp Biol, Autumn Meet, Stirling, UK

Bronk B (1975) Plants consumed by man. Academic Press, New York

Butenko RG, Kuchko AA (1980) Somatic hybridization of *Solanum tuberosum* L. and *Solanum chacoense* Bitt. by protoplast fusion. In: Ferenczy GL (ed) Advances in protoplast research. Acad Kiado, Budapest, pp 293–300

Cassells AC, Deadman ML, Brown CA, Griffin E (1991) Field resistance to late blight (*Phytophthora infestans* (Mont) De Bary) in potato (*Solanum tuberosum* L.) somaclones associated with instability and pleiotropic effects. Euphytica 56: 75–80

Chen HH, Gavinlertavatana P, Li PH (1979) Cold acclimation of stem-cultured plants and leaf callus of *Solanum* species. Bot Gaz 140: 142–147

Cheng J, Bolyard MG, Saxena RC, Sticklen MB (1992) Production of insect resistant potato by genetic transformation with a δ-endotoxin gene from *Bacillus thuringiensis* var. *kurstaki*. Plant Sci 81: 83–91

Cock JH (1985) Cassava: new potential for a neglected crop. Wetview Press, Boulder

Dellweg H (ed) (1983) Biotechnology—a comprehensive treatise, vol 3. Chemie, Weinheim

Dodds JH, Silva-Rodriguez D, Tovar P (1992) Micropropagation of potato (*Solanum tuberosum* L.). In Bajaj YPS (ed) Biotechnology in agriculture and forestry, vol 19. High-tech and micropropagation III. Springer, Berlin Heidelberg New York, pp 91–106

Fabre J, Dereuddre J (1990) Encapsulation-dehydration: a new approach to cryopreservation of *Solanum* shoot tips. Cryo-Lett 11: 413–426

FAO (1992) Production yearbook, vol 46, FAO, Rome

Greenway PJ (1944) Origins of some East African food plants. E Afr Agric J 10: 34–39

Gressel J, Cohen N, Binding H (1984) Somatic hybridization of an atrazine resistant biotype of *Solanum nigrum* with *Solanum tuberosum*. 2. Segregation of plastomes. Theor Appl Genet 67: 131–134

Grout BWW, Henshaw GG (1978) Freeze preservation of potato shoot tip cultures. Ann Bot (Lond) 42: 1227–1229

Harding K, Benson EE, Smith H (1991) The effects of in vitro culture period on the recovery of cryopreserved shoot tips of *Solanum tuberosum*. Cryo-Lett 12: 17–22

Helgeson JP, Hunt GJ, Haberlach GT, Austin S (1986) Somatic hybrids between *Solanum brevidens* and *Solanum tuberosum*: expression of a late blight resistance gene and potato leaf role resistance. Plant Cell Rep 5: 212–214

Hellergren J, Li PH (1981) Survival of *Solanum tuberosum* suspension cultures to $-14\,^{\circ}$C. The mode of action of proline. Physiol Plant 52: 449–453

Howard HW (1958) The storage of potato pollen. Am Potato J 35: 676

Jadari R, Sihachakr D, Rossignol L, Ducreux G (1992) Transfer of resistance to *Verticillium dahliae* Kleb. from *Solanum torvum* SW into potato (*Solanum tuberosum* L.) by protoplast electrofusion. Euphytica 64: 39–47

Jennings DL (1976) Cassava, *Manihot esculenta* (Euphorbiaceae). In: Simmonds NW (ed) Evolution of crop plants. Longman, London, pp 81–84

Kartha KK, Leung NL, Mroginski LA (1982) In vitro growth responses and plant regeneration from cryopreserved meristems of cassava (*Manihot esculenta* Crantz). Z. Pflanzenphysiol 107: 133–140

King JR (1955) Irish potato pollen storage. Am Potato J 32: 460

Lentini Z, Earle ED, Plaisted RL (1990) Insect-resistant plants with improved horticultural traits from interspecific potato hybrids grown in vitro. Theor Appl Genet 80: 95–104

Marin ML, Mafla G, Roca WM, Withers LA (1990) Cryopreservation of cassava zygotic embryos and whole seeds in liquid nitrogen. Cryo-Lett 11: 257–264

Matern U, Strobel G, Shepard J (1978) Reaction to phytotoxins in a potato population derived from mesophyll protoplasts. Proc Natl Acad Sci USA 75: 4935–4939

Montezuma-de-Carvalho J, Guimaraes MLL (1976) Isolation of callus from potato roots. Bol Soc Borteriana (2 a Ser) L: 143–163

Murashige T, Skoog F (1962) A revised medium for rapid growth and bioassays with tobacco tissue cultures. Physiol Plant 15: 473–497

Nasser NMA (1980) Attempts to hybridize wild *Manihot* species with cassava. Econ Bot 34: 13–15

Perl A, Aviv D, Galun E (1991) Protoplast fusion mediated transfer of oligomycin resistance from *Nicotiana sylvestris* to *Solanum tuberosum* by intergenetic hybridization. Mol Gen Genet 225: 11–16

Plucknett DL, Smith NJH, Williams JT, Murthy-Anishetty N (1983) Crop germplasm conservation and developing countries. Science 220: 163–169

Raemakers CJJM, Amati M, Staritsky G, Jacobsen E, Visser RGF (1993) Cyclic somatic embryogenesis and plant regeneration in cassava. Ann Bot 71: 289–294

Roest S, Bokelmann GS (1976) Vegetative propagation of *Solanum tuberosum* L. in vitro. Potato Res 19: 173–178

Rogers DJ (1963) Studies of *Manihot esculenta* Crantz and related species. Bull Torrey Bot Club 90: 43–54

Sales-Andrade AM, Leihner DE (1980) Influence of period and conditions of storage on growth and yields of cassava. In: Weber EJ, Toro JC, Graham M (eds) Cassava cultural practices. Proc Worksh, Salvador, Brazil. Int Dev Res Centre, Ottawa, pp 33–37

Schilde-Rentschler L, Roca WM (1987) Tissue culture for the international exchange of potato and cassava. In: Bajaj YPS (ed) Biotechnology in agriculture and forestry, vol 3. Potato. Springer, Berlin Heidelberg New York, pp 453–465

Schöpke C, Franche C, Bogusze D, Chavarriaga P, Fauquet C, Beachy RN (1993) Transformation in cassava (*Manihot esculenta* Crantz). In: Bajaj YPS (ed) Biotechnology in agriculture and forestry, vol 23. Plant protoplasts and genetic engineering IV. Springer, Berlin Heidelberg New York, pp 273–289

Shepard JF, Bidney D, Shahin E (1980) Potato protoplasts in crop improvement. Science 208: 17–24

Sudarmonowati E, Henshaw GG (1990) Cryopreservation of cassava somatic embryos. VIIth Int Congr Plant tissue and cell culture, Amsterdam. p 378 (Abstr)

Thomson AJ, Gunn RE, Jellis GJ, Boulton RE, Lacey CND (1986) The evaluation of potato somaclones. In: Semal J (ed) Somaclonal variations and crop improvement. Nijhoff, Dordrecht, pp 148–159

Towill LE (1981) *Solanum etuberosum*: a model for studying the cryobiology of shoot-tips in the tuber bearing (*Solanum*) species. Plant Sci Lett 20: 315–324

Towill LE (1984) Survival at ultra-low temperatures of shoot tips from *Solanum tuberosum* groups andigena, phureja stenotomum, tuberosum, and other tuber-bearing *Solanum* species. Cryo-Lett 5: 319–326

Towill LE (1987) Preservation of potato pollen. In: Bajaj YPS (ed) Biotechnology in agriculture and forestry, vol 3. Potato. Springer, Berlin Heidelberg New York, pp 427–440

Weatherhead MA, Grout BWW, Henshaw GG (1978) Advantages of storage of potato pollen in liquid nitrogen. Potato Res 21: 331–334

Wenzel G, Uhrig H (1981) Breeding for nematode and virus resistance in potato via anther culture. Theor Appl Genet 59: 333–340

Wenzel G, Debnath SC, Schuchmann R, Foroughi-Wehr B (1987) Combined application of classical and unconventional techniques in breeding for disease resistant potatoes. In: Ellis GJ, Richardson DE (eds) The production of new potato varieties: technological advances. University Press, Cambridge, pp 277–288

Westcott RJ (1981) Tissue culture storage of potato germplasm. Potato Res 24: 331–352

White PR (1934) Multiplication of viruses of tobacco and aucuba mosaic in growing excised tomato roots. Phytopathology 24: 1003–1011

Yamamoto M (1989) Transformation in potato. In: Bajaj YPS (ed) Biotechnology in agriculture and forestry, vol 9. Plant protoplasts and genetic engineering II. Springer, Berlin Heidelberg New York, pp 122–139

Section VI
Medicinal, Aromatic, and Ornamental Plants

VI.1 Cryopreservation of Germplasm of Medicinal and Aromatic Plants

Y.P.S. Bajaj[1]

1 Introduction

Plants are the most important source of medicines (Balandrin and Klocke 1988); however, due to pressures of human population and economic development, the natural ecosystem is continually disturbed throughout the world. This has resulted in the depletion of naturally occurring genetic resources and germplasm pools from which future pharmaceutical and other useful products might be developed. This has caused international concern for the conservation of germplasm. There is an urgent need to conserve the germplasm of thousands of rare, elite, and endangered species of medicinal plants, especially those of recalcitrant types in which seed cannot be preserved. In this regard in vitro culture has played a significant role, and now there are numerous species of medicinal and aromatic plants belonging to diverse families on which micropropagation studies have been conducted. Bajaj et al. (1988) reviewed the subject that deals with 148 such species belonging to 108 genera.

The storage of cells is important not only for the conservation of germplasm, but for its use in large quantities for the production of medicinal compounds and other pharmaceutical products. Although various in vitro methods that reduce the rate of growth have been employed (see Bajaj 1986), they do not allow long-term conservation. Cryopreservation has the advantage of overcoming this problem. During the last two decades, significant progress has been made and literature has accumulated on the freeze preservation of plant cell, tissue, and organ cultures (Bajaj 1991). Entire plants have been regenerated from cultures cryostored in liquid nitrogen (– 196 °C) for various lengths of time. The technological refinements permit almost 100% survival of retrieved cultures in some plant species (Table 1). In this chapter, which is an update of an earlier contribution by the author (Bajaj 1988a), work on freeze preservation of in vitro cultures of medicinal and aromatic plants is reviewed, and its potential for the retention of biosynthetic capacity of cell cultures and their use in pharmacy is pointed out.

[1]Former Professor of Tissue Culture, Punjab Agricultural University, Ludhiana (Punjab), India. *Present address*: A-137 New Friends Colony, New Delhi 110 065, India

Biotechnology in Agriculture and Forestry, Vol. 32
Cryopreservation of Plant Germplasm I (ed. by Y.P.S. Bajaj)

Table 1. Cryopreservation of in vitro cultures of some medicinal and aromatic plants (also see Chap. I.1, this vol.)

Plant species	Culture	Cryoprotectant	Freezing (°C)	Thawing (°C)	Response	Reference
Allium sativa	Shoot tip	–	– 196	–	–	Niwata (1992)
Anisodus acutangulus	Cell suspension	DMSO or lactalbumin hydrolysate	– 196	37	Hyoscyamine and scopolamine synthesis	Zheng Guang-zhi et al. (1983)
Atropa belladonna	Cell suspension	DMSO (5%)	2/min, – 100, – 196	37	Callus	Nag and Street (1975)
A. belladonna	Pollen embryos	DMSO + sucrose + glycerol	1–3 °C/min, – 196	35–40	Haploid plants	Bajaj (1976b, 1977, 1978)
A. belladonna	Protoplasts	DMSO + mannitol	Exposed to vapors, immersed in LN	35	Plants	Bajaj (1988b)
Beta vulgaris	Meristem	DMSO + sorbitol	– 40, – 196	39	Plantlets	Braun (1988)
B. vulgaris	Root cultures	DMSO (10%)	– 196	25	Biosynthetic activity	Benson and Hamill (1991)
Camellia sinensis	Embryo axis	Not used	Direct – 196	38	Seedlings	Chaudhury et al. (1991) Wesley-Smith et al. (1992)
Cannabis sativa	Cell suspension	DMSO (10%)	2/min, – 50, – 196	40	58% Viability	Jekkel et al. (1989)
Catharanthus roseus	Cell suspension	DMSO (5%) + 1 M sorbitol	0.5 – 3/min to – 35, – 196	40	Callus	Chen et al. (1984)
Chamomilla recutita	Cell culture		– 196			Čellárová et al. (1992)
Chrysanthemum morifolium	Shoot tips	DMSO + glucose	0.2/min, – 40, – 196	25–30	Shoots	Fukai (1990); Fukai et al. (1991)
Cichorium intybus	Shoot tips	–	–	–	Plants	Demeulemeester et al. (1993)
Cinchona ledgeriana	Callus	Mannitol + DMSO	Slow freezing	Rapid thawing	Callus,	Hunter (1986)
Citrus reticulata	Ovule/nucellar embryo	DMSO (7%) + sucrose (7%)	– 196	35–40	Callus, plants	Bajaj (1984)

C. sinensis	Somatic embryos	DMSO (10%)	Slow freezing	Room temperature	Plants	Marin and Duran-Vila (1988)
C. sinensis	Nucellar cells	DMSO + glycerol + glycol	Vitrification, fast freezing	25	80% Survival	Sakai et al. (1990)
Coffea arabica	Somatic embryo	–	–	–	–	Bertrand-Desbrunais et al. (1989)
C. canephora	Somatic embryo and cells	Sucrose 0.85–1 M	-20, -196	40	Secondary embryogenesis	Tessereau et al. (1991)
Coleus blumei	Cell suspension	–	–	–	–	Reuff et al. (1988)
Datura stramonium	Cell suspension	DMSO (7%)	1/min, -196	36–38	40% Cell survival	Bajaj (1976a)
D. innoxia	Cell suspension	1 M sorbitol	1/min, -196	33	Callus	Weber et al. (1983)
D. innoxia	Protoplasts	DMSO + mannitol	Exposed to vapors, immersed in LN	35	Callus	Bajaj (1988b)
Dianthus hybrida	Shoot tips	DMSO + glucose	-196	25–30	Plants, 2-year storage	Fukai (1989)
D. caryophyllus	Shoot tips	DMSO (5%)	Sudden freezing in LN	37	Plants	Seibert (1976)
D. caryophyllus	Shoot tips	DMSO (10%)	-40, -196	–	Plants	Uemura and Sakai (1980)
D. caryophyllus	Shoot apices	High sucrose	Two-step freezing	40	Plants	Dereuddre et al. (1987)
D. caryophyllus	Shoot apices encapsulated	High sucrose	Vitrification, two-step freezing	Room temp.	Plants	Tannoury et al. (1991)
Digitalis lanata	Cell suspension	DMSO + glycerol + sucrose	1/min, -35, -196	40	Callus, biotransformation of β-methyldigitoxin to β-methyldigoxin	Dietterich et al. (1982, 1986) Seitz et al. (1983); Göldner et al. (1991)
Dioscorea deltoidea	Cell suspension	DMSO (7%)	Hardening at 2–10, -30, -70, then -196	20 or 40	Diosgenine synthesis	Butenko et al. (1984)
Haplopappus gracilis	Shoot primordia	DMSO (5%)	0.5/min, -40, -196	37	Plantlets	Taniguchi et al. (1988)

(Contd.)

Table 1. (*Contd.*)

Plant species	Culture	Cryoprotectant	Freezing (°C)	Thawing (°C)	Response	Reference
Lavandula vera	Cell suspension	DMSO (10%) + glucose (20%)	1/min, −40, −196	Rapid	Metabolic potential retained	Watanabe et al. (1983); Kuriyama et al. (1990)
Mentha spp.	Shoot tip	–	−196	–	–	Towill (1990)
Nicotiana tabacum	Cell suspension	DMSO, sucrose or glycerol	1–2/min, or −20, −70, −196	36–38	Haploid plants Callus	Bajaj (1976a) Hauptmann and Widholm (1982); Maddox et al. (1983)
N. tabacum	Pollen embryos	DMSO + sucrose + glycerol	1–3/min, −196	35–40	Plants	Bajaj (1976b, 1977)
N. tabacum	Protoplasts	DMSO + mannitol	Exposed to vapors, immersed in LN	35	Plants	Bajaj (1988b)
N. plumbaginifolia	Cell suspension	DMSO + glycerol	−40 to −70 for	45	Callus	Maddox et al. (1983)
N. sylvestris			4 h, then −196			
N. rustica	Root cultures	DMSO (10%)	−196	25	Biosynthetic activity	Benson and Hamill (1991)
Panx ginseng	Cell suspension	Sucrose/glycerol	Hardening, −30, −70, then −196	20 or 40	Retention of biosynthetic capacity	Butenko et al. (1984); Seitz and Reinhard (1987)
Petunia hybrida	Anthers	DMSO + sucrose + glycerol	−196	35–40	Callusing	Bajaj (1978)
Primula obconica	Androgenic anthers	DMSO (7%) + sucrose (7%)	Fast freezing	36–40	Plantlets	Bajaj (1981)
Theobroma cacao	Immature embryos	DMSO 10% + sucrose 0.5 M	0.4/min, −40, −196	40	Plantlets	Pence (1991)

2 Significance of Cryopreservation for Medicinal Plants

In vitro conservation has demonstrated enormous potential (Bajaj 1986). The reduction in growth rate and nonfrozen storage allow short-term conservation (see Hiraoka 1988). Cryopreservation, however, will enable the conservation of germplasm on a long-term basis. The use of large-scale plant cell cultures in bioreactors for the storage and production of secondary metabolites, medicines, flavors, and other pharmaceutical products has been emphasized (see Hashimoto and Azechi 1988). In this context, the purpose of cryopreservation is threefold.

1. *Retention of Biosynthetic Potential.* In many cell cultures, because too little of the desired product is formed, or there is a decrease in biosynthetic capacity during serial subculture and transfer, cells are needed in large quantities. The cultures therefore have to be collected and stored until sufficient quantities are available for extraction. Although nonfrozen storage is useful, the cultures cannot be maintained for long periods, and there is always the risk of metabolic changes during storage. However, the literature has shown (Table 1) that cultures preserved in liquid nitrogen (LN) retain their biosynthetic potential, therefore, this method seems to be ideal.

2. *Genetic Stability of High-Yielding Somaclones/Somatic Hybrids.* Cell cultures, upon repeated subculturing, undergo genetic erosions that result in somaclonal variation of varied nature and origin, i.e., polyploidy, aneuploidy, gene amplification, chromosome elimination, translocations, etc. (see Bajaj 1990). Most of these variations are of no significance and are therefore discarded. Nevertheless, sometimes mutant cell lines or high-yielding somaclones, e.g., *Coptis, Lithospermum, Anchusa, Epimedium, Taxus*, etc. or somatic hybrids, (Table 2) can be isolated and stored. Since the prime interest is to conserve the clones, cryopreservation holds great promise.

3. *Storage of Genetically Transformed Hairy Roots.* Hairy root cultures are genetically more stable than cell suspensions. *Agrobacterium rhizogenes*-mediated, transformed hairy root cultures that are capable of producing the desired secondary metabolites have now been obtained in a number of medicinal plants (Table 2). Among them, *Lithospermum, Panax, Coreopsis, Swertia*, etc. should be mentioned. Hairy root cultures of tobacco and beet root have been cryopreserved (Benson and Hamill 1991).

3 Cryopreservation Studies

Though considerable progress has been made in the cryopreservation studies on in vitro cultures, the success of the cryoability of a system depends on a number of factors, e.g., the most important are the judicious choice of material, cryoprotectant, and the method of freezing and thawing. The literature on freeze-preservation of various types of in vitro cultures of medicinal and aromatic plants is summarized here.

Table 2. In vitro induced variability in medicinal plants. Some examples of somaclonal cell lines, transformed hairy roots, and the somatic hybrids in which changes in alkaloids/secondary metabolites have been reported

Secondary metabolites	Plant species	Reference
	Somaclonal cell lines	
Nicotine	*Nicotiana tabacum*	Ogino et al. (1978)
Anthocyanin	*Euphorbia millii*	Yamamoto et al. (1982)
Berberine	*Coptis japonica*	Sato and Yamada (1984)
Shikonin	*Lithospermum erythrorhizon*	Fujita et al. (1985)
Rosmarinic acid	*Anchusa officinalis*	De-Eknamkul and Ellis (1988)
Berberine	*Thalictrum* spp.	Ikuta and Itokawa (1988)
Flavonol	*Epimedium diphyllum*	Yamamoto et al. (1992)
Taxol	*Taxus canadensis*	Fett-Neto et al. (1992)
	Transformed hairy roots	
Phenyl glucosides	*Swertia japonica*	Ishimaru et al. (1990)
Phenylpropanoids	*Coreopsis tinctoria*	Reichling and Thron (1990)
Condensed tannins	*Lotus corniculatus*	Morris and Robbins (1992)
Shikonin	*Lithospermum erythrorhizon*	Sim and Chang (1993)
Ginsenoside	*Panax ginseng*	Inomata et al. (1993)
	Somatic hybrids	
Alkaloids	*Datura innoxia* + *Atropa belladonna*	Krumbiegel and Schieder (1979)
Alkaloids	*Atropa belladonna* + *Nicotiana plumbaginifolia*	Gleba et al. (1988)
Alkaloids	*Hyoscyamus muticus* + *Nicotiana tabacum*	Lazar (1989)
Alkaloids	*Atropa belladonna* + *Scopolia carniolica*	Kushnir et al. (1991)
Alkaloids	*Duboisia leichhardii* + *Nicotiana tabacum*	Endo et al. (1991)
Alkaloids	*Rauwolfia serpentina* + *Vinca minor*	Kostenyuk et al. (1991)

3.1 Cryopreservation of Meristems/Shoot Tips

The storage of in vitro cultures of excised meristems has enabled only short- and medium-term maintenance of germplasm (Bajaj 1993), however, cryopreservation has the distinct advantage of allowing long-term conservation (Bajaj 1985a). Since the initial work of Seibert (1976) on meristems of carnations, studies have been conducted on a number of medicinal and aromatic plants (Table 1). Extensive work has been done on various species of *Dianthus*, and the refined technique enables almost 100% survival/regeneration (Seibert 1976; Uemura and Sakai 1980; Dereuddre et al. 1987; Fukai 1989; Tannoury et al. 1991).

The retrieved shoot primordia of *Haplopappus gracilis* ($2n = 4$) were very stable in chromosome number and karyotype (Taniguchi et al. 1988). Meristems of several varieties of sugar beet (*Beta vulgaris*) have survived freezing in LN. Growth recovery resumed anywhere between 2–8 weeks (Braun 1988). In various species of *Chrysanthemum* (Fukai et al. 1991), the shoot regeneration rate of frozen shoot tips (0.5–0.7 mm) varied between 9.4–100% even after 8 months of storage in LN.

3.2 Cryopreservation of Callus/Cell Suspensions

Cryopreservation studies have been conducted on callus/cell suspensions of various species of *Anisodus, Atropa, Cannabis, Catharanthus, Cinchona, Coleus, Coffea, Datura, Digitalis, Dioscorea, Lavandula, Nicotiana, Panax*, etc. The retrieved cultures in most cases retained their morphogenic/biosynthetic potential. These observations form the basis for further investigations on the large-scale storage of cell cultures for commercial purposes.

Bajaj (1976a) successfully revived the cell suspensions of anther-derived haploid tobacco (*Nicotiana tabacum*) frozen to -20, -70, and $-196\,^{\circ}$C, and regenerated complete plants from retrieved cultures. In the initial experiment, it was observed that free cells in filtered cell suspensions survived only occasionally during freezing treatment, whereas cell clumps and small cell aggregates responded well. Thus, in subsequent experiments, unfiltered cell suspensions containing cell clumps and small cell aggregates were employed. Most cells frozen at $-20\,^{\circ}$C for 1 week or 48 h at $-70\,^{\circ}$C survived and resumed growth. When cooled at the rate of $2\,^{\circ}$C/min to $-196\,^{\circ}$C in the presence of various cryoprotectants, some cells survived after 6 months of storage. This is more favorable than the results of Sugawara and Sakai (1974), who reported that tobacco cell suspensions could withstand $-30\,^{\circ}$C, but not $-196\,^{\circ}$C. Due to the low rate of cell survival and the difficulty encountered in determining the accuracy of viability by staining techniques alone, such cultures were grown for 12–20 weeks and their weights recorded. These cultures retained their morphogenic potential and differentiated shoots in 4–5 weeks when grown on an agar-solidified MS medium containing IAA (4 mg/l) and kinetin (2 mg/l). Root tip squashes from these plants showed a normal haploid chromosome number ($n = 24$). The cell suspensions of two other species of tobacco, *N. sylvestris* and *N. plumbaginifolia*, have also been cryopreserved (Maddox et al. 1983).

Datura stramonium cell suspensions frozen at a rate of $1\,^{\circ}$C/min in the presence of 7% DMSO gave the best results (Bajaj 1976a). The cell cultures of periwinkle (*Catharanthus roseus*) subjected to cooling at a rate of $0.5\,^{\circ}$C/min to $-40\,^{\circ}$C, and then immersed in LN showed a viability of about 60% of the controls (Chen et al. 1984). *Dioscorea deltoidea* and *Panax ginseng* are two other species of great medicinal value for the production of diosgenine and saponine, respectively, which have yielded excellent results (Butenko et al. 1984). Preculture and hardening of *Panax* cells in the presence of 7–25% sucrose at 2–10 °C resulted in a considerable increase in their survival. As a cryoprotectant, sucrose was best for *Panax*, whereas 7% DMSO gave optimal results for *Disocorea*, and they retained their biosynthetic potential to produce diosgenine, sitosterol, stigmasterol, and other metabolites (Butenko et al. 1984). Later, Seitz and Reinhard (1987) studied the growth and ginsenoside pattern of cryopreserved cell cultures of *Panax ginseng*.

Cinchona callus precultured for 3 days in a medium containing 5% PVP, and cryoprotected with mannitol + DMSO, frozen slowly to LN vapors, transferred to LN, and then rapidly thawed gave rise to slowly growing calli on agar medium (Hunter 1986). The cell suspensions of hemp (*Cannabis sativa*) first frozen down to −50 °C, and then immersed in LN showed a viability of about 58% (Jekkel et al. 1989).

Callus and cell suspensions of *Anisodus acutangulus* retained their biosynthetic ability to produce hyoscyamine and scopolamine after freezing (Zheng Guang-zhi et al. 1983). Their storage at −196 °C was better than at −20 °C, and callus cultures gave better results than cell suspensions. The use of lactalbumin hydrolysate as a cryoprotectant was similar to DMSO, but glycerine was ineffective.

Extensive studies have been conducted on various aspects of the cryopreservation of *Digitalis lanata.* Up to 50% of their cell cultures survived and the biotransformation potential after cryostorage showed stability. The capacity to transform β-methyldigitoxin to β-methyldigoxin remained unchanged (Diettrich et al. 1982, 1986; Seitz et al. 1983). Göldner et al. (1991) observed that by using trehalose as a cryoprotectant, high post-thaw viability rates were achieved, and the cells resumed growth after a short lag period.

3.3 Cryopreservation of Embryos

The prospects of cryopreservation of various types of embryos, e.g., zygotic, somatic, nucellar, and pollen embryos were discussed earlier (Bajaj 1985b). In the following, work conducted on medicinal and aromatic plants is summarized.

Zygotic Embryos. Large seeds, especially those of recalcitrant species, cannot be preserved under ordinary conditions, thus cryopreservation of zygotic embryos has been suggested as a means to overcome this problem. Early cryopreservation work was done on coconut (*Cocos nucifera*), a large-seeded recalcitrant plantation crop species (Bajaj 1984). The partially desiccated, immature embryos and their segments were subjected to freezing in LN. They had a lag phase of up to 4 months without showing any visible sign of growth. However, subsequently, in some of the cultures, the embryos elongated and proliferated at the cut ends of the segment to form callus. The immature embryos of cacao (*Theobroma cacao*) retained the ability to produce callus and to undergo somatic embryogenesis (Pence 1991). Cryopreservation of desiccated, excised embryonal axes of tea (*Camellia sinensis*) has been attempted with varying degrees of success (Chaudhury et al. 1991; Wesley-Smith et al. 1992). Likewise, embryonal axes of trifoliate orange (*Poncirus trifoliata*) desiccated to a moisture level of 14%, then frozen in LN, revived and formed roots and shoots after 1 week of culture (Radhamani and Chandel 1992).

Somatic Embryos. Since somatic embryos (SE) can now be produced in large numbers in bioreactors (Denchev et al. 1992) and are being employed for mass propagation (Merkle et al. 1991; Bajaj 1991), genetic transformation (Ellis 1993), and for the production of synthetic seed (Redenbaugh et al. 1991), storage methods must be developed so that they can be used when desired. Somatic embryos/embryogenic cultures of a large number of plant species have been successfully cryopreserved and the literature reviewed (see Bajaj 1995).

Cryopreservation studies have been conducted on SEs of citrus (Marin and Duran-Vila 1988) and coffee (Bertrand-Desbrunais et al. 1989; Tessereau et al. 1991). In *Citrus sinensis*, no difference was observed in the viability of SEs frozen with (10% DMSO) or without cryoprotectant. A small number of frozen embryos survived and developed into proliferating cultures that produced whole plants (Marin and Duran-Vila 1988). In *Coffea canephora* (Tesserreau et al. 1991), 100% survival of torpedo-shaped SEs (1–1.5 mm) was obtained by using a high concentration of sucrose (0.85–1 M) as the sole cryoprotectant; their regrowth was expressed by secondary embryogenesis.

Pollen Embryos. The frozen-thawed embryos of *Atropa* (Fig. 1) and tobacco survived (Bajaj 1977) and regenerated plants. Their survival was strongly influenced by the stage of development, the globular embryos withstanding freezing better than the late heart-shaped and differentiated embryos (Bajaj 1978). In cultures, the pollen embryos showed a lag phase proportional to their stage of maturity/size and the extent of freeze injury. Of the various concentrations of DMSO used, 7% gave the best response.

Nucellar Embryos/Tissues. Entire young ovules and the cut micropylar half of the ovules obtained from young unripe fruits of *Citrus* reticulata, frozen in the presence of DMSO and sucrose, showed 28.8 and 24.3% survival, respectively (Bajaj 1984). The retrieved material, like the controls, when cultured on a medium containing casein hydrolysate (500 mg/l), proliferated to form pseudobulbils. In some cases, polyembryonic masses and shoots were formed.

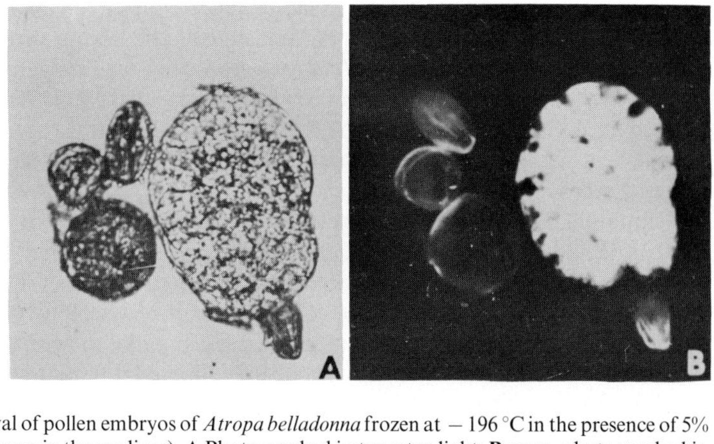

Fig. 1A, B. Survival of pollen embryos of *Atropa belladonna* frozen at −196 °C in the presence of 5% DMSO (+4% sucrose in the medium). **A** Photographed in tungsten light; **B** same, photographed in ultraviolet light; note the fluorescence of pollen embryos in **B** after staining with fluorescein diacetate. (Bajaj 1976b)

Plants have also been regenerated from nucellar cells of navel orange (Kobayashi et al. 1990).

3.4 Cryopreservation of Excised Anthers

Cryopreservation studies have been conducted on anthers of *Atropa belladonna, Nicotiana tabacum, Petunia hybrida* (Bajaj 1977,1978) and *Primula obconica* (Bajaj 1981). The anthers of *Atropa* and *Nicotiana* were cultured for 3–4 weeks to induce callusing/androgenesis (Bajaj et al. 1978), and then subjected to freezing. The viability of such whole anthers was very low. Mostly, the anthers became soft and spongy and the pollen embryos started to abort inside the anthers. This situation, however, was improved by cutting the androgenic anthers into two transverse halves (to expose the pollen embryos) and then freezing them. By doing so, more pollen embryos survived. Anthers of *Primula* cultured for 3 weeks were first treated with an aqueous solution of a mixture of cryoprotectants (7% DMSO and 7% sucrose) for 2 h. The cryoprotectant solution was then poured off, the anthers were blotted on sterile filter paper, and the vial was quickly immersed for 5 min in a cylinder of LN. The retrieved anthers cultured on MS medium supplemented with IAA (0.5 mg/l) + 2,4-D (0.5 mg/l) + zeatin (2 mg/l) showed no sign of growth for 3 weeks. Thereafter, a small patch of white callus cells appeared on about 3% of the browning anthers, which later formed a mass of callus. The cytological study of pollen revealed two modes of development; (1) pollen embryogenesis and (2) the formation of callus which later differentiated into plants (Bajaj 1981).

3.5 Cryopreservation of Protoplasts

The literature on the cryopreservation of plant protoplasts was reviewed earlier (Bajaj 1989). The author's work (Bajaj 1988b) on three alkaloid-producing plants is summarized here. The protoplasts obtained from mesophyll cells of *Nicotiana tabacum* and *Atropa belladonna*, and callus cell suspensions of *Datura innoxia* were subjected to freezing in LN in the presence of DMSO, mannitol, and sucrose, and thawed at 35° C. The degree of survival and regeneration varied a great deal, *Atropa* protoplasts gave the best response, followed by *Nicotiana* and *Datura.*

Initially, the protoplasts were quiescent for 3–5 days in comparison to controls (Bajaj et al. 1978). Afterwards they started to elongate; the chloroplasts became scanty and yellowish and reorganized themselves at the periphery. The first division and the subsequent colony formation were invariably delayed; however, after walls were regenerated, they looked like the control protoplasts. In *Atropa*, a maximum of 44% of the cultures showed differentiation of plantlets in callus. The morphogenetic events were not impaired and cultures showed, as in the control, patterns of differentiation, embryogenesis, and organogenesis (Gosch et al. 1975). The plants thus regenerated, when transferred to pots, continued to grow and formed normal flowers and fruits (Fig. 2). The seeds

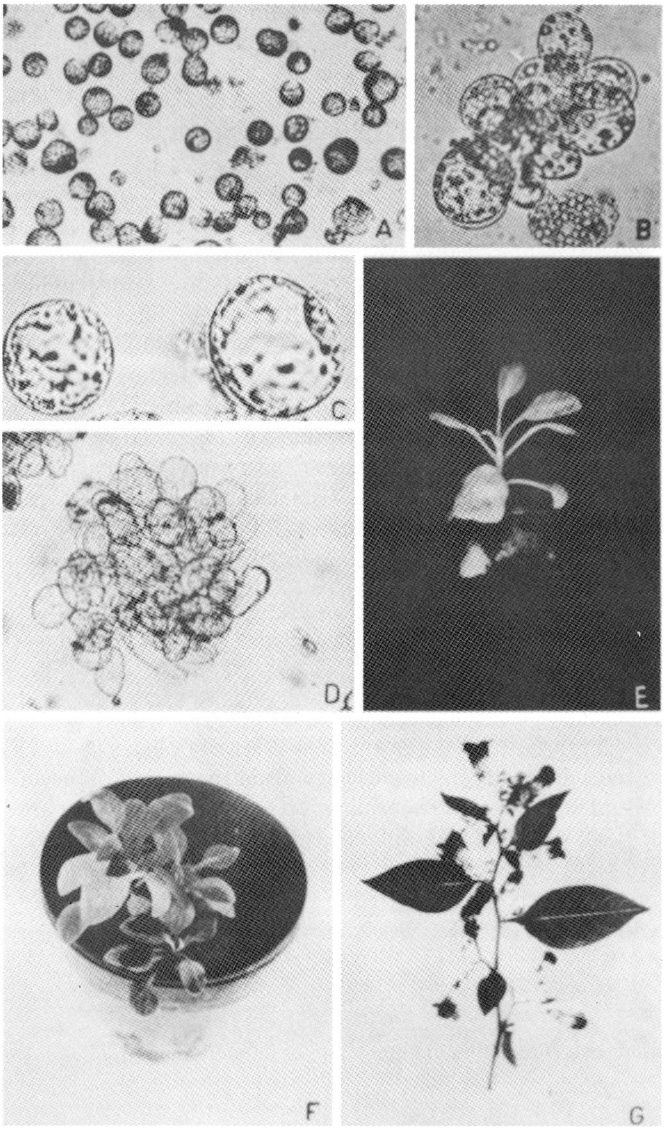

Fig. 2 A–G. Regeneration of plants from protoplasts of *Atropa belladonna, Datura innoxia,* and *Nicotiana tabacum* frozen in liquid nitrogen. **A** Mesophyll protoplasts of *Atropa* after freezing and thawing. **B** Same, 4 weeks after culture in a drop of liquid medium. **C** Retrieved callus cell suspension-derived protoplasts of *Datura* 3 weeks after culture. **D** A callus cell colony obtained from frozen mesophyll protoplasts of tobacco. **E** A plantlet regenerated from frozen protoplast-derived callus (**D**). **F, G** Frozen protoplast-derived plant of *Atropa* 4 weeks after transferring to soil (**F**), and undergoing normal flowering (**G**). (Bajaj 1988b)

obtained from these plants were fertile and, in turn, showed normal germination (Bajaj 1988b).

3.6 Cryopreservation of Hairy Root Cultures

Recently, there has been an upsurge in activity in the genetic transformation of root cultures, by *Agrobacterium rhizogene*, regarding the production of secondary metabolites (Table 2). Transformed hairy roots are fast-growing, more stable in cultures as compared to callus, and are also capable of producing good quantities of medicinal compounds, such as shikonine, ginsenoside, etc. Experiments conducted on the cryopreservation of hairy root cultures of *Nicotiana rustica* and *Beta vulgaris* have yielded positive results (Benson and Hamill 1991). The growth rates, secondary metabolite production, and T-DNA structures were examined after cryopreservation. The secondary metabolite production was essentially unchanged in terms of total betalain (*B. vulgaris*) and alkaloid production (*N. rustica*) and also with regard to their qualitative distribution. Recently hairy root cultures of *Panax ginseng* have also been successfully cryopreserved, and they retained their biosynthetic potential for ginsenoside (Yoshimatsu et al. 1995).

3.7 Cryopreservation of Pollen

The importance of the cryopreservation of pollen was highlighted earlier and the subject reviewed (Bajaj 1987). It suffices to say that the cryopreservation of pollen would facilitate wide hybridization amongst plants that are separated by physiological, geographical, and seasonal limitations, and would help to induce and conserve genetic diversity. Recently, pollen of *Panax ginseng* and *P. quinquefolium* has been reported to retain viability after 11 months of storage in LN (Zhang et al. 1993).

4 Conclusion

For the conservation of germplasm of rare and endangered species and the storage of in vitro induced cell clones yielding increased alkaloids and other metabolites of medicinal and pharmaceutical interest, basic and fundamental cryopreservation experiments have been successfully conducted on the revival of various types of cells, tissues, embryos, anthers, meristems, protoplasts, hairy roots, etc. The retrieved cultures retained their biosynthetic and morphogenetic potential. In view of the recent advances in the large-scale production of cells in bioreactors, the use of cryopreservation technology assumes great importance in the pharmaceutical industry.

References

Bajaj YPS (1976a) Regeneration of plants from cell suspensions frozen at − 20, − 70, and − 196° C. Physiol Plant 37: 263–268

Bajaj YPS (1976b) Gene preservation through freeze storage of cell, tissue and organ cultures.Acta Hortic 63: 75–84

Bajaj YPS (1977) Survival of *Atropa* and *Nicotiana* pollen embryos frozen at − 196° C. Curr Sci 46: 305–307

Bajaj YPS (1978) Effect of superlow temperature on excised anthers and pollen embryos of *Atropa, Nicotiana* and *Petunia.* Phytomorphology 28: 171–176

Bajaj YPS (1981) Regeneration of plants from ultra-low frozen anthers of *Primula obconica.* Sci Hortic 14: 93–95

Bajaj YPS (1984) Induction of growth in frozen embryos of coconut and ovules of citrus. Curr Sci 53: 1215–1216

Bajaj YPS (1985a) Cryopreservation of germplasm of potato (*Solanum tuberosum* L.) and cassava (*Manihot esculenta* Crantz): viability of excised meristems cryopreserved up to four years. Indian J Exp Biol 23: 285–287

Bajaj YPS (1985b) Cryopreservation of embryos. In: Kartha KK (ed) Cryopreservation of plant cells and organs. CRC, Boca Raton, pp 227–242

Bajaj YPS (1986) In vitro preservation of genetic resources – techniques and problems. Int Symp Nuclear techniques and in vitro culture for plant improvement. IAEA/FAO, Vienna, 1985, pp 43–57

Bajaj YPS (1987) Cryopreservation of pollen and pollen embryos, and the establishment of pollen banks. In: Giles KL, Prakash J (eds) Pollen development and cytology. Academic Press, London, pp 397–420

Bajaj YPS (1988a) Cryopreservation and the retention of biosynthetic potential in cell cultures of medicinal and alkaloid-producing plants. In: Bajaj YPS (ed) Biotechnology in agriculture and forestry, vol 4. Medicinal and aromatic plants I. Springer, Berlin Heidelberg New York, pp 169–197

Bajaj YPS (1988b) Regeneration of plants from frozen (− 196 °C) protoplasts of *Atropa belladonna* L., *Datura innoxia* Mill. and *Nicotiana tabacum* L. Indian J Exp Biol 26: 289–292

Bajaj YPS (1989) Cryopreservation of plant protoplasts. In: Bajaj YPS (ed) Biotechnology in agriculture and forestry, vol 8. Plant protoplasts and genetic engineering I. Springer, Berlin Heidelberg New York, pp 97–106

Bajaj YPS (ed) (1990) Biotechnology in agriculture and forestry, vol 11. Somaclonal variation in crop improvement I. Springer, Berlin Heidelberg New York

Bajaj YPS (1991) Storage and cryopreservation of in vitro cultures. In: Bajaj YPS (ed) Biotechnology in agriculture and forestry, vol 17. High-tech and micropropagation I. Springer, Berlin Heidelberg New York, pp 361–381

Bajaj YPS (1993) A suggested method for in vitro long-term storage at 4 °C of Chrysanthemum and *Petunia* germplasm. Plant Tissue Cult 3 (1): 57–58

Bajaj YPS (1995) Cryopreservation of somatic embryos. In: Bajaj YPS (ed) Biotechnology in agriculture and forestry, vol 30. Somatic embryogenesis and synthetic seed I. Springer, Berlin Heidelberg New York, pp 221–229

Bajaj YPS, Gosch G, Ottma M, Weber A, Gröbler A (1978) Production of polyploid and aneuploid plants from anthers and mesophyll protoplasts of *Atropa belladonna* and *Nicotiana tabacum.* Indian J Exp Biol 16: 947–953

Bajaj YPS, Furmanowa M, Olszowska O (1988) Biotechnology of the micropropagation of medicinal and aromatic plants. In: Bajaj YPS (ed) Biotechnology in agriculture and forestry, vol 4. Medicinal and aromatic plants I. Springer, Berlin Heidelberg New York, pp 60–103

Balandrin MF, Klocke JA (1988) Medicinal, aromatic, and industrial materials from plants. In: Bajaj YPS (ed) Biotechnology in agriculture and forestry, vol 4. Medicinal and aromatic plants I. Springer, Berlin Heidelberg New York, pp 3–36

Benson EE, Hamill JD (1991) Cryopreservation and post freeze molecular and biosynthetic stability in transformed roots of *Beta vulgaris* and *Nicotiana rustica.* Plant Cell Tissue Organ Cult 24: 163–172

Bertrand-Desbrunais AJ, Fabre J, Engelmann F, Dereuddre J, Charrier A (1989) Reprise de l'embryogenèse adventive d' embryons somatiques de caféier (*Coffea arabica*) après congélation dans l'azote liquide. CR Acad Sci Paris 307, Ser III: 795–801

Braun A (1988) Cryopreservation of sugarbeet germplasm. Plant Cell Tissue Organ Cult 14: 161–168

Butenko RG, Popov AS, Volkova LA, Chernyak ND, Nosov AM (1984) Recovery of cell cultures and their biosynthetic capacity after storage of *Dioscorea deltoidea* and *Panax ginseng* cells in liquid nitrogen. Plant Sci Lett 33: 285–292

Chaudhury R, Radhamani J, Chandel KPS (1991) Preliminary observations on the cryopreservation of desiccated embryonic axes of tea [*Camellia sinensis* (L.) O. Kuntze] seeds for genetic conservation. Cryo-Lett 12: 31–36

Čellárová E, Cernicka T, Vranova E, Brutovska R, Lapar M (1992) Viability of *Chamomilla recutita* L. Ruaschert cells after cryopreservation. Cryo-Lett 13: 37–42

Chen THH, Kartha KK, Leung NL, Kurz WGW, Chatson KB, Constabel F (1984) Cryopreservation of alkaloid-producing cell cultures of periwinkle (*Catharanthus roseus*). Plant Physiol 75:726–731

De-Eknamkul W, Ellis BE (1988) Rosmarinic acid: production in plant cell cultures. In: Bajaj YPS (ed) Biotechnology in agriculture and forestry, vol 4. Medicinal and aromatic plants I. Springer, Berlin Heidelberg New York, pp 310–329

Demeulemeester MAC, Vandenbussche B, de Profit MP (1993) Regeneration of chicory plants from cryopreserved in vitro shoot tips. Cryo-Lett 14(1): 57–64

Denchev PD, Kuklin AI, Scragg AH (1992) Somatic embryo production in bioreactors. J Biotech 26: 99–109

Dereuddre J, Galerne M, Gazeau C (1987) Effets du saccharose sur la résistance à la congélation dans l'azote liquide (– 196 °C) des méristèmes d'Oeillet (*Dianthus caryophyllus* L.) cultivés in vitro. CR Acad Sci Paris 304: 485–488

Dereuddre J, Fabre J, Bassaglia C (1988) Resistance to freezing in liquid nitrogen of carnation (*Dianthus caryophyllus* L. var. Eolo) axillary shoot-tips excised from different aged in vitro plantlets. Plant Cell Rep 7: 170–173

Diettrich B, Popov AS, Pfeiffer B, Neumann D, Butenko R, Luckner M (1982) Cryopreservation of *Digitalis lanata* cell cultures. Planta Med 46: 82–87

Diettrich B, Haack U, Luckner M (1986) Cryopreservation of *Digitalis lanata* cells grown in vitro. Precultivation and recultivation. J Plant Physiol 126: 63–73

Ellis D (1993) Transformation in spruce (*Picea* species). In: Bajaj YPS (ed) Biotechnology in agriculture and forestry, vol 23. Plant protoplasts and genetic engineering IV. Springer, Berlin Heidelberg New York, pp 315–330

Endo T, Hamaguchi N, Eriksson T, Yamada Y (1991) Alkaloid biosynthesis in somatic hybrids of *Duboisia leichhardtii* F. Muell. and *Nicotiana tabacum* L. Planta 183: 505–510

Fabre J, Dereuddre J (1987) Effects of different sugars (sucrose, glucose, sorbitol and mannitol) on the resistance to deep freezing in liquid nitrogen of meristems from in vitro cultured carnations (*Dianthus caryophyllus* L. var Eolo). CR Acad Sci Paris 304: 507–510

Fett-Neto A, DiCosmo F, Reynolds WF, Sakata K (1992) Cell culture of *Taxus* as a source of the antineoplastic drug taxol and related taxanes. Biotechnology 10: 1572–1575

Fujita Y, Takahashi S, Yamada Y (1985) Selection of cell lines with high productivity of shikonin derivatives by protoplast culture of *Lithospermum erythrorhizon* cells. Agric Biol Chem 49: 1755–1759

Fukai S (1989) Plant regeneration from shoot tips of *Dianthus hybrida* cryopreserved in liquid nitrogen up to 2 years. Plant Tissue Cult Lett 6: 177–178

Fukai S (1990) Cryopreservation of chrysanthemum shoot tips. Sci Hortic 45: 167–174

Fukai S, Goi M, Tanaka M (1991) Cryopreservation of shoot tips of *Chrysanthemum morifolium* and related species native to Japan. Euphytica 54: 201–204

Gleba YY, Hinnisdaels S, Sidorov VA, Kaleda VA, Parakonny AS, Boryshuk NV, Cherep NN, Negrutiu I, Jacobs M (1988) Intergeneric asymmetric hybrids between *Nicotiana plumbaginifolia* and *Atropa belladonna* obtained by gamma-fusion. Theor Appl Genet 76: 760–766

Göldner EM, Seitz U, Reinhard E (1991) Cryopreservation of *Digitalis lanata* Ehrh. cell cultures: preculture and freeze tolerance. Plant Cell Tissue Organ Cult 24: 19–24

Gosch G, Bajaj YPS, Reinert J (1975) Isolation, culture and induction of embryogenesis in protoplasts from cell-suspensions of *Atropa belladonna*. Protoplasma 86: 405–410

Hashimoto T, Azechi S (1988) Bioreactors for the large-scale culture of plant cells. In: Bajaj YPS (ed) Biotechnology in agriculture and forestry, vol 4. Medicinal and aromatic plants I. Springer, Berlin Heidelberg New York, pp 104–122

Hauptmann RM, Widholm JM (1982) Cryostorage of cloned amino acid analog resistant carrot and tobacco suspension culture. Plant Physiol 70: 30–37

Hiraoka N (1988) Nonfrozen storage of plant cell cultures and its effect on metabolites. In: Bajaj YPS (ed) Biotechnology in agriculture and forestry, vol 4. Medicinal and aromatic plants I. Springer, Berlin Heidelberg New York, pp 157–168

Hunter CS (1986) In vitro propagation and germplasm storage of *Cinchona*. In: Withers L, Alderson P, (eds) Plant tissue culture and its agricultural applications. Butterworth, London, pp 291–301

Ikuta A, Itokawa H (1988) Berberine: production through plant (*Thalictrum* spp.) cell cultures. In: Bajaj YPS (ed) Biotechnology in agriculture and forestry, vol 4. Medicinal and aromatic plants I. Springer, Berlin Heidelberg New York, pp 282–293

Inomata S, Yokoyama M, Gozu Y, Shimizu T, Yanagi M (1993) Growth pattern and ginsenoside production of *Agrobacterium*-transformed *Panax* ginseng roots. Plant Cell Rep 12: 681–686

Ishimaru K, Sudo H, Satke M, Shimomura K (1990) Phenyl glucosides from a hairy root culture of *Swertia japonica*. Phytochemistry 29: 3823–3825

Jekkel ZS, Heszky LE, Ali AH (1989) Effect of different cryoprotectants and transfer temperatures on the survival rate of hemp (*Cannabis sativa* L.) cell suspension in deep freezing. Acta Biol Hung 40: 127–136

Kobayashi S, Sakai A, Oiyama I (1990) Cryopreservation in liquid nitrogen of cultured navel orange (*Citrus sinensis* Osb.) nucellar cells and subsequent plant regeneration. Plant Cell Tissue Organ Cult 23: 15–20

Kostenyuk IA, Lubaretz O, Borisyuk N, Voronin V, Cherep N, Stockigt J, Gleba Y (1991) Isolation and characterization of intergeneric somatic hybrids in Apocynaceae family. Theor Appl Genet 82: 713–716

Krumbiegel G, Schieder O (1979) Selection of somatic hybrids after fusion of protoplasts from *Datura innoxia* Mill. and *Atropa belladonna* L. Planta 145: 371–376

Kuriyama A, Watanabe K, Ueno S, Mitsuda H (1990) Effect of post- thaw treatment on the viability of cryopreserved *Lavandula vera* cells. Cryo-Lett 11: 171–178

Kushnir S, Babiychuk E, Bannikova M, Momot V, Komarnitsky I, Cherep N, Gleba Y (1991) Nucleo-cytoplasmic incompatibility in cybrid plants possessing an *Atropa* genome and a *Nicotiana* plastome. Mol Gen Genet 225: 225–230

Lazar (1989) Somatic hybridization in *Hyoscyamus* x *Nicotiana*. In: Bajaj YPS (ed) Biotechnology in agriculture and forestry, vol 8. Plant protoplasts and genetic engineering I. Springer, Berlin Heidelberg New York, pp 356–369

Maddox AD, Gonsalves F, Shields R (1983) Successful preservation of suspension cultures of three *Nicotiana* species at the temperature of liquid nitrogen. Plant Sci Lett 28: 157–162

Marin ML, Duran-Vila N (1988) Survival of somatic embryos and recovery of plants of sweet orange [*Citrus sinensis* (L.) Osb.] after immersion in liquid nitrogen. Plant Cell Tissue Organ Cult 14: 51–57

Merkle SA, Schlarbaum SE, Cox RA, Schwarz OJ (1991) Mass propagation of somatic embryo-derived plantlets of yellow-poplar for field testing. In: Proc 21st Conf Southern Forest tree improvement Knoxville, TN, pp 56–68

Morris P, Robbins MP (1992) Condensed tannin formation by *Agrobacterium rhizogenes* transformed root and shoot organ cultures of *Lotus corniculatus*. J Exp Bot 43: 221–231

Murashige T, Skoog F (1962) A revised medium for rapid growth and bioassays with tobacco tissue cultures. Physiol Plant 15: 473–497

Nag KK, Street HE (1975) Freeze-preservation of cultured plant cells. II. The freezing and thawing phases. Physiol Plant 34: 261–265

Niwata E (1992) Cryopreservation of shoot tips of garlic by vitrification. Jpn J Breed (Suppl 1): 326

Ogino T, Hiraoka N, Tabata M (1978) Selection of high nicotine producing cell lines of tobacco callus by single cell cloning. Phytochemistry 17: 1907–1910

Pence VC (1991) Cryopreservation of immature embryos of *Theobroma cacao*. Plant Cell Rep 10: 144–147

Radhamani J, Chandel KPS (1992) Cryopreservation of embryonic axes of trifoliate orange (*Poncirus trifoliata* L. RAF.). Plant Cell Rep 11: 372–374

Redenbaugh K, Fujii J, Slade D, Viss P, Kossler M (1991) Artificial seeds – encapsulated somatic embryos. In: Bajaj YPS (ed) Biotechnology in agriculture and forestry, vol 17. High-tech and micropropagation I. Springer, Berlin Heidelberg New York, pp 395–416

Reichling J, Thron U (1990) Accumulation of rare phenylpropanoids in *Agrobacterium rhizogenes* transformed root cultures of *Coreopsis tinctoria*. Planta Med 56: 488–490

Reuff I, Seitz U, Ulrich B, Reinhard E (1988) Cryopreservation of *Coleus blumei* suspension and callus cultures. J Plant Physiol 133: 414–418

Sakai A, Kobayashi S, Oyiama I (1990) Cryopreservation of nucellar cells of navel orange (*Citrus sinensis* Osb. var, *brasiliensis* Tanaka) by vitrification. Plant Cell Rep 9: 30–33

Sato F, Yamada Y (1984) High berberine-producing cultures of *Coptis japonica* cells. Phytochemistry 23: 281–285

Seibert M (1976) Shoot initiation from carnation shoot apices frozen to −196 °C. Science 191: 1178–1179

Seitz U, Reinhard E (1987) Growth and ginsenoside pattern of cryopreserved *Panax ginseng* cell cultures. J Plant Physiol 131: 215–223

Seitz U, Alfermann AW, Reinhard E (1983) Stability of biotransformation capacity in *Digitalis lanata* cell cultures after cryogenic storage. Plant Cell Rep 2: 273–276

Sim SJ, Chang HN (1993) Increased shikonin production by hairy roots of *Lithospermum erythrorhizon* in two phase bubble column reactor. Biotechnol Lett 15(2): 145–150

Sugawara Y, Sakai A (1974) Survival of suspension-cultured sycamore cells cooled to the temperature of liquid nitrogen. Plant Physiol 54: 722–724

Taniguchi K, Tanaka R, Ashitani N, Miyagawa H (1988) Freeze preservation of tissue-cultured shoot primordia of the annual *Haplopappus gracilis* (2n = 4). Jpn J Genet 63: 267–272

Tannoury M, Ralambosoa J, Kaminski M, Dereuddre J (1991) Cryoconservation par vitrification d'apex enrobés d'oeillet (*Dianthus caryophyllus* L.) cultivé in vitro. CR Acad Sci, Paris Sér III, 313: 633–638

Tessereau H, Lecouteux C, Florin B, Schlienger C, Petiard V (1991) Use of a simplified freezing process and dehydration for the storage of embryogenic cell lines and somatic embryos. Rev Cytol Biol Veg Bot 14: 297–310

Towill LE (1990) Cryopreservation of isolated mint shoot tips by vitrification. Plant Cell Rep 9: 178–180

Uemura M, Sakai A (1980) Survival of carnation (*Dianthus caryophyllus* L.) shoot apices frozen to the temperature of liquid nitrogen. Plant Cell Physiol 21: 85–94

Watanabe K, Mitsuda H, Yamada Y (1983) Retention of metabolic and differentiation potential of green *Lavandula vera* callus after freeze-preservation. Plant Cell Physiol 24: 119–122

Weber G, Roth EJ, Schweiger HG (1983) Storage of cell suspensions and protoplasts of *Glycine max* (L.) Merr, *Brassica napus* (L.), *Datura innoxia* (Mill.), and *Daucus carota* (L.) by freezing. Z Pflanzenphysiol 109: 29–39

Wesley-Smith J, Vertucci CW, Berjak P, Pammenter NW, Crane J (1992) Cryopreservation of desiccation-sensitive axes of *Camellia sinensis* in relation to dehydration, freezing rate and the thermal properties of tissue water. J Plant Physiol 140: 596–604

Yamamoto Y, Mizuguchi R, Yamada Y (1982) Selection of a high stable pigment-producing strain in cultured *Euphorbia millii* cells. Theor Appl Genet 61: 113–116

Yamamoto H, Ieda K, Tsuchiya SI, Yan K, Tanaka T, Iinuma M, Mizuno M (1992) Flavonol glycoside production in callus cultures of *Epimedium diphyllum*. Phytochemistry 31: 837–840

Yoshimatsu K, Yamaguchi H, Shimomura K (1995) Cold storage and cryopreservation of hairy roots of *Panax ginseng*. Plant Cell Rep (in press)

Zhang LX, Chang WC, Wei YJ, Liu L, Wang YP (1993) Cryopreservation of ginseng pollen. HortScience 28(7): 742–743

Zheng Guang-zhi, He Jing-bo, Wang Shi-ling (1983) Cryopreservation of calli and their suspension culture cells of *Anisodus acutangulus*. Acta Bot Sin 25 (6): 512–517 (in Chinese)

VI.2 Cryopreservation of *Anisodus* Species

G.Z. Zheng and J.P. Luo[1]

1 Introduction

1.1 Important Species and Their Distribution

The genus *Anisodus* (Solanaceae), comprises four species and three varieties: *A. luridus, A. luridus* var. *fischerianus, A. acutangulus, A. acutangulus* var. *breviflorus, A. mairei, A. tanguticus,* and *A. tanguticus* var. *viridulus* (Chen and Chen 1977).

A. luridus is mainly distributed on mountain slopes and grassland between 3800 and 4200 m above sea level in northwest Yunnan province, and on forest land, in mountainous regions, and floodplains between 3200 and 4000 m above sea level in southeast Tibet, Nepal, Bhutan, and northeast India. It has also been introduced in Russia. Roots and seeds are used as a remedy for spasms and pains.

A. acutangulus (C.Y. Wu et C. Chen), a new species, is distributed only on forest land at an elevation of 2750–3100 m in northwest Yunnan province. Roots and seeds are used medicinally as a remedy for fractures, injuries from falls, appendicitis, stomach pains, rheumatism, arthritis, spasmolysis, etc. *A. mairei* is distributed in central and northeast Yunnan province, China. Roots, leaves, and seeds are used medicinally for injuries from falls, rheumatism, arthritis, stomach pains, etc. *A. tanguticus* is mainly distributed in mountain valleys and foothills at an elevation of 2800 m in Gansu province, on riverbanks between 2680 and 3280 m in Sichuan province, on mountain slopes at an elevation of 4200 m in Yunnan province, and on mountain slopes, floodplains, and farmland at an elevation of 3900 m in Tibet, China. It has also been introduced in Russia. Roots and seeds are used medicinally to relieve pain.

The main medicinal compounds in all four species are hyoscyamine (*A. acutangulus, A. mairei*), scopolamine (*A. acutangulus, A. luridus*), and anisodamine (*A. tanguticus*) (Chen and Chen 1977; Cheng and Liang 1978; Chinese Academy of Medical Sciences and Chinghai Institute of Medical Sciences 1976). Hyoscyamine acts on the central nervous system as antispasmodic and analgesic (Chen and Chen 1977). Scopolamine, acting on the parasympatholytic system is used as a remedy for car sickness, sea sickness, vomiting, trembling, paralysis,

[1]Kunming Institute of Botany, Academia Sinica, Kunming, Yunnan 650204, China

Biotechnology in Agriculture and Forestry, Vol. 32
Cryopreservation of Plant Germplasm I (ed. by Y.P.S. Bajaj)

gastrointestinal spasms, mental disease, cerebral abscess, pulmonary abscess, and dementia, etc. (Chen and Chen 1977). Anisodamine, a new kind of choline, is a nerve-blocking medicine. It can expand microarteries and improve microcirculation. It is used to treat acute disturbances of the microcirculation, such as acute appendicitis, shock from poisoning, acute nephritis with cardiac failure, encephalopathy due to arterial hypertension, Meniere's disease, pancreatitis, retinochoroiditis, etc. (Chinese Academy of Medical Sciences and Chinghai Institute of Medical Sciences 1976).

1.2 Need for Germplasm Storage or Cryopreservation

The distribution of the four species of *Anisodus* is limited to an inaccessible mountain area in China and its neighboring region. They belong to these wild plant resources, for which no cultivation or preservation schemes have been set up. In recent years, the wild plant resources have been greatly reduced for a variety of reasons. According to statistics, 20% of these plants will vanish from the earth by the end of this century, especially in China. *A. tanguticus* is distributed in wide areas, compared with *A. acutangulus*, which is found only in a narrow area between 2750 and 3100 m above sea level in northwest Yunnan province. The natural resources of *A. acutangulus* are being rapidly reduced and are in imminent danger from overgathering and digging up. *A. acutangulus* is the raw material for the commercial production of hyoscyamine and scopolamine. Their contents in plants from natural resources are generally much lower (approximately 1%), especially those of scopolamine, whose value and curative effect are higher than those of hyoscyamine, while the amount used medicinally is lower. Therefore, *A. acutangulus* was chosen as a suitable object for the systematic study of tissue and cell culture, and to select excellent variants from calli cultured after irradiation with Co^{60} γ-ray (Zheng and Liang 1976, 1977; Cheng and Liang 1978; Zheng et al. 1980, 1982a,b,c,d, 1983a,b, 1985, 1986; Zheng 1989). By a series of studies and the selection of excellent varieties, the growth rate of cell cultures was increased to 1.5 g dry wt./day/1, which was 15 times the original. The contents of these two alkaloids were 0.63 mg/g dry wt., which was ten times that of the original cell culture particularly in scopolamine content, which was 25 times that of the original. To produce hyoscyamine and scopolamine industrially by mass cell culture, it is not necessary for the cell culture to differentiate and grow to plants. Its proliferation was controlled under artificial conditions by subculturing. Tissue culture and cells which are successively subcultured often show variations in chromosome and genotype, such as polyploidy, aneuploidy, chromosome deletion, and gene translocation (Bajaj 1990). These variations can result in the successive decline of the medicinal contents and the disappearance of their biosynthesis in cell culture. In order to maintain the excellent hereditary features, freeze preservation of tissues and cells is necessary. Various aspects of cryopreservation of medicinal plants have been reviewed (see Bajaj 1988; Jian 1990).

1.3 Methods for Germplasm Preservation of *Anisodus acutangulus*

Three methods are employed:

1. The storage of seed is only allowed for long-lived seeds. When screened seeds of *A. acutangulus* were stored at room temperature, the germination percentage fell to 90% after the first year and 70% after the second. After the third year, the germination percentage fell to 30% and seeds sown in the field grew only very poorly.
2. Micropropagation includes the culture of roots, stems, leaves, flowers, etc. These organs can be cultured and plantlets regenerated.
3. The temperature of freeze preservation is often between -10 and $-269\,^{\circ}$C. For example, the procedure of preservation at low temperatures and freeze storage of calli and suspension culture cells of *A. acutangulus* is given as follows.

2 Preservation of Calli at Low Temperatures

2.1 Method of Preservation at Low Temperatures

The material to be conserved was selected callus variants of *A. acutangulus* from a root callus line that had been mutated by γ-irradiation (Zheng et al. 1983a). Calli were inoculated in LS medium (Linsmaier and Skoog 1965) containing 0.6 mg/l NAA, 2 g/l LH, and 8 g/l agar (Zheng and Liang 1977) before conservation. Calli with the same growth rate, selected according to their increased dry weight after 2 weeks, were divided into groups and preserved in the dark at $0 \pm 1, 5 \pm 1, 10 \pm 1$, and $15 \pm 1\,^{\circ}$C, respectively. After 1 month or more, their increased dry weight and growth rate were determined. Calli were then recultured at $26 \pm 1\,^{\circ}$C in the dark, and the increased dry weight and growth rate determined after 1 week of reculture.

Ten flasks of calli were obtained from culture material preserved or recultured and dried to a constant weight at $-50\,^{\circ}$C in a freeze drier. Thus, the mean dry weight (mg) minus the inoculation quantity equaled the increased dry weight per flask of calli. The growth rate was the increased dry weight of cultures at 1 lit. medium for 1 day.

2.2 Results

When precultured at $26 \pm 1\,^{\circ}$C for 2 weeks before preservation at low temperatures above $0\,^{\circ}$C, calli grew slowly. Preserved at low temperatures, calli could grow continuously. Even if calli were preserved at $0 \pm 1\,^{\circ}$C, their growth rate was as much as 8.5 mg dry wt./day/l. The results showed that calli of *A. acutangulus* to a certain extent possessed resistance to low temperature above $0\,^{\circ}$C. Although calli which were successively subcultured at $26 \pm 1\,^{\circ}$C after preservation at

0 ± 1 °C for 1 month could grow continuously, their growth rate was greatly reduced. The increased dry weight was only 0.5 mg, and the growth rate only 2.8 mg dry wt./day/l (Table 1). Table 1 clearly shows that storage at 0 ± 1 °C can cause freezing injury to calli and is not a suitable preservation temperature.

The growth rates of calli conserved at both 10 ± 1 and 15 ± 1 °C were obviously accelerated, and the growth rates of calli transferred and recultured at 26 ± 1 °C for 7 days were much faster than in those conserved at these low temperatures (Table 2). Preservation at these low temperatures could not slow all metabolic activity, therefore, calli could not be stored at either 10 ± 1 or 15 ± 1 °C over a long period of time.

The calli stored at 5 ± 1 °C grew more slowly, but this temperature prevented freezing injury (Table 1). The growth rate of calli which were transferred and recultured at 26 ± 1 °C increased when stored at 5 ± 1 °C for 30 days. These results indicated that storage at 5 ± 1 °C was suitable for *A. acutangulus* callus. Therefore, the effects of storage period on the growth of calli at 5 ± 1 °C was studied (Table 3).

Table 1. Effects of storage at 0 ± 1 °C and 5 ± 1 °C on callus growth

Temperature (°C) Storage time (days)	26 ± 1 14	10 ± 1 30	→	26 ± 1 7	15 ± 1 30	→	26 ± 1 7
Increased dry wt. (mg)	7.00	6.4		0.5	9.3		6.1
Growth rate (mg dry wt./day/l)	24.9	8.5		2.8	12.4		34.9

Table 2. Effects of storage at 10 ± 1 °C and 15 ± 1 °C on callus growth

Temperature (°C) Storage time (days)	26 ± 1 14	10 ± 1 30	→	26 ± 1 7	15 ± 1 30	→	26 ± 1 7
Increased dry wt. (mg)	7.00	27.3		30.9	44.5		49.0
Growth rate (mg dry wt./day/l)	24.9	36.4		88.4	59.3		284.0

Table 3. The effect of storage period on growth of callus at 5 ± 1 °C

Temperature (°C) Storage time (days)	5 ± 1 30	26 ± 1 21	5 ± 1 60	26 ± 1 21	5 ± 1 120	26 ± 1 21
Increased dry wt. (mg)	9.3	18.3	31.8	96.6	43.2	38.9
Growth rate (mg dry wt./day/l)	12.4	34.9	21.2	184.0	14.4	74.1

During different periods of storage, calli grew slowly. Transferred and recultured at 26 ± 1 °C, calli, excluding the calli stored for 150 days (data not given), grew continuously. On the basis of results, *A. acutangulus* calli could be stored at 5 ± 1 °C for 120 days. In addition, the preservation of *A. acutangulus* seeds at low temperatures was tested. As in results from seeds stored at room temperature, the germination percentage of seeds stored for 3 years was only 30%, and seeds sown in the soil hardly grew.

To conclude, short-lived seeds such as those of *A. acutangulus* could not be stored at room temperature and sown successfully the following year. Conservation of germplasm at low temperatures above 0 °C was suitable for seeds and tissue cultures of *A. acutangulus*, seeds of which could be stored at 5 ± 1 °C, although its calli could be stored only at 5 ± 1 °C for 120 days. Long-term preservation of *A. acutangulus* germplasm, particularly its tissue cultures, under freeze storage, was further investigated.

3 Freeze Preservation of Callus

3.1 Methodology

3.1.1 Freezing Material

A. acutangulus calli for freeze storage and its culture method were as described above. Small pieces of stock selected calli which grew faster were suspended in 100 ml liquid medium in 500-ml flasks and cultured on a rotary shaker with 2.5 cm displacement at 120 rpm at 26 ± 1 °C (Zheng et al. 1982a). The liquid medium was LS supplemented with 0.1 mg/l 2,4-D, 0.2 mg/l NAA, and 0.003 mg/l KT (Zheng et al. 1982b).

3.1.2 Freezing Method

Calli with the same growth rate cultured for 7 weeks were placed in small flasks containing stock liquid under ice-bath conditions, and immediately transferred to the freezer at −20 °C or immersed in liquid nitrogen for freeze storage of different periods. The small frozen flasks were then immediately placed in a water bath at 37 °C and thawed when cell survival was determined or calli were recultured. Suspension cells, suspended for 5 weeks and then filtered, were frozen by the method described above.

3.1.3 Cryoprotectants

Stock solution was LS medium supplemented with different cryoprotectants. The following cryoprotective agents were employed in freezing calli and suspension cells: DMSO (5%), sucrose (15%), glycerine (15%), and LH (5%). All stock solutions were put into freezing flasks (10 ml) with 8 ml stock solution/flask, respectively, sterilized at 120 °C 1 kg/cm for 15 min.

3.1.4 Determination of Cell Survival

Cell survival was determined by the fluorescein diacetate staining (FDA) method. Cells from callus or frozen suspension cells were placed on a slide, washed three times with distilled water, then stained with 1% FDA for 5 min and washed five times with distilled water. The living cells stained with FDA, which showed fluorescence under ultraviolet light, were counted twice with five repeats.

3.1.5 Preculture of Calli and Suspension Cells

Calli and frozen suspension cells were directly inoculated on LS medium at 26 ± 1 °C in the dark. In addition, cultured cells were treated with DMSO, which must be sterile-filtered and washed five times with sterile liquid medium. The increased dry weight, growth rate, and contents of scopolamine and hyoscyamine were determined after 5 weeks of reculture. The increased dry weight and growth rate were measured by the method given above.

3.1.6. Determination of Scopolamine and Hyoscyamine Contents

Centrifuged calli or suspension cells were freeze-dried to constant weight and then crushed. The contents were measured by a TLC spectrophotometer according to the method of The Chinese Academy of Medical Sciences (1972).

3.2 Results

3.2.1 Effects of Cryoprotectants and Temperatures on Freeze Preservation

When calli or suspension cells were frozen without cryoprotectants at −20 °C, cell viability was very low; the longer the freezing period of cultured cells, the lower the cell viability.

DMSO was suitable as a cryoprotectant for storage of calli at −20 °C, but the protective effects of DMSO and LH for storage of suspension cells were similar. Sucrose as a cryoprotectant for calli and suspension cells showed some medium protective effects. When cultured cells were frozen with glycerine as a cryoprotectant, cell survival was very low (Table 4).

These results demonstrated that calli and suspension cells could not be stored at −20 °C without appropriate cryoprotectants. Results in Table 5 show that DMSO and LH had similar effects on calli and suspension cells stored in liquid nitrogen at −196 ° C. Cell viability of calli and cell suspensions frozen at

Table 4. Storage of calli and cell suspensions at $-20\,^{\circ}C$

Cryoprotectant	Viability (%)				
	Calli			Cell suspensions	
Storage time	1 day	1 week	4 weeks	1 week	4 weeks
DMSO	81.21 ± 3.3	76.40 ± 2.6	73.79 ± 1.7	69.44 ± 6.0	67.71 ± 2.3
Sucrose	63.00 ± 3.0	60.28 ± 8.8	58.28 ± 1.8	51.74 ± 7.6	50.69 ± 2.3
Glycerine	40.37 ± 2.2	36.37 ± 6.3	17.71 ± 1.3	31.81 ± 4.0	15.03 ± 1.4
LH	77.43 ± 3.0	74.60 ± 3.9	69.37 ± 4.0	69.40 ± 2.9	67.77 ± 2.4
Control	11.85 ± 3.0	11.40 ± 5.3	7.26 ± 1.0	7.99 ± 2.6	6.72 ± 1.2

Age of stored calli was 9 weeks; age of stored suspension cells 8 weeks.

Table 5. Storage of calli and cell suspensions at $-196\,^{\circ}C$

Cryoprotectant	Viability (%)				
	Calli			Cell suspensions	
Storage time	1 day	1 week	4 weeks	1 week	4 weeks
DMSO	93.28 ± 5.0	90.15 ± 2.0	90.34 ± 2.5	84.67 ± 1.5	84.18 ± 2.5
Sucrose	75.52 ± 5.9	68.43 ± 2.9	68.78 ± 6.7	66.68 ± 2.2	66.20 ± 2.3
LH	92.73 ± 0.7	90.79 ± 1.2	90.10 ± 2.7	83.50 ± 1.9	82.79 ± 2.5

Age of stored calli was 6 weeks; age of suspension culture cells 8 weeks.

$-196\,^{\circ}C$ was higher than in those frozen at $-20\,^{\circ}C$; furthermore, cell viability decreased with prolongation of storage times at -20 and $-196\,^{\circ}C$ (Table 5).

Storage of cell suspensions of *A. acutangulus* at $-196\,^{\circ}C$ had an advantage over storage at $-20\,^{\circ}C$ with the same cryoprotectant. At $-196\,^{\circ}C$, cell suspensions could be stored for 9 weeks. Five percent DMSO was suggested as an excellent cryoprotectant (Bajaj 1979). Our results implied that the effects of LH as a cryoprotectant were similar to those of DMSO, because LH was nutritive to cells and did not need to be washed repeatedly; its effects as a cryoprotectant merit further study.

3.2.2 Effects of the Age of the Stored Material on Cryopreservation

The technology of cryopreservation of plant cells and tissue has been more extensively researched than the physiological states of the materials conserved (Karting et al. 1976; Bajaj 1976; Bajaj and Reinert 1977). This chapter studies the effects of age of calli and cell suspensions of *A. acutangulus* on freeze cryopreservation. Considering the growth of calli and suspension cells which started proliferation after 1 week, the ages of calli and suspension cells as materials for

Table 6. The effect of age of calli and cell suspensions on storage at $-196\,^{\circ}C$

Calli		Suspension culture cells	
Age (weeks)	Viability (%)	Age (weeks)	Viability (%)
3	97.87 ± 2.3	2	93.23 ± 2.5
5	93.64 ± 2.3	4	97.08 ± 1.7
7	88.25 ± 2.0	6	96.09 ± 0.6
9	79.08 ± 2.4	8	84.67 ± 1.2

cryopreservation were 3 and 2 weeks, respectively (Zheng and Liang 1977). The results are listed in Table 6.

In the light of cell survival, 3-week storage was suitable for calli, 4 weeks for cell suspensions. Their cell viabilities were ca. 97%. With increased storage time, cell viabilities of calli and frozen suspension cells decreased progressively (Table 6). The reason for this might be that young cells in calli and suspension cells with a short culture period were abundant. The young cells had a dense protoplast, small vacuole, and good survival, and were able to endure low temperatures. In freeze cryopreservation of *A. acutangulus* germplasm, calli were more successful than cell suspensions.

Calli and cell suspensions of *A. acutangulus* could be stored for 9 weeks at $-196\,^{\circ}C$ in liquid nitrogen. Some calli and suspension cells were frozen for 3 years (1982 to 1985). On thawing, calli and suspension cells retained about 80% cell viability. These cultured cells were used in studies of biotransformation and fermenter culture (Zheng et al. 1985, 1986). These results demonstrated that calli and suspension cells could undoubtedly be stored for more than 2 years in liquid nitrogen.

3.2.3 Effects of Cryopreservation on Growth and Contents of Medicinal Substances

If DMSO was used as a cryoprotectant, *A. acutangulus* calli and frozen suspension cells could not grow successfully, even if washed repeatedly in sterile liquid medium. Table 7 shows results of recultured calli and suspension cells which had been frozen with LH as a cryoprotectant.

As shown in Table 7, the increased dry weight and growth rate of calli and suspension cells frozen after reculture were decreased by 11.1 and 6.3%, respectively, compared with those of normal calli and suspension cells. However, the contents of hyoscyamine and scopolamine in calli and suspension cells frozen after reculture were scarcely lower than those in the normal calli and suspension cells. The results indicated that the ability to synthesize these two alkaloids in calli and suspension cells was retained after they were stored at $-196\,^{\circ}C$. During cryopreservation at $-196\,^{\circ}C$, some cells died or were freeze-injured. The growth of cells recultured after being frozen was slower than that of normal cells. This was identical to the results for *Daucas carota* and *Nicotiana tabacum* cells

Table 7. The effects of cryopreservation on growth and contents of hyoscyamine and scopolamine in calli and suspension cells recultured after revival

Materials	Increased dry wt. (mg)	Growth rate (mg dry wt./day/l)	Hyoscy. content (mg/g dry wt.)	Scopola. content (mg/g dry wt.)
Calli	198.96 ± 0.3	320.90 ± 0.3	0.200 ± 0.005	0.176 ± 0.001
Control	252.58 ± 1.6	360.80 ± 1.6	0.207 ± 0.005	0.177 ± 0.001
Suspension	478.07 ± 2.8	170.80 ± 2.8	0.196 ± 0.006	0.177 ± 0.001
Control	510.00 ± 2.3	182.40 ± 2.3	0.191 ± 0.005	0.179 ± 0.002

Age of stored calli and suspension culture cells was 3 weeks; cryopreservation temperature −196 °C; storage period 4 weeks.

recultured after freezing as reported by Bajaj (1976). To produce medicinal substances by cell culture for industrial purposes, it was important to maintain the potential to synthesize medicinal compounds in cells after cryopreservation. The growth of cells was slow, but it was secondary and transient. During successive subculturing, the growth could be progressively resumed. The growth rates of calli and recultured suspension cells which had been frozen at −196 °C reached 330 mg dry wt./day/l and 175 mg dry wt./day/l at the second subculturing, respectively. At the fifth subculturing, growth reached the standard of normal cells (Zheng et al. 1985, 1986).

3.3 Discussion

Even though the cell viability of calli and frozen suspension cells varied greatly, calli and suspension cells of many plants could be preserved at superlow temperatures. Jian (1990) reviewed the freeze preservation of 33 medicinal plants. Among these, calli of 12 plants and suspension cells of 23 plants were studied, in which cell survival in calli of 7 plants was above 90% and only in *A. acutangulus* was cell survival above 90%. As material for cryopreservation, calli were more suitable than suspension cells, as reported for *A. acutangulus* (Tables 4 and 5).

When calli and suspension cells of *A. acutangulus* were prefrozen step by step before storage, the cell viability was too low to revive. Direct fast freezing could prevent the formation of ice crystals or reduce the rate of ice crystallization and the amount of ice crystallized. Too rapid freezing could make cells vitrify and prevent dehydration, resulting in intracellular freezing and consequent cell death. Cell cytology and physiology were determining factors in rapid freezing, especially for the degree of vacuolation of cells. This was also confirmed by the cryopreservation of calli and suspension cells of *A. acutangulus*, where viability of young cells was higher than that of older vacuolated cells. The choice and concentration of cryoprotectant were also critical factors in cryopreservation. for plant tissues and cells, DMSO is the most extensively used cryoprotectant. Although DMSO is most effective for the cryopreservation of plant calli and suspension cells, it is a permeating cryoprotectant, and its removal from cells is difficult. In our study, the effect of LH as a cryoprotectant was slightly lower than that of DMSO, but LH was an effective nutritive substance (Zheng and Liang 1976, 1977), and frozen calli and suspension cells could be recultured without repeated washes after thawing.

It was shown by the results in this study that when calli and suspension cells were stored at freezing temperatures, their metabolism and growth activities could be delayed or virtually stopped, genetic variations did not occur, and the potential for morphogenesis and synthesis of secondary metabolites could be retained. According to our results, freezing at $-196\,^{\circ}\mathrm{C}$ was suitable for cryopreservation of *A. acutangulus* calli and suspension cells, which could then be stored for more than 2 years.

4 Summary and Conclusions

All four species of *Anisodus* contain important medicinal substances, are of limited distribution, and have not been conventionally cultivated. Particularly, *A. acutangulus* is in imminent danger of extinction, and the preservation of its germplasm is imperative.

The germination percentage of seeds of *A. acutangulus* was reduced by 10, 30, and 70% when stored at room temperature or $5 \pm 1\,^{\circ}\mathrm{C}$ after the first, second, and third year, respectively. Calli could be stored for 15 years at an appropriate constant temperature, but cultured cell lines had to be selected and the culture method improved during successive subculturing. Calli of *A. acutangulus* could be stored for a short time at 0 ± 1, 5 ± 1, 10 ± 1, and $15 \pm 1\,^{\circ}\mathrm{C}$. The period (120 days) of storage at $5 \pm 1\,^{\circ}\mathrm{C}$ was longer than at other temperatures.

Cryopreservation of calli and suspension cells of *A. acutangulus* could maintain the potential of morphogenesis, growth, and synthesis of hyoscyamine and scopolamine. Cell viability of calli and frozen suspension cells was 90 and 95%, respectively. Calli were more suitable than suspension cells as freezing material, and storage at $-196\,^{\circ}\mathrm{C}$ was more suitable than at $-20\,^{\circ}\mathrm{C}$. At $-196\,^{\circ}\mathrm{C}$, calli and suspension cells could be conserved for more than 2 years. Although DMSO was the most effective for cryopreservation of calli and suspension cells of *A. acutangulus* of all cryoprotectants studied in this chapter, calli and suspension cells frozen with DMSO as a cryoprotectant, which is a permeating cryoprotectant, could not be recultured. The effects of LH as a cryoprotectant were similar to those of DMSO; calli and frozen suspension cells could be recultured. Sucrose as a cryoprotectant hardly affected cryoprotection. Cells could not be frozen with glycerine as a cryoprotectant or without any cryoprotectants. For the cryopreservation of *A. acutangulus*, calli and suspension cells should be young, and it was verified that calli cultured for 3 weeks and suspension cells cultured for 4 weeks were suitable as freezing materials. Although the growth and contents of hyoscyamine and scopolamine decreased to a certain extent during freeze cryopreservation of calli and suspension cells, they could resume the level of normal culture cells during successive subculturing.

The following conclusions may be drawn:

1. Seeds could be stored for only 2 years at room temperature or 5 ± 1 °C.
2. Calli could be preserved for 15 years through successive subculturing at 26 ± 1 °C, but the selection of cell lines and improvement of culture conditions should be continuously carried out.
3. Calli could be stored for 120 days at 5 ± 1 °C.
4. Calli and suspension cells could be frozen at -20 or -196 °C. Freeze cryopreservation did not change the potential for morphogenesis and growth of recultured cells. The ability to synthesize hyoscyamine and scopolamine in recultured cells was not lost.
5. For cryopreservation at -196 °C, cell viability of calli was 90% and suspension cells above 95%; both calli and suspension cells could be stored for more than 2 years.
6. For cryopreservation, calli were more suitable than suspension cells as freezing material and storage at -196 °C was better than at -20 °C.
7. Young cells were more suitable for cryopreservation than older cells. Three-week-old calli and 4-week-old suspension cells were suitable for cryopreservation.
8. LH as a cryoprotectant was suitable for cryopreservation of calli and suspension cells. The effect of DMSO as a cryoprotectant was best, however, cells frozen with DMSO as a cryoprotectant could not be recultured after thawing.
9. For cryopreservation of calli and suspension cells, freezing temperatures were dropped directly to -196 or -20 from 0 °C when calli and suspension cells had been cooled from 26 ± 1 °C. For thawing of calli and frozen suspension cells, the temperatures were raised rapidly from -20 or -196 to 37–40 °C.

References

Bajaj YPS (1976) Regeneration of plants from cell suspensions frozen at -20, -70 and -196 °C. Physiol Plant 27: 263–268

Bajaj YPS (1979) Technology and prospects of cryopreservation of germplasm. Euphytica 28: 267–285

Bajaj YPS (1988) Cryopreservation and the retention of biosynthetic potential in cell cultures of medicinal and alkaloid-producing plants. In: Bajaj YPS (ed) Biotechnology in agriculture and forestry, vol 4. Medicinal and aromatic plants I. Springer, Berlin Heidelberg New York, pp 169–187

Bajaj YPS (ed) (1990) Biotechnology in agriculture and forestry, vol 11. Somaclonal variation in crop improvement I. Springer, Berlin Heidelberg New York

Bajaj YPS, Reinert J (1977) Cryobiology of plant cell culture and establishment of gene banks. In: Reinert J, Bajaj YPS (eds) Plant cell tissue and organ culture. Springer, Berlin Heidelberg New York, pp 757–777

Chen C, Chen CL (1977) On the Chinese genera *Scopolia jaco, Anisodus* Link et Otto and *Atropanthe pascher*, Acta Phytotaxon Sin 15(2): 57–68

Cheng KC, Liang Z (1978) Callus culture of the three well-known Chinese herbs and their medicinal contents. Proc Symp Plant tissue culture. Science Press, Peking, pp 469–679

Chinese Academy of Medical Sciences, Institute of Materia Medica (1972) Studies on medicinal components of Chinese herbs, vol 1. People's Hygiene Press, Peking, pp 318–320

Chinese Academy of Medical Sciences and Chinghai Institute of Materia Medica, Institute of Materia Medica (1976) Chemical studies on anisodamine. Acta Chim Sin 34(1): 39–44

Jian LC (1990) Establishment of germplasm banks through super-low temperature storage of cell and tissue cultures. In: Sun CS, Chen WL (eds) Plant biotechnology and crop improvement. Chinese Science and Technology Press, Peking, pp 254–280

Karting T, Russheim U, Maunz B (1976) Beobachtungen über das Vorkommen und Bildung von Cardenoliden in Gewebekulturen, *Digitalis purpurea* und *Digitalis lanata.* Planta Med 29: 275–282

Linsmaier EM, Skoog F (1965) Organic growth factor requirements of tobacco tissue culture. Physiol Plant 18: 100–127

Zheng GZ (1989) *Anisodus acutangulus:* production of scopolamine and hyoscyamine in cell cultures. In: Bajaj YPS (ed) Biotechnology in agriculture and forestry, vol 7. Medicinal and aromatic plants II. Springer, Berlin Heidelberg New York, pp 23–46

Zheng GZ (ed) (1992) Plant cell culture and its secondary metabolism, vol 1. Yunnan University Press, Kunming

Zheng GZ, Liang Z (1976) Studies on tissue culture of medicinal plants I. Callus culture of *Scopolia acutangula* for the production of hyoscyamine and scopolamine. Acta Bot Sin 18(2): 163–169

Zheng GZ, Liang Z (1977) Studies on tissue culture of medicinal plants II. Chemical control of callus growth and synthesis of hyoscyamine and scopolamine by *Scopolia acutangula* callus. Acta Bot Sin 19(3): 209–215

Zheng GZ, He JB, Wang SL (1980) Studies on tissue culture of medicinal plants III. The influence of callus differentiation of *Anisodus acutangulus* on the contents of hyoscyamine and scopolamine. Acta Phytophysiol Sin 6(4): 377–385

Zheng GZ, He JB, Wang SL (1982a) Cell suspension culture of callus from *Anisodus acutangulus* roots. Acta Bot Yunnanica 4(2): 203–206

Zheng GZ, He, JB, Wang SL (1982b) Studies on tissue culture of medicinal plants IV. Hormone regulation on cell suspension culture of *Anisodus acutangulus*. Acta Phytophysiol Sin 8(1): 53–58

Zheng GZ, He JB, Wang SL (1982c) *Anisodus acutangulus* callus strain of high and stable growth rate selected from the callus after irradiation with Co^{60} γ-ray. In: Fujiwara A (ed) Plant tissue culture 1982. Maruzen, Tokyo, pp 339–340

Zheng GZ, He JB, Wang SL (1982d) Studies on tissue culture of medicinal plants VI. The cost question in the cell culture of *Anisodus acutangulus*. Acta Bot Sin 24(3): 209–215

Zheng GZ, He JB, Wang SL (1983a) Studies on tissue culture of medicinal plants V. Variant of *Anisodus acutangulus* with both growth rate and scopolamine content high and stable. Acta Phytophysiol Sin 9(2): 129–134

Zheng GZ, He JB, Wang SL (1983b) Cryopreservation of calli and their suspension culture cells of *Anisodus acutangulus*. Acta Bot Sin 25(6): 512–518

Zheng GZ, He JB, Wang SL (1985) Biotransformation of phenylalanine, ornithine and hyoscyamine to scopolamine in the suspension culture of *Anisodus acutangulus* cells. Acta Bot Yunnanica 7(2): 203–209

Zheng GZ, Wang SL, He JB (1986) Submergence (fermentation) culture of *Anisodus acutangulus* cells and identification of hyoscyamine and scopolamine in the cultured cells. Acta Bot Sin 28(2): 123–31

VI.3 Cryopreservation of Germplasm of Chrysanthemums

S. FUKAI[1]

1 Introduction

1.1 Plant Distribution and Important Species

Chrysanthemum [*Chrysanthemum morifolium* Ramat. Syn. *Dendranthema grandiflorum* (Ramat.) Kitam.] has been cultivated for more than 1500 years. The modern chrysanthemum is supposed to have originated from *C. indicum* var. *procumbens* x *C. zawadskii* var. *latilobum* in China. It is one of the most important cut flowers in Japan, North America, and Europe.

The genus *Chrysanthemum* (family Compositae) consists of about 200 species of annuals, herbaceous perennials and subshrubs, which are widely distributed in Europe, Asia, and North America. Important related species, for example *C. indicum, C. japonense, C. makinoi*, etc. which could be crossed successfully with *C. morifolium*, are mainly distributed in East Asia (Kitamura et al. 1988). These represent important genetic resources for introducing new characters into modern chrysanthemum. A recent example of this is the attractive hybrid Pacificum hybrid chrysanthemum from the hybridization of *C. morifolium* x *C. pacificum* and their back cross to *C. morifolium* (Shibata et al. 1988).

1.2 Various Methods for the Storage of Germplasm of *Chrysanthemum*

Because cultivars of *Chrysanthemum morifolium* are highly heterozygous, they are propagated vegetatively. The plants are usually maintained in the greenhouse or in the field. Cuttings taken from chrysanthemum can be stored for several months at 0 °C, but this is usually a short-term measure.

Many of wild species of *Chrysanthemum* are propagated by seed and these seeds may be stored under low temperature conditions. It is not , however, easy to maintain the phenotypic characteristics of wild species during production because they are easily cross-pollinated with other species in a collection site. Wild species also show a wide range of geographical variations. It is therefore desirable that wild species should also be maintained vegetatively.

[1]Faculty of Agriculture, Kagawa University, Miki-cho, Kita-gun, 761-07, Japan

Biotechnology in Agriculture and Forestry, Vol. 32
Cryopreservation of Plant Germplasm I (ed. by Y.P.S. Bajaj)

It is difficult for breeders and nurseries with a restricted supply of labor and land to maintain so many cultivars and species, and so in vitro storage has been explored as a means of conserving such germplasm. There are two ways of conserving germplasm in vitro; one employs slow growth of plants for short- or medium-term storage, the other uses cryopreservation for long-term storage.

Methods for shoot-tip culture and micropropagation of chrysanthemum have already been established (see Bajaj et al. 1992) and several methods for storage of chrysanthemum under limited growth condition (slow growth) have been reported. Reducing the culture temperature is the most effective way to reduce the growth of plantlets in vitro. Plantlets of chrysanthemum can be maintained successfully in vitro for up to 1 year or more without transfer to fresh medium if cultured on either sucrose-free medium at 10 °C under 1500 lx (Fukai et al. 1988), on the normal medium at 2–3 °C with dim light (Preil and Hoffmann 1985), or at 4 °C (Bajaj 1986). Shoot tips of chrysanthemum which have been precultured on the shoot-tip culture medium for 15 days can be stored for several months at 2 °C in the dark (Fukai and Oe 1989). Recently in vitro cultures of chrysanthemums stored for 6 years at 4 °C survived, and the plants underwent normal flowering (Bajaj 1993).

Modification of the culture medium may also lead to reduced growth of plantlets in vitro. Plantlets of chrysanthemum are usually cultured on medium containing 2 to 3% (w/w) sucrose as a carbon source. If the level of sucrose is reduced to 0.1 or 0%, or mannitol 2–8% is added to the medium, the growth of plantlets of chrysanthemum is significantly reduced (Fukai et al. 1988). The addition of benzyladenine (BA) to the culture medium is effective in delaying the onset of senescence in plantlets during storage at room temperature (Hosoki 1989).

The use of plant growth retardants may also result in slow plant growth. Fukai et al. (1988) found that the growth of plantlets of chrysanthemum was suppressed if they were dipped in 100 mg/l uniconazol for 30 min prior to transplanting to fresh medium. These slow growth techniques may allow laboratory workers to postpone subculture for a year or more.

The cryopreservation of callus cultures of chrysanthemum was first reported in 1976 (Bannier and Steponkus). Successful cryopreservation of the shoot tips has been achieved by using two-step freezing methods (Fukai 1990). Recently, preliminary success in the vitrification of chrysanthemum shoot tips has been reported (Fukai 1992).

Shoot primordia culture method (Tanaka and Ikeda 1983) has also been developed to maintain annual wild species of Compositae. This culture method has been used for in vitro storage and cryopreservation of *Haplopappus* (Taniguchi et al. 1988).

2 Cryopreservation

2.1 Two-Step Freezing Method

In this method, the shoot tips are cooled slowly to subzero temperature in the presence of cryoprotectants prior to immersion into liquid nitrogen (LN).

Preculture. Preculture was essential for the survival of chrysanthemum shoot tips being frozen in LN (Fukai 1990). Shoot tips excised from mother plants were precultured on MS medium (Murashige and Skoog 1962) supplemented with 0.1 mg/l benzyladenine (BA), 1 mg/l α-naphthalene acetic acid (NAA), 20 g/l sucose, and 5% (v/v) dimethylsulfoxide (DMSO) at 25 °C under constant fluorescent lighting 1500 lx for 2 days. Excised shoot tips of chrysanthemum show a large wounded area and tend to turn brown and often die. The effects of preculture are mainly to prevent browning, which leads to death, rather than permeation of cryoprotectants into the shoot tips.

Cooling. After preculture, the shoot tips were put into 0.5 ml plastic straw (3 × 125 mm, bull semen straw Fujihira Industry Co. Ltd. Tokyo) with cryoprotectant solution containing 10% (v/v) DMSO and 3% (w/v) glucose and incubated at 0 °C for 1 h. The straws were then cooled at the rate of 0.1 ~ 1 °C/min from 0°C to − 10 ~ − 40 °C using programmable freezing apparatus (Program freezer ET-1, Fujihira Industry Co. Ltd. Tokyo). The optimum cooling rate and terminal temperature of slow cooling was 0.2 °C/min and − 40 °C, respectively (Figs . 1, 2). During the slow cooling procedure, ice crystals were seeded in the straws at − 3.5 °C by briefly touching the middle part of the straw with tweezers that had been cooled in LN. This seeding procedure minimalized the supercooling of cryoprotectant solution and was necessary for reliable cryopreservation (Fukai 1992).

Thawing. The frozen straws are thawed rapidly after removal from LN. If the shoot tips were thawed slowly in air at 15 °C, survival and shoot regeneration dropped in comparison to shoot tips thawed rapidly in warm water (37 °C) (Table 1).

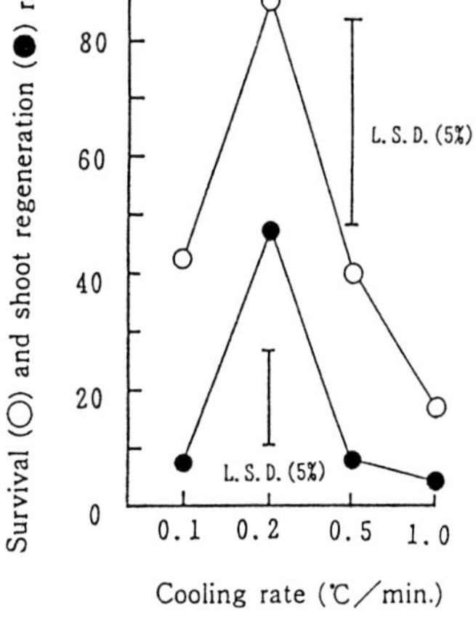

Fig. 1. Effects of cooling rate on the survival and shoot regeneration of chrysanthemum shoot tips (Fukai 1990). The shoot tips were cooled at various rates to − 40 °C after 2 days of preculture on the shoot-tip culture medium with 5% DMSO and were then immersed in LN

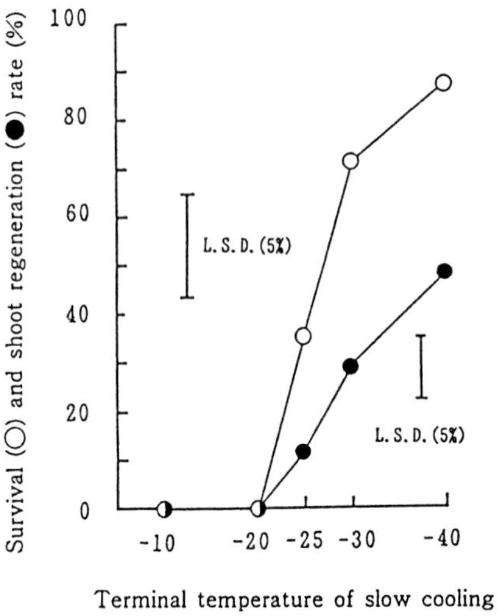

Fig. 2. Effects of various terminal temperatures of slow cooling on the survival and shoot regeneration of chrysanthemum shoot tips (Fukai 1990). The shoot tips were cooled at a rate of 0.2 °C/min down to indicate terminal temperatures after 2 days of preculture on the shoot-tip culture medium with 5% DMSO and were then immersed in LN

Table 1. Effects of thawing methods on the survival and shoot regeneration of carnation and chrysanthemum shoot tips. (Fukai 1992)

Species	Thawing methods	Survival (%)	Shoot regeneration (%)
Dianthus	Hot water (30 °C)	100	96.7
	Ice water (0 °C)	100	100
	Air (15 °C)	35.2	11.4
Chrysanthemum	Hot water (30 °C)	87.5	30.4
	Air (15 °C)	66.6	0

2.2 Shoot Regeneration from Freeze-Thawed Shoot Tips

Culture Conditions After Thawing. Table 2 shows that cytokinin in the culture medium does not affect survival, but greatly affects shoot regeneration in thawed shoot tips. Cultivar differences were observed. Placing filter paper on the medium is effective in suppressing excessive callus growth from thawed shoot tips (data not shown). The light conditions after thawing do not affect survival and shoot regeneration in chrysanthemum (Fukai 1992).

Morphology of Thawed Shoot Tips. Immediately after thawing, no collapse of apical dome and leaf primordia tissue was observed, although shoot tips turned light brown within 24 h of thawing. The leaf primordia of viable shoot tips turned pale green within 7 days of culture. Recovery was observed in the leaf primordium, but rarely in the apical dome. Surviving shoot tips showed differences in size and number of recovering parts and recovery speed. Two patterns of regrowth were observed: a small bump with smooth surface tissue (Fig. 3A) or rough surface callus tissue (Fig. 3B). These surviving tissues grew rapidly and showed a mosaic of smooth and rough surface growth (Fig. 3C). Of the surviving shoot tips, 60–80% regenerated shoots (Fig. 3D), but the origin of the regenerated shoots could not be identified by SEM observation.

Table 2. Effects of cytokinin in the reculture medium on the survival and shoot regeneration of chrysanthemum shoot tips. (Fukai 1992)

cv.	Cytokinin (mg/l)	Survival (%)	Shoot regeneration (%)
Shuhounotikara	BA 0.1	86.6	60.0
	BA 1.0	96.7	96.7
	4PU 0.1	90.0	40.0
	4PU 1.0	83.3	26.7
	Zeatin 0.1	90.0	87.5
	Zeatin 1.0	80.0	30.0
Apricot Marble	BA 0.1	100	71.4
	BA 1.0	93.8	93.8
	4PU 0.1	100	0
	4PU 1.0	98.3	0
	Zeatin 0.1	100	20.0
	Zeatin 1.0	100	14.7

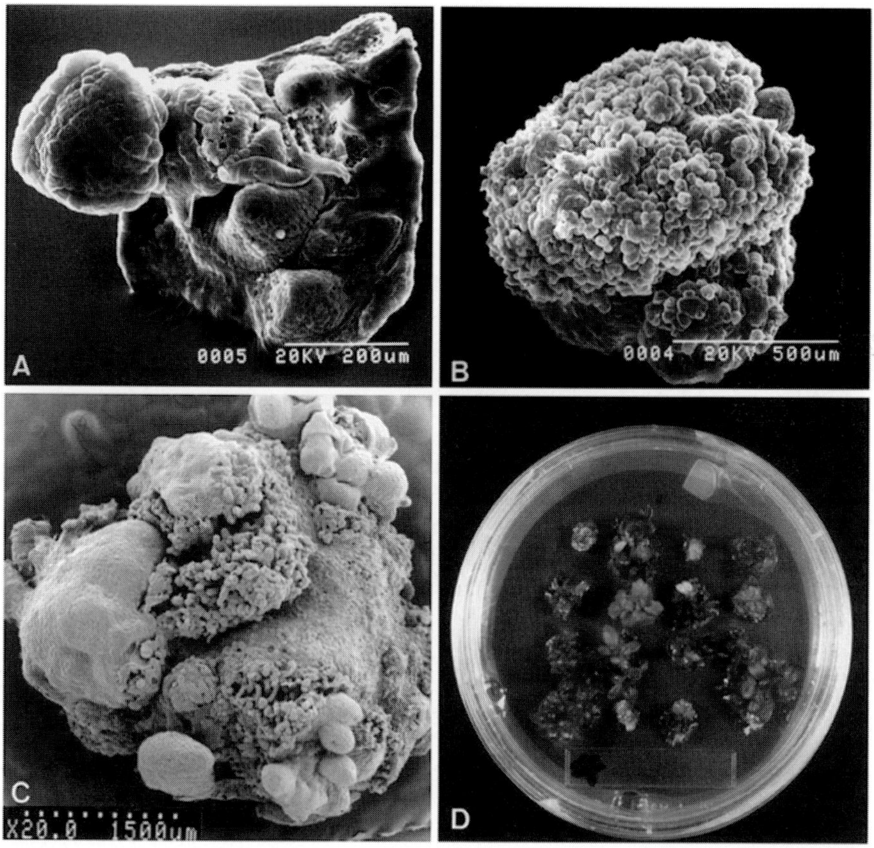

Fig. 3A–D. Regrowth and shoot regeneration from the thawed shoot tips of chrysanthemum (After Fukai and Oe 1990). **A** A viable shoot tip after 20 days of culture. **B** A viable shoot tip after 20 days of culture. **C** Regrowth of a surviving shoot tip. **D** Shoot regeneration from the thawed shoot tip

Table 3. Components of vitrification solution (PVS2). (Sakai et al. 1990)

DMSO	15% w/w
EG	15% w/w
Glycerol	30% w/w
	in MS medium containing 0.5 M sucrose

Table 4. Effects of exposure time to PVS2 on the survival and shoot regeneration of chrysanthemum shoot tips. (Fukai 1992)

Total exposure time (min)	Survival (%)	Shoot regeneration (%)
5	6.7	0
10	9.9	6.5
15	63.9	31.8

Excised shoot tips were precultured on MS medium supplemented with 0.1 mg/l BA, 1 mg/l NAA and 0.5 M sucrose for 2 days and then were treated with PVS2 for 5 to 15 min, followed by immersion into LN.

2.3 Vitrification Method

The shoot tips were treated with a concentrated cryoprotectant solution for a short time in order to osmotically dehydrate the tissue. They were then immersed into LN directly. Ice crystal formation does not occur during rapid cooling. The vitrification method needs a highly concentrated cryoprotectant solution. A vitrification solution for plant tissue (named PVS2, Table 3) is described by Sakai et al. (1990).

Excised shoot tips were precultured on the MS medium supplemented with 0.1 mg/l BA, 1 mg/l NAA, 0.5 M sucrose and 8 g/l agar for 2 days. The shoot tips were then put in 1 ml of PVS2 and incubated at 25 °C for 5 ~ 15 min prior to immersion in LN. After rapid thawing, the shoot tips were rinsed three times with 1.2 M sucrose solution and recultured on the shoot-tip culture medium.

About 64% of the shoot tips treated with PVS2 survived and about half of them regenerated shoots (Table 4). These results are comparable to those of the two-step freezing method.

2.4 Seasonal Variation in Freeze Resistance

The survival and shoot regeneration rates of chrysanthemum shoot tips frozen by two-step freezing methods changed with the season. Higher survival and shoot regeneration was observed in the winter but were lower in mid-summer (Fukai 1992). Storing cuttings at 0 °C for 2 weeks gave rates of survival and regeneration similar to those of plants growing under winter conditions (Table 5).

Table 5. Effects of cold storage of cutting taken in mid-summer on the survival and shoot regeneration of chrysanthemum shoot tips frozen in LN. (Fukai 1992)

Treatment	Survival (%)	Shoot regeneration (%)
Cold storage	95.0	54.4
Control	59.1	28.6

The cuttings were taken in mid-summer and stored at 0°C for 2 weeks. Shoot tips were excised from the cuttings and precultured on the medium containing 5% DMSO for 2 days. The shoot tips were cooled at a rate of 0.2 °C/min to -40 °C prior to immersion in LN.

Therefore cryopreservation during winter or prior to cold storage of cuttings during summer is recommended.

2.5 Estimation of Variation in Regenerated Plants

2.5.1 Growth and Flowering of Cryopreserved Plants

The thawed shoot tips of *Chrysanthemum morifolium* cv. Shuhounotikara were cultured on MS medium supplemented with 0.1 mg/l BA, 1 mg/l NAA, 20 g/l sucrose, and 8 g/l agar. Many of the regenerated shoots had abnormal and vitrified leaves due to the presence of DMSO in the medium (Fukai and Oe 1990). After transplanting into a medium containing 0.3% Hyponex (complete soluble fertilizer, N-P-K = 6.5-6-19), 2% sucrose, and 0.8% agar, however, 4 ± 0.54 normal shoots per thawed shoot tip grew within 60 days of transplantation. The top part of the normal shoots rooted easily in Metro-Mix and were transferred to the greenhouse. There were no differences in the quality of flower between cryopreserved and control plants (Table 6).

2.5.2 Variation in Flower Color in Chimeric Chrysanthemum

Some shoots are derived from callus initiated on the surface of the thawed shoot tips. To clarify the origin of these regenerated shoots, shoot tips of the

Table 6. The cut flower quality of the plants derived from frozen and unfrozen shoot tips of chrysanthemum. (Fukai et al. 1991)

	Date of flowering	Length of flower stem	No. of leaves	Weight of flower (g)	No. of florets	
					R[a]	T[a]
Unfrozen (n = 20)	27.2 ± 0.2 Oct.	56.5 ± 0.9	42.9 ± 0.2	31.6 ± 1.1	192.9 ± 3.2	6.1 ± 0.9
Frozen (n = 18)	27.1 ± 0.4 Oct.	55.0 ± 1.4	41.1 ± 0.4	32.4 ± 1.2	192.4 ± 3.7	5.6 ± 0.8

[a]R: ray floret, T: tuberous floret.

Table 7. Variations of flower color in the plants derived from unfrozen, cryoprotectant loaded, and frozen shoot tips of the periclinally chimeric chrysanthemum cv. Apricot Marble. (Fukai et al. 1994)

Source of shoot tips	BA concentration in the regeneration medium (mg/l)	No. of plants		
		Apricot (%)	Pink (%)	Total
Unfrozen	0.1	86 (100)	0 (0)	86
	1.0	72 (97.3)	2 (2.7)	74
Cryoprotectant loaded	0.1	36 (78.8)	5 (12.2)	41
	1.0	60 (92.3)	5 (7.7)	65
Frozen	0.1	18 (38.3)	29 (61.7)	47
	1.0	15 (26.3)	42 (73.7)	57

periclinally chimeric cultivar Apricot Marble were frozen, and plantlets regenerated from thawed shoot tips were cultured in the greenhouse. Apricot Marble has anthocyanin only in the epidermis (layer I, Ll) and carotinoid only in the mesophyll (layer II, L2). If the periclinally chimeric structure is not retained following cryopreservation, the flower color of the regenerated plants will change. The flower color of plants derived from the conventional shoot tips culture and shoot tips incubated in cryoprotectant did not change. However, 70% of the plants derived from cryopreserved shoot tips had altered flower color from Apricot to Pink (Table 7). This shows that the preculture and cryoprotectant did not affect the chimeric structure of the shoot tips, while indicating that 70% of regenerated shoots were of adventitious origin. The shoot tips suffered damage during freezing and thawing and the chimeric structure of the shoot tips was disturbed during the shoot regeneration process. The Pink flower had anthocyanin in L1, but carotinoids were absent in both L1 and L2. This suggests that these shoots originated in the L1 layer. These results can be attributed to only the L1 surviving after cryopreservation or the L1 having greater regeneration capacity than other tissue surviving in the shoot tips.

2.6 Application to Diverse Genotypes

A standardized cryopreservation protocol applicable to a variety of genotypes is required in order to establish in vitro gene bank. Shoot tips of *C. morifolium* and related species native to Japan were cryopreserved (Table 8). High survival rates were observed in three cultivars, twelve species, and two interspecific hybrids of chrysanthemum. A slightly lower survival rate was observed in three of the related species.

The shoot regeneration responses were divided into three groups as follows: type I, high shoot regeneration of frozen and nonfrozen shoot tips; type II, high shoot regeneration of nonfrozen and relatively low shoot regeneration of frozen shoot tips (*C. morifolium* cultivars, *C. makinoi*, and *C. zawadskii*); type III, low shoot regeneration or poor growth of nonfrozen shoot tips and low shoot

Table 8. Responses of various Japanese chrysanthemum species to cryopreservation. (Fukai et al. 1991)

Species	Frozen shoot tips		Control (unfrozen)	
	S (%)[a]	R (%)[b]	S (%)[a]	R (%)[b]
Chrysanthemum morifolium				
cv. Shuhounotikara	83.3	58.3	100	100
cv. Kenrokukougiku	91.3	52.7	100	100
cv. Parliament	100	62.6	100	100
C. arcticum ssp. *maekawanum*	100	96.5	100	100
C. boreale	93.4	85.0	100	100
C. cuneifolium	95.9	80.0	100	100
C. indicum	96.7	9.4	100	21.0
C. japonense var. *ashizuriense*	96.7	29.3	100	45.0
C. makinoi	96.9	37.5	100	100
C. nipponicum	46.5	28.3	100	100
C. ornatum	87.5	78.2	100	96.9
C. pacificum	63.1	12.3	100	96.9
C. morifolium × *C. pacificum* 22-B	87.5	41.7	100	43.8
22-B × *C. morifolium* Moonlight	89.9	13.0	100	18.8
C. shiwogiku	100	86.7	100	93.8
C. vulgare var. *boreale*	85.9	20.8	100	91.7
C. weyrichii	100	100	100	100
C. yoshinaganthum	90.0	90.0	100	100
C. zawadskii	100	61.5	100	100
C. zawadskii var. *latilobum*	90.0	40.0	100	67.4

[a]S: Survival rate (%) = No. of surviving shoot tips / No. of frozen shoot tips × 100.
[b]R: Shoot regeneration rate (%) = No. of shoots regenerating shoot tips / No. of frozen shoot tips × 100.

regeneration of frozen shoot tips (*C. indicum, C. japonense* var. *shizuriennse, C. pacificum, C. vulgare* var. *boreale,* and *C. zawadskii* var. *latilobum, C. morifolium* x *C. pacificum* 22-B, and 22-B x, *C. morifolium* Moonlight).

The variation in shoot regeneration rates could be attributed to differences in the degree of injury caused by freezing and thawing, the sensitivity to DMSO, and the regeneration ability of the shoot tips. In the case of type III, the unfrozen shoot tips of those species showed low shoot regeneration rates and/or poor growth of the regenerated shoots. This means that the medium is inadequate for those species. Shoot regeneration rates may be improved to some extent, given suitable media. Towill (1984) recognized large variations on survival and shoot regeneration of frozen shoot tips among lines of *Solanum tuberosum* and related species, in response to cryopreservation, and pointed out that a single regeneration medium was inadequate for the cryopreservation of a wide range of species. The determination of a suitable medium for each species is needed to establish the reliable conservation of species and cultivars. Present results show, however, that the two-step freezing method can be applied for the cryopreservation of a wide variety of shoot tips of *Chrysanthemum* species.

3 Summary and Conclusions

A basic protocol for the cryopreservation of chrysanthemum shoot tips is described. This protocol was applicable to a wide range of related species. These reliable in vitro preservation techniques would allow the breeder, researcher, and nurseryman to maintain larger numbers of species and cultivars. Morphological observation and variation of flower color in chimeric cultivars suggest that somaclonal variation in regenerated plants should be taken into consideration when using cryopreserved plants.

References

Bajaj YPS (1986) In vitro preservation of genetic resources – techniques and problems. Int Symp Nuclear techniques and in vitro culture for plant improvement. IAEA/FAO, Vienna 1985, pp 43–57

Bajaj YPS (1993) A suggested method for in vitro long-term storage at 4 °C of *Chrysanthemum* and *Petunia* germplasm. Plant Tissue Cult 3: 57–58

Bajaj YPS, Sidhu MMS, Gill APS (1992) Micropropagation of *Chrysanthemum*. In: Bajaj YPS (ed) Biotechnology in agriculture and forestry, vol 20. High-tech and micropropagation IV. Springer, Berlin Heidelberg New York, pp 69–80

Bannier LJ, Steponkus PL (1976) Cold acclimation of chrysanthemum callus cultures. J Am Soc Hortic Sci 101: 490–512

Fukai S (1990) Cryopreservation of chrysanthemum shoot tips. Sci Hortic 45: 167–174

Fukai S (1992) Cryopreservation of shoot tips of *Dianthus* and *Chrysanthemum*. Mem Fac Agric Kagawa Univ 56: 1–79

Fukai S, Oe M (1989) Cold storage of shoot tips of chrysanthemum [*Dendranthema grandiflorum* (Ramat.) Kitamura]. Plant Tissue Cult Lett 6(1): 10–13

Fukai S, Oe M (1990) Morphological observation of *Chrysanthemum* shoot tips cultured after cryoprotection and freezing. J Jpn Soc Hortic Sci 59: 383–387

Fukai S, Morii M, Oe M (1988) Storage of chrysanthemum [*Dendranthema grandiflorum* (Ramat.) Kitammura] plantlets in vitro. Plant Tissue Cult Lett 5(1): 20–25

Fukai S, Goi M, Tanaka M (1991) Cryopreservation of shoot tips of *Chrysanthemum morifolium* and related species native to Japan. Euphytica 54: 201–204

Fukai S, Goi M, Tanaka M (1994) The chimeric structure of the apical dome of chrysanthemum [*Dendranthema grandiflorum* (Ramat.) Kitam.] is affected by cryopreservation. Sci Hortic 57: 347–351

Hosoki T (1989) In vitro storage of *Chrysanthemum morifolium* at room temperature. Plant Tissue Cult Lett 6(2): 85–87

Kitamura S et al. (1988) *Chrysanthemum* L. sensu ampl. In : Tsukamoto Y (ed) The grand dictionary of horticulture. Shyougakkann, Tokyo, pp 24–30

Murashige T, Skoog F (1962) A revised medium for rapid growth and bioassays with tobacco tissue cultures. Physiol Plant 15: 473–497

Preil W, Hoffmann M (1985) In vitro storage in chrysanthemum breeding and propagation. In: Schäfer-Menuhr A (ed) In vitro techniques. Martinus Nijhoff, Dordrecht, pp 161–165

Sakai A, Kobayashi S, Oiyama I (1990) Cryopreservation of nucellar cells of navel orange (*Citrus sinensis* Osb. var. *brasiliensis* Tanaka) by vitrification. Plant Cell Rep 9: 30–33

Shibata M, Kawata J (1986) Chimerical structure of the Marble sports series in chrysanthemum. In: Kitaura K, Akihama T, Kukimura H, Nakajima K, Horie M, Kozaki I (ed) Development of new technology for identification and classification of tree crops and ornamentals. Fruit Tree Res Stn MAFF Jpn, pp 47–52

Shibata M, Kawata J, Amano M, Kameno T, Yamagishi H, Toyoda T, Yamaguchi T, Okimura M, Uda M (1988) Breeding process and characteristic of Moonlight, and interspecific hybrid between *Chrysanthemum morifolium* Ramat. and *C. pacificum* Nakai (in Japanese with English Summary). Bull Natl Res Inst Veg Ornam Plant Tea Jpn Ser A 2: 257–277

Tanaka R, Ikeda H (1983) Perennial maintenance of annual *Haplopappus gracilis* (2n = 4) by shoot-tip cloning. Jpn J Genet 58: 65–70

Taniguchi K, Tanak R, Ashitani N, Miyagawa H (1988) Freeze preservation of tissue-cultured shoot primordia of the annual *Haplopappus gracilis* (2n = 4). Jpn J Genet 63: 267–272

Towill LE (1984) Survival at ultra-low temperatures of shoot tips from *Solanum tuberosum* group andigena, phureja, stenotomum, tuberosum and other tuber-bearing *Solanum* species. Cryo Lett 5: 319–326

VI.4 Cryopreservation of Germplasm of Carnation (*Dianthus caryophyllus* L.)

J. Dereuddre and M. Tannoury[1]

1 Introduction

1.1 Distribution, Importance, and Propagation of Carnation

The genus *Dianthus* (family Caryophyllaceae) comprises about 75 species which grow in Europe, Asia, and Africa. Some of them (*Dianthus caryophyllus, D. fimbriatus, D. caesius*) have been selected for horticulture, since middle ages. These species are widely used in horticulture due to the beauty and color diversity of their simple or double flowers. Flowers of carnation are regular with five sepals and petals, ten stamens, and one ovary with five styles and stigmas. Seeds contain starchy endosperm (Bonnier 1934).

After rose and gladiolus, carnation is, with tulip, the third flower in importance to be cultivated, slightly before chrysanthemum. (Comité National Interprofessionel de l'Horticulture et des Pépinières, CNIH, pers comm). Three countries appear as leaders in the cultivation of this flower: Italy, Colombia, and Spain (about 2000 acres for each country). However, the economic importance of carnation seems to be decreasing progressively in favor of tulip and chrysanthemum.

Carnation can easily be vegetatively propagated by cutting. However, it is subjected to several pests and diseases: soil nematodes, fungi (*Rhizoctonia solani; Phytophthora* sp.; *Uromyces garyophyllus, Fusarium oxysporum, Bothrytis* sp.), bacteria (*Erwinia chrysanthemi; Pseudomonas caryophilli*), virus (streak, mottle and etched ring, ring spot), and vermin (aphids, thrips, and caterpillars). However, if several pesticides are available, the tendency exists to replace traditional cultures by soil-free cultures in greenhouses. The great problem is the eradication of virus from selected clones, to ensure disease-free plants for commercial distribution.

Considerable work has been done on various aspects of in vitro culture of carnations, and the subject has been reviewed (see Mii et al. 1990; Lu and Chandler 1994; Matern 1994). Carnation can be aseptically micropropagated by nodal cutting (Hackett and Anderson 1967). Virus-free plants can be obtained by meristem culture (for review, see Quak 1977). Absence of virus is usually confirmed by the use of ELISA tests. In vitro propagation by nodal cuttings

[1] Laboratoire de Cryobiologie Végétale, Université Pierre et Marie Curie, 12 rue Cuvier, 75230 Paris Cédex 05, France

Biotechnology in Agriculture and Forestry, Vol. 32
Cryopreservation of Plant Germplasm I (ed. by Y.P.S. Bajaj)

appears suitable for the production of disease-free plants. Providing due precautions are taken, the genotype of plants propagated by shoot-tip culture can be preserved without change, and selected varieties can be subcultured for indefinite period of time.

1.2 Germplasm Storage

For medium-term storage, in vitro cultures can be stored at low temperature ($+4 \pm 1$ °C) in low intensity light (4-h photoperiod). At this temperature, growth of plant material decreases significantly or stops; young rooted in vitro plantlets of carnation can be stored for more than 1 year on standard culture medium (Van Hoof 1980).

For long-term storage, cryopreservation at ultralow temperature (i.e., in liquid nitrogen) can ensure the preservation of material under conditions of good genetic stability for several years (Bajaj 1985). It can be applied to shoot tips of wild species and commercially propagated hybrids. Cryopreservation of wild genes and natural variability may also be ensured by the cryopreservation of seeds and pollen. However, no or little research has been undertaken on tolerance of these organs to deep freezing.

2 Cryopreservation

2.1 Three Cryopreservation Procedures

Cryopreservation of carnation shoot tips was first successfully carried out by Seibert (1976): shoot tips were aseptically excised from plants maintained in a growth chamber, treated with DMSO 5%, plunged in liquid nitrogen (LN) without prefreezing, rapidly rewarmed, and transferred after washing on solid culture medium for reactivation. With this procedure, up to 33% shoot tips survived until cooling in LN, but only 5% of surviving shoot tips developed normal plants. Shoot recovery was considerably improved by culture of carnation donor plants at 4 °C for 3 to 10 days (Seibert and Wetherbee 1977): about 80% shoot recovery was obtained when cooling rates were above 50 °C/min and after rapid rewarming. Conventional procedures described for plant cell suspensions (see Bajaj 1991; Withers 1991) have also been developed for cryopreservation of carnation shoot tips. After several days of cold treatment (0 °C with DMSO 10% as cryoprotectant, stepwise prefreezing to temperatures equal to or lower than -40 °C), 60–80% shoot recovery was achieved after cooling in LN (Uemura and Sakaï 1980). Cryopreserved shoot tips can be stored up to 2 years without decrease in survival rates (Fukai 1989).

Further improvements were obtained using in vitro-cultured donor plants and preculture with high sucrose concentration (0.75 M), in replacement of cold treatments (Dereuddre et al. 1987).

These conventional methods for cryopreservation of plant cells and organs generally involve preculture with sucrose or polyols, loading with DMSO, freeze-induced cell dehydration before plunging in liquid nitrogen, and rapid warming. In order to simplify experimental procedures, two new technologies involving vitrification of both cryoprotective surrounding medium and plant material have been recently proposed for cryopreservation (see Dereuddre 1992).

Vitrification procedures were first applied to cell suspensions and protoplasts (Langis et al. 1989; Uragami et al. 1989; Langis and Steponkus 1990; Sakaï et al. 1991). The usual freeze-induced dehydration was replaced by dehydration at room temperature in extremely concentrated mixtures of permeating and/or nonpermeating cryoprotectants.

In "dehydration" procedures, removal of freezable water from the cells was performed by partial dehydration in air at room temperature (Dereuddre et al. 1990, 1991a; Uragami et al. 1990).

The three protocols which have been successfully applied to the cryopreservation of carnation shoot tips are compared in this chapter; results and protocols are summarized in Table 1.

2.2 Materials and Methods

2.2.1 In Vitro Micropropagation

For in vitro propagation, cuttings were taken from 6 to 8 week-old in vitro plantlets about 12 cm in length (Dereuddre et al. 1987, 1988). Cutting culture medium consisted of the medium of Murashige and Skoog (1962), the vitamin mixture of Morel and Martin (1985), L-glutamine (0.2 g/l), EDTA-Fe (36 mg/l), myoinositol (0.1 mg/l), 0.1 M sucrose, and 1% agar. Cuttings were kept in a culture chamber at 25 °C, with a 16 h/day photoperiod, and irradiance of 25–30 μmol PAR/m^2/s.

2.2.2 Apex Culture

For cryopreservation, explants consisting of apical dome and a pair of leaf primordia were cultured according to the method described previously by Carré et al. (1979). Two kinds of explants were used: axillary and apical shoot tips. Both kinds of shoot tips were cultured on "apex" medium which differed from the "cutting" medium by the presence of nicotinic acid (4.06 mM), pyridoxine hydrochloride (2.43 mM), and glycine (5.33 mM), with agar content 0.7% instead of 1%. Growth regulators were also added to the medium: 1 μM 6-benzylaminopurine and 0.1 μM gibberellic acid (GA_3).

2.2.3 Cryopreservation Procedures

To ensure survival of plant cells after plunging in LN and subsequent rewarming to room temperature, it is necessary to prevent the formation of intracellular ice

Table 1. Summary of freeze-preservation protocols and results for carnation shoot tips. SGM, standard growth medium

Origin of shoot tips	Preculture of shoot tips	Cryoprotective medium	Method of cryopreservation	Cooling and warming procedures	Results/remarks	Reference
Greenhouse-grown plants	4 days in the dark at 26 °C on SGM	5% DMSO 0.1 M sucrose	Conventional	Rapid cooling (> 1000 °C/min) and rapid rewarming	1–2% Shoot recovery	Seibert (1976)
3 days cold acclimation of donor plants (4 °C; 8-h photoperiod)	No	5–10% DMSO 0.1 M sucrose	Conventional	Rapid cooling (> 50 °C/min) and rapid rewarming	Up to 70% shoot recovery	Seibert Wetherbee (1977)
Greenhouse-grown plants stored for several days at 0 °C	No	10% DMSO alone or with 5% sucrose, glucose or sorbitol	Conventional	Slow cooling (0.5 °C/min) to − 40 °C or lower and rapid rewarming	Up to 80% shoot recovery	Uemura and Sakaï (1980)
In vitro plantlets	Overnight on SGM + sucrose (0.5 M)	Preculture medium + 5% DMSO	Conventional	Two-step cooling (0.5 °C/min) to − 40 °C and rapid rewarming	90% Shoot recovery	Dereuddre et al. (1987)
In vitro plantlets	Overnight on SGM (0 or 25 °C) + sucrose, mannitol, sorbitol or glucose (0.5 M)	Preculture medium + 5% DMSO	Conventional	Two-step cooling (0.5 °C/min) to − 40 °C and rapid rewarming	Up to 87% shoot recovery depending on the composition of preculture medium and on temperature of preculture	Fabre and Dereuddre (1987)

(Contd.)

Table 1. (*Contd.*)

Origin of shoot tips	Preculture of shoot tips	Cryoprotective medium	Method of cryopreservation	Cooling and warming procedures	Results/remarks	Reference
In vitro plantlets	Overnight on SGM + sucrose (0.75 M)	Preculture medium + 5% DMSO	Conventional	Two-step cooling (0.5 °C/min) to − 40 °C and rapid rewarming	Survival depends on the position of shoot tips along the stem	Dereuddre et al. (1988)
Greenhouse-grown plants	No	10% DMSO 0.1 M sucrose	Conventional	Two-step cooling (0.5 °C/min) to − 40 °C and then to − 160 °C (33 °C/min) and rapid rewarming	80 to 100% Shoot recovery after several period of storage (up to 2 years)	Fukai (1989)
In vitro plantlets	Encapsulation of shoot tips in alginate beads and overnight preculture in SGM + sucrose (0.75 M)	Vitrification medium (ethylene glycol 60-sucrose 60-water 40) (*P*/*P*/*P*/)	Encapsulation-vitrification	Two-step and rapid cooling slow rewarming	Shoot recovery (73 − 93%) is independent of cooling rate (0.5 to 200 °C/min)	Tannoury et al. (1991)
In vitro plantlets	Encapsulation of shoot tips in alginate beads and overnight preculture in SGM + sucrose (0.75 M)	Increase in sucrose concentration by dehydration	Encapsulation-dehydration	Two-step and rapid cooling slow rewarming	Shoot recovery (79–92%) is independent of cooling rate (0.5 to 200 °C/min)	Tannoury (1994)

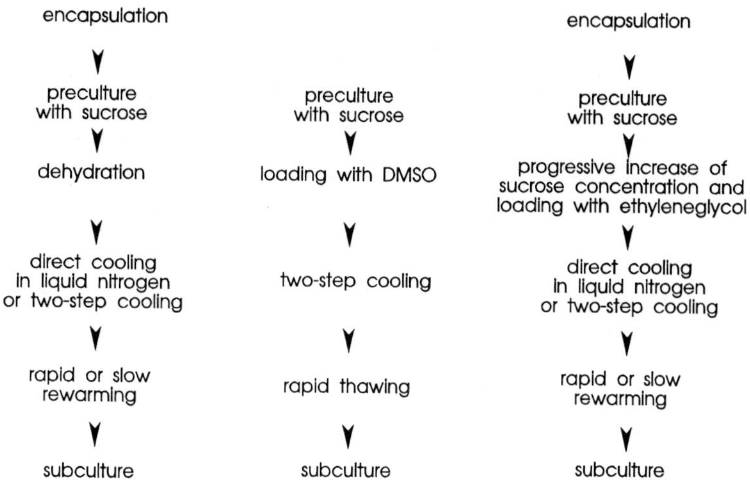

Fig. 1. Procedures used for cryopreservation of carnation shoot tips. In conventional procedures, cell dehydration is ensured during slow cooling from 0 to − 40 °C by outflow of intracellular water towards the cryoprotective medium, where it crystallizes. In vitrification procedure, dehydration is induced by progressive addition of highly concentrated solutions of cryoprotective agents. In dehydration procedure, cellular water is removed by evaporation at room temperature

crystals by dehydrating the cells before immersion in LN. Three procedures have been used to ensure cryopreservation of carnation shoot tips (Fig. 1).

Conventional Procedures. The explants were precultured on a solid "apex" preculture medium supplemented with sucrose at different concentrations (0.1 to 1 M), under the conditions used for micropropagation. DMSO treatment was performed at 0 °C. Apices were suspended in the preculture liquid medium. The solution containing DMSO and the same sucrose concentration as in the preculture medium was added gradually, over 30 min, until the desired DMSO concentration (2.5 to 15%) in the cryopreservative medium was reached. After 2 h incubation, the tips were transferred in cryotubes, containing 1 ml of the cryoprotective solution.

Samples were gradually cooled (0.5 °C/min) to − 40 °C, using a programmable freezer; freezing was done at 3 °C below the freezing temperature of the cryoprotective media with forceps precooled in LN. Once the temperature of − 40 °C was reached, samples were directly immersed in LN, where they were stored for up to 1 week. After rapid thawing for 2 min in a water bath at + 40 °C, shoot tips were transferred to culture medium of same composition as the preculture medium. They were subcultured daily (1 to 4 days) on decreasing sucrose concentrations, until the standard 0.1 M concentration was reached.

Encapsulation-Dehydration Procedure. Beads containing shoot tips were prepared according to the technique of Redenbaugh et al. (1986). Shoot tips were suspended in calcium-free apex medium supplemented with 3% Na-alginate solution and the mixture was dripped into 'apex' medium containing 100 mM calcium chloride.

Beads containing one to three shoot tips were first precultured for 18 h in liquid apex medium supplemented with sucrose at different concentrations (0.1 to 2 M). Dehydration was carried out by placing encapsulated shoot tips in uncovered sterile Petri dishes in a sterile air flow at ambient temperature and humidity (23 ± 1 °C, 55–65% RH) for 0 to 6 h.

Beads were then transferred to cryotubes for cooling in LN for 1 h. Two cooling procedures were used: rapid cooling by direct immersion in liquid nitrogen, and two-step cooling by progressive cooling (0.5 to 20 °C/min) from + 20 to − 150 °C before immersion in LN. After slow rewarming in air at room temperature, beads were transferred for 1 week to Petri dishes containing solid apex medium containing 0.1 M sucrose. Apices were then extracted from the beads and subcultured on new apex culture medium, under the same conditions of cutting culture.

Vitrification Procedure. In the encapsulation-dehydration procedure, the duration of dehydration necessary to give high survival rates for coated plant organs depended on the temperature of the sterile air flow and on its relative humidity. To decrease the effects of these partially uncontrolled parameters, another procedure was proposed (Tannoury et al. 1991). It consisted of vitrification of encapsulated shoot tips after incubation in a vitrification medium of very low water potential.

Encapsulated shoot tips in alginate beads were first precultured for one night in sucrose enriched (0.75 M) apex culture medium. Trapped shoot tips were then transferred into media which were progressively enriched with sucrose until the desired concentration of sucrose (6 g sucrose/4 g water) was reached. After loading for 5 h at 0 °C with ethylene glycol at the concentration of 6 g sucrose/4 g water/6 g ethylene glycol, beads containing one to three shoot tips were transferred in screw-cap vials without cryoprotective medium.

Two cooling procedures were used: direct cooling in liquid nitrogen and two-step cooling. In two-step cooling, beads containing shoot tips were first frozen to − 150 °C at different cooling rates (from 0.5 °C/min to 20 °C/min) and then immersed in liquid nitrogen. After rewarming in air at room temperature, coated shoot tips were subcultured without washing on standard apex culture medium, in Petri dishes.

2.2.4 Thermal Analysis and Survival

Thermal analysis was performed using differential scanning microcalorimetry (DSC) (Mettler DSC 30). By this technique variations of enthalpy due to state change of solutions or live material during cooling as well as during rewarming can be measured . These variations were estimated from temperature differences between two pans cooled or warmed at the same rate. One pan was empty and was used for reference; the second contained the sample to study. This technique could detect crystallization, melting, and glass transitions. It is a useful tool for studying the behavior of cryoprotective solutions and biological material during a freeze-thaw cycle (Dereuddre and Kaminski 1992).

Survival was defined as the percentage of shoot tips recovering normal regrowth after freezing and thawing.

2.3 Results

2.3.1 Conventional Procedure

Resistance of Apical Shoot Tips. In first experiments (Dereuddre et al. 1987), DMSO 5% was used as second cryoprotectant in cryoprotective medium. Shoot recovery after cooling in LN required the presence of high sucrose concentrations during preculture and in the cryoprotective medium. Survival rates above 75% were noted with sucrose concentrations of 0.3 to 0.75 M. Survival decreased to 8% in the presence of 1 M sucrose; this did not appear to be due to the preculture, since all pretreated, but unfrozen, apices developed normally. When 0.75 M sucrose was used in both preculture and cryoprotective media (Dereuddre et al. 1987). Shoot recovery depended on the concentration of DMSO in the cryoprotective medium: only 35% shoot recovery was noted with DMSO-free cryoprotective mixture. Survival increased with DMSO content, reaching 93% with 2.5% DMSO and nearly 100% with 5 to 15% DMSO.

Complementary experiments were performed (unpubl.) to study the effects of both sucrose and DMSO concentrations on the tolerance of apical shoot tips to cooling in LN. Apical shoot tips were first precultured in a range of 0 to 1 M sucrose-supplemented medium. They were then loaded with 0 to 15% DMSO with the same sucrose concentration as in the preculture medium. The results (Table 2) demonstrated that 100% survival and regrowth occurred after precultures with 0.3 to 0.75 M sucrose in the preculture medium.

The concentration of DMSO which allowed 100% regrowth rates depended on the sucrose concentration: 12.5 to 15% after 0.3 M sucrose preculture and 5 to

Table 2. Effects of sucrose (0.1 to 1 M) and DMSO (0 to 15%) concentrations in pretreatment media on survival (% of shoot recovery) of shoot tips after freezing in LN. Values in parentheses represent standard deviation

Sucrose concentration (M)	Shoot tip recovery (% of frozen shoot tips)						
	DMSO concentration (%*v*/*v*)						
	0	2.5	5.0	7.5	10.0	12.5	15.0
0.1	0	0	5	64	72	62	66
0.3	0	63	91	98	100	100	98
0.5	22	70	100	100	100	100	100
0.75	35	93	100	100	100	100	100
1	42	22	—	8	2	0	0

15% after 0.5 or 0.75 M sucrose preculture. Up to 72% regrowth was achieved after preculture of shoot tips in standard culture medium containing only 0.1 M sucrose and 10% DMSO. Up to 35 and 42% survival was obtained with sucrose alone as cryoprotectant (0.75 and 1 M, respectively). Carnation shoot tips tolerated a large range of cryoprotective conditions in experiments using a series of DMSO concentrations versus a series of sucrose concentrations. It may also be noted that sucrose alone allowed tolerance of shoot tips to liquid nitrogen.

2.3.2 *Effects of Other Sugars or Polyols*

In these experiments (Fabre and Dereuddre 1987), sugar and polyol concentrations in preculture and further following loading medium were equal to 0.5 M, and DMSO concentration in the loading medium was 5% (*v/v*). With sucrose as first cryoprotectant, survival rate was equal to 87%. When sucrose was entirely replaced by glucose, survival dropped to 24% (Fig. 2). A similar result (35% regrowth rate) was obtained when sucrose was partially replaced by glucose: in this case the preculture medium contained 0.1 M sucrose and 0.4 M glucose. With

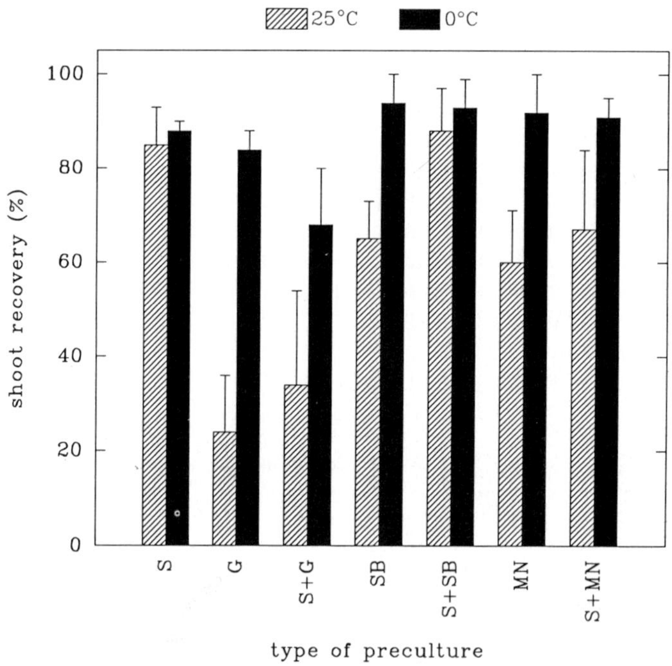

Fig. 2. Recovery rate (%) of carnation shoot tips after freezing in liquid nitrogen, as a function of cryoprotectant and temperature of preculture (25 or 0 °C). Substances were used alone (*S*, *G*, *SB*, and *MN* 0.5 M) or mixed with sucrose: 0.1 M sucrose and 0.4 M glucose (*S* + *G*), sorbitol (*S* + *SB*) or mannitol (*S* + *MN*). *Black bars* Preculture at 25 °C; *hachured bars* preculture at 0 °C. (Fabre and Dereuddre 1987 with permission, editions Lauthier-Villars)

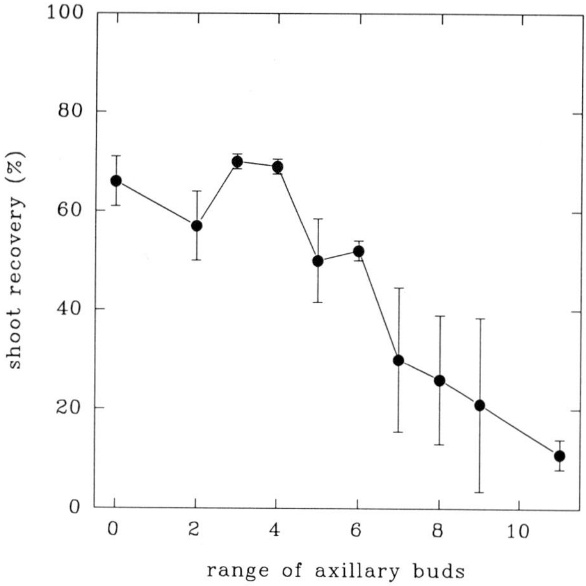

Fig. 3. Changes in the resistance of axillary shoot tips to LN as a function of their range along the stem (range *O* corresponded to apical shoot tip). Sucrose and DMSO concentrations were respectively 5% and 0.75 M, followed by application of cryoprotectant solution containing 5% DMSO and 0.75 M sucrose. These experimental conditions were used to detect differences between shoot tips of different physiological stages. They were applied to study the resistance of axillary shoot tips, before and during their reactivation by nodal cutting culture. (Dereuddre et al. 1988)

sorbitol 0.5 M, regrowth rate was equal to 63%, this value reached 89% when the preculture medium contained 0.1 M sucrose and 0.4 M sorbitol. Similar results were obtained with mannitol, another polyol: 61% regrowth with mannitol alone and 69% for 0.4 M mannitol and 0.1 M sucrose.

2.3.3 *Effect of Preculture Temperature*

Further experiments were performed with shoot tips precultured at 0 °C in the presence of sugars or polyols 0.5 M. No change was observed for sucrose preculture (88% regrowth after preculture at 0 °C against 87% after preculture at 25 °C). When sucrose was partially or completely replaced by glucose, mannitol, or sorbitol, shoot recovery was improved by lowering the temperature of preculture (Fig. 2). With glucose preculture at 0 °C, regrowth rate was raised from 24 to 85%. Similar improvement after cryopreservation was obtained for sorbitol (95 instead of 63%) and mannitol (94 instead of 61%) after preculture at 0 °C.

Tolerance of Axillary Shoot Tips. The physiological problems raised by cryopreservation of meristems relate both to the mechanisms of action of cryoprotectants and to the physiological state of the plant material, which could be one of the main factors for survival after freezing in LN (Dereuddre et al. 1988).

The resistance of axillary shoot tips to LN was studied as a function of their range number from the apical shoot tip on 2-month-old plantlets (Fig. 3). The resistance of axillary shoot tips from range 1 to 4 (90%) was similar to that of apical shoot tips. It decreased gradually from range 5 (68%). Close to the base of the stem (range 11), the survival rate was lower than 10% (6 to 9%, depending on the experiment).

Changes in shoot tip resistance to deep freezing were studied during the first 3 weeks of cutting culture (Dereuddre et al. 1988). Cuttings (range 5 to 9 from the top of the apical shoot tip) were allowed to grow for 0 to 21 days before removal of the apex of the new shoot. During the first 2 days of cutting culture, survival rates increased transitorily to 61%. A decrease was noted on the 3rd day of culture (21%). A similar phenomenon has been noted in tips obtained by the same way (from 3- to 5-day-old cuttings), but kept at 0 °C during preculture on 0.75 M sucrose, all other parameters being the same (DMSO 5% treatment, freezing, thawing, and subculture). Survival rate subsequently increased gradually with the duration of cutting culture; after 14 and 21 days, it reached values similar to those noted for shoot tips excised from 2-month-old plantlets (94 and 98%, respectively).

Axillary shoot tips were more sensitive to deep freezing in LN than apical shoot tips sampled from the same 2-month-old plantlets. This sensitivity of axillary shoot tips increased with distance from the apical shoot tip, and may be closely related with physiological differences caused by correlative inhibitions induced by apical dominance.

Reactivation of axillary buds on cuttings was rapid (less than 2 days), but was not sufficient to increase survival of cryopreserved shoot tips: high resistance to LN in apical shoot tips requires several weeks of cutting culture. This underlines the importance of the physiological stage of organs under study.

2.3.4 *Vitrification Procedure*

Shoot Recovery. Survival rates of the control shoot tips (uncoated shoot tips, coated shoot tips, and shoot tips extracted from the beads after incubation in loading medium) were 100%. After two-step cooling, survival of shoot tips to LN remained high (Fig. 4), whatever the cooling rate: it ranged between 77% (5 °C/min) and 93% (0.5 °C/min). After direct cooling in LN (200 °C/min), survival rate was 83%. One-way analysis of variance showed that survival was independent of cooling rates.

Thermal Analysis. During cooling (Fig. 5, M) as well as during rewarming (Fig. 6, M), thermograms of the loading medium were similar and showed a glass transition without any peak of crystallization or melting. The temperature of the middle point of glass transitions was − 111.0 °C during cooling and − 106.8 °C at rewarming.

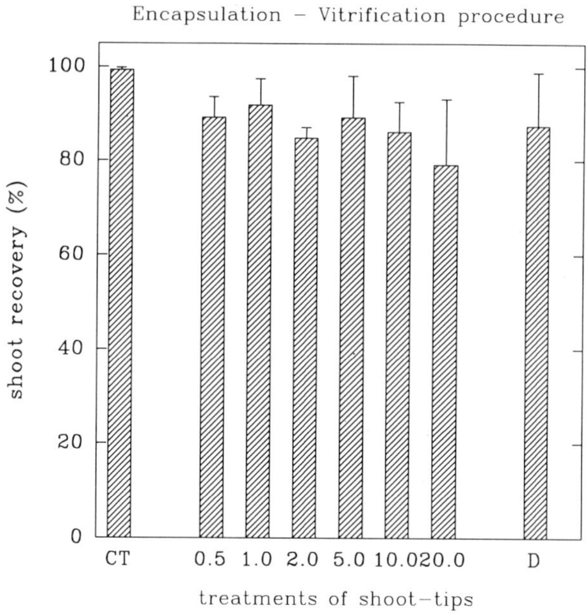

Fig. 4. Percentage of recovery of shoot tips after loading with vitrification (solution medium 6-4-6) (*CT*), direct immersion in liquid nitrogen (*D*) and after two-step cooling from + 20 to − 150 °C (0.5, 1, 2, 5, 10, and 20 °C/min) and immersion in liquid nitrogen. *Vertical bars* Standard deviation. (Tannoury et al. 1991 with permission, editions Lauthier-Villars)

Glass transitions were also obtained with encapsulating material (B) and with shoot tips extracted from the beads (A) after loading for 5 h in cryoprotective medium. The sigmoid shape of cooling and rewarming thermograms was due to the difference in heat capacities of the loading medium before and after the glass transition. Glass transition temperatures of encapsulating material were equal to − 97.6 °C and to − 88.1 °C during cooling and rewarming, respectively. These two temperatures were similar to those obtained with shoot tips extracted from the beads after loading with the vitrification medium: − 94.0 and − 88.1 °C (Tannoury et al. 1991).

2.3.5 Encapsulation-Dehydration Procedure

All control axillary shoot tips excised from in vitro plantlets and subcultured on apex culture medium remained green, and almost 95% of them resumed growth, producing new shoots by direct development of the apical dome. Similar proportions were also obtained from control alginate-encapsulated shoot tips.

Effects of Dehydration. After overnight preculture with 0.75 M sucrose, the water content of beads decreased from 93 to 74%, and continued to decrease during dehydration reaching 23 and 21% after 4 and 6 h of dehydration, respectively (Fig. 7).

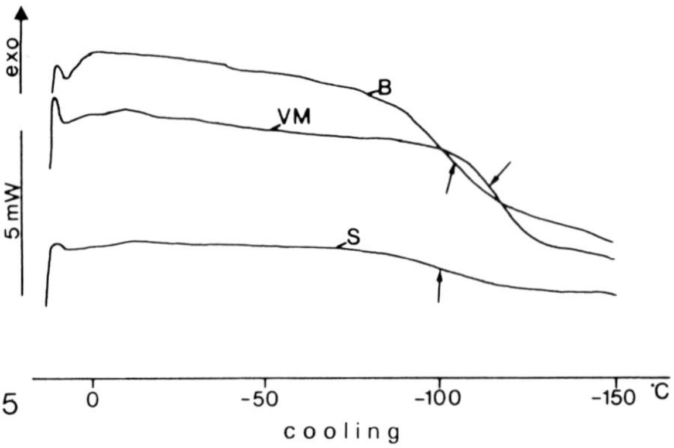

5

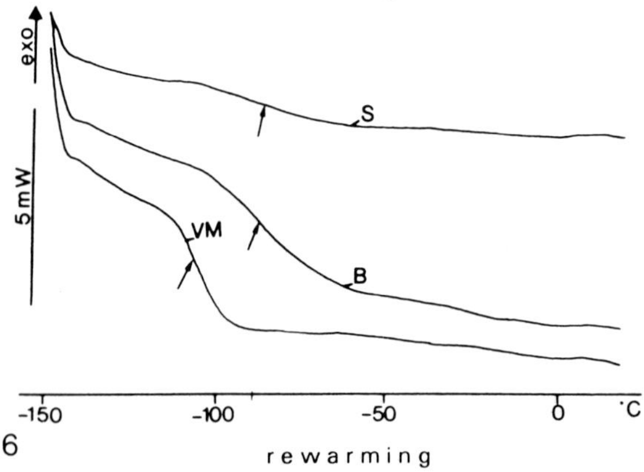

6

Figs. 5, 6. Thermograms of cooling from +20 to −150 °C (**5**) and rewarming from −150 °C to +20 °C (Fig. 6) of loading medium (M), beads treated with loading medium (*B*) and shoot tips extracted from the beads (*A*). Cooling and rewarming rates were 10 °C/min. *Vertical bar* 5 mW; *arrows* middle temperature of glass transitions. (Tannoury et al. 1991 with permission, Lauthier-Villars)

The recovery percentage of dehydrated shoot tips remained high, whatever the duration of dehydration; it decreased progressively to 92% (after 6 h of dehydration). Shoot recovery after exposure to liquid nitrogen was equal to zero without dehydration but increased progressively as a function of the dehydration duration. Best recovery rate (96%) was obtained after 4 h of dehydration, it declined during the two following hours of dehydration.

Effects of Sucrose Concentration. The recovery effect of sucrose concentration on tolerance of encapsulated shoot tips to dehydration and subsequent cooling in liquid nitrogen was studied after overnight preculture with 0.1 to 1.5 M sucrose. After 4 h of dehydration, the water content of the beads was about 23% (±2%). Best results were obtained after 0.75 and 1 M sucrose in the preculture medium (Fig. 8). Tolerance of shoot tips to dehydration increased as a function of sucrose concentration from 0 to 0.75 M, remained constant, and slightly decreased when sugar content of preculture medium was equal to 1.25 or 1.5 M.

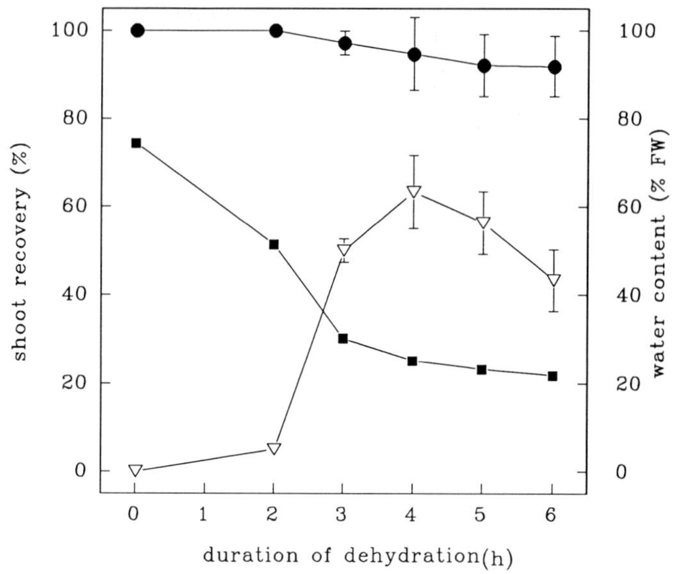

Fig. 7. Effects of the duration of dehydration on the percentage of shoot recovery from shoot tips after dehydration for 0 to 6 h (○) and subsequent direct cooling in liquid nitrogen (△), and on the water content of beads (■). Shoot tips were precultured with 0.75 M sucrose. *Vertical bars* Confidence intervals ($p = 5\%$). Tannoury et al. 1994)

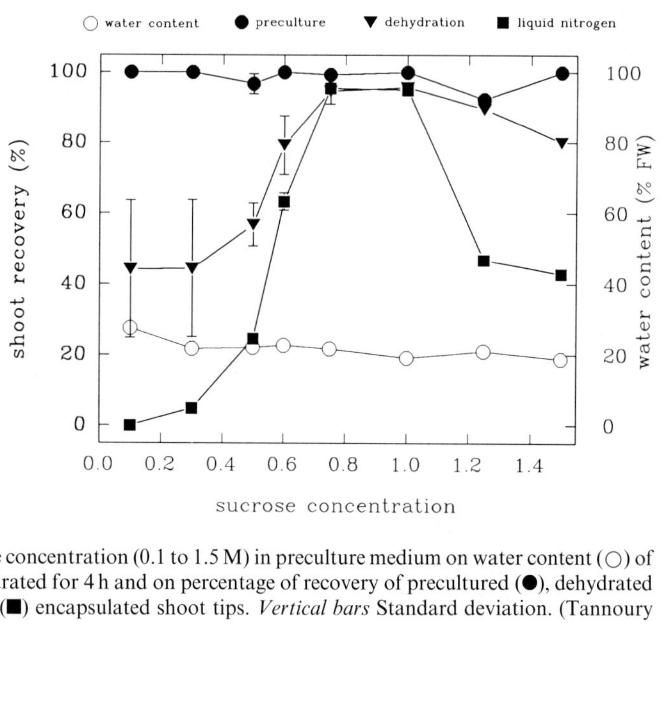

Fig. 8. Effects of sucrose concentration (0.1 to 1.5 M) in preculture medium on water content (○) of the alginate beads dehydrated for 4 h and on percentage of recovery of precultured (●), dehydrated (▼), and cryopreserved (■) encapsulated shoot tips. *Vertical bars* Standard deviation. (Tannoury et al. 1994)

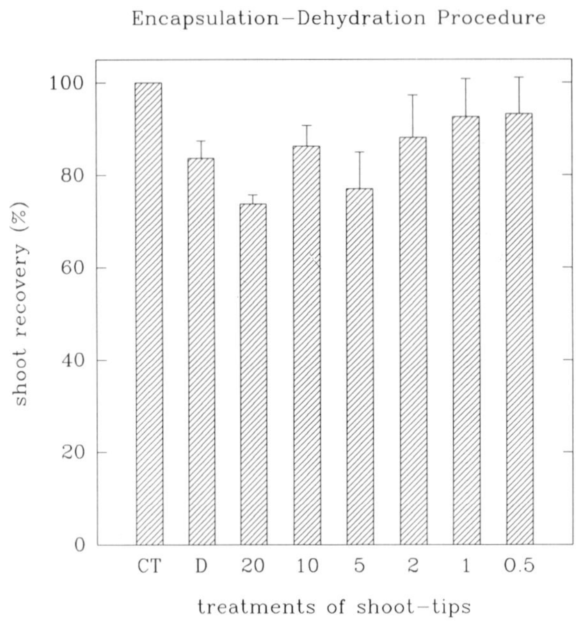

Fig. 9. Effects of cooling rate on recovery of encapsulated shoot tips precultured with 0.75 M sucrose and dehydrated for 4 h *CT* Control of dehydration; *D* direct cooling in liquid nitrogen; 0.5; 1; 5; 10, and 20 °C, cooling rates used for the first step of cooling, from 0 to − 150 °C. *Vertical bar* Standard deviation. (Tannoury et al. 1994)

Effects of Cooling Rate. In these experiments, sucrose concentration in the preculture medium was 0.75 M. Two cooling procedures were used: rapid cooling by direct immersion in LN, and two-step freezing by progressive cooling (0.5 to 20 °C/min) from +20 to −150 °C before immersion in LN (Fig. 9). Shoot recovery was equal to 91% after direct cooling in LN (Fig. 9, D). After two-step cooling, shoot recovery rates ranged from 77% (5 °C/min) to 94% (0.5 °C/min). Whatever the cooling rates, no significant difference in regrowth rates (chi-square test; $p = 5\%$) was noted (Tannoury 1994).

3 Discussion

Survival of shoot tips at liquid nitrogen temperature has been obtained by conventional cryopreservation methods in liquid medium, as well as by more recent procedures involving vitrification of medium and plant material. Intracellular ice crystallization during cooling in LN can be avoided if dehydration is sufficient enough to allow vitrification of the cytoplasm. Survival of plant cells and organs also requires devitrification without recrystallization during warming.

3.1 Conventional Procedure

In conventional procedures, dehydration of cells results from extracellular freezing of the cryoprotective medium during the step of slow cooling to −40 °C. The resistance of carnation apical shoot tips to freezing in LN was first noted for stock plants maintained in a growth chamber (Seibert 1976; Seibert and Wetherbee 1977; Uemura and Sakai 1980). In these experiments, cold treatment of plants for several days was required before sampling of the explants to obtain high regrowth rates after freezing.

With in vitro plantlets as donor plants, the resistance of carnation shoot tips to deep freezing did not require previous cold treatment of donor plants. However, it appeared to be highly dependent on sucrose and DMSO concentrations in the two pretreatment media (preculture and loading). Best results were obtained with high sucrose concentrations when this sugar was employed in mixture with DMSO. However, whereas DMSO appeared to obtain high survival rates, a significant regrowth rate (35%) was achieved with sucrose alone as cryoprotectant. This result confirmed those from immature zygotic embryos of *Capsella bursa pastoris* (Monnier and Leddet 1978), and somatic embryos of *Elaeis guineensis* (Engelmann et al. 1985): sucrose alone may allow tolerance of shoot tips and embryos to liquid nitrogen. This property prefigured further developments when encapsulation was proposed for the cryopreservation of shoot tips (Dereuddre et al. 1990) and somatic embryos (Dereuddre et al. 1991a).

3.2 Vitrification Procedure

In the vitrification procedure which has been successfully applied to carnation shoot tips, dehydration was obtained by placing encapsulated shoot tips in extremely concentrated mixtures of permeating (ethylene glycol or propylene glycol) and low or nonpermeating cryoprotectants (sucrose).

Generally, in vitrification procedures, high survival rates were only achieved if cells and organs were cooled and rewarmed rapidly to avoid recrystallization. This was the case for shoot tips of mint (Towill 1990), white clover (Yamada et al. 1991), pear and apple (Niino et al. 1992b), and mulberry (Niino et al. 1992a). Similar conditions were required for embryogenic cell lines of *Asparagus* (Uragami et al. 1989), *Citrus sinensis* (Sakai et al. 1991), protoplasts (Langis and Steponkus 1990; Sakai et al. 1991), and nonembryogenic cell suspensions (Langis et al. 1989).

After coating in alginate and loading with concentrated medium containing 25% or less water, survival of carnation shoot tips became independent of cooling and warming rates. This property has been related to glass transitions which occurred in shoot tips during cooling and at rewarming. Such a procedure may increase the range of applications of cryopreservation to organs or cells which do not tolerate the presence of ice in the cryoprotective medium or do not tolerate air dehydration.

Thermal analysis of both shoot tips and encapsulating alginate gel led to the hypothesis that independence of survival and cooling or warming procedure was due to their ability to undergo glass transitions during cooling as well as during rewarming. This property appeared when the water content of the cryoprotective media dropped to 30% or less.

3.3 Encapsulation-Dehydration Procedure

In the dehydration procedure, after encapsulation of carnation shoot tips in alginate beads, the loss of water was achieved by evaporation at room temperature. This procedure was first proposed for pear shoot tips (Dereuddre et al. 1990) and successfully applied to shoot tips of *Solanum* (Fabre and Dereuddre 1990), grapevine (Plessis et al. 1991), and eucalyptus (Poissonnier et al. 1991). It has also been applied to somatic embryos (Dereuddre et al. 1991a,b), and to microspore-derived embryos (Uragami et al. 1993).

The water content of the beads (encapsulating material and trapped shoot tips) that allowed best survival after cooling in LN was about 20%. This value was close to those which permitted survival of *Asparagus* shoot tips dehydrated without encapsulation on silicagel (Uragami et al. 1989). It confirmed the range of values (18 to 25%) of residual water content of encapsulated shoot tips of other species (Fabre and Dereuddre 1990; Plessis et al. 1991; Poissonnier et al. 1991; Scottez et al. 1992) and encapsulated somatic embryos (Dereuddre et al. 1991a,b). Thus, the resistance of these explants was closely related to their ability to tolerate dehydration, like zygotic embryos excised from recalcitrant seeds of oil palm (Grout et al. 1983).

After encapsulation and air dehydration, high recovery rates of carnation shoot tips were obtained after direct immersion in LN as well as after two-step cooling, independently of the cooling rate from 0.5 to 200 °C/min. This property, which was also established for pear shoot tips (Scottez et al. 1992), may be due to the ability of encapsulating material and plant organs to vitrify during cooling, to remain in the amorphous solid state at any temperature, and to devitrify without recrystallization during warming whatever cooling and rewarming rates (Dereuddre et al. 1991b).

Tolerance of carnation shoot tips to dehydration and subsequent cooling in LN was obtained after overnight preculture with 0.5 to 1 M sucrose. With this species, previous cold acclimation of in vitro donor plants did not appear to be necessary as for *Asparagus* shoot tips (Uragami et al. 1990). For other species, however, overnight preculture with high sucrose concentrations did not appear sufficient to obtain tolerance of shoot tips to dehydration and subsequent cooling in LN, and complementary pretreatments appeared to be necessary. They consisted in cold acclimation at 0–4 °C of in vitro donor plants (Scottez et al. 1992), stepwise preculture in progressively sucrose enriched preculture medium as for grapevine (Plessis et al. 1991), eucalyptus (Poissonnier et al. 1991), or 7-day preculture in 0.75 sucrose concentrated preculture medium, as was proposed for potato shoot tips (Fabre and Dereuddre 1990).

Hence, the problem of the resistance of cells and organs to LN changed to the problem of tolerance to dehydration or to loading with highly concentrated solutions (Dereuddre 1992). Simplification of the cryopreservation procedure allowed shoot recovery to be independent of cooling and warming rates.

If cells can tolerate the osmotic stress needed for glass transition to occur, they should be protected: phase transitions, and consequently the coexistence of two phases, crystalline and liquid, and consequent phase transition will be avoided.

In conclusion, the ability of carnation (*Dianthus caryophyllus* L.) shoot tips to tolerate cooling in liquid nitrogen depends on their physiological state and on pretreatment conditions. Three cryopreservation procedures are compared. When shoot tips are cryopreserved according to the conventional procedure (overnight preculture with 0.75 M sucrose, treatment with 5% dimethylsulfoxide, two-step freezing, and rapid thawing), tolerance of axillary shoot tips to liquid nitrogen decreased progressively as a function of their distance from the apex. Two other procedures, involving encapsulation of shoot tips in alginate beads and vitrification, can be proposed to ensure good protection of carnation shoot tips against the effects of deep cooling. Vitrification of encapsulated shoot tips can be achieved after loading in vitrification water solutions of sucrose and ethylene glycol (encapsulation-vitrification) or after dehydration in air at room temperature (encapsulation-dehydration). In both cases, shoot tips were first encapsulated in alginate beads and precultured with high sucrose concentration. With these two procedures, shoot recovery became independent of cooling and warming rates; it did not depend on the range of shoot tips along the stem axis.

References

Bajaj YPS (1985) Cryopreservation of germplasm of potato (*Solanum tuberosum* L.) and cassava (*Manihot esculenta* Crantz): viability of excised meristems cryopreserved up to 4 years. Indian J Exp Biol 23: 285–287

Bajaj YPS (1991) Storage and cryopreservation of in vitro cultures. In: Bajaj YPS (ed) Biotechnology in agriculture and forestry, vol 17. High-tech and micropropagation I. Springer, Berlin Heidelberg New York, pp 361–381

Bonnier G (1934) Flore complète de France, Suisse et Belgique. Lib Gén Ens Paris, pp 42–47

Carré M, Martin-Tanguy J, Mussillon P, Martin C (1979) La culture de méristèmes et la multiplication végétative in vitro au service de la pépinière. Bull Petits Fruits 14: 7–65

Dereuddre J (1992) Cryopreservation of plant cells and organs by vitrification and dehydration. In: Dattée Y, Dumas C, Gallais A (eds) Reproductive biology and plant breeding. Springer, Berlin Heidelberg New York, pp 291–300

Dereuddre J, Kaminski M (1992) Applications of thermal analysis in cryopreservation of plant organs. J Therm Anal 38: 1965–1978

Dereuddre J, Galerne M, Gazeau C (1987) Effets du saccharose sur la résistance à la congélation dans l'azote liquide (−196 °C) des méristèmes d'oeillet (*Dianthus caryophyllus* L.) cultivés in vitro. CR Acad Sci Paris 304 Sér III: 485–488

Dereuddre J, Fabre J, Bassaglia C (1988) Resistance to freezing in liquid nitrogen of carnation (*Dianthus caryophyllus* L. var. Eolo) axillary shoot tips excised from different aged in vitro plantlets. Plant Cell Rep 7: 170–173

Dereuddre J, Scottez C, Arnaud Y, Duron M (1990) Résistance d'apex axillaires de poirier (*Pyrus communis* L.) à une déshydratation puis à une congélation dans l'azote liquide: effet d'un traitement au froid des vitroplants. CR Acad Sci Paris 310 Sér III: 317–323

Dereuddre J, Blandin S, Hassen N (1991a) Resistance of alginate-coated somatic embryos of carrot (*Daucus carota* L.) to freezing in liquid nitrogen: 1. effects of preculture. Cryo Lett 12: 125–134

Dereuddre J, Hassen N, Blandin S, Kaminski M (1991b) Resistance of alginate-coated somatic embryos of carrot (*Daucus carota* L.) to desiccation and freezing in liquid nitrogen. 2. Thermal analysis. Cryo Lett 12: 135–148

Engelmann F, Duval Y, Dereuddre J (1985) Survie et prolifération d'embryons somatiques de palmier à huile (*Elaeis guineensis* Jacq.) après congélation dans l'azote liquide. CR Acad Sci 301 Sér III: 111–116

Fabre J, Dereuddre J (1987) Effets de différentes substances (saccharose, glucose, sorbitol et mannitol) sur la résistance à la congélation dans l'azote liquide de méristèmes d'oeillet (*Dianthus caryophyllus* L., var. Eolo) cultivés in vitro. CR Acad Sci Paris 304 Sér III: 507–510

Fabre J, Dereuddre J (1990) Encapsulation-dehydration: a new approach to cryopreservation of *Solanum* shoot tips. Cryo Lett 11: 413–426

Fukai S (1989) Plant regeneration from shoot tips of *Dianthus hybrida* cryopreserved in liquid nitrogen up to 2 years. Plant Tissue Cult Lett 6: 177–178

Grout BWW, Shelton K, Pritchard HW (1983) Orthodox behaviour of oil palm seed and cryopreservation of the exised embryo for genetic conservation. Ann Bot 52: 381–384

Hackett WP, Anderson JM (1967) Aseptic multiplication and maintenance of differentiated carnation shoot tissue derived from shoot apices. Proc Am Soc Hortic Sci 90: 365–369

Langis R, Steponkus PL (1990) Cryopreservation of rye protoplasts by vitrification. Plant Physiol 92: 666–671

Langis R, Schnabel B, Earle ED, Steponkus PL (1989) Cryopreservation of *Brassica campestris* L. cell suspensions by vitrification. Cryo Lett 10: 421–428

Lu CY, Chandler SF (1995) Transformation of carnation (*Dianthus caryophyllus*). In: Bajaj YPS (ed) Biotechnology in agriculture and forestry, vol 34. Plant protoplasts and genetic engineering VI. Springer, Berlin Heidelberg New York (in press)

Matern U (1994) *Dianthus* species (Carnation): in vitro culture and the biosynthesis of dianthalexin and other secondary metabolites. In: Bajaj YPS (ed) Biotechnology in agriculture and forestry, vol 28. Medicinal and aromatic plants VII. Springer, Berlin Heidelberg New York, pp 170–184

Mii M, Buiatti M, Gimelli F (1990) Carnation. In: Ammirato PV, Evans DA, Sharp WR, Bajaj YPS (eds) Handbook of plant cell culture, vol 5. Ornamental species. McGraw-Hill, New York, pp 284–318

Monnier M, Leddet C (1978) Sur l'acquisition au froid des embryons immatures de *Capsella bursa-pastoris*. CR Acad Sci Paris 287 Sér III: 615–618

Morel G, Martin C (1985) Guérison de pommes de terre atteintes de maladies à virus. CR Acad Agric 41:472–475

Murashige T, Skoog F (1962) A revised medium for rapid growth and bioassays with tobacco tissue cultures. Physiol Plant 15: 473–497

Niino T, Sakai A, Enomoto S, Magosi J, Kato S (1992a) Cryopreservation of in vitro-grown shoot tips of mulberry by vitrification. Cryo Lett 13: 303–312

Niino T, Sakai A, Yakuwa H, Nojiri K (1992b) Cryopreservation of in vitro-grown shoot tips of apple and pear by vitrification. Plant Cell Tissue Organ Cult 28: 261–266

Plessis P, Leddet C, Dereuddre J (1991) Résistance à la déshydratation et à la congélation dans l'azote liquide d'apex enrobés de vigne (*Vitis vinifera* L. cv Chardonnay). CR Acad Sci Paris 313 Sér III: 373–380

Poissonnier M, Monod V, Pâques M, Dereuddre J (1991) Cryoconservation dans l'azote liquide d'apex d'*Eucalyptus gunnii* (Hook. F.) cultivé in vitro après enrobage et déshydratation. Ann Rech Sylv: 5–13

Quak F (1977) Meristem culture and virus-free plants. In: Reinert J, Bajaj YPS (eds) Applied and fundamental aspects of plant cell, tissue, and organ culture. Springer, Berlin Heidelberg New York, pp 598–615

Redenbaugh K, Paasch BD, Nichol JW, Kossle ME, Viss PR, Walker KA (1986) Somatic seeds: encapsulation of asexual plant embryos. Biotechnology 4: 797–801

Sakai A, Kobayashi S, Ooyama I (1991) Survival by vitrification of nucellar cells of navel orange (*Citrus sinensis* var. *brasiliensis* Tanaka) cooled to − 196 °C. J Plant Physiol 137: 465–470

Scottez C, Chevreau E, Godard N, Aranaud Y, Duron M, Dereuddre J (1992) Cryopreservation of cold-hardened shoot tips of pear (*Pyrus communis* L. cv. Beurré Hardy) in vitro cultures after encapsulation-dehydration. Cryobiology 29: 691–700

Seibert M (1976) Shoot initiation from carnation shoot apices frozen to −196 °C. Science 191: 1178–1179
Seibert M, Wetherbee PJ (1977) Increased survival and differentiation of frozen herbaceous plant organ cultures through cold treatment. Plant Physiol 59: 1043–1046
Tannoury M, Ralambosoa J, Kaminski M, Dereuddre J (1991) Cryoconservation par vitrification d'apex enrobés d'oeillet (*Dianthus caryophyllus* L.) cultivé in vitro. CR Acad Sci Paris 313 Sér III: 633–638
Tannoury M, Vintejoux C, Dereuddre J (1994) Cryoconservation d'apex d'oeillet (*Dianthus caryophyllus* L.) et d'embryons somatiques de carottte (*Daucus carota* L.) par les procédés d'enrobage-deshydratation et d'eurobage-vitrification. Thesis, Univ Paris VI, 200 pp
Towill LE (1990) Cryopreservation of isolated mint shoot tips by vitrification. Plant Cell Rep 9: 178–180
Uemura M, Sakai A (1980) Survival of carnation (*Dianthus caryophyllus* L.) shoot apices frozen to the temperature of liquid nitrogen. Plant Cell Physiol 21: 85–94
Uragami A, Sakai A, Nagai M, Takahashi T (1989) Survival of cells and somatic embryos of *Asparagus officinalis* cryopreserved by vitrification. Plant Cell Rep 8: 418–421
Uragami A, Sakai A, Nagai M (1990) Cryopreservation of dried axillary buds from plantlets of *Asparagus officinalis* L. grown in vitro. Plant Cell Rep 9: 328–331
Uragami A, Lucas MO, Ralambosoa J, Renard M, Dereuddre J (1993) Cryopreservation of microspore embryos of oilseed rape (*Brassica napus* L.) by dehydration in air with or without alginate encapsulation. Cryo Lett 14: 83–90
Van Hoof P (1980) Conservation et multiplication des oeillets in vitro. Rev Agric 30: 79–84
Withers LA (1991) In vitro conservation. Biol J Linn Soc 43: 31–42
Yamada T, Sakai A, Matsumura T, Higushi S (1991) Cryopreservation of apical meristems of white clover (*Trifolium repens* L.) by vitrification. Plant Sci 78: 81–87

VI.5 Cryopreservation Studies on *Digitalis lanata* (Foxglove)

U. SEITZ[1]

1 Introduction

Plants of the genus *Digitalis* (foxglove, Scrophulariaceae) are biennial or perennial herbs, and belong to the Eurasian flora. *D. lanata* is found especially in southeast Europe, for example in Hungary and in the Balkan countries. It has become naturalized in North America. *D. purpurea* is not only a medicinal plant; as an ornamental it is widespread in western and central Europe. Like *D. lanata*, it has become naturalized in North America. Other important species are *D. grandiflora* and *D. lutea*.

The foliage of foxglove consists of a rosette of leaves with inflorescences of about 1 m height. Secondary products from species of the genus *Digitalis* are used in the treatment of heart diseases. The major therapeutic category of these substances is defined as the cardiotonic glycosides (cardenolides).

Attempts to establish in vitro cultures of *Digitalis* species date back to the 1950s. Gautheret already reported on these first results in his handbook published in 1959. Since then, investigations have been carried out in numerous laboratories, and many types of in vitro techniques have been applied to *Digitalis*, including meristem culture and clonal multiplication of elite plants, protoplast culture and regeneration, induction of morphogenetic and embryogenetic processes, and, last but not the least, large-scale culture techniques. It is not within the scope of this chapter to compile details of this topic; it has been reviewed comprehensively by Rücker (1988), and Luckner and Diettrich (1992).

Cultivated cells of *Digitalis lanata* do not synthesize cardenolides de novo, but they are able to transform added cardenolides (biotransformation). Accordingly, it becomes possible to transform a substance of lower into one of greater scientific or commercial value. A typical reaction is the 12β-hydroxylation of β-methyldigitoxin, which results in the formation of β-methyldigoxin (Alfermann et al. 1983). High-yield cell lines used in such systems are often the product of a long phase of optimization and selection processes. Difficulties can arise when it becomes necessary to maintain such cultures for future use. Continuous subcultivation is expensive and time-consuming, and it carries the risk of accidental loss and of somaclonal variation.

The fact that cardenolide formation depends on the degree of differentiation stimulated work on embryogenic *Digitalis* cultures. Kuberski et al. (1984)

[1]University of Tübingen, Botanical Institute, 72076 Tübingen, Germany

Biotechnology in Agriculture and Forestry, Vol. 32
Cryopreservation of Plant Germplasm I (ed. by Y.P.S. Bajaj)

observed that by increasing the cytokinin/auxin ratio a culture strain established from filament callus could be induced to form large numbers of adventive embryos from which plantlets regenerated in hormone-free media. Upon irradiation, the globular stage embryos turned green and formed cardenolides in amounts comparable to those produced by plantlets or shoots. From a biotechnological point of view, such cultures are desirable because they are easier to handle than the more complex structures, and also they can be used to study the in vitro formation of cardenolides (Scheibner et al. 1987). The ability of cultures to form somatic embryos is restricted to a limited number of cell strains and, furthermore, this capacity can be lost after serial subculturing.

Thus, the necessity has arisen for efficient storage methods. For short- and medium-term storage, the cells are normally maintained in the form of callus cultures on solid or semisolid media. A reduction in temperature may be beneficial in prolonging subculture intervals. For long-term storage, all biochemical processes must be completely suppressed in the cells. The only method that meets this demand is the cryopreservation technique. The cryopreservation of medicinal plants has been reviewed by Bajaj (1988).

2 Cryopreservation

Several authors have reported on the cryopreservation of *Digitalis* cells, especially of *D. lanata* and *D. purpurea*, and best results were obtained with suspension cultures. Studies on the cryopreservation of *Digitalis* started relatively early (Diettrich et al. 1982; Seitz et al. 1983). In the meantime, a broad knowledge has become available, not only with respect to the establishment of freezing protocols, but also about the response of the cells to various pretreatments, the characteristics of frozen-thawed cells, and the events during freezing and thawing on a cellular level. The work by various authors is summarized in Table 1.

2.1 Cell Cultures

It has been documented that cells from the early growth phase, which have a relatively high cytoplasm-to-vacuole ratio, especially those that have recently passed through mitosis, are the most tolerant to subsequent freezing. Sepasgosarian (1987) established cell cultures from *D. purpurea* that were characterized by small cell sizes and vacuoles that were hardly visible under the light microscope. This may be the reason why, without exception, *D. purpurea* cells showed very high post-thaw viability rates and good growth after being thawed (Seitz, unpubl.). Exactly the same correlation was observed with *Coleus blumei* cells (Reuff et al. 1988).

Like the vast majority of cell cultures, *Digitalis* requires preculture in a modified medium where the cells become adapted to high osmotic pressures. Medium additives may be sugar alcohols (mannitol, Diettrich et al. 1982; Seitz

Table 1. Summary of cryopreservation work on *Digitalis lanata*. PA: preculture additives, C: cryoprotectants

No.	Reference	Culture type, specimens	Freezing method	Results; comments
1	Diettrich et al. (1982)	Cell cultures	PA: mannitol; C: sucrose/glycerol and other mixtures; slow cooling	Optimization of procedure; about 50% survival; no selection during freezing and thawing
2	Seitz et al. (1983)	Cell cultures (shake cultures and bioreactors)	PA: mannitol; C: DMSO/glycerol/sucrose; slow cooling	Preservation of biotransformation capacity
3	Diettrich et al. (1985)	Embryogenic cell cultures	PA: mannitol; C: sucrose/glycerol; slow cooling	Morphogenetic potential unchanged
4	Seitz et al. (1985)	Cell cultures	As no. 2	50–75% viability; freezing resistance and cardenolide content
5	Diettrich et al. (1986)	Cell cultures	PA: mannitol; C: glycerol/sucrose; DMSO and proline also tested	Behavior of cells during pretreatments, cooling and recultivation: osmotic value, plasmolysis viability
6	Diettrich et al. (1987a)	Cell cultures and protoplasts	PA and C as no. 3; low and high cooling rate	Cryomicroscopy; intracellular ice formation; high cooling rate causes low viability
7	Diettrich et al. (1987b)	Shoot tips (cold-hardened)	C: DMSO or glycerol/sucrose; ultrarapid cooling; slow cooling	Survival about 70%; shoot regeneration from newly formed callus
8	Diettrich et al. (1991)	Cell cultures and protoplasts	PA: mannitol; C: glycerol/sucrose; slow and rapid cooling	Ultrastructure; slow cooling: increase of the number of vacuoles and vesicles; rapid cooling: damage of cytological structures
9	Göldner et al. (1991)	Cell cultures	Various preculture additives C: DMSO/glycerol/sucrose; slow cooling	Effect of preculture additives on freeze tolerance: trehalose > mannitol > sucrose, melibiose > proline, sorbitol; viability during preculture; stress response

et al. 1983) or sugars (sucrose, trehalose, Göldner et al. 1991). A reduction in cell viability was observed after the cells were transferred into preculture media. In the most favorable cases this reduction was found to be transient (Göldner et al. 1991). Mannitol seems to be taken up only to a negligible extent (Diettrich et al. 1986), but an accumulation of sugars, especially of sucrose, has been observed when the cells are cultivated in the presence of mannitol or trehalose (Seitz, unpubl.). By means of a plasmolysis test, Diettrich et al. (1986) noticed an enhancement of the osmotic values of the cells during preculture in the presence of mannitol (0.15 M). The cellular changes, as far as they have been investigated, resemble those found in cells undergoing cold or other stress-hardening.

2.1.1 Cryoprotection

In the case of suspension cultures, a mixture containing DMSO, glycerol, and sucrose seems to be widely applicable and has also been used with *Digitalis* cultures. Sucrose, the third component, may be replaced by proline, trehalose, or sugar alcohols. There are, however, indications that it may be sufficient to use sucrose as the only additive (0.3 M during preculture, 1 M as a cryoprotective agent). This was shown for several species: *D. purpurea, Panax ginseng, Tabernaemontana coronaria, Plumeria rubra*, and *Sterculia* sp. (Seitz, unpubl.). Bhandal et al. (1985) were able to demonstrate successful cryopreservation using trehalose as the sole cryoprotectant. Cryoprotection treatment normally lasts for 1 h at 0–4 °C.

2.1.2 Freezing, Storage, and Thawing

For freezing, cell suspensions are transferred into suitable containers. Presterilized, screw-cap ampules with a capacity of 2 ml are the most commonly used. A slow freezing rate of 1 to 2 °C per min is imperative for most cell cultures because it allows protective dehydration, which is a prerequisite for avoiding the formation of ice crystals within the cells, thus reducing freeze damage. Cooling is usually programmed to the range of − 30 to − 40 °C. Finally, the ampules are plunged into liquid nitrogen, where they are then stored.

After the chosen storage period, cell suspensions are thawed rapidly by agitating the ampules in warm water. Immediately after thawing, cell suspensions are spread on semisolid culture medium. Diettrich et al. (1982) noted postthaw viability rates of about 50%, Seitz et al. (1985) of 50–75% with *D. lanata* treated with the protocol described above. To improve the growth response of cryopreserved cells, a filter paper layer is recommended, with which the cells can easily be transferred to fresh medium after several hours. This makes the removal of cryoprotectants a gradual process, and problems associated with deplasmolysis can be reduced.

2.2 Conservation of Biosynthetic Potentials

Before a specific storage method is used to preserve important cell lines, it is vital to have clear proof that the culture type does not change due to the treatment itself and that no selection will occur. In the history of cryopreservation, cell viability was the first criterion for evaluating the experiments. Later, regrowth was examined and, finally, the characteristics of frozen-thawed cells. Bajaj (1976) and Dougall and Whitten (1980) reported anthocyanin accumulation in carrot cell cultures. Seitz et al. (1983) showed that the capacity of *Digitalis lanata* cells to transform β-methyldigitoxin to β-methyldigoxin remained unchanged after cryogenic storage. This result was obtained in 300-ml shake cultures as well as in 20-l bioreactors after a scale-up process. A typical experiment is shown in Fig. 1. The final yields of β-methyldigoxin were not reduced after storage periods of 2 months, 6 months, 1 year, and 2 years, but, similar to the findings of Dougall and Whitten (1980), an increase in the standard deviation was observed (Seitz, unpubl.). The freeze tolerance was not influenced significantly by the presence of cardenolides (Seitz et al. 1985). Another biotransformation reaction, namely the glucosylation of digitoxin added to *D. lanata* cells, was also preserved during cryogenic storage. This finding could be confirmed after a 3-year storage period (Diettrich et al. 1985).

2.3 The Freezing Process

Diettrich et al. (1987a) investigated *Digitalis lanata* cells and protoplasts under the cryomicroscope during the processes of freezing and rewarming, and simulta-

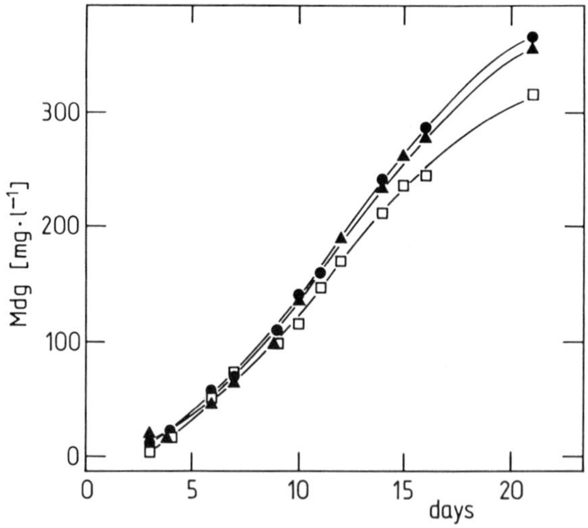

Fig 1. Biotransformation of β-methyldigitoxin to β-methyldigoxin (*Mdg*) in a 300-ml shake culture of *Digitalis lanata* (strain 287). Glucose was fed from day 8 to maintain a glucose level of between 2 and 5 g/l. An aliquot of culture medium was diluted with methanol and analyzed by HPLC (Seitz et al. 1983). ●, ▲ Frozen-thawed cells; □ control

neously monitored cell viability under the respective conditions. Cells frozen at rates $< 3\,^{\circ}$C/min in the presence of an adequate mixture of cryoprotectants remained viable. Focusing on protoplasts, the authors pointed out that cell volume decreased continuously during cooling. The cell volume can be used to estimate the degree of dehydration, which is a consequence of the osmotical equilibrium with the surrounding medium during freezing. The maximum dehydration is 80%; at that point, the cell volume remains stable, thus indicating that free water is no longer available inside the cells. Maximum dehydration takes about 60 min. This can be achieved either by cooling to $-40\,^{\circ}$C or by cooling to $-20\,^{\circ}$C and maintaining that temperature for a certain period of time before the samples are transferred into liquid nitrogen.

For a closer look at the above-mentioned process of cell shrinkage, Diettrich et al. (1991) investigated the ultrastructure of cells and isolated protoplasts. Their spherical shape makes protoplasts ideal subjects for morphometric studies despite their vulnerability. Under the appropriate conditions the ultrastructure appeared undamaged. The area of the plasma membrane was considerably reduced when cells (or protoplasts) were cooled slowly in the presence of osmotically active cryoprotectants. Concomitantly, the authors observed an increase in the number of vacuoles and membrane-coated vesicles. A kind of endocytosis seemed to be involved in vesicle formation. However, if protoplasts were treated in the presence of gold granules, only a small, yet significant percentage of vesicles was found to contain gold particles. The formation of vesicles is thought to be an important mechanism by which the area reduction of the plasma membrane and the tonoplast can be achieved.

Intracellular ice is generally regarded as being detrimental for living cells. Its formation can be documented under the cryomicroscope using the so-called flash phenomenon, a sudden reduction of the transparency of the cytoplasm that accompanies the growth of ice crystals (Steponkus et al. 1983). Diettrich et al. (1987a) reported the flash phenomenon to occur between -20 and $-25\,^{\circ}$C if the samples were subjected to higher cooling rates ($> 3\,^{\circ}$C/min). Viability was severely reduced in such cells and they were not able to resume growth.

2.4 Meristem and Shoot Tip Culture

Digitalis lanata plants that have been found to produce cardenolides in large amounts are used for breeding and propagation programs. The stability of the selected lines, however, is not sufficiently high. Furthermore, plants cultivated in the field will die after flowering in the second vegetation period. The possibility of using in vitro techniques for the multiplication and conservation of elite plants is therefore of great value.

The rapid clonal multiplication of *D. lanata* axillary buds has been reported by Erdei et al. (1981). Since then, several papers have been published on this subject. The literature is accumulated in the recent review by Luckner and Diettrich (1992) which comprises information on all steps of the procedure, including the adaptation to greenhouse and field conditions. In field tests the clonal plants showed normal development and the seeds were fertile. The cardenolide contents of the clonal plants showed a certain variation, but were comparable with those of the mother plants. On the basis of these results it would be safe to say that the propagation of *D. lanata* by shoot tip culture is possible and that the rapid multiplication of elite plants is an established procedure.

Depending on material and objectives, one can envisage various storage methods. Once in vitro plantlets have developed from isolated meristems, the axenic shoots can be transferred to media with enhanced osmolarity and then kept at low temperatures (+4 °C) and dim light. Under these conditions, a substantial growth reduction can be achieved, making this a suitable medium-term storage method.

For the long-term storage of organized structures, the cryopreservation of shoot tips is the method of choice. Diettrich et al. (1987b) have presented a method for preserving shoot tips of *Digitalis lanata* in liquid nitrogen. It includes the following steps: (1) hardening of 4-week-old plantlets at 4 °C for 8 weeks, (2) preparation of meristems from axillary buds, size of the explants about 0.5 mm, (3) treatment with 2 M DMSO for 2 h, (4) slow cooling (rate 0.5 °C/min), initiation of crystallization at −5 °C, cooling to −40 °C, transfer of ampules into liquid nitrogen, (5) storage at −196 °C, (6) ultrarapid warming in a warm water bath, (7) recovery growth on solidified medium, and (8) determination of survival after 14 days. It has been found that the cold-hardening step considerably increased the survival of *Digitalis* meristems, as well as the shoot regeneration capacity. Besides the method of controlled slow cooling, Diettrich et al. (1987b) performed experiments using the rapid cooling technique. Treatment of meristems with 0.7 M DMSO for 18 h followed by direct quenching in liquid nitrogen resulted in viability rates that were similar to those obtained with the slow cooling method. Regeneration of shoots was also comparable. The advantage of this technique lies in its simplicity and in the fact that expensive freezing equipment is not necessary. Its limits result from high contamination rates observed especially after longer storage periods.

3 Summary and Conclusions

The cryopreservation work performed with *D. lanata* comprises all types of culture systems, although most of the results have been obtained with cell cultures. It has been shown by several authors that pretreatment with mannitol improves the freeze tolerance markedly. Disaccharides, especially sucrose and trehalose, are also highly effective. Cryoprotectants can be the mixtures of DMSO/ glycerol/sucrose, glycerol/sucrose, or sucrose as the sole additive. Under optimum conditions, viability rates of more than 50% are normal and reproducible. Cold-hardening works well with shoot-tips of *D. lanata*; this material can then be frozen successfully by either slow or ultrarapid cooling.

Cellular and biochemical events during the cryopreservation process, such as the increase in the osmotic value, stress responses, and changes in the ultrastructure, have been investigated. The area of the plasma membrane is reduced considerably when cells or protoplasts are cooled slowly in the presence of cryoprotectants. Intracellular ice formation does not occur and the cells are able to retain their viability. *Digitalis lanata* is one of the species where the biochemical capacities of frozen-thawed cells were studied very early. The rate at which added cardenolides are biotransformed, i.e., glucosylated or hydroxylated, is fully maintained. Cryopreserved shoot tips retain their morphogenetic potential. Thus, for all practical purposes, the recovered material is indistinguishable from the one not subjected to cryopreservation.

References

Alfermann AW, Bergmann W, Figur C, Helmbold U, Schwantag D, Schuller I, Reinhard E (1983) Biotransformation of β-methyldigitoxin to β-methyldigoxin by cell cultures of *Digitalis lanata*. In: Mantell SH, Smith H (eds) Plant biotechnology. Cambridge Univ Press, Cambridge, pp 67–74

Bajaj YPS (1976) Regeneration of plants from cell suspensions frozen at −20, −70, and −196 °C. Physiol plant 37: 263–268

Bajaj YPS (1988) Cryopreservation and the retention of biosynthetic potential in cell cultures of medicinal and alkaloid-producing plants. In: Bajaj YPS (ed) Biotechnology in agriculture and forestry, vol 4. Medicinal and aromatic plants I. Springer, Berlin Heidelberg New York, pp 169–187

Bhandal IS, Hauptmann RM, Widholm JM (1985) Trehalose as cryoprotectant for the freeze preservation of carrot and tobacco cells. Plant Physiol 78: 430–432

Diettrich B, Popov AS, Pfeiffer B, Neumann D, Butenko R, Luckner M (1982) Cryopreservation of *Digitalis lanata* cell cultures. Planta Med 46: 82–87

Diettrich B, Haack U, Popov AS, Butenko RG, Luckner M (1985) Long-term storage in liquid nitrogen of an embryogenic cell strain of *Digitalis lanata*. Biochem Physiol Pflanz 180: 33–43

Diettrich B, Haack U, Luckner M (1986) Cryopreservation of *Digitalis lanata* cells grown in vitro. Precultivation and recultivation. J Plant Physiol 126: 63–73

Diettrich B, Haack U, Thom F, Matthes G, Luckner M (1987a) Influence of intracellular ice formation on the survival of *Digitalis lanata* cells grown in vitro. Cryo Lett 8: 98–107

Diettrich B, Wolf T, Bormann A, Popov AS, Butenko RG, Luckner M (1987b) Cryopreservation of *Digitalis lanata* shoot tips. Planta Med 53: 359–363

Diettrich B, Neumann D, Luckner M (1991) Cryopreservation of *Digitalis lanata* cells. Membrane area reduction of the protoplast during cryoprotector treatment and cooling. J Plant Physiol 139: 212–217

Dougall DK, Whitten GH (1980) The ability of wild carrot cell cultures to retain their capacity for anthocyanin synthesis after storage at −140 °C. Planta Med (Suppl): 129–135

Erdei I, Kiss Z, Maliga P (1981) Rapid clonal multiplication of *Digitalis lanata* in tissue culture. Plant Cell Rep 1: 34–35

Gautheret RJ (1959) La culture des tissus végétaux. Techniques et réalisations. Masson, Paris

Göldner EM, Seitz U, Reinhard E (1991) Cryopreservation of *Digitalis lanata* Ehrh. cell cultures: preculture and freeze tolerance. Plant Cell Tissue Organ Cult 24: 19–24

Kuberski C, Scheibner H, Steup C, Diettrich B, Luckner M (1984) Embryogenesis and cardenolide formation in tissue cultures of *Digitalis lanata*. Phytochemistry 23: 1407–1412

Luckner M, Diettrich B (1992) Micropropagation of woolly foxglove (*Digitalis lanata*) by shoot tip culture. In: Bajaj YPS (ed) Biotechnology in agriculture and forestry, vol 19. High-tech and micropropagation III. Springer, Berlin Heidelberg New York, pp 400–424

Reuff I, Seitz U, Ulbrich B, Reinhard E (1988) Cryopreservation of *Coleus blumei* suspension and callus cultures. J Plant Physiol 133: 414–418

Rücker W (1988) *Digitalis* spp.: in vitro culture, regeneration, and the production of cardenolides and other secondary products. In: Bajaj YPS (ed) Biotechnology in agriculture and forestry, vol 4. Medicinal and aromatic plants I. Springer, Berlin Heidelberg New York, pp 388–418

Scheibner H, Björk L, Schulz U, Diettrich B, Luckner M (1987) Influence of light on cardenolide accumulation in somatic embryos of *Digitalis lanata*. J Plant Physiol 130: 211–219

Seitz U, Alfermann AW, Reinhard E (1983) Stability of biotransformation capacity in *Digitalis lanata* cell cultures after cryogenic storage. Plant Cell Rep 2: 273–276

Seitz U, Reuff I, Reinhard E (1985) Cryopreservation of plant cell cultures. In: Neumann KH, Barz W, Reinhard E (eds) Primary and secondary metabolism of plant cell cultures. Springer, Berlin Heidelberg New York, pp 323–333

Sepasgosarian J (1987) 16β-Hydroxylierung durch Zellkulturen von *Digitalis purpurea* L. Thesis, Univ Tübingen, pp 6–8

Steponkus PL, Dowgert MF, Gordon-Kamm WJ (1983) Destabilization of the plasma membrane of isolated plant protoplasts during a freeze-thaw cycle: the influence of cold acclimation. Cryobiology 20: 448–465

VI.6 Cryopreservation of Germplasm of *Dioscorea deltoidea* (Medicinal Yam)

A.S. Popov, L.A. Volkova, and R.G. Butenko[1]

1 Introduction

1.1 *Dioscorea* Species of Practical Importance

The genus *Dioscorea* (family Dioscoreaceae) comprises about 700 species. All species are climbers, their underground organs – rhizomes and tubers – are perennial. Eleven species are used as a food source. They contain a good deal of starch, vitamin C, and other nutrient substances. These species are common in the humid tropics, particularly in West Africa near the Gulf of Guinea, providing the traditional food for the local population (Ng 1992). The most widespread species is *D. alata.*

Nine species are medicinal plants, which accumulate steroid saponins in their rhizomes. The genin of these saponins – diosgenin – can serve as a raw material for steroid hormone preparation and for oral contraceptives (Purseglove 1972). The highest productivity is shown by the Mexican species *D. floribunda* Mart. and Gal., *D. composita* Hemsl. and *D. mexicana* Guillem (Onwueme 1978), with a diosgenin content as high as 10% of the dry weight of the rhizomes, together with the strains *D. elephantipes* (L. Her.) Engl. (South Africa) and *D. deltoidea* Wall.

D. deltoidea has its origin in Kashmir, northern India, where it grows in the Himalayan foothills at 1400–3000 m altitude. In diosgenin content, this species is second to those from Mexico. It accumulates 3–5% diosgenin under moderate temperatures in Kashmir (Chandra et al. 1955) and 7–8% diosgenin in tropical Bangalore (Purseglove 1972), the economically feasible concentration being considered to be 3%. The other northern species are *D. balcanica* Košan. from southwest Yugoslavia and Northern Albany, and *D. caucasica* Lipsky from South Russia, where it occurs in the southern foothills of the Caucasus (Kichenko and Panina 1965; Grubishich et al. 1991). Both species are included in the same section Stenophora as *D. deltoidea*, and are characterized by 2n = 20 chromosomes, both being endemic relicts of the Eocene *Dioscorea* (Burkill 1960).

One more species of *Dioscorea, D. nipponica* Makino (Shreter et al. 1965), is widespread in the far east of Russia, but its limited abundance and low diosgenin content make it unsuitable for pharmaceutics. Attempts to introduce some

[1]K.A. Timiryazev Institute of Plant Physiology, Russian Academy of Sciences, Botanical Str. 35, Moscow 127276, Russia

Biotechnology in Agriculture and Forestry, Vol. 32
Cryopreservation of Plant Germplasm I (ed. by Y.P.S. Bajaj)

Dioscorea species (*D. balcanica, D. caucasica, D. nipponica, D. villosa, D. deltoidea*) showed that they were sufficiently tolerant to Moscow winter conditions. Nevertheless, their diosgenin content was too low and they were not economically feasible (Kichenko 1961; Kichenko and Panina 1965). Simultaneously, some difficulties emerged with raw material import as intensive exploitation resulted in depletion of the natural resources of the medicinal species of this genus in India, South Africa, and Mexico.

1.2 Methods for the Storage of *Dioscorea* Germplasm

The main and conventional method of propagation of all *Dioscorea* species is vegetative multiplication by tubers or their segments, and rhizome cuttings. The tubers and the rhizomes survive in the soil until the next rainy season . In regions where dioscoreas (yams) are common agricultural crops, their tubers are stored for several months at normal temperatures if well ventilated (Onwueme 1978; Purseglove 1988).

Many investigators have studied *Dioscorea* tissue and cell cultures in vitro. Much attention was paid to *D. deltoidea* (Kaul and Staba 1967). Adventitious shoots and callus culture of this plant were obtained in the Timiryazev Institute of Plant Physiology in Moscow in 1969 (Abroshnikova (Sarkisova) et al. 1971), from which suspension cultures and mutant cell strains were derived. Various studies of cell physiology, secondary metabolism, cultivation on an industrial scale, etc. were performed (Butenko et al. 1992). Also Furmanowa and coworkers obtained adventitious shoots in vitro for micropropagation (Furmanowa et al. 1984; Furmanowa and Guzewska 1989). In the course of 1 year, one aseptic *D. floribunda* plantlet may give 2.5×10^6 plantlets by rooting each one-node leaf microcutting. Recently, microtuberization of *D. alata* (two cultivars) and *D. abyssinica* was performed in vitro, and organ regeneration from vine-nodal explants of *D. bulbifera* was studied (Martine and Cappadocia 1989; Okonkwo et al. 1990). In future, the microtubers and the microbulbils obtained from healthy mericlones of the species in question may be very useful for storage and for the industrialization of yam planting (Malaurie et al. 1993).

2 Cryopreservation

Cryopreservation of in vitro culture of medicinal plants has attracted much attention, because of its potential for the long-term conservation of high-yielding cell clones (see Bajaj 1988). Our work on the freeze preservation of cell cultures of *Dioscorea deltoidea* is discussed here.

2.1 Materials and Methods (Protocol)

2.1.1 Culture of Various Cell Strains

Dioscorea deltoidea strain D-1 (wild original strain) and derived mutants DM-0.5, DM-1, and DM-8 were grown as suspension cultures at 25 °C in darkness at

70% relative humidity. Cultivation was performed in flasks on a hanging shaker (the rotation rate 95–98 rpm, rotation radius 20 mm) on MS medium (Murashige and Skoog 1962) supplemented with vitamins (Kaul and Staba 1967), casein hydrolysate 500 mg/l, sucrose 30 g/l, 2,4-D 1 mg/l, and kinetin 0.1 mg/l. The hormone-independent strain M-8 was grown on a medium without casein and phytohormones. Inoculation density was 3×10^5 cells/ml.

2.1.2 Growth, Viability, and Productivity of Various Strains

Phenosafranine staining was used for monitoring the cell viability, at least 200 cell aggregates were examined in each sample (Widholm 1972; Popov 1985). The level of aggregation was estimated by viewing at least 1000 clusters and determining the number of cells in each. The growth index was estimated as P/P_0 ratio, where P was the number of cells by the 14th day of cultivation, and P_0 the number of cells at subculturing (at day 0). The number of chromosomes was estimated after fixation with ethanol-acetic acid (3:1) and staining with acetocarmine. At least 100 metaphase plates were analyzed on squashed preparations. The content of intracellular sugars was determined by phenolic method (Dubois et al. 1956).

Steroids were extracted with chloroform-methanol (2:1) from air-dry mass of the 14-day-old cultures, hydrolyzed for 15 h with 1 N HCl in 50% ethanol at 85 °C, and then extracted with diethyl ether (Volkova et al. 1986). The intracellular ice formation was monitored under the light microscope with cryoinstallation. Addition of some ice crystals to the preparation at −2 and −3 °C was used to initiate ice crystallization.

2.1.3 Freezing, Cryopreservation, and Recovery of Cell Cultures

Cells from cultures in the exponential growth phase were used for freezing. Originally unstable on freezing, the strain D-1 needed special precultivation. To increase stability, the cells were grown for 5–6 days with addition of one of the following amino acids: asparagine, alanine, proline, or serine, with a final concentration of 0.02, 0.05, 0.02, and 0.01 M, respectively.

In some cases, pretreatment was carried out by cell suspension incubated at lowering temperature (9–10 °C) for 18–20 h. During incubation, the suspension was not disturbed, and layer depth was about 1 cm.

After 4–6 days' cultivation under normal conditions (mutant strains) or following precultivation or incubation at lowering temperature as described above (strain D-1), the cell suspensions were settled at about 0 °C and preliminary cooled double strength solution of cryoprotectant in nutrient medium was added to cell pellet in the same volume. The addition was made in four portions within 20 min. Dimethylsulfoxide (DMSO) with a final concentration of 7% and freezing point −2.82 °C was used (Butenko et al. 1983; Popov 1985). The suspension was then poured into plastic ampules, diameter 10–12 mm, volume 1.2–1.8 ml. The ampules were placed in an ice bath and then in a freezer, which was cooled to about 0 °C. The following freezing program was used: the ampules were cooled to -4 to $-5\,^{\circ}C$ and after short temperature stabilization, ice crystals were initiated either by touching the ampule bottoms onto liquid nitrogen for 0.5–1.5 s, or by submerging all the cassettes with ampules into LN for 0.4–0.8 s, or by using the special device of Popov et al. (1988). Temperature was maintained at the same level for a further 18–20 min, followed by a temperature decrease to $-30\,^{\circ}C$ at a rate of 0.3–0.5 °C/min and then down to $-70\,^{\circ}C$ at a rate of 9–10 °C/min. Finally, the ampules were immersed rapidly in LN.

Ampules were thawed in a water bath at 40 °C with shaking until visible ice crystals disappeared. The cells were rinsed to remove DMSO, and their viability was estimated as described above.

The rinsed cells were concentrated by sedimentation and plated on agar medium surface in Petri dishes sealed by parafilm and placed in darkness at 25 °C. The callus was formed after 4–6 weeks, then transferred into a small amount of liquid medium to renew the suspension culture.

2.2 Results

Cell viability after cryopreservation with 7% DMSO, was six times higher for mutant strains than for the original strain D-1 (Table 1). Such high viability enabled us to regenerate the mutant cell strains of *D. deltoidea* even without special pretreatment. The level of cell viability of the original strain D-1 cells was insufficient to renew the culture (Volkova et al. 1982). To prepare these cells for deep freezing, they were cultivated for 5–6 days with addition of one of the following amino acids: asparagine, alanine, proline, or serine at a concentration of 0.02, 0.05, 0.02, and 0.01 M, respectively. By 7–10 days of cultivation cell viability dropped to 50% or lower. Hence, 5–6-day-old cultures were used for freezing.

Cells grown in the presence of amino acids differed from the controls in some morphological, physiological, and cytogenetical properties. By the 5th day of subculture with addition of asparagine or alanine, starch grains could either be

Table 1. The viability of *Dioscorea deltoidea* cells following deep freezing and preservation at $-196\,^{\circ}C$

Treatment	Cell strain			
	D-1	DM-0.5	DM-1	DM-8
	Cells uncolored with phenosafranine, % of total cells			
Control	70.6 ± 6.5 n = 16	75.3 ± 5.7 n = 4	77.0 ± 8.1 n = 7	74.5 ± 7.8 n = 6
Cryopreservation	5.2 ± 0.5 n = 49	29.4 ± 2.3 n = 14	29.3 ± 2.7 n = 7	33.6 ± 7.9 n = 31

Note: Both treatments were carried out in the presence of DMSO 7%. n, Total number of viability determinations.

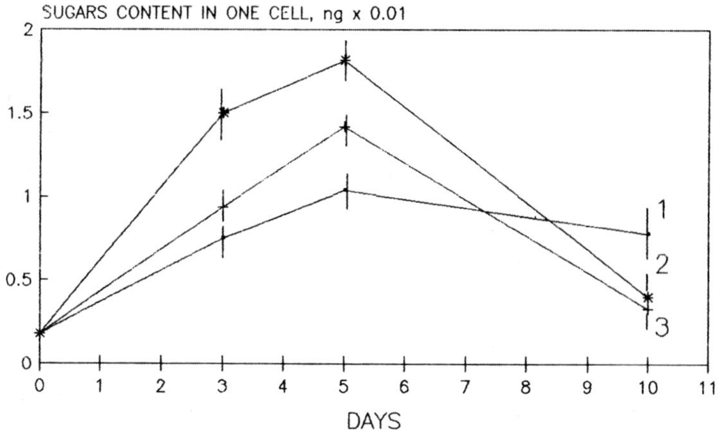

Fig. 1. Content of soluble sugars in D-1 cells grown with addition of various amino acids to the nutrient media. *1* Control culture (without amino acids); *3* + proline (0.02 M); *2* + asparagine (0.02 M)

seen only as the dispersion of small particles instead of the large ones in the control, or they disappeared altogether. The intracellular sugar content increased simultaneously by 80% in the cells grown on a medium with asparagine, by 60% with alanine, and by 40% with proline in comparison to the control level (Fig. 1). The drop in sugar content by the 10th day may be explained by depletion of starch storage under treatment with amino acids at the above concentrations.

In mutant cells, the soluble sugar content was 1.5–2 times higher than in the original strain. This coincides with our data on the double value of cell sap osmotic pressure in mutants (Volkova et al. 1986).

Suspension cultures of *Dioscorea* cells are aneuploid and reveal morphological and genetic heterogeneity. The number of chromosomes for *D. deltoidea* varied from 9 to 80, with an average of 20. Subculturing with amino acids led to an increase in diploid cell contribution, whereas in the control culture the number of tetraploid cells exceeded that of diploid cells (Fig. 2). The number of triploid cells did not change. The similar pattern of cell ploidy is characteristic of mutant strains DM-0.5, DM-1, DM-8 (Fig. 3), their cryoresistance being higher than that of the original strain. Various strains of *Dioscorea* differed also by their degree of aggregation (Fig. 4). More distinct aggregation was observed in the strain DM-8, its population comprising over 60% of large clusters with more than 20 cells. Strains D-1 and DM-1 showed the lowest aggregation. These peculiarities of strain aggregation were retained for all phases of culture cycle.

The growth index for separate strains calculated on the basis of cell number varied only insignificantly (Table 2).

Viewing with the cryomicroscope enabled observation of the most dangerous changes – the appearance of intracellular ice crystals, which depends directly on the rate of temperature decrease (Table 3). Under cryoprotection with DMSO, the rate of cooling may be 2.5 times higher. Preculturing with amino acids further augments cryoprotection.

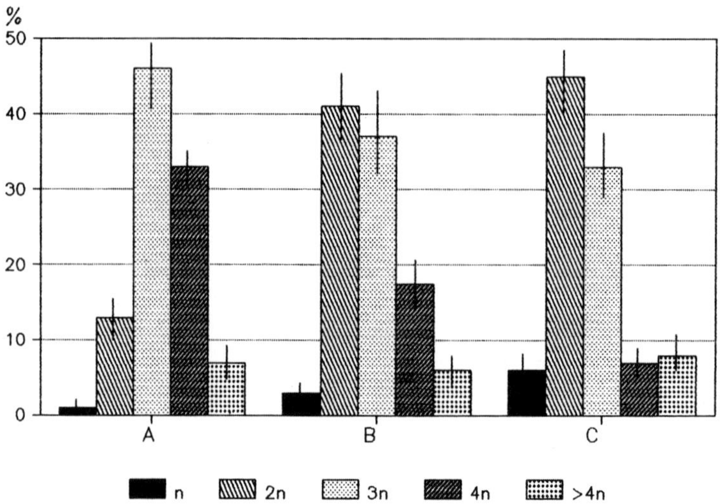

Fig. 2. Pattern of ploidy distribution in *Dioscorea deltoidea* cell population. *A* Control cells; *B* after subculturing with addition of asparagine (0.02 M); *C* the second subculture after thawing (without asparagine); *n* ploidy

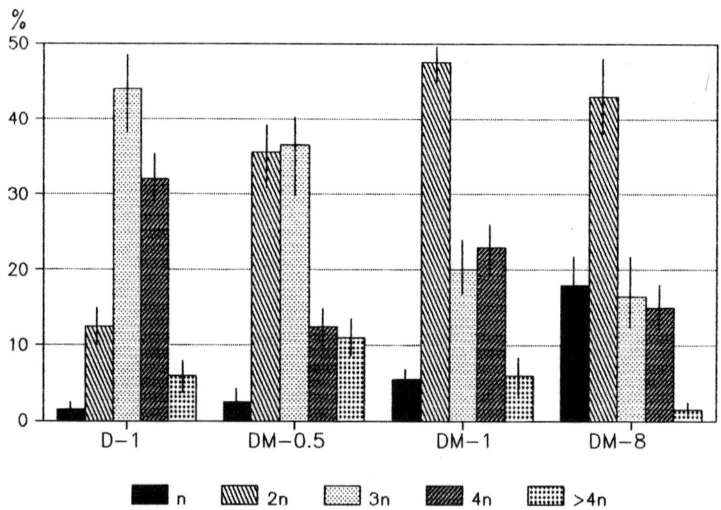

Fig. 3. Pattern of ploidy distribution in various cell strain population of *Dioscorea deltoidea* (2n = 20). *D. deltoidea* strains: D-1, DM-0.5, DM-1, DM-8; *n* ploidy

Another technique to prepare for freezing the cells of strain D-1 is their exposure at 9–10 °C for 18–20 h (Table 4). This resulted in a 2.5-fold increase in viability for the strain D-1, and 1.5-fold for the mutant DM-0.5. Combined treatment of these factors, amino acid addition followed by lowering

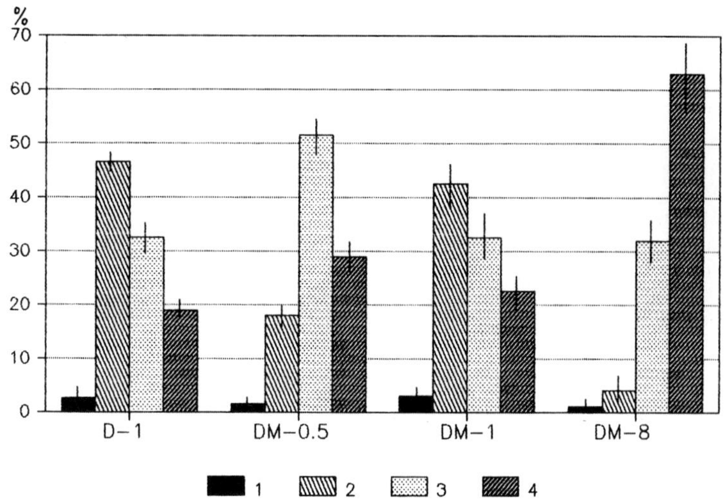

Fig. 4. Degree of cell aggregation in suspension cultures of various strains of *Dioscorea deltoidea* Number of cells in aggregate: *1* One cell; *2* 2–5 cells; *3* 5–20 cells; *4* > 20 cells

Table 2. Growth of the suspension cultures of *D. deltoidea* cell strains. Growth index on cell number basis, rel. values (P/P_0), see Section 2.1.2

Treatment	Cell strain			
	D-1	DM-0.5	DM-1	DM-8
Treatment	Rel. value			
Control	6.8 ± 0.3	7.9 ± 0.5	6.9 ± 0.4	7.1 ± 0.3
After cryopreservation	6.9 ± 0.4	8.1 ± 0.3	6.8 ± 0.6	7.9 ± 0.5

temperature, resulted in a synergetic increase in viability of up to 50% of the cell population. Corresponding to the synergetic increase in viability on combined treatment, D-1 cell exposure at 9–10 °C for 18–20 h did not lead to any rise in intracellular sugar content. Repeated measuring in widely separated subcultivated cycles gave average means 61.6 and 60.8 μg of intracellular sugars/1 ml of cell suspension for control and cooled cultures correspondingly during the first experiment and 54.9 and 54.1 μg/ml during the second.

After cryopreservation, *Dioscorea* cell strains completely retained their biosynthetic properties. Neither the composition (Volkova et al. 1982) nor the quantity of steroids (Table 5) changed significantly. The growth index of cell strains also did not shift (Table 2), neither did the ploidy pattern (Fig. 5).

Table 3. Influence of amino acids and DMSO on the maximal rate of cooling without intracellular ice crystallization

Treatment	Rate of cooling, °C/min
Asparagine (0.02 M) + DMSO (7%)	4
Asparagine (0.02 M)	2
Alanine (0.05 M) + DMSO (7%)	4
Alanine (0.05 M)	2
Proline (0.02 M) + DMSO (7%)	3
Proline (0.02 M)	1.5
DMSO (7%)	2.5
Control	1

Table 4. Influence of low-temperature (9–10 °C) preincubation on the viability (cells uncolored with phenosafranine, % of total cells) of *D. deltoidea* cell strains after cryopreservation

Cell strain	Conditions	9–10 °C	Control
D-1	No precultivation	26.0 ± 3.2	10.4 ± 2.4
D-1	Precultivation with asparagine	50.0 ± 2.2	39.0 ± 3.1
DM-0.5	No precultivation	48.5 ± 3.9	31.0 ± 2.8
DM-0.5	Precultivation with asparagine	54.0 ± 2.1	38.0 ± 2.7

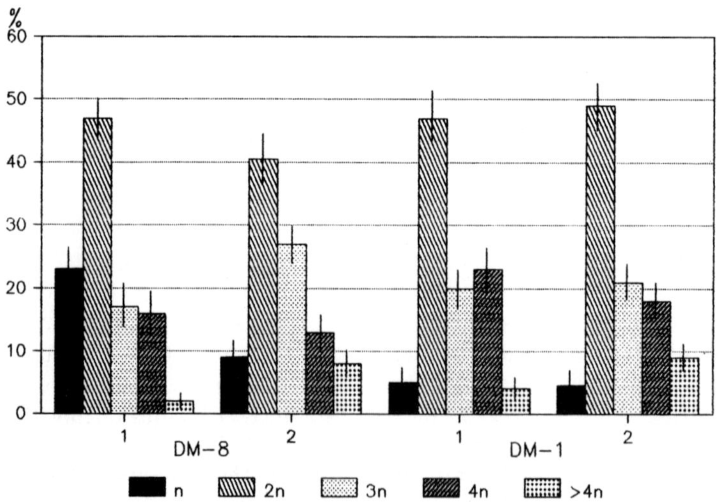

Fig. 5. Retaining of ploidy of *Dioscorea deltoidea* cells in suspension cultures after cryopreservation in liquid nitrogen. *1* Initial cultures; *2* renewing cultures, third subculture

Table 5. Steroid content and composition in various cell strains of *Dioscorea deltoidea*

Steroids		Cell strain			
		D-1 mg/g dry wt.	DM-0.5	DM-1	DM-8
Diosgenine	C	6.07 ± 0.04	12.7 ± 1.0	4.4 ± 0.4	5.5 ± 0.2
	R	6.12 ± 0.21	13.3 ± 0.4	3.8 ± 0.2	5.7 ± 0.2
Sitosterol	C	1.33 ± 0.06	0.44 ± 0.04	0.41 ± 0.01	0.57 ± 0.01
	R	1.20 ± 0.13	0.48 ± 0.04	0.37 ± 0.04	0.60 ± 0.04
Stigmasterol	C	0.43 ± 0.01	0	1.03 ± 0.03	0.46 ± 0.01
	R	0.45 ± 0.02	0	0.95 ± 0.08	0.48 ± 0.02
Campesterol	C	0.59 ± 0.02	0.36 ± 0.08	0.48 ± 0.03	0.40 ± 0.04
	R	0.63 ± 0.01	0.38 ± 0.08	0.54 ± 0.05	0.47 ± 0.02

Note: C, control culture; R, culture renewed after cryopreservation.

3 Discussion

The critical factors responsible for lethal cell injuries under freezing are vigorous dehydration and ice formation inside the cells. Dehydration is inevitable. At the same time, some dehydration could even be beneficial as it prevents the formation of large intracellular ice crystals. On growth, such crystals destroy all intracellular structures, primarily the membranes; but some specific properties of plant cells with active metabolism (large size, large vacuoles, high water content) make dehydration itself a principal cause of injuries. Excessive dehydration causes irreversible damage to plasmalemma (Steponkus 1984) and consequently leads to cell death.

The acceptable rate of dehydration depends on the water content in the cells and the cell wall permeability for water. In this work, as well as in many other investigations on cryopreservation, dehydration, as a component of the freezing process, occurs due to a phenomenon well known in cryophysics: the vapor pressure over the ice is lower than that over the water in the overcooled liquid state at the same temperature. This results in removal of water from the cell and its freezing on the surface as extracellular ice crystals. Thus, the freezing rate determines the rate of dehydration. The greater the water content of the cell, the less the rate of dehydration must be. The initiation of extracellular ice crystallization and the optimal rate of temperature decrease retard ice formation inside the cells and increase their revival after cryopreservation (Butenko et al. 1983; Popov 1985).

When added prior to freezing, chemical cryoprotectors at a high concentration (5–10%) affect the cells via various mechanisms. First of all, they decrease ice formation and expansion in the external solution, lowering freezing temperature, and hindering growth in the preformed crystals. The freezing of water in a solution results in an increased concentration of the cryoprotector and, as a consequence, augments its penetration into the cells. This is especially the case with easily penetrating DMSO. Solutions of this substance have the ability to solidify as an amorphous mass even at a moderate negative temperature, i.e., vitrification is possible without inevitably requiring overfast deep freezing. Thus, our experiments performed in 1980 showed that the freezing procedure with the initiation of extracellular ice crystallization and temperature decrease at a rate of 0.7–1 °C/min avoids the appearance of ice crystals inside the majority of cells even at −27 °C (Volkova et al. 1984). Microscope observation at further temperature decrease was impossible, but −27 °C means an overcooling equal to 24.2 °C as against the freezing point of the cryoprotector (7% DMSO in nutrition medium). In other words, these experiments showed an almost complete absence of water available to freezing at −27 °C in accordance with the DMSO properties considered above, which means more or less complete dehydration. These results coincide to a great extent with data obtained by Nuclear Magnetic Spectroscopy. In these experiments, *Catharanthus roseus* cells submerged in 5–10% DMSO solutions and frozen at the same rate of temperature change down to −30 °C showed a water amount near the lowest possible value (the plateau being at −40 °C), meaning an almost complete loss of free water (Chen et al. 1984).

The degree of dehydration and the resultant plasmalemma damage decrease with the lowering of the water potential of the cells due to accumulation of osmoticum. Our experiments show a significant increase of cryoresistance, enhancement of starch hydrolysis, and an increase in soluble sugar content in cells of *Dioscorea deltoidea* D-1 as a result of precultivation in medium supplemented with amino acids in subtoxic concentration (asparagine, alanine, proline, or serine). Interestingly, the cells of cryoresistant *Dioscorea* strains accumulated 1.5–2 times more sugars than the cells of the nonresistant strain D-1. Also the cells of one of the strains of *Panax ginseng*, supplemented with a rather high amount of sucrose, showed an increase in both endogenous sugar content and cryoresistance due to cold hardening (Fedorovsky et al. 1993). The sugars are known to increase the water-binding ability of cell, which leads to reduced shrinkage during plasmolysis and thus prevents plasmalemma destruction.

Many authors have concentrated on the important role of plasmalemma in cell resistance to damage under freezing (Steponkus 1984; Steponkus and Lynch 1989; Popov 1993). It is possible that the osmotically acting mechanism is not the only factor in increase in plasmalemma cryoresistance during precultivation on medium supplemented with amino acids. Thus, in particular, some amino acids of the second amino group are shown to interact directly with the phosphate groups of the plasmalemma phospholipids (Anchordoguy et al. 1988). Some carbohydrates can also stabilize membranes due to interaction with polar heads of phospholipids (Crowe et al. 1988). Evidently, similar processes could have taken place in our experiments, which showed 1.5–2-fold sugar content increase and, at the same time, five- to sixfold increase in cell viability.

Our results on the kinetics of cryo- and osmodamage of plasmalemma at deep freezing-thawing obtained using the same cells and under the same conditions confirm these suggestions. For these measurements, we used a new fluorimetric method which revealed lethal damage to plasmalemma structure (Popov and Fedorovsky 1992). Precultivation of cells of strain D-1 in the presence of asparagine resulted in a 2.5-fold increase of cryoresistant cells. For the cryoresistant mutant lines, resistant cells showed the same value (Fedorovsky and Popov 1992).

In the pattern of ploidy distribution in a population of strain D-1 cells, changes after precultivation in the presence of amino acids can be considered as nonspecific, emerging under any stress factors. Such precultivation resulted in increased cryoresistance and in spectral shifts from ploidy to diploid forms. Amino acid removal caused the slow (after 14 cycles) restoration of the ploidy pattern and, simultaneously, a decrease in cryoresistance. Both features are characteristic for this strain when grown without amino acids. It is possible that diploid cells of *Dioscorea deltoidea* are more cryoresistant than cells with a higher degree of ploidy. The ploidy degree higher than di- or triploid was not inherent to mutant strains with comparatively resistant cells.

Resistance in cells of strain D-1, initially rather sensitive to freezing, can be enforced also by another technique: cell exposure at 9–10 °C for 18–20 h. The effects of both methods on cells, and the absence of a sugar increase after cooled exposure illustrate the different mechanisms of action of these methods. It cannot be ruled out that cell incubation at lowering temperature directly influences cell membranes. At least such a phenomenon was demonstrated on the plasmalemma of *Rauwolfia serpentina* cells (Yamada et al. 1980). The cultivation of these cells at 10 °C resulted in an increase in nonsaturation of fatty acids and, hence, plasmalemma membrane fluidity, that in turn contributed to an increase in cryoresistance (Popov 1993).

4 Summary and Conclusions

For 20 years, *Dioscorea deltoidea* germplasm has been maintained in the Department of Cell Biology and Biotechnology of the K.A. Timiriazev Institute of Plant Physiology. At first, the original cell strain D-1 was obtained and then the mutant cell strains DM-0.5, DM-1, and the hormone-independent DM-8 were selected. All the strains have a similar growth index and number of viable cells in suspension cultures, but differ from the initial line in having a higher number of diploid cells and degree of cell clustering. The largest clusters are observed in strain DM-8. The mutant cells differ in the amount and composition of their steroid substances. Strain DM-0.5 is of great importance, possessing a considerable amount of diosgenin.

Methods were developed for cryopreservation in liquid nitrogen (– 196 °C) for all of these cell strains. Cryoresistance in mutant lines was five to six times greater than in the initial strain. They were successfully deep frozen with 7% DMSO only, and renewed after thawing; but for the original strain D-1, two techniques of preparation for cryopreservation were elaborated. The first used preliminary cultivation on the medium supplemented with amino acids for 5 days. Such cultivation resulted in a decrease in the size of starch grains, an increase in the amount of endogenous sugars, and in a ploidy shift towards cell diploidy. The best D-1 cell survival after cryopreservation was found if asparagine (0.02 M) had been employed. Intracellular ice crystals did not emerge at temperatures as low as − 27 °C under optimal conditions of culture preparation, cryoprotection, and freezing. The second method used incubation at 9–10 °C for 18–20 h. Applied together, both techniques had a synergetic effect. The rate of survival of cells after storage for 6–9 years was the same as after 3 months. All strains were renewed, and retained their morphologic, cytogenetic, growth, and biosynthetic characteristics after cryopreservation.

References

Abroshnikova (Sarkisova) MA, Butenko RG, Singukhin AM (1971) Tissue culture of *Dioscorea deltoidea* is a producent of steroid saponins and genins. Rastit Resour 7: 517–524 (in Russian)

Anchordoguy TJ, Carpenter JF, Loomis SH, Crowe JH (1988) Mechanisms of interaction of amino acids with phospholipid bilayers during freezing. Biochim Biophys Acta M 946: 299–306

Bajaj YPS (1988) Cryopreservation and the retention of biosynthetic potential in cell cultures of medicinal and alkaloid-producing plants. In: Bajaj YPS (ed) Biotechnology in agriculture and forestry, vol 4. Medicinal and aromatic plants I. Springer, Berlin Heidelberg New York, pp 169–187

Burkill IH (1960) The organography and evolution of *Dioscoreaceae*: the family of yams. J Linn Soc 56: 319–412

Butenko RG, Popov AS, Volkova LA, Chernyak ND, Nosov AM (1984) Recovery of cell cultures and their biosynthetic capacity after storage of *Dioscorea deltoidea* and *Panax ginseng* cells in liquid nitrogen. Plant Sci Lett 33: 285–292

Butenko RG, Popov AS, Volkova LA, Diettrich B, Lückner M (1983) Plant cell viability at different programs of deep freezing. Tsitologia (Cytology) 25: 1191–1196 (in Russian with English Summary)

Butenko RG, Vorobyov AS, Nosov AM, Knajzkov IE (1992) Synthesis, accumulation and location of steroid glycosides in the cells of different strains of *Dioscorea deltoidea* Wall. Sov Plant Physiol 39: 1146–1153 (pp of Russian edition)

Chandra J, Handa K, Kapoor L (1955) A short note on *Dioscorea deltoidea* growing in Jammu and Kashmir. Indian J Pharm 17: 142–144

Chen THH, Kartha KK, Constabel F, Gusta LV (1984) Freezing characteristics of cultured *Catharanthus roseus* (L.) G. Don. cells treated with DMSO and sorbitol in relation to cryopreservation. Plant Physiol 75: 720–725

Crowe LM, Carpenter JF, Rudolf AS, Wistrom CA, Spargo BJ, Anchordoguy TJ (1988) Interactions of sugars with membranes. Biochim Biophys Acta Rev Biomembr 947 (MR17): 367–384

Dubois M, Gilles KA, Hamilton JK, Rebers PA, Smith F (1956) Colorimetric method for determination of sugars and related substances. Anal Chem 28: 350–356

Fedorovsky DN, Popov AS (1992) Plasmalemma injuries in different strains of *Dioscorea deltoidea* during cryopreservation. Sov Plant Physiol 39: 381–386

Fedorovsky DN, Chernyak ND, Popov AS (1993) Investigation of plasmalemma disturbances of *Panax ginseng* cells during cryopreservation. Sov Plant Physiol 40: 94–98

Furmanowa M, Guzewska J (1989) *Dioscorea*: In vitro culture and the micropropagation of diosgenin-containing species. In: Bajaj YPS (ed) Biotechnology in agriculture and forestry, vol 7. Medicinal and aromatic plants II. Springer, Heidelberg New York, pp 162–184

Furmanowa M, Guzewska J, Beldowska B (1984) Organ regeneration in callus of *Dioscorea deltoidea* Wall. In: Novak FJ et al (eds) Proc Int Symp Plant tissue cell cult, application to crop improvement. Acad Sci Prague, pp 167–168

Grubishich D, Chulafich L, Boevich-Tsvetich D (1991) Regeneration of yam plants (*Dioscorea caucasica* Lipsky and *D. balcanica* Košanin) in in vitro culture. Sov Plant Physiol 38: 1018–1022 (pp of Russian edition)

Kaul B, Staba EJ (1967) *Dioscorea deltoidea* Wall. tissue cultures: I. Production and isolation of diosgenin. Lloydia 30: 283–291

Kichenko VI (1961) *Dioscorea* is raw material for a synthesis of steroid hormones. USSR Med Indust No. 3: 17–22 (in Russian)

Kichenko VI, Panina VV (1965) Diosgenin level in rhizomes of some *Dioscorea* species introduced near Moscow. Rastit Resour 1: 397–401 (in Russian)

Malaurie B, Pungo O, Dumont R, Trouslot M-F (1993) The creation of an in vitro germplasm collection of yams (*Dioscorea* spp.) for genetic resources preservation. Euphytica 65: 113–122

Martine J, Cappadocia M (1989) In vitro tuberization of *Dioscorea abyssinica* Hoch. and *D. alata* L. cv. Brazo fuerte and Florido. Am J Bot 76 (Suppl): 39–40

Murashige T, Skoog F (1962) A revised medium for rapid growth and bio-assays with tobacco tissue cultures. Physiol Plant 15: 473–497

Ng SYC (1992) Micropropagation of white yam (*Dioscorea rotundata* Poir.) In: Bajaj YPS (ed) Biotechnology in agriculture and forestry, vol 19. High-tech and micropropagation III. Springer, Berlin Heidelberg New York, pp 133–159

Okonkwo SNC, Nwoke FIO, Okezie CEA (1990) The effects of photoperiod and growth regulators on organ regeneration from vine nodal explants of *Dioscorea bulbifera* L. cultured in vitro. In: Abstr VII Int Congr Plant Tissue Cell Cult, Amsterdam, June 1990. Kluwer, Dordrecht, 121 pp

Onwueme IC (1978) Tropical tuber crops: yams, cassava, sweet potato and cocoyams. Wiley, New York

Popov AS (1985) Cryoconservation and plant cell bank. In: Butenko RG (ed) Plant cell culture. MIR, Moscow, pp 175–196

Popov AS (1993) Mechanisms of in vitro cryoinjury in plant cell cryopreservation. Russian J Plant Physiol 40: 421–431

Popov AS, Fedorovsky DN (1992) Plasmalemma disturbances during cryopreservation of cultured in vitro *Dioscorea* cells. Sov Plant Physiol 39: 211–216

Popov AS, Isakov NA, Butenko RG (1988) A device for freezing of live biological objects in containers. USSR Inventor Certificate 1440137

Purseglove JW (1972) Tropical crops, monocotyledons. Longmans Green, London

Shreter AI, Pimenov MG, Vasil'eva VD (1965) Systematics, distribution and resources of Soviet Far East *Dioscorea*. Rastit Resour 1: 390–396 (in Russian)

Steponkus PL (1984) Role of the plasma membrane in freezing injury and cold acclimation. Annu Rev Plant Physiol 35: 543–584

Steponkus PL, Lynch DV (1989) Freeze-thaw-induced destabilization of the plasma membrane and the effects of cold acclimation. J Bioenerg Biomembr 21: 21–41

Volkova LA, Popov AS, Nosov AM, Butenko RG (1982) Preservation of biosynthetic potential of dioscorea (*Dioscorea deltoidea* Wall.) cells after cryogenic storage. Proc USSR Acad Sci 265: 504–506 (in Russian)

Volkova LA, Popov AS, Samygin GA (1984) Effect of exogenous amino acids on cultivation of *Dioscorea deltoidea* cell suspension and recultivation after storage at −196 °C. Sov Plant Physiol 31: 632–638 (pp of Russian edition)

Volkova LA, Gorskaya NV, Popov AS, Paukov VN, Urmantseva VV (1986) Invariability of main characteristics of mutant *Dioscorea* strains during storage at superlow temperature. Sov Plant Physiol 33: 779–787 (pp of Russian edition)

Widholm J (1972) The use of fluorescein diacetate and phenosafranine for determining viability of cultured plant cells. Stain Technol 47: 189–194

Yamada Y, Hara Y, Katagi H, Senda M (1980) Protoplast fusion. Effect of low temperature on the membrane fluidity of cultured cells. Plant Physiol 65: 1099–1102

VI.7 Preservation of Cell Cultures of *Lavandula vera* (Lavender)

A. TANAKA[1] and J. TAKAHASHI[2]

1 Introduction

1.1 Importance of Cell Cultures of *Lavandula vera*

Plant cell cultures are very attractive biocatalysts and are now used to produce different useful compounds (Parr 1989; Tanaka 1989; Bajaj 1994). Cultured green cells of *L. vera* grown under illumination were found to accumulate the greatest amount of free biotin of the cultured plant cells examined (Watanabe et al. 1982). Gamma-irradiation followed by repeated selection resulted in cell lines which synthesize a large amount of free biotin ranging from 0.097 to 0.909 μg/g fresh weight (Watanabe and Yamada 1982a, b). Thus, these selected cell lines are very important in producing free, true biotin, because free biotin accumulated in microbial cells is usually composed of desthiobiotin to a great extent, with true biotin in a small quantity.

Furthermore, the cultured green cells of *L. vera* can produce extracellular blue pigments in the presence of *S*-containing compounds, such as L-cysteine, L-cystine, DL-homocysteine, and DL-homocystine (Watanabe et al. 1985a). These pigments can be used as coloring agents in the food and cosmetic industries. These facts strongly indicate the importance of the cell cultures of *L. vera* for the production of useful plant-specific metabolites (Segura and Calvo 1991).

1.2 Need for Preservation of Cultured Cells

In the production of blue pigments by *Lavandula vera* cells, immobilization by entrapment in adequate polymer gels has been found to be effective in stabilizing the pigment productivity of the cells for a long period (Nakajima et al. 1985, 1986). Although a fine suspension culture is essential for immobilization of the cells, such cells are generally more sensitive to the surrounding conditions than a callus culture, and without subculturing, the suspended cells could not remain alive for more than 2–3 weeks at 26 °C. It is difficult and time-consuming to obtain a stable fine suspension of *L. vera* cells, because the cells tend to form

[1]Department of Synthetic Chemistry and Biological Chemistry, Faculty of Engineering, Kyoto University, Yoshida, Sakyo-ku, Kyoto 606-01, Japan
[2]Discovery Research Laboratories I, Discovery Research Division, Takeda Chemical Industries, Ltd., 10 Wadai, Tsukuba, Ibaraki 300-42, Japan

Biotechnology in Agriculture and Forestry, Vol. 32
Cryopreservation of Plant Germplasm I (ed. by Y.P.S. Bajaj)

aggregates in a liquid medium. Addition of calcium ion at a relatively high concentration was found to be effective in keeping the suspension fine. However, several successive subcultures in the presence of calcium ion are essential before stabilization of the culture. This means that a long time is needed to prepare a fine suspension of *L. vera* cells from the callus when the cells become inactive in producing the pigments. Therefore, development of a simple preservation method of the suspended cultured cells is desired.

1.3 Methods for Preservation of Cultured Cells

Freeze preservation of cells has been successfully used for several species of medicinal and aromatic plants (see Bajaj 1988). Green callus of *L. vera* was preserved successfully in liquid nitrogen under frozen condition for up to 3 weeks (Watanabe et al. 1983). Storage of the cells cultured in a liquid medium was also examined in liquid nitrogen, the freeze-thaw procedures causing no damage to the biosynthetic capability for biotin and the differentiation potentials of *L. vera* (Watanabe et al. 1985b). However, this method is generally unsuitable for routine work. Therefore, nonfreeze preservation has also been applied to the preservation of callus cultures (Hiraoka and Kodama 1982; Hiraoka 1988).

We have developed a new, simple and labor-saving method of preserving suspended plant cell cultures at a low but nonfreezing temperature, which needs only short-term incubation to recover the viability and the metabolic activity of the cells (Takahashi et al. 1991).

2 Preservation of Cell Cultures of *L. vera*

2.1 Nonfreezing Method

2.1.1 Methodology

Strains and Cultivation. Lavandula vera strain L_{10} 4–2154, which had been selected as a potent biotin-producing strain and found to be a pigment-producer (Watanabe et al. 1982, 1985a), was cultivated in LS medium (Linsmaier and Skoog 1965) containing 3% (w/v) sucrose, 10^{-5} M indole-3-butyric acid, and 10^{-6} M 6-benzyladenine with shaking (100 strokes per min) at 26 °C under illumination (ca. 6000 lx) (Nakajima et al. 1985). The liquid medium was added by 50 mM calcium chloride to keep the suspension fine.

Immobilization. L. vera cells from suspended cultures (6 ml) at the stationary growth phase, corresponding to about 2.4 g wet weight cells, were immobilized by entrapment with several gel materials, which had been sterilized at 120 °C for 10 min before use (Nakajima et al. 1985, 1986). Typical procedure is as follows. Calcium ion in the cell suspension was removed by washing the cells with 3%

sucrose solution prior to entrapment. The cell suspension in the LS medium (6 ml) was mixed with 7.2 ml of 3% (w/v) sodium alginate solution in deionized water at room temperature. Calcium alginate beads were prepared by dropping the mixture from a wide-mouthed pipette into 0.1 M calcium chloride solution at room temperature and keeping it for 15 min. The beads thus obtained (diameter ca. 5 mm) were washed with 3% sucrose solution.

Preservation. The free cells obtained by filtration from 6 ml of a suspended culture at the stationary phase were suspended in 20 ml of the LS medium with or without agar purified (DIFCO) and left at 4 °C in the dark in 40-ml sealed glass bottles. The entrapped cells were washed with deionized water and preserved in 20 ml of 3% sucrose solution or the LS medium, or without any aqueous phase.

Analyses. Oxygen uptake was measured with 1–3 ml of the cell suspension or 1–2 g of wet gels entrapping the cells in 3% sucrose solution (total volume, ca. 12 ml) by using a Beckman Fieldlab oxygen analyzer at 30 °C (Nakajima et al. 1985). Triphenyl tetrazolium chloride (TTC) reduction activity was measured as follows. The free or entrapped cells were collected by filtration, rinsed with deionized water, immersed in 0.6% (w/v) TTC solution in 0.05 M KH_2PO_4–NaOH buffer (pH 7.5), and left at 26 °C for 20 h in the dark. Thereafter, TTC solution was removed and the free cells and the cell-entrapping gels cut into small pieces (ca. 1 × 1 × 1 mm) were rinsed with distilled water. Triphenylformazan formed was extracted from the cells with 99% (v/v) ethanol at 50–60 °C for 2 to 3 h. Cell viability was expressed as the absorbance at 530 nm of the ethanol extract (Steponkus and Lampher 1967).

2.1.2 *Results and Discussion*

When suspended cells were left statically at 4 °C, they died within a week, whereas they remained alive for 3 weeks in the LS medium at 4 °C with shaking. These results suggest that cell sedimentation might make the diffusion of certain essential materials contained in air or in the medium difficult. It seemed, therefore, that, if the cells were dispersed during storage, the cell viability might be increased.

When the cells were preserved at 4 °C in the LS medium with 0.3% agar (the minimal agar concentration to keep cells suspended), about 20% of the oxygen uptake activity was retained after 30 days (Table 1). A high concentration of agar (0.6%, the maximum concentration to keep agar solution fluid) was not so effective on the cell viability in spite of a similar ability to keep the cells suspended. If the suspension conditions were of prime importance for the survival of the cells, such a difference in agar concentration might not be observed. These results suggested that diffusion of certain essential materials, which was inhibited by agar, could play an important role in survival. Entrapment of cells in suitable gels seems to keep the cells suspended. Therefore, entrapment in calcium alginate beads was examined for the preservation of cells. In this case, the entrapped cells were preserved in 3% sucrose solution (an

Table 1. Oxygen uptake of free and entrapped *L. vera* cells after storage for 30 days. (Takahashi et al. 1991)

Cell		Oxygen uptake (μmol/h/g wet cells)	Survival (%)
Free	In LS medium	0	0
	In 0.3% agar[a]	0.96	19.4
	In 0.6% agar[a]	0.62	12.5
Ca-alginate-entrapped	In 3% sucrose	0.36	7.6
	In LS medium	0.49	10.3
	In 0.3% agar[a]	0.59	12.4
	Without aqueous phase	3.12	65.8

[a]0.3% or 0.6% agar in the LS medium.
In each case, 2.4 g wet cells were stored at 4 °C in the dark in sealed glass bottles. Survival (%) is expressed by the oxygen uptake rate of the preserved cells compared with that of the free cells (4.96 μmol/h/g wet cells) or the entrapped cells (4.74 μmol/h/g wet cells) before storage.

isotonic sucrose solution) in the LS medium, to examine the effect of the medium components. However, little effect of the medium components on the entrapped cells was observed (Table 1). On the contrary, the viability of the entrapped cells after preservation for 30 days without any aqueous phase was considerably higher than that of both entrapped and free cells with aqueous phases (Table 1). These results strongly suggested that contact with air was most important in the preservation of cells, since air seems to pass easily through the gel beads when the entrapped cells are stored without a liquid solution.

The results demonstrated a possible important function of water around the cells. The medium was removed from the suspended cells by filtration and the different volume of the isotonic sucrose solution was added to simplify the effect of water. When the amount of water was small, the cells could not survive for a long period, probably due to dryness. A large amount of water was also not effective for cell preservation, indicating that a proper volume of water around the cells was essential during the preservation period. These results can be explained by the efficiency of the air supply. As the volume of water was decreased, the air supply toward the cells was enhanced and cell survival increased. Excess water seemed to hinder the cells from contact with air, resulting in their death.

Considering that the air supply is significant for cell preservation, entrapment seems to be an attractive treatment, because entrapped cells have a larger surface in contact with air than their free counterparts sedimented in the bottom of a bottle. Furthermore, gels can serve as a water reservoir for a long storage period. Therefore, it is important to find a suitable gel material. Table 2 shows that entrapment in calcium alginate, κ-carrageenan, and a photosensitive resin [a derivative of polyvinyl alcohol, PVA-SbQ (Ichimura 1984; Nakajima et al. 1986)] were all effective in cell preservation. The entrapped cells, except for those in κ-carrageenan, retained the ability to produce the blue pigments (Nakajima et al. 1985, 1986) even after storage for 30 days. κ-Carrageenan gel was not stable when incubated after storage. This instability might affect the viability of the cells

Table 2. Activities of *L. vera* cells entrapped in different gels after storage for 30 days. (Takahashi et al. 1991)

Gel material	Survival (%) based on		Growth after storage	Pigment production
	Oxygen uptake	TTC reduction		
None	32.8	49.5	+	+
Ca-alginate	65.7	59.0	+	++
κ-Carrageenan	70.0	66.8	–	–
PVA-SbQ	44.3	77.6	+	+

In each case, 2.4 g wet cells were entrapped and stored at 4 °C in the dark in sealed glass bottles without any aqueous phase (entrapped cells). In the case of the free cells (none), the medium was removed by decantation (the amount of water corresponding to 1.5 ml/system) and the cells were stored as mentioned above. Oxygen uptake and TTC reduction of the stored cells were compared with those of the corresponding cells before preservation. After storage, the cells were rinsed with the LS medium and cultivated in 25 ml of the medium (for growth) or the medium containing L-cysteine [for pigment production (Nakajima et al. 1985)] for 10 days at 26 °C.

during cultivation in the medium for 10 days. In spite of these facts, the cells entrapped in the gels were successfully preserved irrespective of the gel materials, indicating again the importance of efficient contact of cells with air for cell preservation.

Calcium alginate gel was stable in the LS medium, and could be solubilized in a phosphate buffer without damaging the cells after being cultivated for activation. Therefore, the suspended cells can be easily obtained as required. Moreover, the calcium alginate-entrapped cells showed especially high growth activity and pigment productivity even after preservation, enabling direct application of the preserved entrapped cells to the pigment production. From these results, entrapment in calcium alginate was selected for long-term preservation. Even after preservation for 50 days, the calcium alginate-entrapped cells still recovered a high oxygen uptake activity after only 10 days cultivation (Table 3). In this case, extra water on the gel surface must be wiped off with filter paper before storage to avoid the death of the cells dipped in water accumulated at the bottom of the bottle.

This method was also applicable to the preservation of tobacco cells. *Nicotiana tabacum* cv. Samsun NN cells (Sato et al. 1988) were sufficiently viable after preservation for 30 days, despite the fact that the conditions were not optimized for tobacco cells. Although the optimization of preservation conditions is essential, depending on cell species, this nonfreezing method is thought to be widely applicable to plant suspension cultures.

In these experiments, *L. vera* cells were stored in sealed glass bottles. When stored in plastic bottles, which permit air to pass through more easily than glass, the cells could not survive for more than 2 weeks. Metabolite gases, such as carbon dioxide, might play a significant role in the preservation of cells, as suggested by Hammer et al. (1990) and Rushing (1990).

Table 3. Preservation of calcium alginate-entrapped *L. vera* cells. (Takahashi et al. 1991)

Preservation period (days)	Activity after preservation		Activity after cultivation	
	Oxygen uptake[a]	TTC reduction[b]	Oxygen uptake[a]	TTC reduction[b]
0	11.4	39.1	34.7	417
30	7.5	23.0	34.2	213
50	5.4	7.9	28.3	253

[a] μmol/h/system.
[b] A_{530}/system.
In each case, 2.4 g wet cells were entrapped in calcium alginate beads and stored under the same conditions as in Table 1, without an aqueous phase, for the indicated periods. Oxygen uptake and TTC reduction were measured before and after cultivation for 10 days in the LS medium.

2.2 Cryopreservation

Bajaj (1988) has summarized the purposes of cryopreservation of cultured plant cells as follows: retention of biosynthetic potentials, genetic stabilization of somaclones, and germplasm conservation. In fact, the cell lines, which were selected by various methods as useful ones, have to be maintained for a long period without undergoing any change.

The potent biotin-producing cell lines of *L. vera* were preserved in liquid nitrogen with keeping their synthetic ability for biotin and differentiation potentials (Watanabe et al. 1983). Green *L. vera* callus was cultivated at 27 °C under illumination (6000 lx) on an LS agar medium containing 1% (w/v) agar, 10^{-5} M indole-3-butyric acid, and 10^{-6} M 6-benzyladenine. Small cell aggregates ($\leqq$ 1 mm) were cut from the callus at exponential growth phase and suspended in the liquid medium of the same composition, followed by cooling on ice. The cryoprotectant solution, 10% (v/v) dimethylsulfoxide and 20% (w/v) glucose in distilled water, which had been sterilized and cooled on ice, was added gradually to an equal volume of the cooled cell suspension over a period of 1h. The resulting cell suspension was dispensed into sterile polypropylene tubes on ice (0.5 ml settled cells + 0.5 ml liquid in 2-ml tube). The cell suspension was cooled at 1 °C/min to − 40 °C in a Cryo-Med programmed freezer, then transferred into liquid nitrogen. After 6 to 21 days of storage in liquid nitrogen, the cell suspension was thawed rapidly by shaking the tube in a water bath at 60 °C for 40 to 60 s just before the ice in the tube melted completely, then the tube was kept on ice. After thawing, the cell suspension was slowly diluted with the fresh medium at room temperature. The cells recovered were cultivated on the agar medium.

The biosynthetic capability for biotin, together with the differentiation potentials, were retained well after the freeze preservation. Repeated freeze-thaw procedures were also possible for cultured green cells of *L. vera* (Watanabe et al. 1985b).

3 Summary and Conclusions

A simple procedure to preserve cultured plant cells under nonfreezing conditions has been demonstrated. *Lavandula vera* cells were stored under dark and cool (4 °C) conditions in sealed glass bottles. When medium was removed from the system with the cells entrapped in calcium alginate gels, enhanced viability after long-term preservation was observed. The cells recovered from the gels were able to grow successfully after storage for 30 days, and incubation for only 10 days after storage was enough to recover the oxygen uptake activity to the level of the cells before storage. *L. vera* cells retained the ability to synthesize the blue pigments, specific secondary metabolites of the cells, even after storage for 30 days. Entrapment in calcium alginate beads was effective to preserve the cells for more than 50 days. The importance of air supply in preservation was strongly suggested.

Cell aggregates from the callus of *L. vera* were also cryopreserved successfully in liquid nitrogen. The cells recovered after thawing retained the biosynthetic capability for biotin and the differentiation potentials as before preservation.

References

Bajaj YPS (1988) Cryopreservation and the retention of biosynthetic potential in cell cultures of medicinal and alkaloid-producing plants. In: Bajaj YPS (ed) Biotechnology in agriculture and forestry, vol 4. Medicinal and aromatic plants I. Springer, Berlin Heidelberg New York, pp 169–187

Bajaj YPS (1994) (ed) Biotechnology in agriculture and forestry, vols 26, 28. Medicinal and aromatic plants VI, VII. Springer, Berlin Heidelberg New York

Hammer PE, Yang SF, Marois JJ (1990) Postharvest control of *Botrytis cinerea* infections on cut roses using fungistatic storage atmospheres. J Am Soc Hortic Sci 115: 102–107

Hiraoka N (1988) Nonfrozen storage of plant cell cultures and its effect on metabolites. In: Bajaj YPS (ed) Biotechnology in agriculture and forestry, vol 4. Medicinal and aromatic plants I. Springer, Berlin Heidelberg New York, pp 157–168

Hiraoka N, Kodama T (1982) Effects of non-frozen cold storage on the growth, organogenesis, and secondary metabolite production of callus cultures. In: Fujiwara A (ed) Plant Tissue Culture 1982, Maruzen, Tokyo, pp 359–360

Ichimura K (1984) A convenient photochemical method to immobilize enzymes. J. Polym Sci Polym Chem Ed 22: 2817–2828

Linsmaier EM, Skoog F (1965) Organic growth factor requirements of tobacco tissue cultures. Physiol Plant 18: 100–127

Nakajima H, Sonomoto K, Usui N, Sato F, Yamada Y, Tanaka A, Fukui S (1985) Entrapment of *Lavandula vera* cells and production of pigments by entrapped cells. J Biotechnol 2: 107–117

Nakajima H, Sonomoto K, Morikawa H, Sato F, Ichimura K, Yamada Y, Tanaka A (1986) Entrapment of *Lavandula vera* cells with synthetic resin prepolymers and its application to pigment production. Appl Microbiol Biotechnol 24: 266–270

Parr AJ (1989) The production of secondary metabolites by plant cell cultures. J Biotechnol 10: 1–26

Rushing JW (1990) Cytokinins affect respiration, ethylene production, and chlorophyll retention of packaged broccoli florets. HortScience 25: 88–90

Sato F, Shigematsu Y, Yamada Y (1988) Selection of an atrazine-resistant tobacco cell line having a mutant *psbA* gene. Mol Gen Genet 214: 358–360

Segura J, Calvo MC (1991) *Lavandula* spp. (Lavendula): In vitro culture, regeneration of plants, and the formation of essential oils and pigments. In: Bajaj YPS (ed) Biotechnology in agriculture and forestry, vol 15. Medicinal and aromatic plants III. Springer, Berlin Heidelberg New York, pp 283–310

Steponkus PL, Lampher FO (1967) Refinement of the triphenyl tetrazolium chloride method of determining cold injury. Plant Physiol 42: 1423–1426

Takahashi J, Nakajima H, Sonomoto K, Sato F, Ichimura K, Yamada Y, Tanaka A (1991) Development of simple non-freezing method to preserve cultured plant cells. J Ferment Bioeng 72: 71–73

Tanaka A (1989) Cultured plant cells as the catalysts for bioreactor. In: Fiechter A, Okada H, Tanner RD (eds) Bioproducts and bioprocesses. Springer, Berlin Heidelberg New York, pp 3–12

Watanabe K, Yamada Y (1982a) Selection of high biotin producing strains of cultured *Lavandula vera* cells. In: Fujiwara A (ed) Plant Tissue Culture 1982, Maruzen, Tokyo, pp 357–358

Watanabe K, Yamada Y (1982b) Selection of variants with high levels of biotin from cultured green *Lavandula vera* cells irradiated with gamma rays. Plant Cell Physiol 23: 1453–1456

Watanabe K, Yano S, Yamada Y (1982) The selection of cultured plant cell lines producing high level of biotin. Phytochemistry 21: 513–516

Watanabe K, Mitsuda H, Yamada Y (1983) Retention of metabolic and differentiation potentials of green *Lavandula vera* callus after freeze preservation. Plant Cell Physiol 24: 119–122

Watanabe K, Sato F, Furuta M, Yamada Y (1985a) Induction of pigment production by *S*-containing compounds in cultured *Lavandula vera* cells. Agric Biol Chem 49: 533–534

Watanabe K, Yamada Y, Ueno S, Mitsuda H (1985b) Change of freezing resistance and retention of metabolic and differentiation potentials in cultured green *Lavandula vera* cells which survived repeated freeze-thaw procedures. Agric Biol Chem 49: 1727–1731

Subject Index

Springer-Verlag and the Environment

We at Springer-Verlag firmly believe that an international science publisher has a special obligation to the environment, and our corporate policies consistently reflect this conviction.

We also expect our business partners – paper mills, printers, packaging manufacturers, etc. – to commit themselves to using environmentally friendly materials and production processes.

The paper in this book is made from low- or no-chlorine pulp and is acid free, in conformance with international standards for paper permanency.

Printing: Druckerei Zechner, Speyer
Binding: Buchbinderei IVB, Heppenheim